# Sustainable Water Resources: Development and Management

# Sustainable Water Resources: Development and Management

Editor: Herbert Lotus

R CALLISTO REFERENCE

www.callistoreference.com

**Callisto Reference,**
118-35 Queens Blvd., Suite 400,
Forest Hills, NY 11375, USA

Visit us on the World Wide Web at:
www.callistoreference.com

ISBN: 978-1-63239-838-3 (Hardback)

### Cataloging-in-publication Data

Sustainable water resources : development and management / edited by Herbert Lotus.
    p. cm.
Includes bibliographical references and index.
ISBN 978-1-63239-838-3
1. Water-supply. 2. Water resources development. 3. Water-supply--Management. 4. Water use.
5. Sustainable development. I. Lotus, Herbert.
TD345 .S87 2017
628.1--dc23

# Table of Contents

**Permissions**

**List of Contributors**

**Index**

# Preface

This book elucidates the concepts and innovative models around prospective developments with respect to sustainable water resources. Water crisis is a major concern of today's world. Therefore, the need for sustainable water resources is an essential one. This book will give insights into the practice of managing, planning and distribution techniques required for the optimum utilization of water resources. This book includes some of the vital pieces of work being conducted across the world, on various topics related to this area. It discusses the fundamentals as well as modern approaches of this field. It unfolds the innovative aspects revolving around sustainable management of water resources which will be crucial for the progress of this area in the future. Scientists and students actively engaged in this field will find this text full of crucial and unexplored concepts.

In my initial years as a student, I used to run to the library at every possible instance to grab a book and learn something new. Books were my primary source of knowledge and I would not have come such a long way without all that I learnt from them. Thus, when I was approached to edit this book; I became understandably nostalgic. It was an absolute honor to be considered worthy of guiding the current generation as well as those to come. I put all my knowledge and hard work into making this book most beneficial for its readers.

I wish to thank my publisher for supporting me at every step. I would also like to thank all the authors who have contributed their researches in this book. I hope this book will be a valuable contribution to the progress of the field.

**Editor**

# The Potential Connectivity of Waterhole Networks and the Effectiveness of a Protected Area under Various Drought Scenarios

**Georgina O'Farrill**[1]*, **Kim Gauthier Schampaert**[3], **Bronwyn Rayfield**[2], **Örjan Bodin**[4], **Sophie Calmé**[3,5], **Raja Sengupta**[6], **Andrew Gonzalez**[2]

1 Ecology and Evolutionary Biology Department, University of Toronto, Toronto, Ontario, Canada, 2 Biology Department, McGill University, Montreal, Quebec, Canada, 3 Département de géomatique (KGS), Département de biologie (SC), Université de Sherbrooke, Sherbrooke, Québec, Canada, 4 Stockholm Resilience Centre, Stockholm University, Stockholm, Sweden, 5 Departamento de conservación de la biodiversidad, El Colegio de la Frontera Sur, Chetumal, Quintana Roo, Mexico, 6 Geography Department, McGill University, Montreal, Quebec, Canada

## Abstract

Landscape connectivity is considered a priority for ecosystem conservation because it may mitigate the synergistic effects of climate change and habitat loss. Climate change predictions suggest changes in precipitation regimes, which will affect the availability of water resources, with potential consequences for landscape connectivity. The Greater Calakmul Region of the Yucatan Peninsula (Mexico) has experienced a 16% decrease in precipitation over the last 50 years, which we hypothesise has affected water resource connectivity. We used a network model of connectivity, for three large endangered species (Baird's tapir, white-lipped peccary and jaguar), to assess the effect of drought on waterhole availability and connectivity in a forested landscape inside and adjacent to the Calakmul Biosphere Reserve. We used reported travel distances and home ranges for our species to establish movement distances in our model. Specifically, we compared the effects of 10 drought scenarios on the number of waterholes (nodes) and the subsequent changes in network structure and node importance. Our analysis revealed that drought dramatically influenced spatial structure and potential connectivity of the network. Our results show that waterhole connectivity and suitable habitat (area surrounding waterholes) is lost faster inside than outside the reserve for all three study species, an outcome that may drive them outside the reserve boundaries. These results emphasize the need to assess how the variability in the availability of seasonal water resource may affect the viability of animal populations under current climate change inside and outside protected areas.

**Editor:** Cédric Sueur, Institut Pluridisciplinaire Hubert Curien, France

**Funding:** This study was carried out with the support of the Consejo Nacional de Ciencia y Tecnologia-Mexico (CONACYT- doctoral fellowship) and Fonds Québécois de la Recherche sur la Nature et les Technologies (FQRNT- postdoctoral fellowship) to GO, the National Science and Engineering Research Council from Canada (NSERC) Discovery Grants program to AG and RS, the Canada Research Chair Program (to AG), the Global Environmental and Climate Change Center (GEC3), the Swedish Foundation for Strategic Environmental Research (MISTRA), the strategic research program Ekoklim at Stockholm University, and the NASA-LCLUC program(NAG5-6046 and NAG5-11134) to the Southern Yucatán Peninsular Region(SYPR) project. The Walter Hitschfeld Geographic Information Centre in the Geography Department at McGill University provided the authors with the Quick Bird images. The funders had no role in study design, data collection and analysis, decision to publish, or preparation of the manuscript.

**Competing Interests:** The authors have declared that no competing interests exist.

* E-mail: georgina.ofarrill@gmail.com

## Introduction

The synergistic effects of land use change, habitat loss and climate change are expected to affect species persistence by altering the distribution and connectivity of resources and habitat. These effects will have significant consequences for biodiversity conservation and management [1,2]. Landscape connectivity allows species movement and dispersal, influencing the distribution of genes, resources and populations of many species [3,4]. Landscape connectivity analysis encompasses both the ease with which an animal can move from one resource patch to another (the animal perspective, [5]), and the location and quality of resources (the landscape perspective) that will determine the species' motivation to move [6,7].

Changes in temperature and precipitation regimes predicted by climate change models [8] are likely to influence resource availability through changes in their abundance (e.g. fruits, water).

Resource connectivity analyses are particularly important when resources found within habitat patches vary in space and over short (within years) and long (between years) time scales (e.g. [9]). Hence highly variable fluctuations in temporal resource availability make connectivity within a resource network dynamic and stochastic in space and time [10], which is expected to influence species movement patterns [11].

Resource connectivity studies should incorporate the temporal variability in the availability and connectivity of resources given the current predictions of climate change worldwide, the species' differential use and accessibility to resources, and the amount of suitable habitat remaining after landscape connectivity is lost [12–14]. This is particularly relevant when areas with low or declining connectivity may not be able to support viable populations of some species over long periods of time [15]. Given current observations on the long-term effects of climate change, longer-term fluctuations in precipitation than the ones presented currently may affect water networks resulting in multi-annual trends in network connectivity [10,11].

In the Yucatan Peninsula of Mexico, climate change is causing an increase in drought periods [16], which seems to be influencing the availability of resources and the movement patterns of animals (e.g. [17]). The Greater Calakmul Region of the Yucatan Peninsula of Mexico is a continuous forested area in a highly seasonal tropical climate where, over the last 50 years, annual precipitation decreased by 16% while drought frequency increased [16]. Yearly fluctuations in precipitation show a decreasing pattern despite reports of years with high precipitation (Comisión Nacional del Agua, unpublished data). According to regional climate models, this area will increasingly suffer from extreme droughts in future years [8]. This area is a karstic upland area, where freshwater is only available to wildlife and people in superficial waterholes and small seasonal streams. Therefore, water is a scarce and dynamic resource in the area. Many small waterholes dry up during the dry season, and if the reduction in precipitation continues, we hypothesize this will further influence the availability of waterholes. In addition, we hypothesize that if drought events affect surface water availability (i.e., waterhole presence), waterhole connectivity will decrease given that waterholes that remain in the landscape will be located beyond species maximum travel capabilities or home ranges.

We used a network (graph theoretical) analysis to test our hypotheses about the change in the spatial distribution of waterholes in this network [18]. We treated these waterholes and short seasonal streams as nodes, and we defined a link between any two if they fell within the range of our study species' movement distances [19,20]. We modelled the movement of Baird's tapir (*Tapirus bairdii* Gill, 1865), white-lipped peccary (*Tayassu pecari* Link, 1795) and jaguar (*Panthera onca* Linnaeus, 1758). We used these species because they are of significant conservation concern, they rely on waterholes, and data about their movement are available. We assessed the connectivity of waterholes for each species given the temporal and spatial changes of water availability within and between years inside and outside the Calakmul Biosphere Reserve considering actual observations and climate change predictions of severe droughts for the area. We used reported travel distances for our study species, rather than measuring the actual movement of individuals; our results therefore represent the potential connectivity network of waterholes. Although other studies have evaluated the importance of seasonal water resources for species survival [11,12,21], none have addressed the connectivity of water resource under scenarios of climate change and how resource availability interacts with species movement capacities to modify the functional connectivity of a landscape.

## Materials and Methods

### Study Area

The waterhole network is situated in an area of approximately 750 km$^2$ located in the north-eastern part of the Greater Calakmul Region to the south of the Yucatan Peninsula of Mexico (19°15' to 17°50'N and 90°20' to 89°00'W). Fieldwork activities inside and in the areas surrounding the Calakmul Biosphere Reserve were authorized by the Director of the Reserve (Fernando Durand Siller- Permit no. D-RBC-020-10-07). Fieldwork outside the reserve was carried out in the Ejido of Nuevo Becal (communal land) and the Community commissioner authorized field activities after the assembly of the community was notified and agreed on allowing our visit to their land. Even though this study considers

movement capacities of endangered species, data were collected from the literature and field activities did not involve any of the endangered species studied. The Greater Calakmul Region (Fig. 1) is part of the Selva Maya, the second largest area of tropical forests in the Americas. Approximately 13% of the forest cover is disturbed by human activities in the region [22]. Of our study area, 30–40% lies within the Calakmul Biosphere Reserve, while the eastern section represents a continuous forested landscape that corresponds to communal land and is mostly a managed forest reserve. Forest cover within and outside the Calakmul Biosphere Reserve does not show any large difference [22]; however, there are human settlements outside the borders of the reserve, which represent a threat to species, i.e. hunting and logging activities.

Between 1953 and 1998 precipitation decreased by almost 16% in the Calakmul Region; mean precipitation declined from 1,300 mm in 1950s to 790 mm in 1990s [23] following the same pattern until 2005 (1955–2005; Comisión Nacional del Agua, unpublished data; [23]). In addition, climate models predict a further decline in precipitation and warmer temperatures, and suggest that most severe droughts in Mexico would occur during el Niño years affecting 50% of the area covered by deciduous tropical forest [8]. Due to its geomorphologic conditions (karst topography), the region does not have any rivers and the majority of the superficial water is stored in small depressions on the landscape: waterholes (locally called "*aguadas*") and small seasonal streams [24]. No waterholes are formed by underground water in this area. Therefore, during the dry season these superficial water bodies represent the only available water source for many animal species.

### Study Species

In the Greater Calakmul Region, species such as the Baird's tapir, white-lipped peccary and jaguar are endangered [25] [26] [27]. These species are highly associated with water bodies for water consumption, to regulate their body temperature (tapirs and peccaries [28]) or to find their prey (jaguars [27]). Given that water is only present in waterholes and short streams, species such as these depend on seasonal water bodies and the associated habitat surrounding them [27,29–31].

The number of water bodies throughout the landscape used by each species depends on their home ranges and daily movements. For individuals of Baird's tapir, reported yearly mean home ranges encompass approximately 1.3 km$^2$ ($\pm 0.73$ km$^2$, SD) with a maximum home range of 2.3 km$^2$ in Costa Rica [32]. These authors reported that even though home ranges did not vary between seasons, during the wet season individual tapirs shared 26% of their annual home range with other tapirs, whereas overlap was usually null in the dry season; these observations suggest that water availability influences tapir use of the landscape. For white lipped peccaries, Reyna-Hurtado [30] estimated herd annual home ranges from 43.9 to 97.5 km$^2$ in the tropical semi-dry forest of the Calakmul Biosphere Reserve (home range estimates based on VHF data and 95% fixed kernel). White-lipped peccaries visit waterholes disproportionately more often during the dry season than in the wet season [17]. During this study, when water became scarce and was only available at the larger waterhole in the area, the groups remained at this waterhole and foraged in a radius of 6 km (mean distance <600 m). In the Calakmul Region, jaguars show differential habitat use by season based on the availability of water bodies, which affects the density and location of prey species [27]. Data obtained from satellite collars showed that the activity area for two males was about 1000 km$^2$ [27].

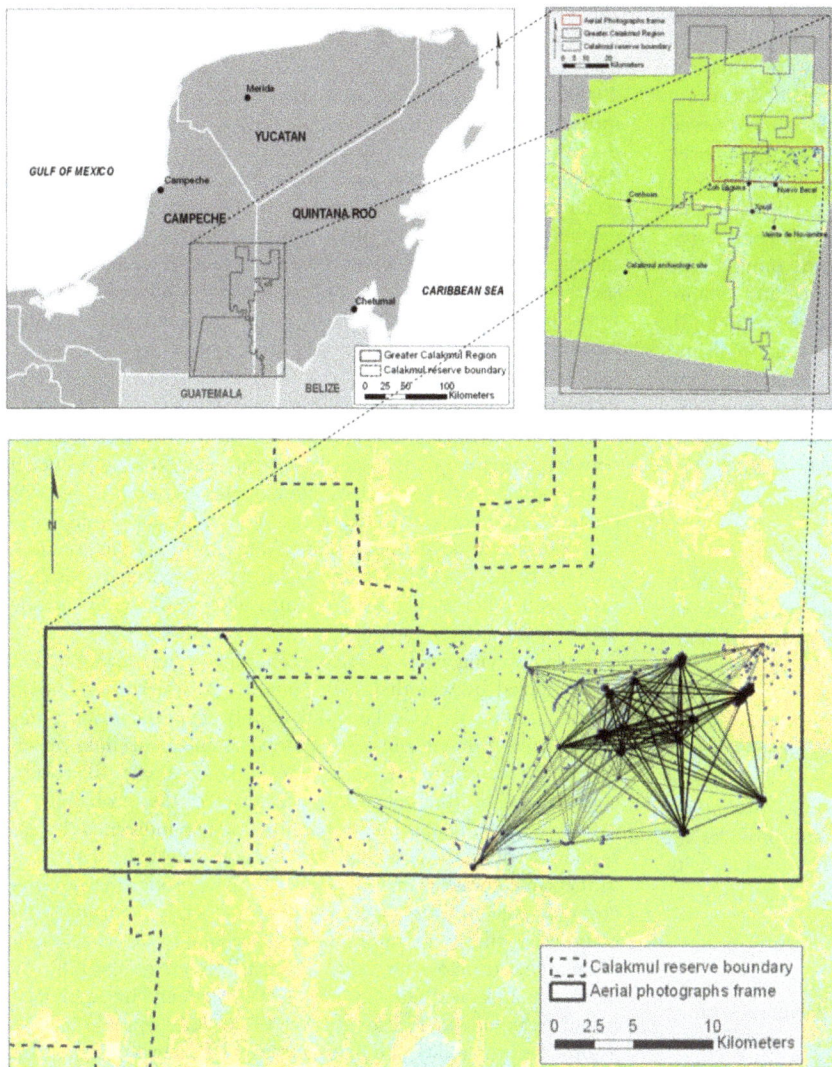

**Figure 1. Greater Calakmul Region and study area (upper figures).** This figure shows a representation of drought scenario D (grey links) and E (black links) for the 13 km travel distance. Links are lost inside and adjacent to the reserve and a narrow stepping-stone strip of waterholes remains, connecting the interior of the reserve with the smaller sub-network to the east, outside the reserve. The circle shows the waterhole that maintains the connectivity between the sub-networks.

In addition to home ranges, maximum travel capacities provide information about the ability of a species to reach distant water bodies if required. In Peru, Tobler [33] reported lowland tapir (*Tapirus terrestris*) individuals moving up to 13 km over a 24-hour period (GPS radio-collared data), with a mean movement distance in a 24-hr period of 5.2 km (range 3.6–6.7 km). This is the only formal study documenting tapir movements with GPS collars. When considering tapir movements, we used the travel distances reported for this species of tapir; however, we are confident these observations can be used for the Baird's tapir in our region as all tapir species depend on water for their survival [25]. Reyna-Hurtado *et al.* [17] found that white-lipped peccaries require visits to water bodies on an almost daily basis in our study region, performing search patterns at two spatial scales: they search one area intensively by moving no more than 3 km every day and occasionally perform long displacements (9 to >16 km) in a single direction that take them out of the previous searched area over the course of one to three days. For jaguars, the mean daily travelled

distance was 2.24 km with a maximal daily distance travelled of 10 km based on radio collar data [34].

In summary our model species have been reported to move distances between 3 and 13 km (minimum and maximum reported travel distances by tapirs), between 3 and 16 km (minimum and maximum travel distances by white-lipped peccaries) and between 2.24 and 10 km (minimum and maximum travel distances by jaguars). This range (3–16 km) provides the minimum and maximum potential movement capacities of these species. This suggests that these species have the ability to reach waterholes located further away than their daily home ranges.

## Remotely Sensed Image Interpretation and Ground Truthing

For our analysis, we used remotely sensed imagery (orthophotographs from March 1998 and 2001) to obtain the most accurate locations of the waterholes in our study area (see File S1). During fieldwork, we observed that most of the waterholes smaller than 400 m$^2$ were dry at the beginning of the dry season, therefore we only digitized water bodies >400 m². We visited 15 waterholes observed in the orthophotos, confirming the presence of 80% of them in the field. All misidentified waterholes were small (< 700 m²) and were subsequently excluded from the model. From 2006 to 2009, we repeatedly visited 15 waterholes and observed that large waterholes did not dry or were dry only during the peak of the dry season of very dry years. For example, in May 2006 a waterhole of 90,000 m² was observed to go dry for the first time since people re-colonized the area 40 years earlier (Nicolas Arias, comm. pers). These observations support climate models that suggest an increase in drought conditions in the area and capture the situation during a very dry year (2005) in the region [8]. Animal tracks (e.g., tapir, deer, peccaries) and dung near small and large waterholes are found at lower densities when waterholes are dry, suggesting that animals use waterholes mainly for water [35]. The intervening matrix surrounding waterholes was comprised of accessible forest for all species; no major physical barriers (e.g. roads, mountains, rivers) occur in the study area.

## Waterhole Networks

Water resource connectivity was assessed for each of the three model species using species-specific waterhole networks. Various drought scenarios were simulated on these networks through node deletion sequences that reflected waterhole-drying patterns. Deletion sequences were based on observed waterhole drying during the dry season of 4 consecutive years (2006–2009). The implications of these drought scenarios on species-specific water resource connectivity were assessed by quantifying the structure of sub-networks, network-wide probability of connectivity, and access to waterhole-associated habitat. Our analysis thus provides an indication of potential future changes in waterhole connectivity based on current waterhole distribution and drying pattern.

## Waterhole Network Delineation

The three species' networks consisted of the same set of nodes (waterholes), but each had a unique set of links reflecting the range of reported movement abilities of the three species, varying from 3 to 16 km (tapirs: 3–13 km; white-lipped peccaries: 3–16 km; and jaguars: 2.24–10 km). We identified links between waterhole centroids rather than waterhole edges due to high variability in waterhole edges with rainfall. Links that were longer than the maximum movement ability of each species were removed from their respective networks. Although we do not expect animals to necessarily travel their maximal distances in a day or constantly, these distances represent a reasonable estimate of the distance that these species might be able to travel to find water when it becomes limiting, e.g. to locate a new waterhole. Furthermore, in contrast to most resource networks where nodes represent patches of suitable habitat surrounded by an inhospitable matrix, in our study, nodes (waterholes) are separated by suitable habitat as the matrix landscape represents a continuous forested area. All waterhole networks were delineated using complete graph extraction in SELES (A Spatially Explicit Landscape Even Simulator; [36]).

## Waterhole Network Deletion Sequences Based on Drought Scenarios

The impact of drought on the connectivity of the waterhole network was modelled by removing nodes in increasing order of waterhole area. We made this assumption based on field observations that showed the largest waterholes were the last to become dry at the peak of the dry season from 2006 to 2009. We created a base scenario (A) that included all waterholes larger than

700 m², and 10 drought scenarios (B–K) excluding waterholes smaller than or equal to (B) 1,000 m², (C) 2,500 m², (D) 5,625 m²; (E) 10,000 m², (F) 15,625 m², (G) 22,500 m², (H) 30,600 m², (I) 40,000 m², (J) 50,600 m² and (K) 62,500 m². In the field, we measured the perimeter to evaluate the area of several waterholes that were monitored for water availability over the four years. The distribution of waterhole sizes allowed us to select our size class scenarios by providing maximum and minimum size limits. After deleting nodes in each scenario, only links shorter than the species-specific maximum movement distance were maintained within the species' waterhole network. These pruned waterhole networks were used in the calculation of connectivity for each drought scenario.

## Waterhole Network Connectivity Analyses

Prior to modelling drought scenarios, we analysed the sub-network structure of the species-specific waterhole networks by enumerating the connected components (clusters). Network clusters represent connected areas of a network within which individuals can move among nodes (waterholes) via direct links or indirect paths [20,37]. Different clusters are effectively isolated from each other as no links or paths allow individuals to move between them.

After applying each drought scenario, we recorded the total area covered by waterholes, the number of waterholes, and the density of links. We calculated link density (L) as the proportion of actual links present out of the total number of links in the equivalent complete network (L≤1). To evaluate the overall connectivity of the network we calculated the probability of connectivity defined as the probability that two individuals located randomly within the landscape are found at waterholes that belong to the same component [38]. Given the low proportion of water-covered area with respect to the total study area, we only considered the numerator of the probability of connectivity calculations to allow for better comparisons of results [38]. We derived a link attribute for dispersal probability by applying a negative exponential dispersal kernel to link lengths (m). We assumed a probability of 0.05 that species could move further than their maximum movement distance (straight line) to parameterize the negative exponential dispersal kernel [20]. We used the patch removal technique [39] to assess the contribution of each node in maintaining network connectivity for each species and under each drought scenario, which allowed us to rank waterholes based on their node importance to the network. This was measured as delta probability of connectivity [38,40]. We calculated the area covered by the network or sub-networks using ArcGIS, and analysed network connectivity using the igraph package in R (R Development Core Team, 2005).

## Waterhole-associated Habitat

Our model species are strongly associated with water bodies in the study area. Therefore, we assumed that their habitat must contain at least one waterhole and defined the suitable habitat as the area surrounding that waterhole. We assume that species are in general unwilling to travel long distances in search for resources on a daily basis, thus their movement patterns will likely occur mostly within a small area surrounding a waterhole as has been observed for the white-lipped peccary [17]. Therefore, to evaluate the species suitable habitat, we created 1 and 2 km radius buffers around the edge of each waterhole in each scenario using ArcGIS 9.3 Spatial Analyst Tool (ESRI 2009). By considering these radii, we assume species will move short distances during the day <3 km (minimum daily travel distance reported by all species) around a waterhole if all their requirements are fulfilled in that area. This analysis allowed us to estimate the loss of suitable habitat for the species and to complement the network connectivity analysis where we focus on the ability for the species to undertake long and rare movements. In addition, we quantified the percentage of total area covered by the suitable habitat in each scenario.

## Results

The initial waterhole network (scenario A) included 187 waterholes that represented a total area covered by water of 3.14 km$^2$ and included 17,391 potential links (when no threshold distance was considered; Fig. 2). For all travel distances, link density abruptly decreased when all small waterholes (2,500 m$^2$) were deleted showing a threshold type response at an early stage of the model. The frequency of waterholes for each size class declines as a power function with many small waterholes (n = 115 of ≤ 2,500 m$^2$) and a few large waterholes (n = 15 of ≥40,000 m$^2$; Fig. S1). Our base scenario consisted of one connected network for all travel distances, except for the 3 km distance. This suggests that only when individuals move more than 3 km they will be able to reach the remaining waterholes in each scenario. As travel distance decreases, the number of clusters (2 or more linked waterholes) increases (Fig. 3).

### Waterhole Network Structure

Drought dramatically influenced waterhole network structure and therefore the potential connectivity of waterholes and potential suitable habitat for our model species. An abrupt decrease in link density showing a threshold was observed when small waterholes (≤2,500 m$^2$) were removed, followed by stabilization in link density curve after scenario E, i.e., when medium size waterholes were removed (10000 m$^2$), with similar patterns for the different travel distances considered (Fig. 2). For example, for a travel distance of 13 km (e.g., within tapir and peccaries daily travel capacities), the number of links in the base scenario (scenario A-that included all waterholes larger than 700 m$^2$) was 7,424 (42.7% of all potential links, i.e., without a threshold distance). For this same example, scenario C would correspond to a rapid decrease in link density with only 209 links left, which corresponds to 16.3% of the links present in the base scenario. In scenario F (when the link density curve stabilizes) we observed only 94 links left, i.e., 1.3% of all links found in the base scenario.

The removal of waterholes ≤2,500 m$^2$ (scenario C), which represents the most frequent waterhole size in our study area, resulted in a loss of 61.5% of the total number of waterholes included in the network (Fig. 2). Most of the waterholes of this size dry fast during the early dry season and are not a reliable source of water. The removal of 90.9% of the waterholes (scenario F) caused a loss of only 15% of the total area covered by water; however, this caused a drastic reduction of more than 95% of the links for all travel distances (Fig. 2, scenario F). For scenario C, network connectivity was maintained inside and outside the reserve when a minimum of 5 km travel distance was considered- a distance within all species movement range (Fig. 3); however, these waterholes were typically observed to dry every year in the field.

A travel distance of 5 km (within all species travel capacities) did not ensure connectivity for all drought scenarios; for example, the sub-network observed inside the reserve had a sparser density of waterholes linked compared to the sub-network outside the reserve, and its connectivity was maintained only when waterholes smaller than or equal to 10,000 m$^2$ persisted in the landscape (Fig. S2). In a conservative scenario of waterhole deletion (scenario E) and for a travel distance of 13 km (maximal reported travel distance by tapirs), the sub-network within the reserve collapsed, leaving only three waterholes within the boundary of the reserve linked to the sparse network outside the reserve (Fig. 3). In the

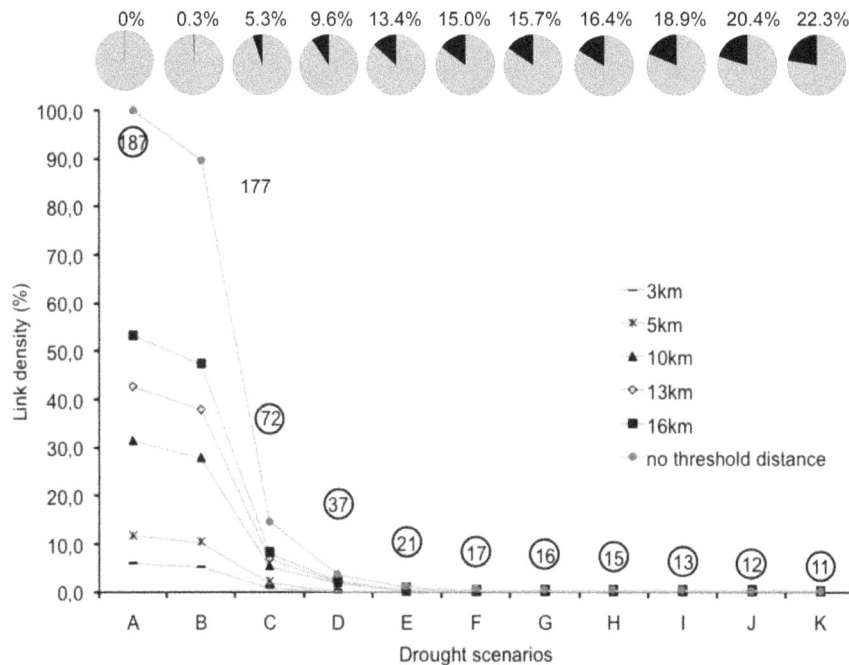

**Figure 2. Number of waterholes (embedded in circle) and link density in each drought scenario by travel distances of species: tapirs (3–13 km), white-lipped peccaries (3–16 km) and jaguars (3–10 km).** Drought scenarios correspond to waterhole removal based on waterhole size: A (all), B (≤1000 m$^2$), C (≤2500 m$^2$), D (≤5625 m$^2$), E (≤10000 m$^2$), F (≤15625 m$^2$), G (≤22500 m$^2$), H (≤30600 m$^2$), I (≤40000 m$^2$), J (≤50600 m$^2$), K (≤62500 m$^2$). Pie charts show the amount of waterhole area lost in each drought scenario. Lines between points are included only as reference to observe point-decreasing pattern.

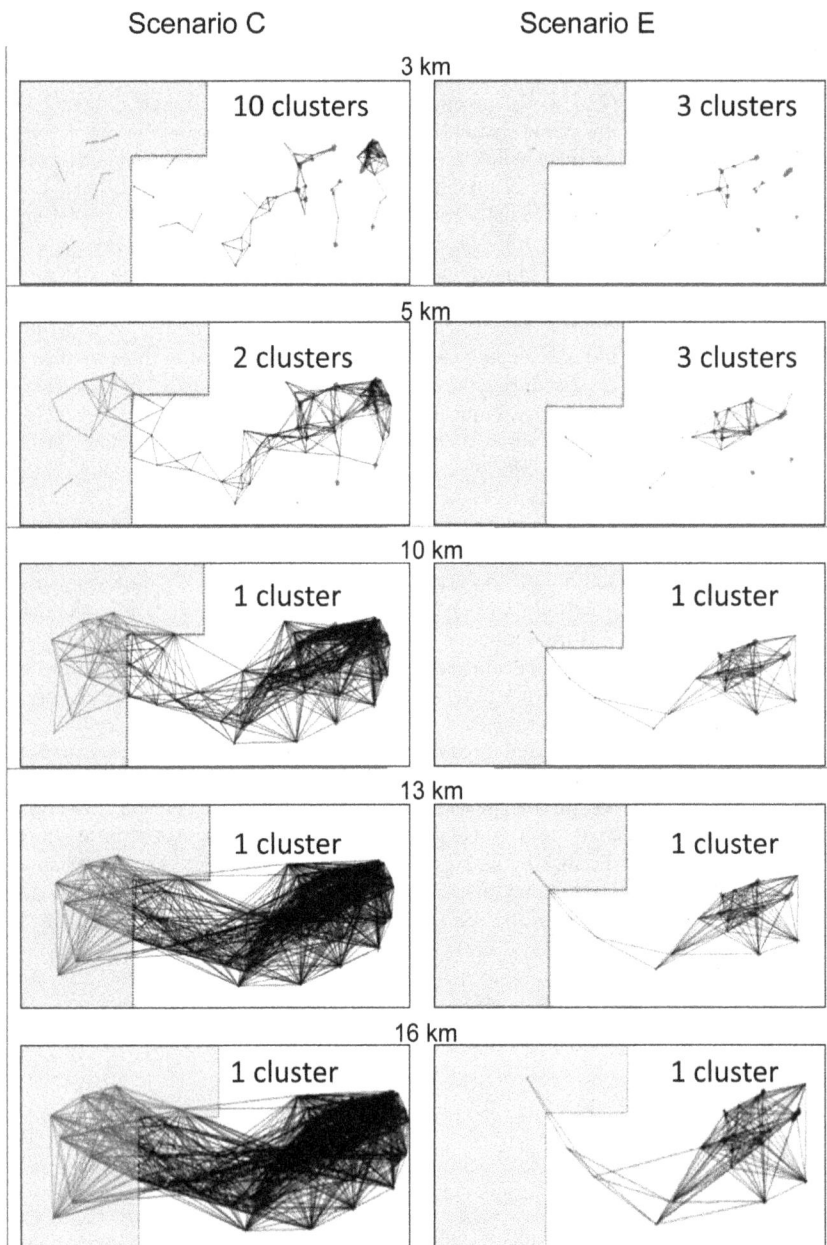

**Figure 3. Representations of the network graphs showing the changes in network structure for Scenario C (removal of waterholes ≤2500 m²) and E (removal of waterholes ≤10000 m²) at 3, 5, 10, 13 and 16 km which represent a range of travel distances for our study species: tapirs (3–13 km), white-lipped peccaries (3–16 km), and jaguars (3–10 km).** The two scenarios presented show abrupt changes in connectivity.

most severe drought scenario (K) waterholes became very sparsely connected outside the reserve and non-existent inside the reserve.

### Probability of Connectivity and Node Importance

We observed a decrease in the probability of connectivity as waterholes were removed from the network for each travel distance. The probability of connectivity increased with increasing travel distance. The probability of connectivity for all scenarios was lower at the lower range of travelled distances of all three species (3 and 5 km; Fig. 4). However, we found that even though node importance (given by the probability of connectivity of each node) decreased with waterhole area (smaller waterholes were less

important than large waterholes), the location of waterholes further influenced their node importance; therefore, large waterholes located far from other waterholes had lower node importance than smaller waterholes located close to other waterholes. For example, a waterhole of 172,500 m² was ranked 9th with respect to its node importance even though it was larger than waterholes ranked 8th given its location (Fig. S3).

### Suitable Habitat

The waterhole-associated area, or "suitable habitat" using a buffer radius of 1 km around waterholes, corresponded to 52% of the total area in the base scenario. The removal of waterholes ≤

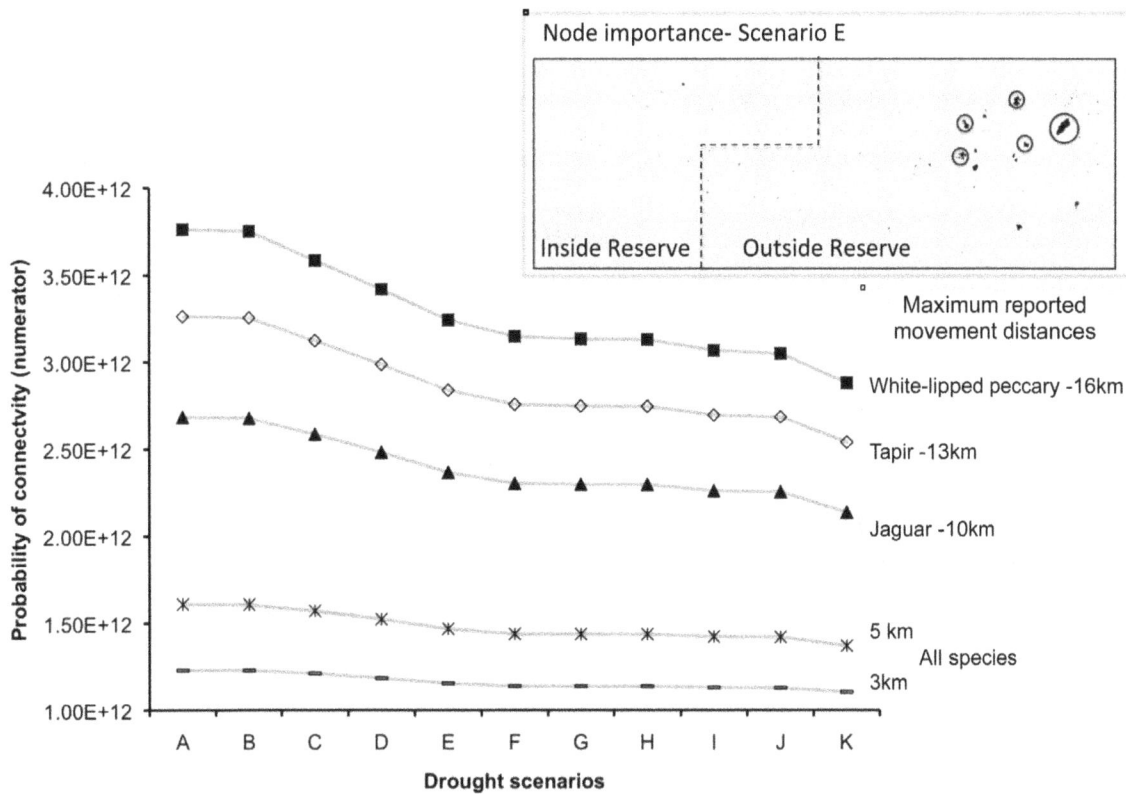

**Figure 4. Changes in the probability of connectivity with different travel distances.** Inset node importance map illustrates the five waterholes (encircled) with higher ranking based on their contribution to maintaining the overall probability of connectivity of the waterhole network in scenario E for all species (added probability of connectivity of each waterhole by all 5 distance thresholds used).

10,000 m$^2$ (scenario E) caused a large reduction with only 21% of initial suitable habitat remaining (waterhole-associated area in the base scenario; Fig. S4). If we set our buffer radius to 2 km, the removal of waterholes $\leq$10,000 m$^2$ still caused a large reduction of suitable habitat with only 29% of the suitable area remaining. Of the total area covered by our network, only 9% was found within a radius of 1 km and 25% within a radius of 2 km from the existing waterholes in scenario F (simulating the peak of the dry season). Given a buffer radius of 1 km, we found that the number of connected patches of suitable habitat (components of buffered patches) decreased by half (from 13 to 6 connected patches) from scenario E to K. The same pattern was observed when considering buffers of 2 km; only three connected patches of suitable habitat remained in our last scenario, K (Fig. S4).

## Discussion

Our results emphasize that: 1) waterhole number is seasonal and very sensitive to dry seasons, 2) changes in waterhole availability may not sustain a functionally connected network of waterholes for our endangered study species under present and future drought scenarios, and 3) network analysis can improve our understanding of reserve functioning and potential habitat connectivity in highly seasonal landscapes. Our analysis revealed that the potential connectivity of the waterhole network is very sensitive to drought for jaguars, white-lipped peccaries and Baird's tapirs. By using a range of reported travelled distances, we were able to model the effects of both daily movement patterns around waterholes (suitable habitat analysis) and long, rare movements (connectivity analysis). Our analyses highlight the potential negative effects for

three endangered species of observed trends of decreasing precipitation and future projections of changes in drought conditions in the area. The availability of water outside the protected area might result more attractive for water-dependant species, demonstrating the need for further species conservation programs in such a human dominated landscape.

### Loss of Connectivity

Based on our models, maintaining waterholes smaller than 10,000 m$^2$ (scenarios A–D) is especially important for the potential connectivity of the landscape both inside and adjacent to the reserve. The distribution and abundance of these small waterholes ensure accessibility to water without large and costly movements. In addition, animals might use these small waterholes as stepping-stones en route to large waterholes, especially to move between sub-networks inside and outside the reserve (Fig. 3). Removing small waterholes caused a non-linear decrease in the connectivity of the network showing that link density abruptly decreases after a threshold of waterhole size removal. When only large waterholes remain in the landscape these are connected by few links (Fig. 3), showing a fragile network of waterholes available for species. Our results can be explained by percolation theory, which suggests that when random habitat loss occurs across more than 60% of the landscape, the largest habitat patch size decreases abruptly and no longer spans the landscape [41]. In our landscape, more than 60% of the waterholes are lost at an early stage of our deletion model (scenario C), which causes an abrupt decrease in link density and connectivity, with only a few large and clumped waterholes remaining.

The results from our network analyses do not imply that tapirs, white-lipped peccaries and jaguars move between waterholes on a daily basis but suggest that movement may be constrained when the species are forced to move to new waterholes due to seasonal or permanent waterhole drying (e.g., due to climate change) or to disturbance (e.g., logging). In addition, we do not expect species to walk in straight lines, but straight-line movements are general considerations relevant to all network connectivity analysis and represent the shortest distance our study species could cover to move between waterholes; therefore, our results showing changes in potential connectivity using maximal distances should be seen as representing the upper limit of potential connectivity. Additional information on species' water requirements would improve the assessment of the distances species move between waterholes; however this information is not available for these species. Our approach can be applied to test the potential functional resource connectivity of any other temporally and spatially dynamic resource [11,21,42,43].

The potential connectivity of waterholes in the area was severely reduced for our focal species, despite their body size and dispersal capabilities, given drought scenarios that represent the peak of the dry season in the driest years during the study period. Even though a small portion of the network remained connected (Fig. 3), this remnant network fell outside the reserve. Our study emphasizes the importance of considering the spatiotemporal dynamics of resources inside and around protected areas [13,44]. Even though our study only corresponds to a portion of the Greater Calakmul Region, our fieldwork suggests that these patterns may be common throughout the region.

## Habitat Area vs. Quality?

Our results further emphasize that the spatio-temporal distribution of resources will likely determine the functional connectivity of the landscape. If resources become increasingly rare and then isolated, the chance that they fall outside the movement range of species will increase [45,46]. In this study, matrix habitat between waterholes is a homogenous-forested landscape, which allows free movement between waterholes. Our movement scenarios were not sensitive to estimates of the resistance to movement of the forested matrix, as is often the case in least-cost connectivity analyses in heterogeneous landscapes. Suitable habitat for the study species considers waterhole use; therefore habitat quality decreases as waterholes dry up causing species to perform longer and unusual movements. In our study, the changes in resource (waterhole) network connectivity in different drought scenarios show dynamic connectivity that will likely not be identified if only habitat area is considered in connectivity analyses; a result that suggests the importance of evaluating habitat quality in addition to habitat extent [11,47,48].

Tapirs, white-lipped peccaries, and jaguars possibly require more access to water when temperatures rise during the dry season, either to lower body temperature or, in the case of tapirs and white-lipped peccaries to compensate for the lower water content of their food items [28]. This shows that in very dry scenarios species are likely to be forced to move beyond their usual daily travel distances and in some cases even the maximum reported movement capacities of our model species might be insufficient for individuals to reach water resources. If species regularly perform short distance movements (no more than 2 km) suitable habitat may be seasonally limited in this continuous forested area. In addition, the number of connected patches of suitable area (overlapping buffer areas) decreases suggesting limited movement between suitable areas.

Even if our three model species are able to perform such long distance movements in less than three days [17,33,34], these movements are costly and performed rarely. Given the species' requirements, forested areas with only small waterholes, irrespective of the quality of the forest itself, are thus of decreasing quality as the dry season progresses. As observed in Figure 4, only a small percentage of suitable habitat remains during the most severe drought scenarios we studied. Such a situation was actually observed by Reyna-Hurtado et al. [17] in the Calakmul Biosphere Reserve. The four groups of white-lipped peccary they studied behaved like central-place foragers around, and foraged close to, the only remaining waterhole in their 240 km$^2$ study area. In addition, the loss of waterholes will not only affect jaguar habitat directly but it will indirectly influence prey availability near waterholes, which will have an overall effect on habitat quality. Therefore, habitat quality must be assessed in terms of resource availability for each species. More generally, although large reserves often contain critical areas of high quality habitat, this may not always be the case and analyses of the sort reported here can inform reserve design and management [49].

## Reserve and Corridor Design and Management

The effect of drought was not uniform across the study region and was particularly apparent within the Calakmul Biosphere Reserve. Climate models predict an increase in temperature and a reduction in precipitation for the area [8]; therefore our study, based on patterns of waterhole drying observed in the field between 2006 and 2009, assumed a realistic drought scenario. Our study allowed us to show that the potential connectivity of the waterhole network is dependent on movement capabilities and is dynamic in space and time. Additionally, only one waterhole keeps the reserve connected to the network outside the reserve (Fig. 1), the loss of which would interrupt a potential resource corridor for individuals moving east to the waterhole network beyond the reserve boundary. The loss of waterholes of this size was observed in the field at the peak of the dry season during a very dry season (e.g., 2006).

Our study region is an important part of the Mesoamerican Biological Corridor and is of considerable conservation value because of its high diversity and area [50]. The Greater Calakmul Region contains large waterholes that ensure water availability throughout the dry season even in years of very low precipitation (e.g. 2005; personal observation). Our study area is represented by a homogenous-forested matrix, which allows free access to waterholes for most species. However, the network analysis revealed higher node importance of large waterholes and documents that the spatial distribution of these large waterholes was heterogeneous and aggregated beyond the boundary of the reserve (Fig. 3). Available habitat outside the reserve might represent a better habitat for species in terms of water resources. Habitat outside the protected area experiences threats caused by human activities such as hunting or habitat disruption by logging and agriculture which can hinder the survival of species. Waterholes located outside the reserve, and with a larger area, showed higher node importances in all scenarios, which emphasize their contribution to maintaining connectivity of the remaining network (Fig. 4). These large waterholes are all located outside the reserve are found close to each other, which suggests that this area might become a refuge for species if drought conditions continue as predicted.

Our results suggest that resource connectivity should be at the centre of reserve network design [51]. This will be especially important if trends of climate or land use change directly impact resource availability. In addition, human populations in the area also depend on water bodies. Critical waterholes important for humans and fauna in the region currently lack a sustainable management plan. The approach we used in this paper can be used to rapidly assess landscape viability and vulnerability for a range of species specifically considering variations across space and time. Our approach can also be used to initially prioritize habitat and resources for landscape management programs and to target further data collection and monitoring [47,49] in areas where critical resources are spatially and temporally dynamic. Our results emphasize the need to better understand the availability of water inside the reserve and the consequences for species survival in this protected area. Conservation actions are needed outside the reserve not only to ensure the survival of species in areas with low waterhole abundance, but also to identify areas of potential human-animal conflicts if animals move outside the reserve to find water (or other resources). A higher rate of hunting and crop-raid events might be expected outside the reserve; therefore, further studies on these topics are needed to inform conservation actions.

Given the rapid effects of climate change, which in some areas has now translated into altered precipitation regimes [21], we require new approaches to create dynamic reserve and corridor network designs that incorporate the temporal and spatial dynamics of resources [43,51]. This study considered the effect of climate change on species persistence by evaluating the effect of changes in precipitation in water resource availability and the connectivity of resources. We have provided science-based information that can inform future conservation programs in the area. These programs can be established for the protection of key water bodies inside and outside the reserve, for the conservation of areas where water bodies with higher node importance value were found, and to promote further studies on the movement capacities of these endangered species, their water requirements, and the potential consequences of a higher abundance of these species outside the protected area.

## Future Considerations

Our analysis suggests that even though waterholes may remain connected during the wet season, resource connectivity is abruptly affected during the dry season. If current trends of precipitation continue, drought periods are expected to be longer and affect the waterhole network. In particular, our results suggest that the loss of a small number of water holes has a large effect on the network's structure and connectivity. These effects will increase in strength when dry years occur consecutively and may force species to move to unprotected areas beyond the reserve. Currently this effect is seasonal but it may reflect future scenarios with longer and more severe dry seasons. A permanent shift in conditions would have consequences for species persistence in and around the reserve. Climate and land use changes will dramatically alter the functional connectivity of this region, and hence the conservation capacity of the Calakmul Biosphere Reserve.

Our network approach allowed us to link field and GIS data to analyse the potential connectivity of a waterhole network for three

large mammal species of conservation concern. Although this modelling approach has been applied to study the impacts of habitat fragmentation [4,19] few results are available for water resource networks [52] and their dynamics [53]. We recognize that detailed data and habitat use patterns are still missing from our model. However, our modelling approach is easily updated as movement data (e.g., from GPS collars) becomes available. Next steps will involve modelling demography to allow the identification of key features of the network that are critical for metapopulation persistence under climate change [11,20]. Furthermore, changing human land use in the area might influence species' movements in the future, so the inclusion of land use change (e.g. logging roads or human settlement) will also be important to evaluate change in landscape connectivity.

## Supporting Information

**Figure S1　Power relationship between waterhole size and frequency, $R^2 = 0.64$.** Each dot corresponds to a drought scenario described in the text. Note the log scale of both axes.

**Figure S2　Network graphs showing changes in network structure when considering a 5 km travel distance for scenarios A (all waterholes considered), C (waterholes $\leq 2500$ m$^2$ removed), E (waterholes $\leq 10000$ m$^2$ removed), and J (waterholes $\leq 50600$ m$^2$ removed).** The grey area corresponds to the reserve.

**Figure S3　Area of patches with the ten highest node importance values.** The symbols correspond to each of the 11 drought scenarios A to K.

**Figure S4　Percentage of suitable habitat lost in each drought scenario.** The upper figures show the buffer analysis graphs with waterholes remaining in drought scenario E and K and their 1 km and 2 km buffers.

## Acknowledgments

We thank Marie-Josée Fortin, Patrick Leighton and Martin Lechowicz for their valuable comments on a previous version of this manuscript.

## Author Contributions

Analyzed the data: GO KGS BR. Contributed reagents/materials/analysis tools: KGS OB RS AG. Wrote the paper: GO. Manuscript editing: BR SC RS AG.

## References

1. Crooks KR, Sanjayan M (2006) Connectivity conservation: Cambridge University Press, Cambridge.
2. With KA, Crist TO (1995) Critical thresholds in species responses to landscape structure Ecology 76: 2446–2459.
3. Bodin O, Norberg J (2007) A network approach for analyzing spatially structured populations in fragmented landscape. Landscape Ecol 22: 31–44.
4. Kadoya T (2009) Assessing functional connectivity using empirical data. Popul Ecol 51: 5–15.
5. Hetherington DA, Miller DR, Macleod CD, Gorman ML (2008) A potential habitat network for the Eurasian lynx Lynx lynx in Scotland. Mamm Rev 38: 285–303.

6. Clobert J, Le Galliard JF, Cote J, Meylan S, Massot M (2009) Informed dispersal, heterogeneity in animal dispersal syndromes and the dynamics of spatially structured populations. Ecol Lett 12: 197–209.
7. Stevens VM, Baguette M (2008) Importance of habitat quality and landscape connectivity for the persistence of endangered natterjack toads. Conserv Biol 22: 1194–1204.
8. IPCC (2007) Climate Change 2007: Synthesis Report. Contribution of Working Groups I, II and III to the Fourth Assessment Report of the Intergovernmental Panel on Climate Change [Core Writing Team, Pachauri, R. K. and Reisinger, A. (eds.)]. IPCC, Geneva, Switzerland. 104.

9. Larson B, Sengupta R (2004) A spatial decision support system to identify species-specific critical habitats based on size and accessibility using US GAP data. Environ Model 19: 7–18.

10. Wright CK (2010) Spatiotemporal dynamics of prairie wetland networks: power-law scaling and implications for conservation planning. Ecology 91: 1924–1930.

11. Fortuna MA, Gomez-Rodriguez C, Bascompte J (2006) Spatial network structure and amphibian persistence in stochastic environments. Proc R Soc Lond, Ser B: Biol Sci 273: 1429–1434.

12. Chamaillé-Jammes S, Fritz H, Valeix M, Murindagomo F, Clobert J (2008) Resource variability, aggregation and direct density dependence in an open context: the local regulation of an African elephant population. J Anim Ecol 77.

13. Hansen AJ, DeFries R (2007) Ecological mechanisms linking protected areas to surrounding lands. Ecol Appl 17: 974–988.

14. Andren H (1994) Effects of habitat fragmentation on birds and mammals in landscapes with different proportions of suitable habitat: a review Oikos 71: 355–366.

15. Gonzalez A, Rayfield B, Lindo Z (2011) The disentangled bank: how habitat loss fragments and disassembles ecological networks. Am J Bot 98: 503–516.

16. Márdero S, Nickl E, Schmook B, Schneider L, Rogan J, et al. (2012) Sequías en el sur de la península de Yucatán: análisis de la variabilidad anual y estacional de la precipitación. Investigaciones Geográficas, Boletín del instituto de Geografía, UNAM, Mexico 78: 19–33.

17. Reyna-Hurtado R, Chapman CA, Calmé S, Pedersen EJ (2012) Searching in heterogeneous and limiting environments: foraging strategies of white-lipped peccaries (*Tayassu pecari*). J Mammal 93: 124–133.

18. Dale MRT, Fortin MJ (2010) From graphs to spatial graphs. Annu Rev Ecol, Evol Syst 41: 21–38.

19. Andersson E, Bodin O (2009) Practical tool for landscape planning? An empirical investigation of network based models of habitat fragmentation. Ecography 32: 123–132.

20. Urban D, Keitt T (2001) Landscape connectivity: A graph-theoretic perspective. Ecology 82: 1205–1218.

21. Valeix M, Fritz H, Chamaillé-Jammes S, Bourgarel M, Murindagomo F (2008) Fluctuations in abundance of large herbivore populations: insights into the influence of dry season rainfall and elephant numbers from long-term data. Anim Conserv 11: 391–400.

22. Vester HFM, Lawrence D, Eastman JR, Turner BL, Calmé S, et al. (2007) Land change in the southern Yucatan and Calakmul biosphere reserve: Effects on habitat and biodiversity. Ecol Appl 17: 989–1003.

23. Martínez E, Galindo-Leal C (2002) La vegetación de Calakmul, Campeche, México: clasificación, descripción y distribución. Boletín de la Sociedad Botánica de México 71: 7–32.

24. García-Gil G, Palacio JL, Ortiz MA (2002) Reconocimiento geomorfológico e hidrográfico de la Reserva de la Biosfera Calakmul, México. Investigación Geográficas Boletín del Instituto de Geografía, UNAM 48: 7–23.

25. Castellanos A, Foerster CR, Lizcano DJ, Naranjo E, Cruz-Aldan E, et al. (2008) *Tapirus bairdii*. IUCN 2009. IUCN Red List of Threatened Species: Version 2009.2.

26. Reyna-Hurtado R, Taber A, Altrichter M, Fragoso JMV, Keuroghlian A, et al. (2008) *Tayassu pecari*. *IUCN 2012*. IUCN Red List of Threatened Species. Version 2012.1.

27. Chávez Tovar JC (2010) Ecology and conservation of jaguar (*Panthera onca*) and puma (*Puma concolor*) in the Calakmul Region, and its implications for the conservation of the Yucatan peninsula: Ph.D Thesis. Departamento de Biologia Animal, Universidad de Granada.

28. Owen-Smith RN (1992) Megaherbivores: The influence of very large body size on ecology: Cambridge University Press.

29. Naranjo E, Bodmer RE (2002) Population ecology and conservation of Baird's tapir (*Tapirus bairdii*) in the Lacandon Forest, Mexico. Newsletter of the IUCN/SSC Tapir Specialist Group 11: 25–33.

30. Reyna-Hurtado R, Rojas-Flores E, Tanner GW (2009) Home range and habitat preferences of white-lipped peccaries (*Tayassu pecari*) in Calakmul, Campeche, Mexico. J Mammal 90: 1199–1209.

31. Caso A, Lopez-Gonzalez C, Payan E, Elzirik E, de Oliveira T, et al. (2008) *Panthera onca*. In: IUCN 2012. IUCN Red List of Threatened Species. Version 2012.1.

32. Foerster CR, Vaughan C (2002) Home range, habitat use, and activity of Baird's tapir in Costa Rica. Biotropica 34: 423–437.

33. Tobler MW (2008) The ecology of the lowland tapir in Madre de Dios, Peru: using new technologies to study large rainforest mammals: Ph. D. Thesis. Texas A&M University.

34. Colchero F, Conde DA, Manterola C, Chavez C, Rivera A, et al. (2011) Jaguars on the move: modeling movement to mitigate fragmentation from road expansion in the Mayan Forest. Anim Conserv 14: 158–166.

35. Perez-Cortez S, Enriquez PL, Sima-Panti D, Reyna-Hurtado R, Naranjo EJ (2012) Influence of water availability in the presence and abundance of *Tapirus bairdii* in the Calakmul forest, Campeche, Mexico. Revista Mexicana de Biodiversidad 83: 753–761.

36. Fall A, Fall J (2001) A domain-specific language for models of landscape dynamics. Ecol Model 141: 1–18.

37. Galpern P, Manseau M, Fall A (2011) Patch-based graphs of landscape connectivity: A guide to construction, analysis and application for conservation. Biol Conserv 144: 44–55.

38. Saura S, Pascual-Hortal L (2007) A new habitat availability index to integrate connectivity in landscape conservation planning: Comparison with existing indices and application to a case study. Landscape Urban Plann 83: 91–103.

39. Bodin O, Saura S (2010) Ranking individual habitat patches as connectivity providers: Integrating network analysis and patch removal experiments. Ecol Model 221: 2392–2405.

40. Pascual-Hortal L, Saura S (2006) Comparison and development of new graph-based landscape connectivity indices: towards the priorization of habitat patches and corridors for conservation. Landscape Ecol 21: 959–967.

41. Swift TL, Hannon SJ (2010) Critical thresholds associated with habitat loss: a review of the concepts, evidence, and applications. Biological Reviews 85: 35–53.

42. Gomez-Rodriguez C, Diaz-Paniagua C, Serrano L, Florencio M, Portheault A (2009) Mediterranean temporary ponds as amphibian breeding habitats: the importance of preserving pond networks. Aquat Ecol 43: 1179–1191.

43. Telleria JL, Ghaillani HEM, Fernandez-Palacios JM, Bartolome J, Montiano E (2008) Crocodiles *Crocodylus niloticus* as a focal species for conserving water resources in Mauritanian Sahara. Oryx 42: 292–295.

44. Pringle CM (2001) Hydrologic connectivity and the management of biological reserves: a global perspective. Ecol Appl 11: 981–998.

45. Loarie SR, Van Aarde RJ, Pimm SL (2009) Fences and artificial water affect African savannah elephant movement patterns. Biol Conserv 142: 3086–3098.

46. Sitters J, Heitkonig IMA, Holmgren M, Ojwang GSO (2009) Herded cattle and wild grazers partition water but share forage resources during dry years in East African savannas. Biol Conserv 142: 738–750.

47. Saura S, Rubio L (2010) A common currency for the different ways in which patches and links can contribute to habitat availability and connectivity in the landscape. Ecography 33: 523–537.

48. Metzger JP, Decamps H (1997) The structural connectivity threshold: An hypothesis in conservation biology at the landscape scale. Acta Oecologica-International Journal of Ecology 18: 1–12.

49. Calabrese JM, Fagan WF (2004) A comparison-shopper's guide to connectivity metrics. Front Ecol Environ 2: 529–536.

50. Neeti N, Rogan J, Christman Z, Eastman JR, Millones M, et al. (2012) Mapping seasonal trends in vegetation using AVHRR-NDVI time series in the Yucatan Peninsula, Mexico. Remote Sens Lett 3: 433–442.

51. Herbert ME, McIntyre PB, Doran PJ, Allan JD, Abell R (2010) Terrestrial reserve networks do not adequately represent aquatic ecosystems. Conserv Biol 24: 1002–1011.

52. Pereira M, Segurado P, Neves N (2011) Using spatial network structure in landscape management and planning: A case study with pond turtles. Landscape Urban Plann 100: 67–76.

53. Cote D, Kehler DG, Bourne C, Wiersma YF (2009) A new measure of longitudinal connectivity for stream networks. Landscape Ecol 24: 101–113.

# 2

# Interactive Effects of Elevated CO$_2$ Concentration and Irrigation on Photosynthetic Parameters and Yield of Maize in Northeast China

**Fanchao Meng**[1,2], **Jiahua Zhang**[1,3]*, **Fengmei Yao**[4], **Cui Hao**[1]

**1** Institute of Eco-Environment and Agro-Meteorology, Chinese Academy of Meteorological Sciences, Beijing, China, **2** College of Atmospheric Science, Nanjing University of Information Science & Technology, Nanjing, China, **3** Key Laboratory of Digital Earth Science, Institute of Remote Sensing and Digital Earth, Chinese Academy of Sciences, Beijing, China, **4** Key Laboratory of Computational Geodynamics, Chinese Academy of Sciences, Beijing, China

## Abstract

Maize is one of the major cultivated crops of China, having a central role in ensuring the food security of the country. There has been a significant increase in studies of maize under interactive effects of elevated CO$_2$ concentration ([CO$_2$]) and other factors, yet the interactive effects of elevated [CO$_2$] and increasing precipitation on maize has remained unclear. In this study, a manipulative experiment in Jinzhou, Liaoning province, Northeast China was performed so as to obtain reliable results concerning the later effects. The Open Top Chambers (OTCs) experiment was designed to control contrasting [CO$_2$] i.e., 390, 450 and 550 μmol·mol$^{-1}$, and the experiment with 15% increasing precipitation levels was also set based on the average monthly precipitation of 5–9 month from 1981 to 2010 and controlled by irrigation. Thus, six treatments, i.e. C$_{550}$W$_{+15\%}$, C$_{550}$W$_0$, C$_{450}$W$_{+15\%}$, C$_{450}$W$_0$, C$_{390}$W$_{+15\%}$ and C$_{390}$W$_0$ were included in this study. The results showed that the irrigation under elevated [CO$_2$] levels increased the leaf net photosynthetic rate ($P_n$) and intercellular CO$_2$ concentration ($C_i$) of maize. Similarly, the stomatal conductance ($G_s$) and transpiration rate ($T_r$) decreased with elevated [CO$_2$], but irrigation have a positive effect on increased of them at each [CO$_2$] level, resulting in the water use efficiency (*WUE*) higher in natural precipitation treatment than irrigation treatment at elevated [CO$_2$] levels. Irradiance-response parameters, e.g., maximum net photosynthetic rate ($P_{nmax}$) and light saturation points (*LSP*) were increased under elevated [CO$_2$] and irrigation, and dark respiration ($R_d$) was increased as well. The growth characteristics, e.g., plant height, leaf area and aboveground biomass were enhanced, resulting in an improved of yield and ear characteristics except axle diameter. The study concluded by reporting that, future elevated [CO$_2$] may favor to maize when coupled with increasing amount of precipitation in Northeast China.

**Editor:** Gerrit T.S. Beemster, University of Antwerp, Belgium

**Funding:** This work was supported by Global Change Research Key Project of MOST 973 Program (No. 2010CB951302), the Hundred Talents Program of CAS (Y24002101A), the Social Commonweal Meteorological Research Project (No. GYHY201106027), 1-3-5 Innovation Project of CAS-RADI (Y3ZZ15101A) and CAS-TWAS project for drought monitoring and assessment. The funders had no role in study design, data collection and analysis, decision to publish, or preparation of the manuscript.

**Competing Interests:** The authors have declared that no competing interests exist.

* E-mail: jhzhang@ceode.ac.cn

## Introduction

The CO$_2$ concentration ([CO$_2$]) in the atmosphere is about 390 μmol·mol$^{-1}$ as a consequence of fossil fuel combustion and deforestation, which is predicted to reach 550 μmol·mol$^{-1}$ by the middle of this century [1]. Elevated [CO$_2$] is an important abiotic factor, and has significant fertilization effects on crops. Extensive previous studies have reported that elevated [CO$_2$] significantly improved water use efficiency, lower transpiration rate, shorten maize growth period, and increased plant height, leaf number, leaf area, growth rate and yield [2–12]. In addition, the increasing of atmospheric [CO$_2$] affects precipitation balance, which can change the seasonal precipitation distribution [13]. It has been estimated that this effect would bring about a 10% increase or decrease in water resources at different areas [14]. The global annual average precipitation increase is about 2% since the beginning of the 20$^{th}$ century [15–16], and this rise over the area of 30°–85°N has shown a 7%–12% increase [17]. It has been predicted that the rainfall decrease will be noticed in middle-and-lower regions of Yangtze River (24°N–34°N,108°E–122°E), while the rain belts are likely to move towards north of China and precipitation would increase in Northeast China in the future [18]. The crop growth of Northeast China will likely be affected by both elevated [CO$_2$] and increasing precipitation, which are important abiotic factors that directly or indirectly affect crop growth, physiological processes and productivity. Thus, it is necessary to understand the interactive effects of elevated [CO$_2$] and increasing precipitation on crop growth in Northeast China under future climate change.

In fact, lots of studies have been focused on the interactive effects of elevated [CO$_2$] and other environmental factors on plant growth. The study of the interactive effects of elevated [CO$_2$] and temperature indicated that the effects on photosynthesis and growth in C$_4$ species are obvious [19–22]. FACE (Free Air Carbon-dioxide Enrichment) and chamber experiment have demonstrated that the interactive effects of elevated [CO$_2$] and drought stress have an increase in the leaf water-use efficiency [23–24], and more recent evidence shows that maize will benefit from the increase in [$CO_2$] under drought condition [25–29]. Also, the studies of the interactive effects of elevated [$CO_2$] and light on plant found that high light have a great effect on net photosynthesis in condition of elevated [$CO_2$] [30–31]. Regarding the interactive effects of elevated [$CO_2$] and Ozone ($O_3$), studies showing that elevated [$CO_2$] inhibits adverse effects of $O_3$ and increased trees seedling stem diameters at low $O_3$ [32–35]. Moreover, the interactive effects of elevated [$CO_2$] and soil nutrition have been investigated. For example, the studies of the interactive effects of elevated [$CO_2$] and nitrogen (N) indicated that there is a positive $CO_2 \times N$ interaction for grain yield of rice [36–38], while the research on the interactive effects of elevated [$CO_2$] and potassium (K) found that plants grown under elevated [$CO_2$] are more sensitive to K deficiency with higher leaf critical K levels [39]. Further, there are lots of studies involving the interactive effects of elevated [$CO_2$] and other factors (e.g., Nacl-salinity, plant diversity) have been reported [40–41]. However, the interactive effects of elevated [$CO_2$] and increasing precipitation on photosynthesis and yield of maize are not well understood. In particular, there has been no detailed study evaluating the interactive effects of elevated [$CO_2$] and increasing precipitation on photosynthesis efficiency, water use efficiency and yield of maize in Northeast China.

Northeast China (38°N–56°N,120°E–135°E) is located in the middle-high latitudes and east of the Eurasian continent, which has a cultivated land area of 21.53 million hm$^2$, accounted for 16.6% of the country's total cultivated areas [42]. The summer is warm and short, and the annual precipitation is 400–800 mm. The precipitation from July to September is accounting for 60% of the annual precipitation. Moreover, Northeast China has fertile black soil, belonging to one of the three pieces of black soil in the world [43], hence, which is the biggest commercial grain production base and provides 30–35 million tons of commercial grain to country every year [44]. Therefore, it plays an important role to stabilize the grain market and keep sustainable development of China's national economy. In Northeast China, maize (*Zea mays L.*) is the major cultivated crop, and its yield has accounted for about 1/3 of the national total maize yield [45]. The growth of maize requires more water, which yields tend to decrease if water deficit occurred during the key growing stages (e.g., silking stage) [46]. The precipitation in Northeast China can meet the water requirements of maize in most of the years, but slight drought has been discovered to occur in some of the past years. Therefore, for the rain-fed maize in Northeast China, precipitation is a very important climatic factor. If the water deficit occurred at silking stage of maize will cause disaccord flowering season, and then affect the pollination and seed formation, resulting in maize yield reduction.

To examine the interactive effects of elevated [$CO_2$] and increasing precipitation on maize in Northeast China, we conducted an Open Top Chambers (OTCs) experiment under the combined effects of elevated [$CO_2$] and precipitation in Jinzhou, Liaoning province during maize growing season (May to September) in 2013. Firstly, we tested the response of leaf gas exchange parameters (e.g., $P_n$, $T_r$) and irradiance-response parameters (e.g., $P_{nmax}$, *LSP*) to the combined elevated [$CO_2$] and increasing precipitation. Secondly, we examined the change in the growth parameters of maize (e.g., leaf area, aboveground biomass), yield and ear characters (e.g., ear length, ear diameter). The results of this study would be crucial for evaluating the possible consequences of climate change on crop photosynthetic

capacity and yield in Northeast China, and may help inform regulatory policies to cope with the future climate change.

## Materials and Methods

### Experimental site

This study site is located at Jinzhou Ecological and Agricultural Meteorological Experiment Center (41°09′N,121°10′E,27.4 m a.s.l.) in Liaoning province of China, which is a warm temperature monsoon climate zone. The mean annual precipitation is 568.8 mm, and the mean annual temperature is 9.1°C. The annual frost-free period is approximately 180 d in duration, with an annual accumulated activity temperature is 3700°C·d. The site has typical brown soil, and the soil pH is approximately 6.3. The soil organic matter and total N are 6.41–9.43 g/kg and 0.69 g/kg, respectively [47].

### Open Top Chamber design

Three pairs of Open Top Chambers (OTCs), each 3.5 m high with an octagonal ground surface area of 11.73 m$^2$ were constructed. An inclined plane of 45° (inward on upper side of chambers) was provided for reducing gas escape from the top. The set up was completed in May 2011 (Fig. 1).

Additionally, the OTCs were constructed with a 5.5-m-wide buffer zone between them to prevent mutual shading. Carbon dioxide was supplied to the chambers through a pipe with pinholes connected to industrial carbon dioxide cylinders (liquid carbon dioxide, purity was 99.99%, supplied by Anjin Gas Corporation) outside the chambers. There was an exchange fan of each chamber, which mixed the entered carbon dioxide and fresh air from outside, then transported by pipe and well distributed in the entire chamber by the octagonal-pipe with holes, and the gas would discharge from the opening on top and put in air circulation. Carbon dioxide was supplied for 24 hours a day and the [$CO_2$] was monitored by taking constant measurements with an infrared gas analyzer.

### Experimental design

The effects of elevated [$CO_2$] and precipitation on photosynthesis, growth, yield and ear characteristics of maize were examined in the chamber experiments. Considering the present ambient [$CO_2$] and projected increasing [$CO_2$] levels in next several decades from IPCC (2007) [1], three [$CO_2$] levels were conducted with the OTC experiments including 390 μmol·mol$^{-1}$ ($C_{390}$), 450 μmol·mol$^{-1}$ ($C_{450}$) and 550 μmol·mol$^{-1}$ ($C_{550}$).

**Figure 1. View of Open Top Chambers on 21 July 2013 at Liaoning province, China.**

According to 5–9 month average monthly precipitation in Jinzhou during 1981–2010 (see Fig. 2) and the prediction of increasing rainfall in Northeast China [18,48–49], two levels of watering including natural precipitation ($W_0$) and precipitation increased by 15% ($W_{+15}$) were designed. With this, maize, grown from seeds, was subjected to 6 different combined treatments (Table 1).

Three pairs of OTCs were used, and within each pair one was randomly assigned to receive the control (390 $\mu mol \cdot mol^{-1}$) and the others as the elevated [$CO_2$] (450 and 550 $\mu mol \cdot mol^{-1}$). Every chamber was designed with two watering gradient (0 and +15%), and each gradients with 7 pots (50.5 cm (diameter)×32.5 cm (height)), producing 14 plots per treatment with a total of 84 plots. Based on the local long-term (1981–2010) monthly mean precipitation, watering volumes of each pot were 20.43 L, 35.43 L, 32.21 L and 11.96 L in the $W_{+15\%}$ treatment plots from June to September, respectively. It was divided into everyday average irrigation, directed into the pots in the morning and evening daily, and covering the raining days too.

The maize cultivar used in this study was Danyu 39, and now it is widely planted in Northeast China. Three maize seeds were planted in the pots using field soil on 10 May 2013, with final seeding at four-leaf stage. On 1 June, the 84 pots containing plants were moved into the chambers randomly, and each chamber contained 14 pots. Meanwhile, the levels of [$CO_2$] and water supplied were under monitored and controlled until harvest time (15 September).

## Rainfall date and relative soil moisture measurements

The rainfall data were obtained from the Jinzhou weather station. The relative soil moisture (RSM) was calculated by equation (1) given below [50]:

$$RSM(\%) = \frac{soil\ moisture\ content}{soil\ field\ capacity} \times 100\% \qquad (1)$$

The soil moisture content was measured at a 0–20 cm soil depth. The soil samples of each treatment was collected and recorded as fresh weights and the samples were dried in an oven at 105°C for at least 48 hours. The soil moisture content was then measured using equation (2) given below:

$$Soil\ moisture\ content(\%) =$$
$$\frac{fresh\ weight\ of\ soil - dry\ weight\ of\ soil}{dry\ weight\ of\ soil} \times 100\% \qquad (2)$$

And the soil field capacity in this study was used the value of 21%, which according to the average value of soil field capacity in the last few years at this study site [51].

## Gas exchange measurements of maize

Measurements were made between 8:30 a.m. and 11:30 a.m. (local time) from 24–26 July, 2013 (at the maize silking stage). Three representative plants were chosen for per treatment and the middle of the ear-leaves was measured, then the averages were taken. Gas exchange measurements were conducted using portable gas exchange systems (LI-6400, LI-Cor, Lincoln, NE, USA). The [$CO_2$] in the leaf chamber was controlled by the LI-Cor $CO_2$ injection system, and the built-in LED lamp (red/blue) supplied the irradiance. Temperature in leaf chamber was set at 30°C, and the actual temperature of the leaf chamber ranged from 29 to 33°C. The vapour pressure deficit on the leaf surface (VpdL) was between 2.9 and 3.4 kPa, and the flow control was at 500 $\mu mol \cdot s^{-1}$. The lamp settings across the series of 2000, 1600, 1200, 1000, 800, 600, 400, 200, 150, 100, 80, 50, 20 and 0 $\mu mol \cdot m^{-2} \cdot s^{-1}$, and the measurements were recorded after equilibrium was reached. Each individual curve took approximately 30 min to complete.

$P_n$ curve fitting and analyzed of parameters were performed using the modified rectangular hyperbolic model by Ye and Yu [52–53], and its expression and correlative equations is as given below:

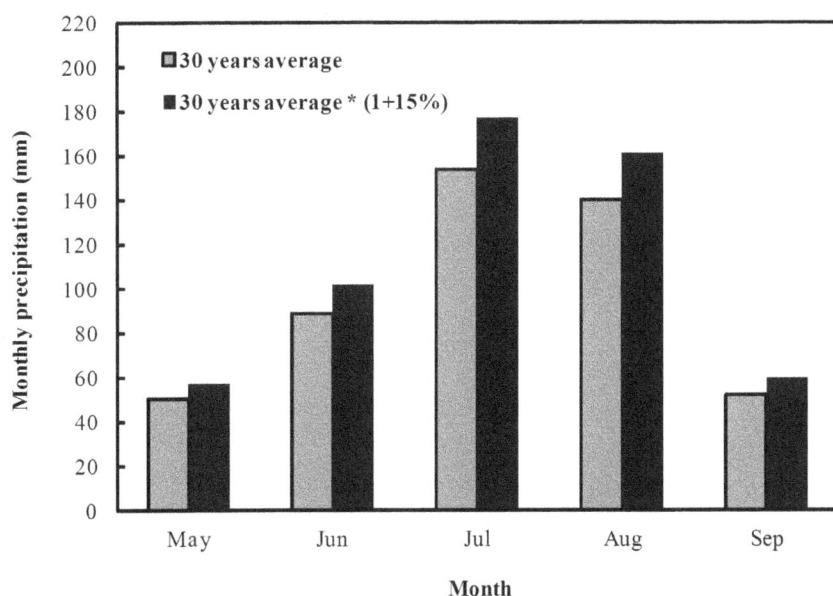

**Figure 2. Monthly average precipitation during 1981–2010 years in the growing season of Maize in study area.** Gray bars indicate the regional monthly averages precipitation from 1981 to 2010, and black bars indicate increased 15% precipitation based on gray bars.

**Table 1.** Treatments performed in OTCs.

| Treatments | Description |
|---|---|
| $C_{550}W_{+15\%}$ | Elevated $[CO_2]$ concentration (550 $\mu mol \cdot mol^{-1}$) and Increased 15% of precipitation |
| $C_{550}W_0(CK)$ | Elevated $[CO_2]$ concentration (550 $\mu mol \cdot mol^{-1}$) and Natural precipitation |
| $C_{450}W_{+15\%}$ | Elevated $[CO_2]$ concentration (450 $\mu mol \cdot mol^{-1}$) and Increased 15% of precipitation |
| $C_{450}W_0(CK)$ | Elevated $[CO_2]$ concentration (450 $\mu mol \cdot mol^{-1}$) and Natural precipitation |
| $C_{390}W_{+15\%}$ | Ambient $[CO_2]$ concentration (390 $\mu mol \cdot mol^{-1}$) and Increased 15% of precipitation |
| $C_{390}W_0(CK)$ | Ambient $[CO_2]$ concentration (390 $\mu mol \cdot mol^{-1}$) and Natural precipitation |

$$P_n(PAR) = \alpha \frac{1 - \beta PAR}{1 + \gamma PAR} PAR - R_d \qquad (3)$$

where $PAR$ is irradiance, $\alpha$ is the initial slope of irradiance-response curve of photosynthesis when irradiance approaches to zero, $\beta$ and $\gamma$ are coefficients which are independent of $PAR$, $R_d$ is dark respiration ($\mu mol \cdot m^{-2} \cdot s^{-1}$). From these parameters, we can calculate $P_{nmax}$ (maximum net photosynthetic rate, $\mu mol \cdot m^{-2} \cdot s^{-1}$), $LSP$ (light saturation point, $\mu mol \cdot m^{-2} \cdot s^{-1}$), $LCP$ (light compensation point, $\mu mol \cdot m^{-2} \cdot s^{-1}$) and $\varphi_c$ (quantum efficiency of the light compensation point, $mol \cdot mol^{-1}$) using equations (4)–(7) as described by Han et al. [54].

$$P_{nmax} = \alpha \left( \frac{\sqrt{\beta + \gamma} - \sqrt{\beta}}{\gamma} \right)^2 - R_d \qquad (4)$$

$$\varphi_c = \alpha \frac{1 - 2\beta LCP - \beta\gamma LCP^2}{(1 + \gamma LCP)^2} \qquad (5)$$

$$LCP = \frac{\alpha - \gamma R_d - \sqrt{(\gamma R_d - \alpha)^2 - 4\alpha\beta R_d}}{2\alpha\beta} \qquad (6)$$

$$LSP = \frac{\sqrt{(\beta + \gamma)/\beta} - 1}{\gamma} \qquad (7)$$

$T_r$ (transpiration rate, $mol \cdot m^{-2} \cdot s^{-1}$), $G_s$ (stomatal conductance, $\mu mol \cdot m^{-2} \cdot s^{-1}$) and $C_i$ (intercellular $CO_2$ concentration, $\mu mol \cdot mol^{-1}$) were also measured at the same irradiance, temperature and vapour pressure when the measurements of $P_n$ were conducted. Additionally, $WUE$ (water use efficiency, $\mu mol \cdot mol^{-1}$) was calculated as $P_n / T_r$.

### Growth and harvesting of maize

Maize growth stages were recorded throughout the growing season. The plant height, ear height, stem diameter, leaf area and aboveground biomass were measured at the silking stage of the maize. Leaf area of each plant was determined with long-width coefficient method (length × width × 0.75). Aboveground biomass

was obtained by dry weight. Before weighing, three plants from each treatment were separated into stem, leaf and grain. This was as a result of the need to shrivel it in oven at 105°C for 45 min and drying to constant weight at 85°C for at least 48 h.

At maturity stage, ten plants of each treatment were harvested for the yield components. The grains from each ear of maize were threshed by hand after air dried and weighed. The measured ear characteristics include: ear length, ear diameter, ear weight, 100-kernel weight, rows per ear, kernel number, shriveled kernels, bare-tip length and axle diameter. The yield of each plant was calculated by 14% moisture content of grain.

During sampling, three representative plants of each treatment were randomly selected for measurement.

### Statistical analysis

In this study, statistical significance of growth and yield components was tested at 0.05 probability level ($P < 0.05$) following the DUNCAN test, performed using DPS 7.05 (Data-processing System, Zhejiang University, China). Irradiance-response curve fitting and parameter analysis based on modified rectangular hyperbolic model. The two-way analysis of variances (ANOVAS) were used to examine the interactive effects of elevated $[CO_2]$ and irrigation on growth parameters and yield of maize, and statistical significance were set at $P < 0.05$ and $P < 0.01$. These statistical analyses were conducted with SPSS 17.0 software (SPSS Institute Incorporated, Chicago, Illinois, USA). The standard deviation (S.D.) was calculated to compare the treatment means.

## Results

### Rainfall data and relative soil moisture

The variations of 5–9 month average monthly precipitation in Jinzhou during 1981–2010 are as shown in this work (Fig. 2). The work indicated that the amount of rainfall was maximum in July (176.93 mm) and less in May (57.43 mm).

The relative soil moisture of irrigation treatment was higher than that of the natural precipitation treatment at each $[CO_2]$ level (Fig. 3). The range of relative soil moisture in irrigation treatment was about 68.55%–69.84% and that of natural precipitation was about 56.68%–58.32%. According to the national standard of the Classification of Meteorological Drought (GB/T20481-2006) [50], we can see that the natural precipitation treatments are in-fact under slight drought.

### Photosynthetic gas exchange parameters of maize

According to previous studies, the relationship between photosynthetic rate and irradiance could be well described by modified rectangular hyperbolic model [52]. This model was used to obtain the irradiance-response curve (Fig. 5 A), with $R^2 > 0.99$

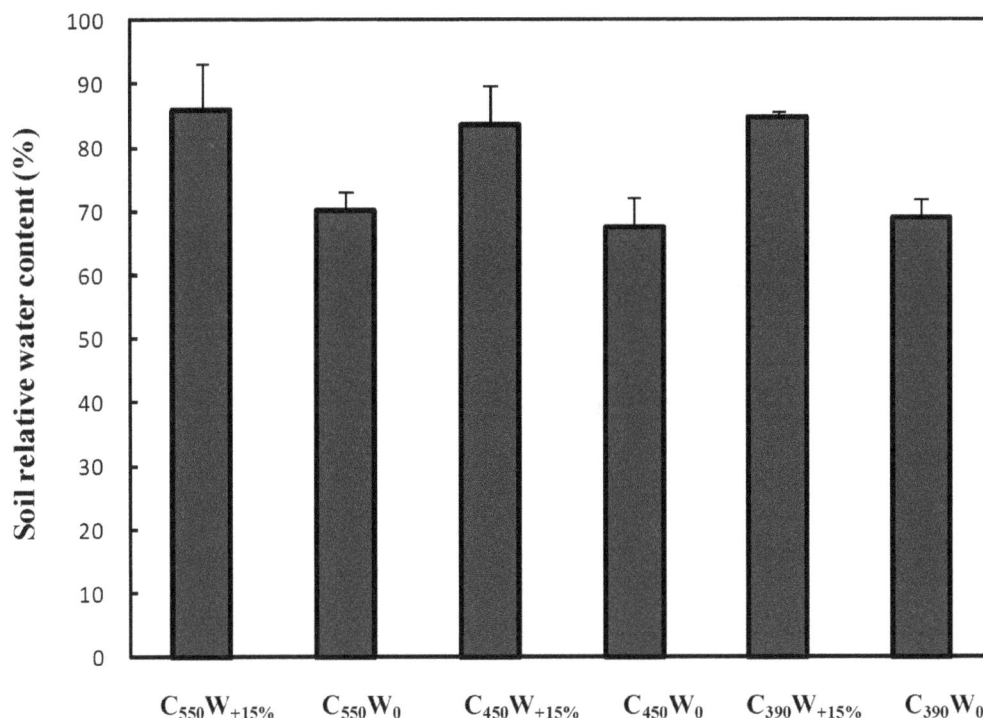

**Figure 3. The relative soil moisture (means ± SD) (n = 3) in six treatments (C$_{550}$W$_{+15\%}$, C$_{550}$W$_0$, C$_{450}$W$_{+15\%}$, C$_{450}$W$_0$, C$_{390}$W$_{+15\%}$ and C$_{390}$W$_0$) in silking stage of maize under effects of elevated [CO$_2$] and irrigation.**

for all treatments, meanwhile the characteristic parameters were calculated from table 2. From the comparison between $P_n$ of experimental value and model predicted value (Fig. 4), it could be seen that there was a good agreement between them.

Figure 5 showed the dynamic changes of $P_n$, $T_r$, $WUE$, $G_s$ and $C_i$ with $PAR$ increased during silking stage of maize under interactive effects of elevated [CO$_2$] and irrigation. With increased of $PAR$, $P_n$ curves of six treatments increased with elevated [CO$_2$] and irrigation. Similarly, when $PAR$ is above 1500 $\mu$mol·m$^{-2}$·s$^{-1}$, $P_n$

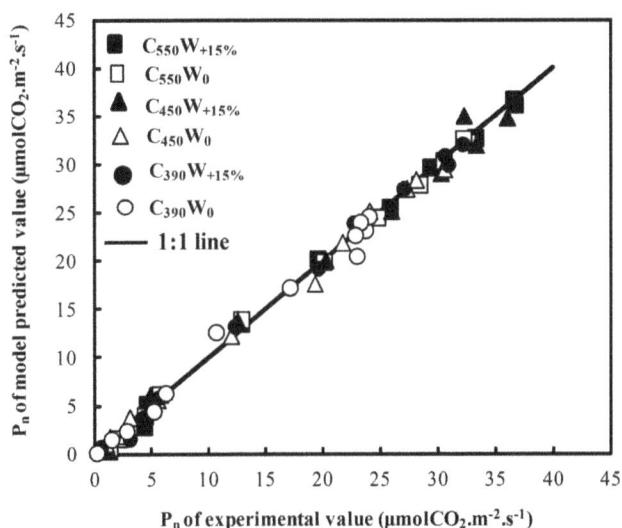

**Figure 4. $P_n$ of experimental value comparing with $P_n$ of model predicted value.** The 1:1 line indicates $P_n$ of experimental value equals to the predicted value.

curves closed to saturation and becomes stable. The order of six treatments were: C$_{550}$W$_{+15\%}$>C$_{450}$W$_{+15\%}$>C$_{550}$W$_0$>C$_{390}$W$_{+15\%}$ >C$_{450}$W$_0$>C$_{390}$W$_0$ (Fig. 5A); Whereas all the $T_r$ curves decreased with elevated [CO$_2$], and were much lower in natural precipitation than irrigation treatment at each [CO$_2$] level (Fig. 5B). $WUE$ showed higher values with elevated [CO$_2$], and natural precipitation treatment showed higher than irrigation treatment at elevated [CO$_2$] levels (Fig. 5C). However, $G_s$ curves showed opposite trend. The $G_s$ curves were lower with elevated [CO$_2$] and rose with irrigation at the same [CO$_2$] levels (Fig. 5D). There were high trends of $C_i$ under elevated [CO$_2$], and the irrigation treatments were higher with the increased amounts in $PAR$ at each [CO$_2$] level (Fig. 5E).

The $P_n$ curves were fitted and the parameters were generated by the modified rectangular hyperbolic model. The parameter $P_{nmax}$ of the ear-leaves for maize was increased by 12.43%, 19.80% and 29.70% under irrigation conditions above the natural precipitation at 390, 450 and 550 $\mu$mol·mol$^{-1}$ [CO$_2$] levels, respectively. $\varphi_c$ was not affected by irrigation and elevated [CO$_2$] (Table 2). Irrigation increased $LSP$ by 8.25%, 7.60% and 9.97% at 390, 450 and 550 $\mu$mol·mol$^{-1}$ [CO$_2$] levels, respectively, but $LCP$ increased by 38.73%, 15.81% and 42.92%, and $R_d$ increased by 42.71%, 33.93% and 41.81% under the same conditions (Table 2).

## Growth and development of maize

Plant height, ear height, stem diameter, leaf area and aboveground biomass of maize all had increasing trend under elevated [CO$_2$] and irrigation, while C$_{390}$W$_0$ always at its lowest value (Fig. 6). plant height of C$_{550}$W$_{+15\%}$, C$_{450}$W$_{+15\%}$ and C$_{390}$W$_{+15\%}$ were significantly higher than C$_{550}$W$_0$, C$_{450}$W$_0$ and C$_{390}$W$_0$ by 5.28%, 4.60% and 5.86%, respectively ($P<0.05$; Fig. 6A), while ear height were higher by 5.69%, 3.34% and 3.50%, respectively, and the same trend as plant height ($P<0.05$;

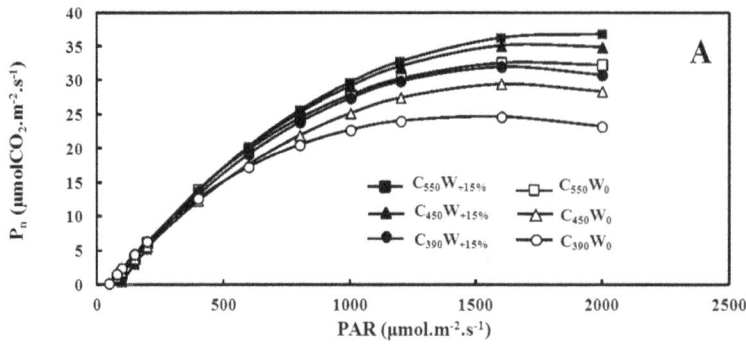

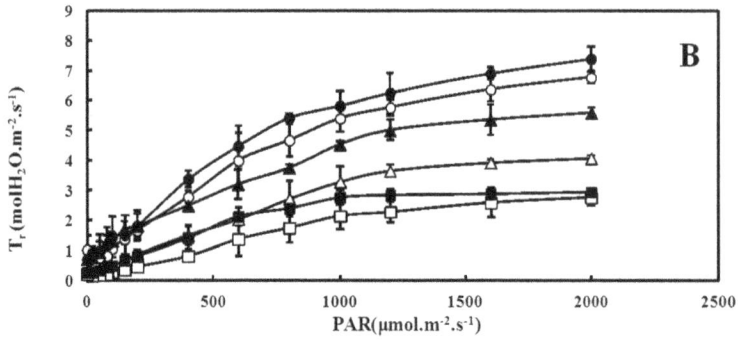

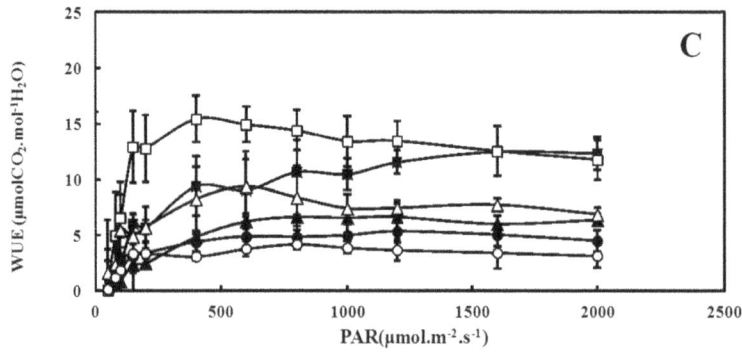

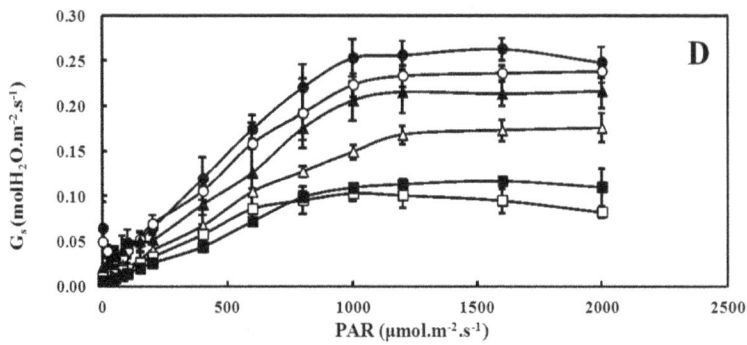

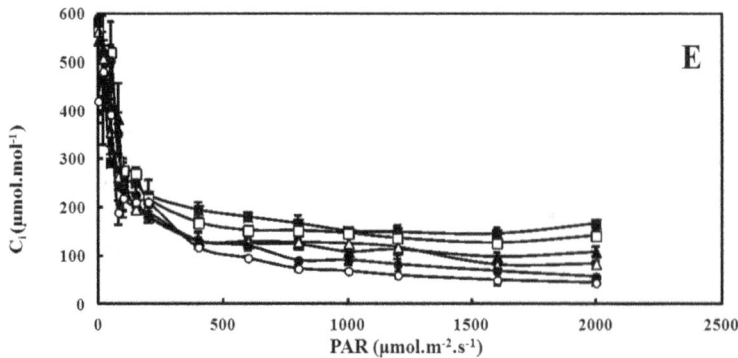

**Figure 5. Dynamic curves of (A) net photosynthetic rate ($P_n$), (B) transpiration rate ($T_r$), (C) water use efficiency (*WUE*), (D) stomatal conductance ($G_s$) and (E) intercellular $CO_2$ concentration ($C_i$) (means ± SD) (n = 3) in six treatments ($C_{550}W_{+15\%}$, $C_{550}W_0$, $C_{450}W_{+15\%}$, $C_{450}W_0$, $C_{390}W_{+15\%}$ and $C_{390}W_0$) in silking stage of maize under effects of elevated [$CO_2$] and irrigation.**

Fig. 6B). There were no significant differences of stem diameter for all treatments at silking stage, but $C_{390}W_0$ was observed to be low ($P<0.05$; Fig. 6C). There was a significant interactive effects for elevated [$CO_2$] and irrigation in plant height ($P<0.05$; Table 4), but no significant interactive effects in ear height and stem diameter ($P<0.05$; Table 4). Irrigation significantly increased the leaf area by 12.04%, 9.90% and 7.75% at 390, 450 and 550 μmol·mol$^{-1}$ [$CO_2$] levels, respectively ($P<0.05$; Fig. 6D). In addition, irrigation significantly increased aboveground biomass by 16.66%, 10.75% and 7.65% at 390, 450 and 550 μmol·mol$^{-1}$ [$CO_2$] levels, respectively ($P<0.05$; Fig. 6E). The interactive effects of elevated [$CO_2$] and irrigation were highly significant in leaf area ($P<0.05$; Table 4) and aboveground biomass ($P<0.05$; Table 4).

### Yield and ear characteristics of maize

The study also revealed that when irrigation was compared with the natural precipitation at 390, 450 and 550 μmol·mol$^{-1}$ [$CO_2$] levels, significantly increased the seed yield by 17.48%, 14.91% and 10.59%, respectively ($P<0.05$; Table 3). There was also a significant increase in biological yield by 12.39%, 9.30% and 8.39%, respectively ($P<0.05$; Table 3). Following from the above, economic coefficient showed a significant increase that was higher in irrigation than natural precipitation at each [$CO_2$] level ($P<0.05$; Table 3). There were significant interactive effects of elevated [$CO_2$] and irrigation on maize seed yield ($P<0.05$; Table 4) and biological yield ($P<0.01$; Table 4).

The maize ear characteristics showed significant differences in six treatments under elevated [$CO_2$] and irrigation ($P<0.05$; Table 5). Irrigation increased 100-kernel weight by 10.59%, 8.20% and 5.19% at 390, 450 and 550 μmol·mol$^{-1}$ [$CO_2$] respectively, whereas that of shriveled kernels decreased by 70.33%, 74.70% and 73.68%, respectively. Kernels per row and kernel number of $C_{390}W_{+15\%}$, $C_{450}W_{+15\%}$ and $C_{550}W_{+15\%}$ were significantly increased more than $C_{390}W_0$, $C_{450}W_0$ and $C_{550}W_0$ ($P<0.05$). Also, rows per ear increased, but there was no difference among the six treatments. Maize ear length, ear diameter and ear weight were all significantly increased under irrigation, whereas axle diameter showed increasing trend at 390, 450 and 550 μmol·mol$^{-1}$ [$CO_2$] levels ($P<0.05$). However, there was no noteworthy difference for bare-tip length ($P<0.05$; Table 5).

## Discussion

### Interactive effects on photosynthetic parameters of maize

The response of plant photosynthesis to each of the environmental variables (e.g., water availability, temperature, nitrogen) associated with the elevated [$CO_2$] has not been sufficiently understood [55–58]. The present study indicated that leaf $P_n$ of maize improved with both elevated [$CO_2$] and irrigation, and the curves of $C_{550}W_{+15\%}$, $C_{550}W_0$, $C_{450}W_{+15\%}$, $C_{450}W_0$ and $C_{390}W_{+15\%}$ showed higher values than $C_{390}W_0$ curve (Fig 5A). Similar results have been found in studies of *Stipa breviflora*, which reported a significant increase in $P_n$ under elevated [$CO_2$] and increased precipitation of 15% [59], and wall *et al.* (2001) also observed that elevated [$CO_2$] increased $P_n$ of sorghum by 9% in wet condition in FACE [60]. Leakey (2006) indicated that elevated [$CO_2$] can increase plant photosynthetic capacity and yield by adjusting its water state, so elevated [$CO_2$] will have positive effect in water deficit condition [61]. In this experimental site, the relative soil moisture of natural precipitation is lower as compared with irrigation (Fig. 3) and in-fact under slight drought, and it implied that water deficit is a key factor limiting maize growth. In contrast to $P_n$, $T_r$ decreased with elevated [$CO_2$], and much lower in natural precipitation treatment than irrigation treatment at each [$CO_2$] level (Fig 5B). A number of other controlled environment studies [62–63] all observed the $T_r$ decreased at elevated [$CO_2$]. It is worth mentioning, that the decrease of $T_r$ was associated with decrease of $G_s$, when elevated [$CO_2$] decreased leaf $G_s$, and caused increasing resistance from intrinsic leaf to outside, resulting in the decrease of $T_r$ [64]. Additionally, the studies of elevated [$CO_2$] and drought on plant reported that elevated [$CO_2$] declined $T_r$ in low soil moisture than high soil moisture [25,62], which in accordance with our findings. This suggested that elevated [$CO_2$] may increased stomatal resistance and leaded to the reduce of $T_r$ [25], but irrigation have a positive effect on stomatal opening and $T_r$ have a increased trend in irrigation treatment compare with natural precipitation treatment at each [$CO_2$] level. Moreover, our data indicated that *WUE* increased with elevated [$CO_2$], and higher in natural precipitation treatment than irrigation treatment at elevated [$CO_2$] levels (Fig 5C). This is in agreement with the results of previous studies, which reported elevated [$CO_2$] have an increase trend in *WUE* under drought condition than well-watered condition [27,61,62,65]. Specifically, we found that the *WUE* have

**Table 2.** Parameters of irradiance-response curves of maize under effects of elevated [$CO_2$] and different irrigation.

| Treatments | $P_{nmax}$ /(μmol·m$^{-2}$·s$^{-1}$) | *LSP*/(μmol·m$^{-2}$·s$^{-1}$) | *LCP* /(μmol·m$^{-2}$·s$^{-1}$) | $\varphi_c$ /(mol·mol$^{-1}$) | $R_d$ /(μmol·m$^{-2}$·s$^{-1}$) | $R^2$ |
|---|---|---|---|---|---|---|
| $C_{550}W_{+15\%}$ | 36.8702 | 1884.3777 | 94.8387 | 0.0544 | 5.1268 | 0.997 |
| $C_{550}W_0$(CK) | 32.7945 | 1740.8274 | 68.3083 | 0.0530 | 3.5925 | 0.998 |
| $C_{450}W_{+15\%}$ | 35.3561 | 1768.8612 | 62.8339 | 0.0483 | 3.0519 | 0.994 |
| $C_{450}W_0$(CK) | 29.5133 | 1650.8724 | 54.2541 | 0.0417 | 2.2788 | 0.997 |
| $C_{390}W_{+15\%}$ | 32.0309 | 1643.9243 | 66.9802 | 0.0469 | 3.1668 | 0.993 |
| $C_{390}W_0$(CK) | 24.6966 | 1501.2016 | 46.8651 | 0.0482 | 2.2332 | 0.992 |

Abbreviations are: $P_{nmax}$ - maximum net photosynthetic rate, *LSP* - light saturation point, *LCP* - light compensation point, $\varphi_c$ - quantum efficiency of the light compensation point, $R_d$ - dark respiration.

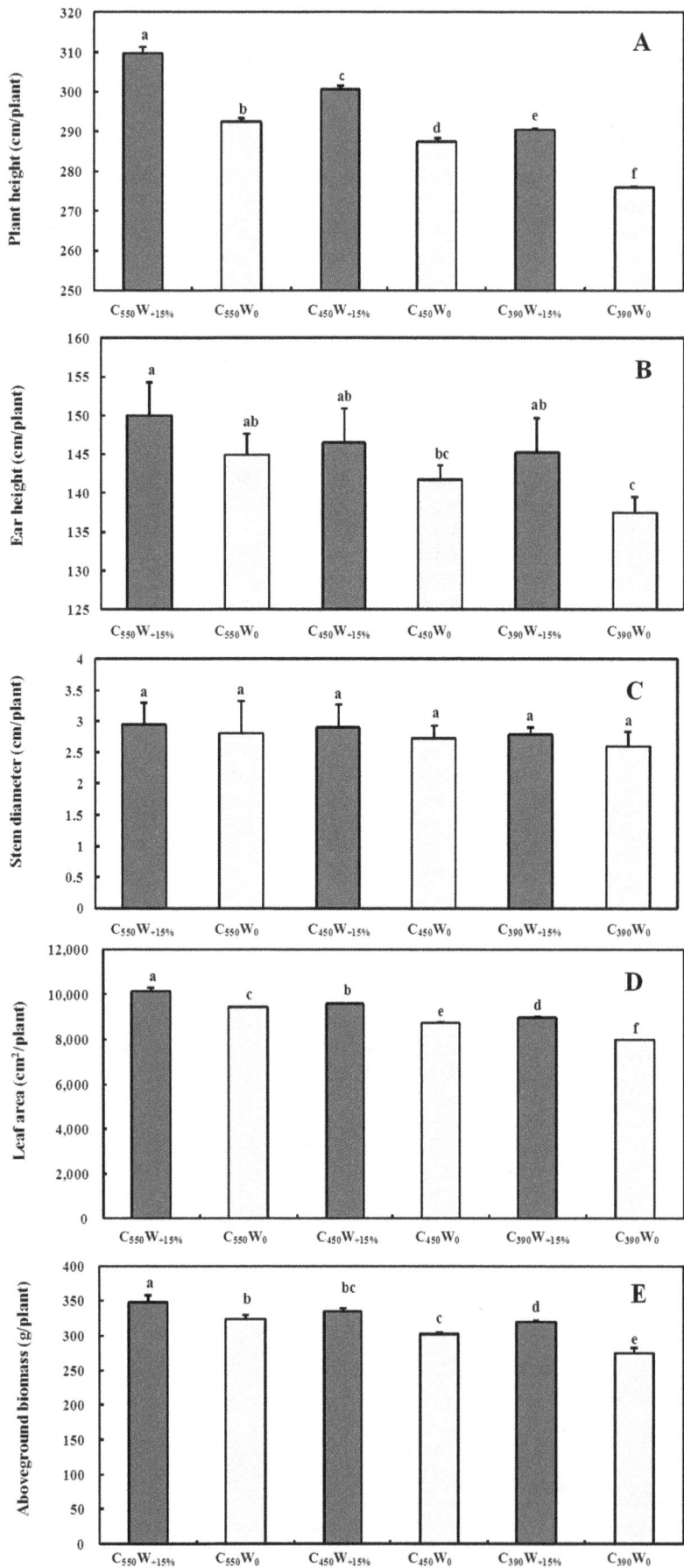

**Figure 6. Changes of (A) plant height, (B) ear height, (C) stem diameter, (D) leaf area and (E) aboveground biomass (means ± SD) (n = 3) in six treatments ($C_{550}W_{+15\%}$, $C_{550}W_0$, $C_{450}W_{+15\%}$, $C_{450}W_0$, $C_{390}W_{+15\%}$ and $C_{390}W_0$) in silking stage of maize under effects of elevated [CO$_2$] and irrigation.**Different lower cases letters indicated significant difference (P<0.05).

a decrease trend in irrigation treatment than natural precipitation treatment at elevated [CO$_2$] levels, whereas the yield of that increased. This is mainly due to the increase of leaf area and biomass of crop community under elevated [CO$_2$] and irrigation, resulting in the increase of *WUE* of crop community, at last performance increase in yield of maize [25,66–69]. In our study, $C_i$ increased with elevated [CO$_2$], whereas irrigation caused a small increase of $C_i$ at each [CO$_2$] level (Fig 5E). However, $G_s$ decreased with elevated [CO$_2$], and irrigation increased it greatly at each [CO$_2$] level (Fig 5D). Some controlled environment studies [70–72] also showing that elevated [CO$_2$] could cause a decrease in plant stomatal conductance ($G_s$) and partly closing of the stomata. Curtis *et al.* (1998) found that doubling [CO$_2$] average reduced $G_s$ by 11% [73]. Further, the decreased $G_s$ might be associated with the increase in $C_i$, because a prior work showing that, $C_i$ increased with rising [CO$_2$], wheat adjusted stomata opening width will bring about a decrease in $C_i$ so as to keep the intercellular CO$_2$ partial pressure to be always lower than atmospheric CO$_2$ partial pressure [74].

Photosynthetic parameter represented photosynthetic capacity and efficiency [75], which was usually obtained from irradiance-response model. Ye and Yu (2008) indicated that the modified rectangular hyperbolic model fitting results were quite close to the real values compare with the other models by series verification test, and the main photosynthetic parameters can be directly generated without any assumption by the model [52–53]. The model mentioned above was used and the $P_n$ curves were fitted, and producing the main parameters with $R^2 > 0.99$ (Table 2). Irrigation increased $P_{nmax}$ by 12.43%, 19.8% and 29.70% at 390, 450 and 550 µmol·mol$^{-1}$ [CO$_2$] levels, respectively (Table 2). The rise in $P_{nmax}$ has shown an improvement concerning the photosynthetic electron transport rate and photophosphorylation activity. It has also shown an improvement in the photosynthetic capacity of maize. In addition, *LSP* increased by 9.51%, 7.15% and 8.25% under irrigation more than natural precipitation at 390, 450 and 550 µmol·mol$^{-1}$ [CO$_2$] levels, respectively (Table 2). The emergence of *LSP* is actually as a result of dark reactions which are able to keep up with light reactions under the intense radiation, thus restricting the increase of photosynthetic rate. However, additional irrigation under elevated [CO$_2$] might partly alleviate the negative effect and could promote photosynthetic capacity of maize under high light intensity. Normally, the

photosynthesis reaction in a plant is very weak when in *LCP*, thus there was no significant effect on maize photosynthesis, even after there was a change in *LCP* with elevated [CO$_2$] and irrigation (Table 2). In addition, irrigation increased $R_d$ at each [CO$_2$] level, which is the limiting factor for maize photosynthesis. Generally, elevated [CO$_2$] can cause an increase in temperature, and make plant respiration quickened, leading to increase in consumption. Thus, the net effect is an increase in $R_d$ [70]. Irrigation might enhance this effect, but further researches are needed to explore the reasons.

## Interactive effects on growth of maize

The plant height increased by 5.28%, 4.60% and 5.86% under conditions of irrigation as compared with natural precipitation at 390, 450 and 550 µmol·mol$^{-1}$ [CO$_2$] levels, respectively (Fig. 6A). Similar results could be found for ear height (Fig. 6B), whereas stem diameter did not show any significant difference (Fig. 6C). Significant interactive effects of elevated [CO$_2$] and irrigation were observed in plant height, whereas there was no significant interactive effects of elevated [CO$_2$] and irrigation in ear height and stem diameter of maize ($P < 0.05$; Table 4). Some studies reported that leaf area of maize significantly increased under elevated [CO$_2$] [76–77], and this research work is in agreement with those findings (Fig. 6D). In addition, irrigation increased leaf area by 12.04%, 9.90% and 7.75% at 390, 450 and 550 µmol· mol$^{-1}$ [CO$_2$] levels, respectively (Fig. 6D). A similar conclusion could be found in *Stipa breviflora*, in which, the leaf area increased to a maximum under conditions of elevated [CO$_2$] and increased precipitation of 15% [78]. Chamber studies showed that changing [CO$_2$] has no effect on biomass of maize and sorghum under adequate moisture conditions [79–81], and even the negative effects [82]. However, this study found that aboveground biomass of maize significantly increased by 16.66%, 10.75% and 7.65% under irrigation more than natural precipitation at 390, 450 and 550 µmol·mol$^{-1}$ [CO$_2$] levels, respectively (Fig. 6E). This result is consistent with Cure's findings [83], who reported that maize aboveground biomass increased by 7% under well-watered more than dry conditions at 550 µmol· mol$^{-1}$ [CO$_2$] in chamber (n = 4). Shi *et al.* (2013) observed that the aboveground biomass of *Stipa breviflora* significantly increased with elevated [CO$_2$] and increasing precipitation [78]. Moreover, There were significant interactive effects of elevated

**Table 3.** Multiple comparison on yield of maize under effects of elevated [CO$_2$] and irrigation.

| Treatments | Seed yield /(g/plant) | Increase compare with its CK % | Biology yield /(g/plant) | Increase compare with its CK % | Economical coefficient |
|---|---|---|---|---|---|
| $C_{550}W_{+15\%}$ | 336.76±2.75[a] | 10.59 | 617.34±5.23[a] | 8.39 | 0.55±0.01[a] |
| $C_{550}W_0$(CK) | 304.52±1.01[c] | | 569.55±1.19[c] | | 0.53±0.00[a] |
| $C_{450}W_{+15\%}$ | 314.34±4.22[b] | 14.91 | 585.81±3.84[b] | 9.30 | 0.54±0.00[a] |
| $C_{450}W_0$(CK) | 273.55±5.52[e] | | 535.98±2.10[e] | | 0.51±0.01[b] |
| $C_{390}W_{+15\%}$ | 281.53±6.35[d] | 17.48 | 554.18±3.78[d] | 12.39 | 0.51±0.01[b] |
| $C_{390}W_0$(CK) | 239.64±1.09[f] | | 493.09±5.54[f] | | 0.49±0.00[c] |

The data are means ± SD (n = 3).
Different lower cases letters (a,b,c,d,e,f) indicated significant difference (P<0.05).

**Table 4.** The two-way analysis of variances on growth and yield of maize between different elevated [$CO_2$] and irrigation treatment.

| Variable | $CO_2$ concentration | | | Irrigation | | | Interaction | | |
|---|---|---|---|---|---|---|---|---|---|
| | df | F | P | df | F | P | df | F | P |
| Plant height | 2 | 536.048 | 0.000 | 1 | 1105.743 | 0.000 | 2 | 6.507 | 0.012* |
| Ear height | 2 | 9.315 | 0.002 | 1 | 26.167 | 0.000 | 2 | 0.726 | 0.498 |
| Stem diameter | 2 | 0.988 | 0.392 | 1 | 2.454 | 0.135 | 2 | 0.020 | 0.980 |
| Leaf area | 2 | 622.766 | 0.000 | 1 | 777.613 | 0.000 | 2 | 4.821 | 0.029* |
| Aboveground biomass | 2 | 58.838 | 0.000 | 1 | 137.036 | 0.000 | 2 | 4.339 | 0.038* |
| Seed yield | 2 | 663.180 | 0.000 | 1 | 806.085 | 0.000 | 2 | 5.113 | 0.017* |
| Biology yield | 2 | 945.868 | 0.000 | 1 | 1627.130 | 0.000 | 2 | 9.953 | 0.001** |

\* and ** indicated $P<0.05$ and $P<0.01$, respectively.

**Table 5.** Change on ear characteristics of maize under effects of elevated [$CO_2$] and irrigation.

| Treatments | Ear length (cm) | Ear diameter (cm) | Ear weight (g) | 100-kernel weight (g) | Kernels per row (kernels/row) | Kernel number (No.) | Rows per ear (row/ear) | Shriveled kernels (kernels/ear) | Bare-tip length (cm) | Axle diameter (cm) |
|---|---|---|---|---|---|---|---|---|---|---|
| $C_{550}W_{+15\%}$ | 24.17±0.35[a] | 6.16±0.04[a] | 362.45±0.93[a] | 43.97±0.21[a] | 44.67±1.15[a] | 820.67±2.31[a] | 18.67±1.15[a] | 5.00±1.00[e] | 4.51±0.04[bc] | 3.80±0.31[a] |
| $C_{550}W_0$(CK) | 22.21±0.05[c] | 5.97±0.04[b] | 332.28±16.86[b] | 41.80±0.40[b] | 41.33±0.58[b] | 757.67±1.53[b] | 17.33±1.15[a] | 19.00±1.00[c] | 4.45±0.04[c] | 3.72±0.37[ab] |
| $C_{450}W_{+15\%}$ | 23.05±0.45[b] | 6.03±0.04[b] | 344.69±4.08[b] | 42.60±0.50[b] | 42.33±0.58[b] | 762.00±1.39[b] | 18.00±0.01[a] | 7.00±1.00[d] | 4.42±0.04[c] | 3.64±0.28[bc] |
| $C_{450}W_0$(CK) | 21.02±0.12[d] | 5.57±0.01[d] | 305.56±4.07[c] | 39.37±0.84[c] | 38.33±0.58[c] | 614.67±3.06[d] | 17.33±1.15[a] | 27.67±2.08[b] | 4.62±0.01[b] | 3.61±0.31[bc] |
| $C_{390}W_{+15\%}$ | 21.25±0.03[d] | 5.69±0.17[c] | 312.01±1.77[c] | 40.40±0.75[c] | 39.00±1.00[c] | 676.00±1.58[c] | 17.33±1.15[a] | 9.00±1.00[d] | 4.44±0.17[c] | 3.55±0.26[c] |
| $C_{390}W_0$(CK) | 19.28±0.35[e] | 5.17±0.04[e] | 254.25±1.45[d] | 36.53±1.31[d] | 35.67±0.58[d] | 594.33±2.89[e] | 16.67±1.15[a] | 30.33±1.15[a] | 4.87±0.04[a] | 3.54±0.28[c] |

The data are means ± SD (n=3).
Different lower cases letters (a,b,c,d,e) indicated significant difference ($P<0.05$).

[$CO_2$] and irrigation in leaf area ($P<0.05$; Table 4) and aboveground biomass ($P<0.05$; Table 4).

## Interactive effects on yield and ear characteristics of maize

The results of that elevated [$CO_2$] increased maize yield have been indicated in previous studies. Cure *et al.* (1986) reported that maize average yield increased by 27% (n = 3) by doubling [$CO_2$] [83], while Guo (2003) indicated that maize yield might be increased by 22.88% when [$CO_2$] levels rise up to 700 $\mu$mol· mol$^{-1}$ [84]. Long *et al.* (2006) used comprehensive observation data of chamber studies (n = 14) and found out that grain yield of maize and sorghum increased by an average of 18% when [$CO_2$] were elevated to 550 $\mu$mol·mol$^{-1}$ [85]. This study has exhaustively shown that the maize yield of Northeast China increased by 14.15% and 27.07% with 450 and 550 $\mu$mol·mol$^{-1}$ [$CO_2$] levels, respectively (Table 3). Additionally, irrigation significantly increased maize yield by 10.59%, 14.91% and 17.48% as compared with natural precipitation at 390, 450 and 550 $\mu$mol·mol$^{-1}$ [$CO_2$] levels, respectively (Table 3). Similar results have been reported in previous studies [86–87], but others have reported decreases [82] or no significant change at all [79–80]. Allen (2011) predicted that management of irrigation water in a future high [$CO_2$] world could potentially increase overall $C_4$ crop yield (in water-limited areas) [62], we agree with this idea. Moreover, biological yield has also been increased by 8.39%, 9.30% and 12.39% under irrigation more than natural precipitation at 390, 450 and 550 $\mu$mol·mol$^{-1}$ [$CO_2$] levels, respectively (Table 3). The work also revealed that there has been significant interactive effects of elevated [$CO_2$] and irrigation on maize seed yield ($P<0.05$; Table 4) and biological yield ($P<0.01$; Table 4). Consequently, maize economic coefficient increased in irrigation more than natural precipitation. Economic coefficient reflects the transport and store ability of crop "source" to "sink" form photosynthetic products [88]. Kirschbaum (2010) reported that plant growth response to elevated [$CO_2$] increase with a plant's sink capacity and nutrient status [89]. In present study, maize photosynthetic capacity was enhanced while the ear capacity was expanded under elevated [$CO_2$] and irrigation, resulting in a more dry matter accumulation and yield increase.

Maize ear characteristics significantly changed under the interactive effects of elevated [$CO_2$] and irrigation. 100-kernel weight and kernal number were all significantly increased under irrigation than natural precipitation at each [$CO_2$] level (Table 5). Additionally, ear length, ear diameter and ear weight were all increased in accordance with yield, and the increase of 100-kernel weight and ear length was consistent with the study done by Wang et al. (1996) [90]. There is a study reported that kernels per row is the main factor that will increase maize yield [91]. In this study, irrigation increased kernels per row by 9.34%, 10.44% and 8.08% at 390, 450 and 550 $\mu$mol·mol$^{-1}$ [$CO_2$] levels respectively,

whereas rows per ear change was not remarkable (Table 5). The decreased of shriveled kernels was beneficial to yield increase. In our study, irrigation decreased shriveled kernels by 70.33%, 74.70% and 73.68% at 390, 450 and 550 $\mu$mol·mol$^{-1}$ [$CO_2$] levels, respectively (Table 5). Optimization of the ear characteristics were reflected in an increase in yield under elevated [$CO_2$] and irrigation. Therefore, it is necessary to irrigate with additional water for the elevated [$CO_2$] plots of maize to compensate for photosynthesis resistance from lacking of water under elevated [$CO_2$] condition. Moreover, bare-tip length of maize made no significant difference in all treatments, but axle diameter increased within a narrow range. The increase was by 0.28%, 0.83% and 2.15% in irrigation more than natural precipitation at 390, 450 and 550 $\mu$mol·mol$^{-1}$ [$CO_2$] levels (Table 5). This change implied that elevated [$CO_2$] and irrigation increased ear length, ear diameter as well as axle diameter. Thus, it can be estimated that maize yield could not be increased unlimitedly with elevated [$CO_2$] and irrigation, but increasing the axle diameter serves as a limiting factor.

## Conclusions

Our results demonstrated the following: 1) With elevated [$CO_2$] and irrigation, leaf $P_n$ and $C_i$ of maize significantly increased, while elevated [$CO_2$] brought a notable decrease in $G_s$ and $T_r$, but irrigation have a positive effect on them, thereby increasing *WUE* in natural precipitation treatment than irrigation treatment at elevated [$CO_2$] levels. 2) Irradiance-response parameters $P_{nmax}$ and *LSP* increased more under irrigation conditions than natural precipitation at each [$CO_2$] level. $R_d$ also increased under the same conditions, and this is a limiting factor. 3) Irrigation under elevated $CO_2$ increased plant height, ear height, stem diameter, leaf area and aboveground biomass, resulting in the increase of yield. In addition, ear characteristics of maize were all superior except axle diameter. However, further study should be taken to ensure the contribution rate of elevated [$CO_2$] and irrigation.

## Acknowledgments

We would like to appreciate Drs. Zhenzhu Xu and Yueming Bai for their valuable advice on the experiment design, Jinzhou Ecological and Agricultural meteorological experiment center in Liaoning province of China for providing experiment facilities and field assistance, Zhengming Zhou, Pengcheng Qin, Hui Li, Xiliang Chen and Xiang Wu for their help with the fieldwork and laboratory analysis, and academic editor and additional editor for their valuable comments in the manuscript.

## Author Contributions

Conceived and designed the experiments: JZ. Performed the experiments: FM CH. Analyzed the data: FM FY. Contributed reagents/materials/analysis tools: FM FY. Wrote the paper: FM JZ.

## References

1. IPCC (2007) IPCC fourth assessment report: climate change 2007. Cambridge University Press.
2. Leadley PW, Drake BG (1993) Open top chambers for exposing plant canopies to elevated $CO_2$ concentration and for measuring net gas exchange. Vegetatio 104–105(1): 3–15.
3. Wang XL, Xu SH, Liang H (1998) The experimental study of the effects of $CO_2$ concentration enrichment on growth, development and yield of $C_3$ and $C_4$ crops. Sci Agric Sin 31(1): 55–61. (in Chinese).
4. Ziska LH, Sicher RC, Bunce JA (1999) The impact of elevated carbon dioxide on the growth and gas exchange of three $C_4$ species differing in $CO_2$ leak rates. Physiol Plant 105(1): 74–80.

5. Ghannoum O, Von Caemmerer S, Ziska LH, Conroy JP (2000) The growth response of $C_4$ plants to rising atmospheric $CO_2$ partial pressure: a reassessment. Plant Cell Environ 23(9): 931–942.
6. Shaw MR, Zavaleta ES, Chiariello NR, Cleland EE, Mooney HA, et al. (2002) Grassland responses to global environmental changes suppressed by elevated $CO_2$. Science 298(5600): 1987–1990.
7. Long SP, Ainsworth EA, Rogers A, Ort DR (2004) Rising atmospheric carbon dioxide: plants FACE the future. Annu Rev Plant Biol 55: 591–628.
8. Ainsworth EA, Long SP (2005) What have we learned from 15 years of free-air $CO_2$ enrichment (FACE)? A meta-analytic review of the responses of photosynthesis, canopy properties and plant production to rising $CO_2$. New Phytol 165(2): 351–372.

9. Ainsworth EA, Rogers A (2007) The response of photosynthesis and stomatal conductance to rising [$CO_2$]: mechanisms and environmental interactions. Plant Cell Environ 30(3): 258–270.

10. Onoda Y, Hirose T, Hikosaka K (2007) Effect of elevated $CO_2$ levels on leaf starch, nitrogen and photosynthesis of plants growing at three natural $CO_2$ springs in Japan. Ecol Res 22(3): 475–484.

11. Wall GW, Garcia RL, Wechsung F, Kimball BA (2011) Elevated atmospheric $CO_2$ and drought effects on leaf gas exchange properties of barley. Agr Ecosyst Environ 144(1): 390–404.

12. Ziska L (2013) Observed changes in soyabean growth and seed yield from *Abutilon theophrasti* competition as a function of carbon dioxide concentration. Weed Research 53(2): 140–145.

13. Easterling DR, Evans JL, Groisman PY, Karl TR, Kunkel KE, et al. (2000) Observed variability and trends in extreme climate events: a brief review. Bull Amer Meteor Soc 81: 417–425.

14. Wallace JS (2000) Increasing agricultural water use efficiency to meet future food production. Agric Ecosyst Environ 82(1): 105–119.

15. Jones PD, Hulme M (1996) Calculating regional climatic time series for temperature and precipitation: methods and illustrations. Int J Climatol 16: 361–377.

16. Hulme M, Osborn TJ, Johns TC (1998) Precipitation sensitivity to global warming: comparison of observations with HadCM2 simulations. Geophys Res Lett 25(17): 3379–3382.

17. Houghton RA (2001) Counting terrestrial sources and sinks of carbon. Clim Change 48(4): 525–534.

18. Wu JD, Wang SL, Zhang JM (2000) A numerical simulation of the impacts of climate change on water and thermal resources in northeast china. Resources Sci 22(6): 36–42. (in Chinese).

19. Kim SH, Gitz DC, Sicher RC, Baker JT, Timlin DJ, et al. (2007) Temperature dependence of growth, development, and photosynthesis in maize under elevated $CO_2$. Environ Exp Bot 61(3): 224–236.

20. Wang D, Heckathorn SA, Barua D, Joshi P, Hamilton EW, et al. (2008) Effects of elevated $CO_2$ on the tolerance of photosynthesis to acute heat stress in $C_3$, $C_4$, and CAM species. Amer J Bot, 95(2): 165–176.

21. Hamilton EW, Heckathorn SA, Joshi P, Wang D, Barua D (2008) Interactive effects of elevated $CO_2$ and growth temperature on the tolerance of photosynthesis to acute heat stress in $C_3$ and $C_4$ species. J Integr Plant Biol 50(11): 1375–1387.

22. Xu ZZ, Shimizu H, Ito S, Yagasaki Y, Zou CJ, et al. (2014) Effects of elevated $CO_2$, warming and precipitation change on plant growth, photosynthesis and peroxidation in dominant species from North China grassland. Planta 239(2): 421–435.

23. Liang N, Maruyama K (1995) Interactive effects of $CO_2$ enrichment and drought stress on gas exchange and water-use efficiency in Alnus firma. Environ Exp Bot 35(3): 353–361.

24. Burkart S, Manderscheid R, Weigel HJ (2004) Interactive effects of elevated atmospheric $CO_2$ concentrations and plant available soil water content on canopy evapotranspiration and conductance of spring wheat. Eur J Agron 21(4): 401–417.

25. Kang SZ, Zhang FC, Hu XT, Zhang JH (2002) Benefits of $CO_2$ enrichment on crop plants are modified by soil water status. Plant Soil 238(1): 69–77.

26. Leakey ADB, Ainsworth EA, Bernacchi CJ, Rogers A, Long S P, et al. (2009) Elevated $CO_2$ effects on plant carbon, nitrogen, and water relations: six important lessons from FACE. J Exper Bot 60(10): 2859–2876.

27. Manderscheid R, Erbs M, Weigel HJ (2014) Interactive effects of free-air $CO_2$ enrichment and drought stress on maize growth. Eur J Agron 52: 11–21.

28. Zong YZ, Wang WF, Xue QW, Shangguan ZP (2014) Interactive effects of elevated $CO_2$ and drought on photosynthetic capacity and PSII performance in maize. Photosynthetica 52(1):63–70.

29. Markelz RJC, Strellner RS, Leakey ADB (2011) Impairment of $C_4$ photosynthesis by drought is exacerbated by limiting nitrogen and ameliorated by elevated [$CO_2$] in maize. J Exp Bot 62(9): 3235–3246.

30. Allen SG, Idso SB, Kimball BA (1990) Interactive effects of $CO_2$ and environment on net photosynthesis of water-Lily. Agric Ecosystems Environ 30(1): 81–88.

31. Idso SB, Wall GW, Kimball BA (1993) Interactive effects of atmospheric $CO_2$ enrichment and light intensity reductions on net photosynthesis of sour orange tree leaves. Environ Exp Bot 33(3): 367–375.

32. Olszyk DM, Wise C (1997) Interactive effects of elevated $CO_2$ and $O_3$ on rice and *flacca* tomato. Agric Ecosystems Environ 66(1): 1–10.

33. Olszyk DM, Johnson MG, Phillips DL, Seidler RJ, Tingey DT, et al. (2001) Interactive effects of $CO_2$ and $O_3$ on a ponderosa pine plant/litter/soil mesocosm. Environ Pollut 115(3): 447–462.

34. Ainsworth EA (2008) Rice production in a changing climate: a meta-analysis of responses to elevated carbon dioxide and elevated ozone concentrations. Global Change Biol 14(7): 1642–1650.

35. Ainsworth EA, Rogers A, Leakey ADB (2008) Targets for crop biotechnology in a future high-$CO_2$ and high-$O_3$ world. Plant Physiol 147(1): 13–19.

36. Kim HY, Lieffering M, Kobayashi K, Okada M, Mitchell MW, et al. (2003) Effects of free-air $CO_2$ enrichment and nitrogen supply on the yield of temperate paddy rice crops. Field Crop Res 83(3): 261–270.

37. Kim HY, Lieffering M, Miura S, Kobayashi K, Okada M (2001) Growth and nitrogen uptake of $CO_2$-enriched rice under field conditions. New Phytol 150(2): 223–229.

38. Rogers A, Ainsworth EA, Leakey ADB (2009) Will elevated carbon dioxide concentration amplify the benefits of nitrogen fixation in legumes? Plant Physiol 151(3): 1009–1016.

39. Reddy KR, Zhao D (2005) Interactive effects of elevated $CO_2$ and potassium deficiency on photosynthesis, growth, and biomass partitioning of cotton. Field Crop Res 94(2): 201–213.

40. Geissler N, Hussin S, Koyro HW (2009) Interactive effects of NaCl salinity and elevated atmospheric $CO_2$ concentration on growth, photosynthesis, water relations and chemical composition of the potential cash crop halophyte *Aster tripolium* L. Environ Exp Bot 65(2): 220–231.

41. Milcu A, Paul S, Lukac M (2011) Belowground interactive effects of elevated $CO_2$, plant diversity and earthworms in grassland microcosms. Basic Appl Ecol 12: 600–608.

42. Zhou HQ, Wang CJ (2006) Analysis on comprehensive capability of food production in Northeast area. J Northeast Agric Univ 4(1): 5–8. (in Chinese).

43. Hu G, Wu YQ, Liu BY, Yu ZT, You ZM, et al. (2007) Short-term gully retreat rates over rolling hill areas in black soil of Northeast China. Catena, 71(2): 321–329. (in Chinese).

44. Liu XT, Tong LJ, Wu ZJ, Liang WJ (1998) Analysis and prediction of grain production potential in Northeast region. Sci Geogr Sin 18(6): 501–509. (in Chinese).

45. Ma SQ, Wang Q, Luo XL (2008) Effect of climate change om maize (*Zea mays*) growth and yield based on stage sowing. Acta Ecologica Sinica 28(5): 2131–2139. (in Chinese).

46. Bai XL, Sun SX, Yang GH, Liu M, Zhang ZP, et al. (2009) Effect of water stress on maize yield during different growing stages. J Maize Sci 17(2): 60–63. (in Chinese).

47. Han GX, Zhou GS, Xu ZZ, Y Y, Liu JL, et al. (2007) Soil temperature and biotic factors drive the seasonal variation of soil respiration in a maize (*Zea mays* L.) agricultural ecosystem. Plant Soil 291(1): 15–26.

48. Panel on the Northeast Regional Climate Change Assessment Report (2013) Summary for policy-makers and implementation of Northeast Regional Climate Change Assessment Report 2012. Beijing: China Meteorological Press. (in Chinese).

49. Wigley TML, Raper SCB (2001) Interpretation of high projections forglobal-mean warming. Science 293: 451–454.

50. Zhang Q, Zou XK, Xiao FJ, Lu HQ, Liu HB, et al. (2006) GB/T20481-2006, Classification of meteorological drought. People's Republic of China National Standards (in Chinese). Beijing: Standardization Press of China.

51. Liu JL, Shi KQ, Liang T, Zhang LH (2008) Relationship between underground water level and precipitation, soil moisture in maize field in Jinzhou. Anhui Agric Sci Bull 14(20): 126–130. (in Chinese).

52. Ye ZP (2007) A new model for relationship between irradiance and the rate of photosynthesis in *Oryza sativa*. Photosynthetica 45(4): 637–640.

53. Ye ZP, Yu Q (2008) A coupled model of stomatal conductance and photosynthesis for winter wheat. Photosynthetica 46(4): 637–640.

54. Han RF, Li JM, Hu XH, Da HG, Bai RF (2012) Research on dynamic characteristics of photosynthesis in muskmelon seedling leaves. Acta Ecologica Sinica 32(5): 1471–1480. (in Chinese).

55. Lee JS (2011) Combined effect of elevated $CO_2$ and temperature on the growth and phenology of two annual $C_3$ and $C_4$ weedy species. Agr Ecosyst Environ 140(3): 484–491.

56. Reddy AR, Rasineni GK, Raghavendra AS (2010) The impact of global elevated $CO_2$ concentration on photosynthesis and plant productivity. Current Science 99(1): 46–57.

57. Hättenschwiler S, Zunbrunn T (2006) Hemiparasite abundance in an alpine treeline ecotone increases in response to atmospheric $CO_2$ enrichment. Oecologia 147: 47–52.

58. Temperton V, L ard PM, Jarvis P (2003) Does elevated atmospheric carbon dioxide affect internal nitrogen allocation in the temperate trees *Alnus glutinosa* and *Pinus syvestris*? Global Change Biol 9: 286–294.

59. Wang H, Zhou GS, Jiang YL, Shi YH, Xu ZZ (2012) Interactive effects of changing precipitation and elevated $CO_2$ concentration on photosynthetic parameters of *Stipa breviflora*. Chin J Plant Ecol 36(7): 597–606. (in Chinese).

60. Wall GW, Brooks TJ, Adam NR, Cousins AB, Kimball BA (2001) Elevated atmospheric $CO_2$ improved *sorghum* plant water status by ameliorating the adverse effects of drought. New Phytol, 152(2): 231–248.

61. Leakey ADB, Uribelarrea M, Ainsworth EA, Naidu SL, Rogers A, et al. (2006) Photosynthesis, productivity, and yield of maize are not affected by open-air elevation of $CO_2$ concentration in the absence of drought. Plant Physiol 140, 779–790.

62. Allen Jr LH, Kakani VG, Vu JCV, Boote KJ (2011) Elevated $CO_2$ increases water use efficiency by sustaining photosynthesis of water-limited maize and sorghum. J Plant Physiol 168(16): 1909–1918.

63. Bernacchi CJ, Kimball BA, Quarles DR, Long SP, Ort DR (2007) Decreases in stomatal conductance of soybean under open-air elevation of [$CO_2$] are closely coupled with decreases in ecosystem evapotranspiration. Plant Physiol 143(1): 134–144.

64. Xu L, Zhao TH, Hu YY, Shi Y (2008) Effects of $CO_2$ Enrichment on Photosynthesis and Grain Yield of Spring Wheat. J Triticeae Crops 28(5): 867–872. (in Chinese).

65. Kang SZ, Zhang FC, Liang YL, Ma QL, Hu XT (1999) Effects of soil water and the atmospheric $CO_2$ concentration increase on evapotranspiration, Photosynthesis, growth of Wheat, Maize and Cotton. Acta Agron Sin 25(1): 55–63.

66. Samarakoon AB, Müller WJ, Gifford RM (1995) Transpiration and leaf area under elevated $CO_2$: effects of soil water status and genotype in wheat. Aust J Plant Physiol 22: 33–44.

67. Roden JS, Ball MC (1996) The effect of elevated $[CO_2]$ on growth and photosynthesis of two eucalyptus species exposed to high temperatures and water deficits. Plant Physiol 111(3): 909–919.

68. Newton PCD, Clark H, Bell CC, Glasgow EM (1996) Interaction of soil moisture and elevated $CO_2$ on the above-ground growth rate, root length density and gas exchange of turves from temperate pasture. J Exp Bot 47(299): 771–779.

69. Robredo A, Pérez-López U, de la Maza HS, González-Moro B, Lacuesta M, et al. (2007) Elevated $CO_2$ alleviates the impact of drought on barley improving water status by lowering stomatal conductance and delaying its effects on photosynthesis. Environ Exp Bot 59(3): 252–263.

70. Murray DR (1995) Plant responses to carbon dioxide. Amer J Bot 82(5): 690–697.

71. Ainsworth EA, Rogers A (2007) The response of photosynthesis and stomatal conductance to rising $[CO_2]$: mechanisms and environmental interactions. Plant Cell Environ 30: 258–270.

72. Ainsworth EA, Long SP (2005) What have we learned from 15 years of free-air $CO_2$ enrichment (FACE)? A meta-analytic review of the responses of photosynthesis, canopy properties and plant production to rising $CO_2$. New Phytol 165(2): 351–372.

73. Curtis PS, Wang XZ (1998) A meta-analysis of elevated $CO_2$ effects on woody plant mass, form and physiology. Oecologia 113: 299–313.

74. Wang RJ, Zhang XC, Gao SM, Yu XF, Ma YF (2010) Effects of atmospheric $CO_2$ enrichment and nitrogen application rate on photosynthetic parameters and water use efficiency of spring wheat. Agric Res Arid Areas 28(5): 32–37. (in Chinese).

75. Wang JL, Wen XF, Zhao FH, Fang QX, Yang XM (2012) Effects of doubled $CO_2$ concentration on leaf photosynthesis, transpiration and water use efficiency of eight crop species. Chin J Plant Ecol 36(5): 438–446. (in Chinese).

76. Samarakoon AB, Gifford RM (1995) Soil water content under plants at high $CO_2$ concentration and interactions with the direct $CO_2$ effects: a species comparison. J Biogeogr 22(2/3): 193–202.

77. Maroco JP, Edwards GE, Ku MSB (1999) Photosynthetic acclimation of maize to growth under elevated levels of carbon dioxide. Planta 210(1): 115–125.

78. Shi YH, Zhou GS, Jiang YL, Wang H, Xu ZZ (2013) Effects of interactive $CO_2$ concentration and precipitation on growth characteristics of *Stipa breviflora*. Acta Ecologica Sinica 33(14): 4478–4485. (in Chinese).

79. Marc J, Gifford RM (1984) Floral initiation in wheat, sunflower, and sorghum under carbon dioxide enrichment. Can J Bot 62(1): 9–14.

80. Mauney JR, Fry KE, Guinn G (1978) Relationship of photosynthetic rate to growth and fruiting of cotton, soybean, sorghum, and sunflower. Crop Sci 18(2): 259–263.

81. Ziska LH, Bunce JA (1997) Influence of increasing carbon dioxide concentration on the photosynthetic and growth stimulation of selected $C_4$ crops and weeds. Photosynth Res 54(3): 199–208.

82. Ellis RH, Craufurd PQ, Summerfield RJ, Roberts EH (1995) Linear relations between carbon dioxide concentration and rate of development towards flowering in sorghum, cowpea and soybean. Ann Bot 75(2): 193–198.

83. Cure JD, Acock B (1986) Crop responses to carbon dioxide doubling: a literature survey. Agr Forest Meteorol 38(1/3): 127–145.

84. Guo JP (2003) Responses of mainly plants in northern China to $CO_2$ enrichment and soil drought. Beijing: China Meteorological Press. (in Chinese).

85. Long SP, Ainsworth EA, Leakey ADB, Nösberger J, Ort DR (2006) Food for thought: lower-than-expected crop yield stimulation with rising $CO_2$ concentrations. Science 312(5782): 1918–1921.

86. Amthor JS, Mitchell RJ, Runion GB, Rogers HH, Prior SA, et al. (1994) Energy content, construction cost and phytomass accumulation of *Glycine max* (L.) Merr. and *Sorghum bicolor* (L.) Moench grown in elevated $CO_2$ in the field. New Phytol 128(3): 443–450.

87. Reeves DW, Rogers HH, Prior SA, Wood CW, Runion GB (1994) Elevated atmospheric carbon dioxide effects on sorghum and soybean nutrient status. J Plant Nutr 17(11): 1939–1954.

88. Mason TG, Maskell EJ (1928) Studies on the transport of carbohydrates in the cotton plant: II. The factors determining the rate and the direction of movement of sugars. Ann Bot 42(3): 571–636.

89. Kirschbaum MUF (2011) Does enhanced photosynthesis enhance growth? Lessons learned from $CO_2$ enrichment studies Plant Physiol 155(1): 117–124.

90. Wang CY, Bai YM, Wen M (1996) Effects of $CO_2$ concentration increase on yield and quality of corn. Acta Sci circumst 16(3): 331–336. (in Chinese).

91. Yang JH, Mao JC, Li FM, Ran LG, Liu J, et al. (2003) Correlation and path analysis on agronomic traits and kernal yield of maize hybrids. Chin Agric Sci Bull 19(4): 28–30. (in Chinese).

# 3

# Aberrant Water Homeostasis Detected by Stable Isotope Analysis

**Shannon P. O'Grady**[1]*, **Adam R. Wende**[2], **Christopher H. Remien**[3], **Luciano O. Valenzuela**[1,4], **Lindsey E. Enright**[1], **Lesley A. Chesson**[1,4], **E. Dale Abel**[2], **Thure E. Cerling**[1,4,5], **James R. Ehleringer**[1,4]

1 Department of Biology, University of Utah, Salt Lake City, Utah, United States of America, 2 Division of Endocrinology, Metabolism, and Diabetes, Salt Lake City, Utah, United States of America, 3 Department of Mathematics, University of Utah, Salt Lake City, Utah, United States of America, 4 IsoForensics, Inc., Salt Lake City, Utah, United States of America, 5 Department of Geology and Geophysics, University of Utah, Salt Lake City, Utah, United States of America

## Abstract

While isotopes are frequently used as tracers in investigations of disease physiology (i.e., $^{14}C$ labeled glucose), few studies have examined the impact that disease, and disease-related alterations in metabolism, may have on stable isotope ratios at natural abundance levels. The isotopic composition of body water is heavily influenced by water metabolism and dietary patterns and may provide a platform for disease detection. By utilizing a model of streptozotocin (STZ)-induced diabetes as an index case of aberrant water homeostasis, we demonstrate that untreated diabetes mellitus results in distinct combinations, or signatures, of the hydrogen ($\delta^2H$) and oxygen ($\delta^{18}O$) isotope ratios in body water. Additionally, we show that the $\delta^2H$ and $\delta^{18}O$ values of body water are correlated with increased water flux, suggesting altered blood osmolality, due to hyperglycemia, as the mechanism behind this correlation. Further, we present a mathematical model describing the impact of water flux on the isotopic composition of body water and compare model predicted values with actual values. These data highlight the importance of factors such as water flux and energy expenditure on predictive models of body water and additionally provide a framework for using naturally occurring stable isotope ratios to monitor diseases that impact water homeostasis.

**Editor:** Adrian Vella, Mayo Clinic College of Medicine, United States of America

**Funding:** S.P.O. is funded by an Intelligence Community Postdoctoral Fellowship, A.R.W. is funded by a Juvenile Diabetes Research Foundation Advanced Postdoctoral Fellowship, and E.D.A. by U01HL087947. L.O.V., L.A.C., T.E.C. and J.R.E. are partially funded by IsoForensics, Inc. The funders had no role in study design, data collection and analysis, decision to publish, or preparation of the manuscript.

**Competing Interests:** L.O.V., L.A.C., T.E.C. and J.R.E. are partially funded by IsoForensics, Inc.

* E-mail: ogrady@biology.utah.edu

## Introduction

The stable hydrogen ($\delta^2H$) and oxygen ($\delta^{18}O$) isotopes of body water have been used as tracers of movement and migration in the fields of anthropology [1,2], ecology [3,4], and forensics [5]. This is possible because the isotopic composition of body water ($\delta_{bw}$) is positively correlated with the isotopic composition of local precipitation (*e.g.* drinking water; $\delta_{dw}$) [6–8] and the isotopic composition of meteoric water varies geographically [9,10]. The isotopic signature of body water is a composite of quantitative and qualitative influx and efflux factors, such as fluid and food intake, urine output, and sweat production [6,7,11,12]. Although drinking water is the major contributing factor to the body water pool, the oxidation of hydrogen and oxygen atoms contained in food and water elevate the isotopic composition of body water above that of drinking water (Fig. 1). In addition to the impact that the $\delta^2H$ and $\delta^{18}O$ values of drinking water can have on the isotopic signature of body water, the volume of water processed by the body may also be a major determinant of body water signature.

The combined volume of influx and efflux factors impacting the body water pool is defined as total water flux (TWF). In healthy humans drinking water, food water, and atmospheric oxygen are the primary input factors, contributing 43%, 30%, and 20% to the body water pool, respectively. Liquid water in the form of urine and sweat are the primary outputs, constituting ~63% of total water loss (Podlesak et al., in review). All of these factors result in an offset between the isotopic composition of the major drinking water input and the isotopic composition of body water (Fig. 1). While relatively few studies have measured TWF in humans [13], studies examining body water from animals known to have high TWF, such as dairy cows and hippopotamids [14,15], suggest that as TWF increases, the isotopic contribution of drinking water also increases, and the $\delta^2H$ and $\delta^{18}O$ of body water approach that of drinking water.

Untreated diabetes mellitus is a state of elevated water flux. Glycosuria and polyuria ensue when blood glucose levels exceed the renal threshold [16]. Eventually, untreated diabetes results in dehydration, thirst, and increased fluid intake. We hypothesized that increased water flux in animals with untreated diabetes mellitus would yield a body water isotope signature more similar to drinking water isotope values than control animals. To test this hypothesis, we measured water intake, urine output, body water stable isotope ratios, and blood glucose values utilizing a model of streptozotocin (STZ)-induced diabetes mellitus. Treatment with streptozotocin destroys pancreatic islet cells, thus inducing insulin-dependent diabetes mellitus [17]. Additionally, we applied an isotope mass balance model to estimate the $\delta^2H$ and $\delta^{18}O$ values of mouse body water for different levels of TWF.

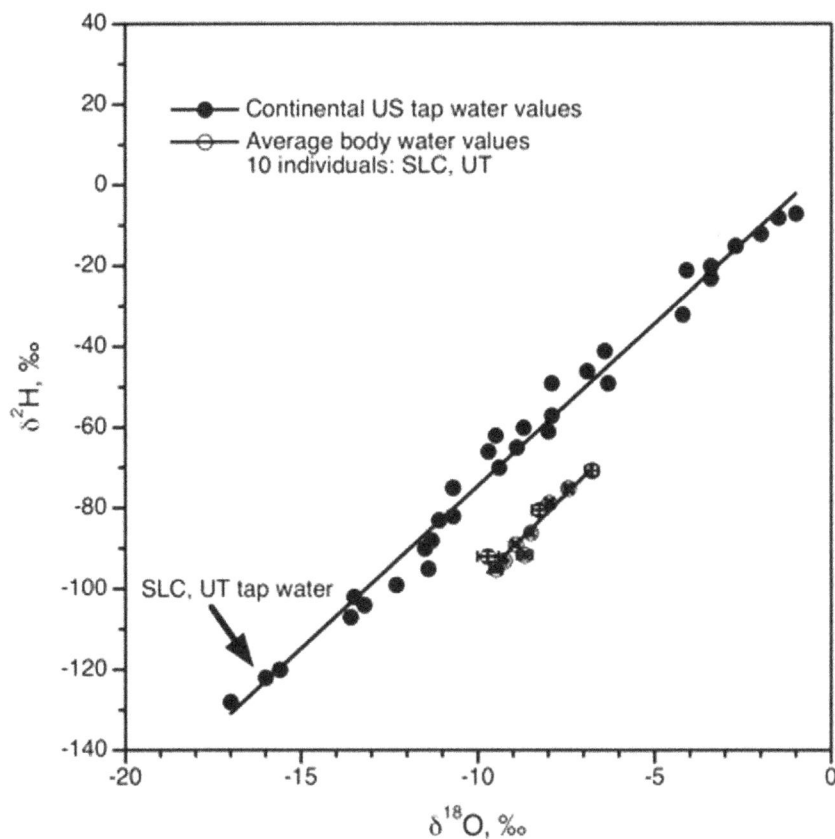

**Figure 1. Isotopic values of continental tap water relative to body water.** Measured $\delta^2H$ and $\delta^{18}O$ values of tap water samples collected from 18 states across the continental United States (●) and body water (urine) samples collected within Salt Lake City, Utah (○). The simple linear regression between the hydrogen and oxygen of tap water ($\delta^2H = 8.06^{18}O + 6.15$, $r^2 = 0.992$) and body water ($\delta^2H = 8.91\delta^{18}O - 9.72$, $r^2 = 0.962$) are also shown.

## Results

### TWF is higher in diabetic STZ mice relative to vehicle-treated (VEH) controls

Water intake of STZ mice (week 1: 33.6±6.0 ml s.d.; week 4: 44.6±8.4 ml s.d.) was five to six times that of VEH treated mice (week 1: 5.5±1.1 ml s.d.; week 4: 9.7±6.0 ml s.d.), at 1 and 4 weeks post-injection (two tailed t-test: $\alpha = 0.05$, $p < 0.0001$ and $p < 0.0001$, respectively; Fig. 2). Additionally, average daily urine output was greater in STZ mice (week 1: 25±4.1 ml s.d.; week 4: 33.2±5.2 ml s.d.) relative to VEH treated mice (<5ml) at both week 1 and 4.

### Varying only TWF, model accurately predicts body water isotope composition

The $\delta^2H$ and $\delta^{18}O$ values of body water from STZ and VEH animals exhibited significant covariance ($r^2 = 0.99$, $p < 0.0001$ and $r^2 = 0.31$, $p < 0.0001$; Fig. 3a). Varying only the modeled influx of drinking water, the model accurately predicts both the actual range and concavity of body water $\delta^2H$ and $\delta^{18}O$ values in STZ and VEH mice as average drinking water intake increases ($r^2 = 0.83$ and $r^2 = 0.79$, respectively; Fig. 3b and c). The accuracy of model predictions increases for both $\delta^2H$ and $\delta^{18}O$ in STZ and VEH mice if the total energy expenditure (TEE) of animals is increased from 37.4 KJ/day to 65.0 KJ/day ($r^2 = 0.96$ and $r^2 = 0.96$, respectively; Fig. 3b and c)

### Body water of diabetic STZ mice is isotopically distinct from that of VEH mice

There was no significant difference between STZ and VEH animals in the isotopic composition of body water or blood glucose readings at the start of the study (week 0, Fig 4). Additionally, neither the isotopic composition of body water or blood glucose of VEH animals changed significantly during the course of the study (Fig. 4). By week 1, relative to VEH animals, STZ animals had significantly lower $\delta^2H$ and $\delta^{18}O$ body water values (ANOVA: $\alpha = 0.05$, $p < 0.0001$; $-88.6±2.3‰$ s.d. vs. $-101.9±5.4‰$ s.d. and $-7.9±.5‰$ s.d. vs. $-10.6±1.0‰$ s.d., respectively; Fig. 4) and significantly higher blood glucose (ANOVA: $\alpha = 0.05$, $p < 0.0001$; 232±47 mg/dL s.d. vs. 465±66 mg/dL s.d., respectively; Fig. 4). The average isotopic composition of body water of STZ animals was significantly different at week 2 relative to week 1 (ANOVA: $\alpha = 0.05$, $p < 0.0001$; week 2: $\delta^2H = -114.8±2.2‰$ s.d., $\delta^{18}O = -13±.7‰$ s.d.), but there was no difference in the isotopic composition of body water of STZ animals between weeks 2, 3, and 4 (Fig. 4). This was also true for blood glucose values of STZ animals, with week 2 values significantly increased relative to week 1 (ANOVA: $\alpha = 0.05$, $p < 0.0001$; week 2: 680.9±141 mg/dL s.d.), but no difference between weeks 2, 3, and 4 (Fig. 4).

### Body water of STZ mice approaches isotopic signature of drinking water

By week 4, the $\delta^2H$ and $\delta^{18}O$ values of body water from STZ animals were similar to those of drinking water ($\delta^2H = -121‰$,

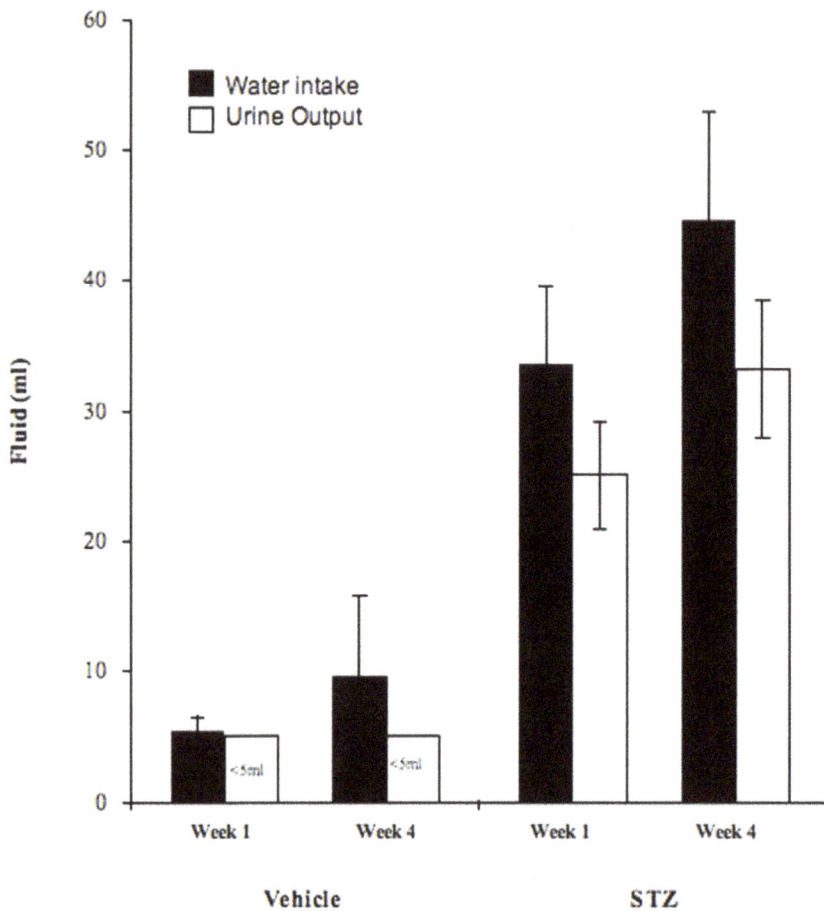

**Figure 2. Water flux of VEH and STZ mice.** Average water intake (■) and urine output (□) from VEH treated (n = 6) and STZ treated mice (n = 6), at 1 week and 4 weeks post-injection. Bars delineate standard deviation within each group. Urine output from VEH treated mice was less than 5ml/day and below measurement precision.

$\delta^{18}O = -16‰$), with approximately a 6‰ and 3‰ difference between body water and drinking water H and O isotope ratios, respectively (Fig. 4). In contrast, relative to drinking water, the $\delta^2H$ and $\delta^{18}O$ values of body water from VEH animals were elevated by approximately 34‰ and 8‰, respectively (Fig. 4).

## Discussion

Water intake and urine output data (Fig. 2) indicate that TWF is higher in diabetic STZ mice relative to VEH treated controls. Consistent with our hypothesis, the $\delta^2H$ and $\delta^{18}O$ values of body water from diabetic STZ mice resembled the isotopic signature of drinking water much more closely than did the body water values of VEH mice (Fig. 4). Our results suggest that as TWF increases, the isotopic input of drinking water becomes a major determinant of the isotopic signature of body water. We hypothesize that altered water flux due to hyperglycemia, and not metabolism, is the mechanism underlying the similarity between the isotopic signatures of diabetic STZ body water and drinking water. Further, supporting the impact of water flux on the isotopic composition of body water, the mathematical model used here accurately predicts the isotopic range of body water as determined by changes only in water flux, with all other model parameters held constant (Fig. 3). Both the time series data (Fig. 4) and the data generated by the body water model (Fig. 3) indicate that the isotopic composition of body water reaches a threshold level of

water flux at which point the changes in the $\delta^2H$ and $\delta^{18}O$ values of body water cannot be determined. Intuitively, this threshold is isotopically indistinguishable from the isotopic value of the drinking water input (for this study, $\delta^2H = -121‰$ and $\delta^{18}O = -16‰$). While this study provides convincing evidence that the mechanism behind this isotopic distinction is water flux, we plan to conduct similar studies in a model of diabetes insipidus. As opposed to diabetes mellitus, which is characterized by low insulin production or resistance and can result in hyperglycemia, diabetes insipidus is a disease state caused by either the deficiency of or insensitivity to arginine vasopressin, also known as antidiuretic hormone, resulting in elevated water flux with no impact on blood glucose. Such a study would enable us to isolate the impact of elevated water flux on body water stable isotope ratios, independent of hyperglycemia.

Disease state can significantly alter both metabolism and behavior, which in turn, will directly impact total energy expenditure (TEE). Food consumption and activity levels were not measured in this study and it is possible that these factors differed between treatment groups as hyperglycemia and resulting diabetes progressed in the STZ group. Although previous research has demonstrated no difference in the metabolic rates of VEH and STZ treatment groups [18,19], the issue that either altered food consumption or activity may impact the isotopic composition of body water should be considered. As it would be expected for a change in mass to occur if food consumption or metabolism

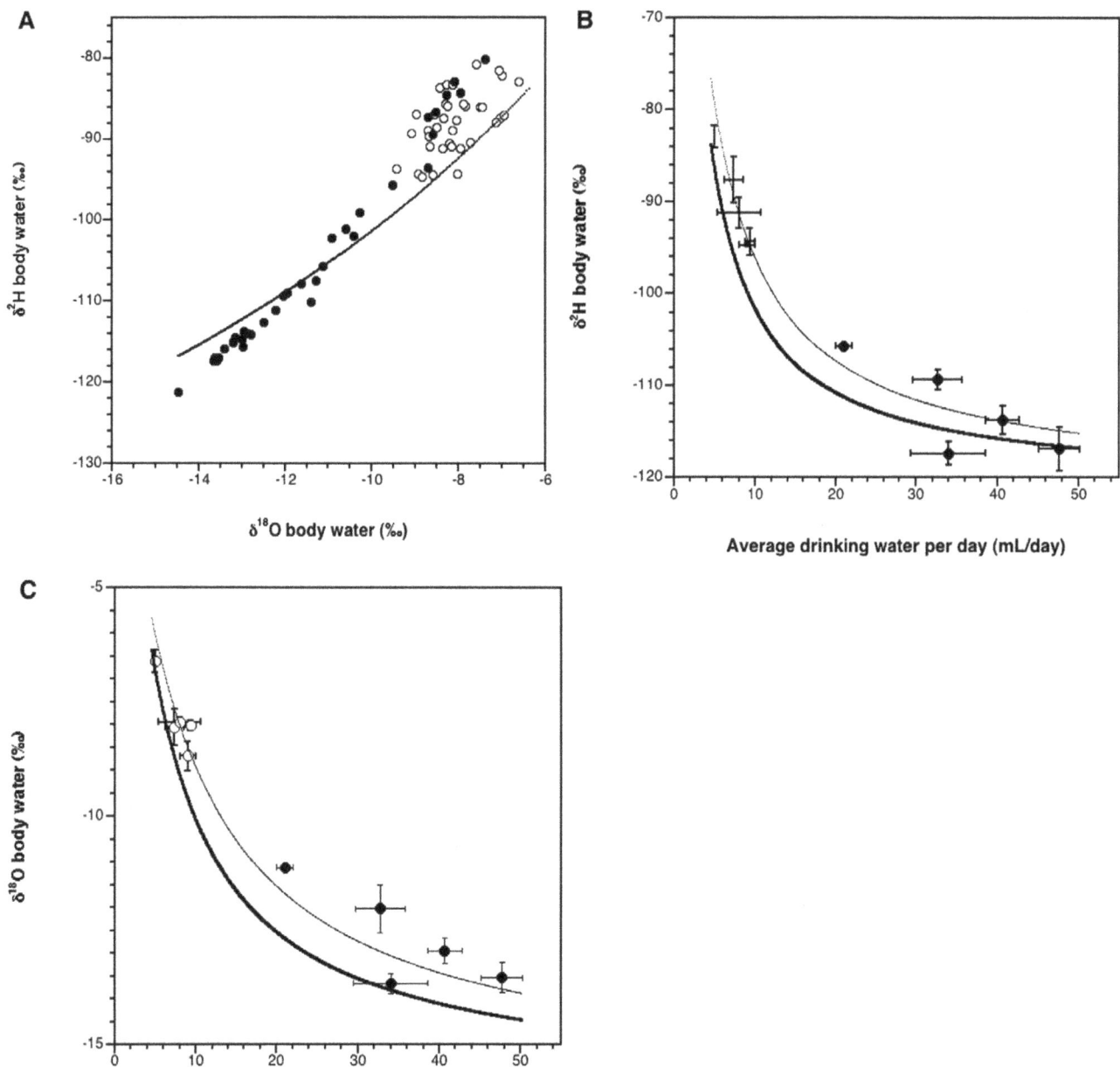

**Figure 3. Isotopic values of body water relative to model predictions.** Isotopic signature of body water of VEH ($\bigcirc$) and STZ ($\bullet$) mice relative to body water model predictions in which the only model variant is water influx. (A) Measured $\delta^2H$ relative to $\delta^{18}O$ of body water for all animals over the entire 5-week study period. The curve depicts model predictions for $\delta^2H$ and $\delta^{18}O$ values over a range of water influx values. (B) Measured $\delta^2H$ values of body water and (C) $\delta^{18}O$ values of body water relative to average drinking water per day of 9 individual animals (mL/day). The bold curve on each panel indicates values predicted by our mathematical model of the isotopic composition of body water using a total energy expenditure (TEE) of 37.4 KJ/day. Correlations between the model and measured $\delta^2H$ and $\delta^{18}O$ of body water yielded an $r^2 = 0.83$ and $r^2 = 0.79$, respectively. The lighter curve on each panel indicates values predicted by our mathematical model of the isotopic composition of body water using a TEE of 65.0 KJ/day. Correlations between the model and measured $\delta^2H$ and $\delta^{18}O$ of body water yielded an $r^2 = 0.96$ and $r^2 = 0.96$, respectively. Bars delineate standard deviation for sample analyses.

changed significantly, we can approach this issue indirectly by examining the respective mass changes of each group throughout the progression of the study. There was no significant change in either the average mass of the VEH group (mass = 26.5± 0.6g; ANOVA: $\alpha = 0.05$, p = 0.52) or the STZ group (mass = 25.0±0.8g; ANOVA: $\alpha = 0.05$, p = 0.24), and thus we can surmise that the changes observed in the isotopic composition of body water are likely the sole result of altered water flux due to hyperglycemia.

The model accurately predicts both the actual range and concavity of body water $\delta^2H$ and $\delta^{18}O$ values in STZ and VEH mice as average drinking water intake increases ($r^2 = 0.83$ and $r^2 = 0.79$, respectively; Fig. 3b and c). While all parameters were either measured directly, or estimated from directly measurable parameters, the authors acknowledge that variation in the values assigned to these parameters may alter the accuracy of model predictions. For example, in addition to values of water flux, the model is also sensitive to values assigned TEE. The value used for

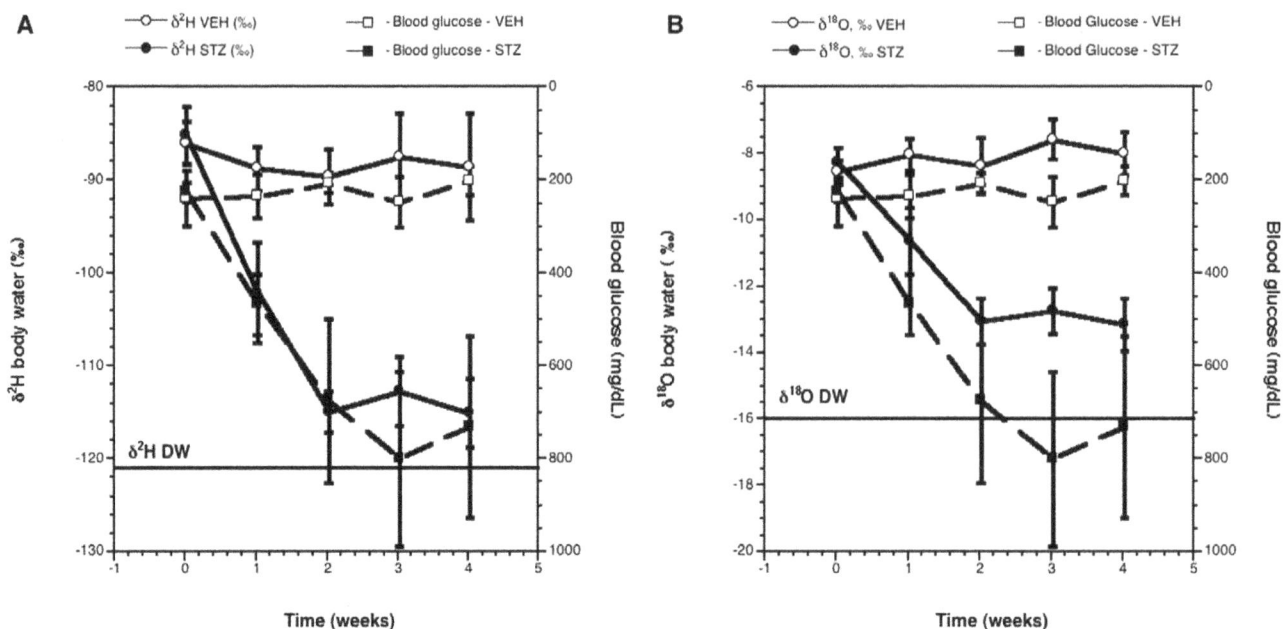

**Figure 4. Isotopic values of body water relative to blood glucose.** Isotopic signature of body water (y1 axis) and blood glucose values (y2 axis) over the 5 week study period relative to the isotopic signature of drinking water (−DW). Time 0 is prior to STZ treatment. (A) $\delta^2H$ of body water and blood glucose values of VEH treated (○ and □, respectively; n = 7) and STZ treated (● and ■, respectively; n = 6) mice. (B) $\delta^{18}O$ of body water and blood glucose values of VEH treated (○ and □, respectively ; n = 7) and STZ treated (● and ■, respectively; n = 6) mice. Bars delineate standard deviation for duplicate sample analyses.

TEE in this exercise was 37.4 KJ/day and was calculated based on the hourly consumption of oxygen ($VO_2$) by a 25-gram control mouse housed independently. This value may be an underestimate of the actual TEE of animals housed in groups (as in this study), where between animal interactions may elevate energy expenditure. To demonstrate the models' sensitivity to TEE, we nearly doubled the value of TEE used in this study (increased from 37.4 KJ/day to 65.0 KJ/day) and applied the model over the same range of water influxes used in the previous exercise. In this case, doubling the value of TEE actually increases the accuracy with which the model predicts the isotopic composition of body water ($r^2 = 0.96$ for both $\delta^2H$ and $\delta^{18}O$; Fig. 3b and c). These exercises emphasize the importance of factors such as TWF and TEE in determining the isotopic composition of body water, particularly in reconstructive or predictive applications.

## Materials and Methods

All experiments were approved by the University of Utah Institutional Animal Care and Use Committee (IACUC protocol 09-08011). Water flux measurements (n = 5/treatment) were taken 1, 3, and 4 weeks following a 5-day low dose STZ injection protocol [20]. We determined the stable hydrogen and oxygen isotope ratios of water cryogenically extracted from blood (hereafter referred to as body water) of STZ (n = 6) and VEH (n = 7) treated controls prior to STZ injections and weekly for the 4 weeks following STZ treatment. Samples were analyzed in duplicate at the University of Utah Stable Isotope Ratio Facility for Environmental Research (SIRFER; http://sirfer.net) on a ThermoFinnigan-MAT Delta Plus XL isotope ratio mass spectrometer (Bremen, Germany) with a high temperature conversion elemental analyzer (TC/EA) attached. Samples were pyrolyzed at 1400°C to produce $H_2$ and CO gas. Resultant gases were separated on a 1-m, 0.25 in (outer diameter) molecular sieve

5Å gas chromatography column (Costech Analytical, Valencia, CA, USA). Water samples were introduced to the pyrolysis column using a PAL autosampler (LEAP Technologies, Carrboro, NC, USA) and analyzed alongside a set of three laboratory water reference materials previously calibrated to the Vienna Standard Mean Ocean Water (VSMOW) scale. The analytical precision for samples was ±1.55‰ and ±0.17‰ for H and O, respectively.

## Stable Isotope Notation

Stable isotope abundances are reported in *d*-notation as parts per thousands (‰), where

$$\delta_{sample} = (R_{sample}/R_{STD} - 1) * 1000$$

and $R_{sample}$ and $R_{standard}$ are the molar ratios of the rare to abundant isotope (e.g., $^2H/^1H$) in the sample and standard, respectively. The international standard for both hydrogen and oxygen stable isotope analysis is VSMOW.

## Modeling

We applied a mechanistic model describing the isotopic composition of body water to determine if we could accurately predict the H and O isotope ratios of body water of STZ and VEH mice. Similar models have been applied to reconstruct paleoclimate from the $\delta^{18}O$ value of fossil biogenic phosphate [6,21,22] and to determine energy expenditure via the doubly labeled water technique [11,23].

Assuming an animal is in isotopic equilibrium, the general mass balance equation can be written as

$$R_{dw}^O F_{dw}^O + R_{fw}^O F_{fw}^O + R_{food}^O F_{food}^O + R_{wvg}^O F_{wvg}^O + R_{air}^O F_{air}^O$$
$$= \alpha_{co2}^O R_{bw}^O F_{co2}^O + \alpha_{bwv}^O R_{bw}^O F_{bwv}^O + \alpha_{twv}^O R_{bw}^O F_{twv}^O + R_{bw}^O F_{rw}^O$$

**Table 1.** Values used in body water modeling.

| Influx | Oxygen Flux (TEE = 37.4 KJ/day, 65.0 KJ/day) | Hydrogen Flux (TEE = 37.4 KJ/day, 65.0 KJ/day) | Source |
|---|---|---|---|
| dw – drinking water | 0.125–1.39 | 0.250–2.78 | measured |
| fw – food water | 0 | 0 | measured |
| food – bound O and H | 0.0315, 0.0547 | 0.0734, 0.128 | [18,21] |
| wvg – water vapor gain | 0.00602, 0.0105 | 0.0120, 0.0209 | [21] |
| air – atmospheric O | 0.0761, 0.132 | n/a | [18,21] |
| **Efflux** | **Oxygen Flux (TEE = 37.4 KJ/day, 65.0 KJ/day)** | **Hydrogen Flux (TEE = 37.4 KJ/day, 65.0 KJ/day)** | **Source** |
| $CO_2$ – carbon dioxide | 0.0709, 0.123 | n/a | [18,21] |
| bwv – breath water vapor | 0.0577, 0.100 | 0.115, 0.201 | [6] |
| Twv – transcutaneous water vapor | 0.0463 | 0.0925 | [23] |
| rw – remaining water | 0.0637–1.39 | 0.127–2.78 | calculated |
| **Fractionation Factors** | **Oxygen** | **Hydrogen** | **Source** |
| $\alpha_{co2}$ | 1.038 | n/a | [24] |
| $\alpha_{bwv}$ | 0.991 | 0.946 | [23,25] |
| $\alpha_{twv}$ | 0.981 | 0.935 | [23,25] |
| **Isotope Ratio** | **Oxygen ($\delta^{18}O$)** | **Hydrogen ($\delta^2H$)** | **Source** |
| dw – drinking water | −16 | −121 | measured |
| fw – food water | n/a | n/a | n/a |
| food – bound O and H | 23.9 | −97.9 | measured |
| wvg – water vapor gain | −25.5 | −165 | [6,26] |
| air – atmospheric O | 15.1 | n/a | [6] |

Where applicable, values for modeling exercises completed with a total energy expenditure of 37.4 and 65.0 KJ/day are included.

for $^{18}O$, and similarly

$$R_{dw}^H F_{dw}^H + R_{fw}^H F_{fw}^H + R_{food}^H F_{food}^H + R_{wvg}^H F_{wvg}^H$$
$$= \alpha_{bwv}^H R_{bw}^H F_{bwv}^H + \alpha_{twv}^H R_{bw}^H F_{twv}^H + R_{bw}^H F_{rw}^H$$

for $^2H$. Isotope abundances are expressed as ratios ($R$) of heavy to light isotopes, fluxes ($F$) are in units mole/day, and isotopic fractionation relative to body water is expressed as $\alpha$. Oxygen and hydrogen influxes and effluxes considered include drinking water (*dw*), free water in food (*fw*), oxygen and hydrogen bound in food (*food*), water vapor gain (*wvg*), atmospheric oxygen (*air*), carbon dioxide ($CO_2$), breath water vapor (*bwv*), transcutaneous water vapor (*twv*), and unfractionated remaining water loss (*rw*) which includes urine, fecal water, and sweat.

Solving for the isotope ratio of body water yields

$$R_{bw}^O = \frac{R_{dw}^O F_{dw}^O + R_{fw}^O F_{fw}^O + R_{food}^O F_{food}^O + R_{wvg}^O F_{wvg}^O + R_{air}^O F_{air}^O}{\alpha_{co2}^O F_{co2}^O + \alpha_{bwv}^O F_{bwv}^O + \alpha_{twv}^O F_{twv}^O + F_{rw}^O}$$ for oxygen, and $$R_{bw}^H = \frac{R_{dw}^H F_{dw}^H + R_{fw}^H F_{fw}^H + R_{food}^H F_{food}^H + R_{wvg}^H F_{wvg}^H}{\alpha_{bwv}^H F_{bwv}^H + \alpha_{twv}^H F_{twv}^H + F_{rw}^H}$$ for hydrogen.

Influxes and effluxes of oxygen and hydrogen fall into two categories, those associated with free water (*dw, fw, wvg, bwv, twv, rw*) and those derived from the metabolism of carbohydrates, fats, and protein (*food, air, co2*). Oxidation reactions for glucose,

palmitic acid, and alanine are considered first order representatives of carbohydrate, fat, and protein substrates [21]. The flux terms associated with metabolism are therefore functions of diet composition, specifically the fraction of energy derived from carbohydrates, fats, and protein, as well as total energy expenditure. All parameters were either measured directly or estimated from directly measurable parameters (Table 1). Animals were housed in a temperature controlled facility and facility-recorded averages were used for temperature (20°C) and humidity (20%) inputs in the model. Modeling exercises were performed using two values of total energy expenditure (TEE): 37.4 KJ/day and 65.0 KJ/day. For each TEE value, the drinking water influx ($F_{dw}$) was manipulated to simulate the measured range (5–50 mL/day), while all other parameters were held constant.

## Acknowledgments

We thank J. Barnette and M. Lott for technical assistance. We also wish to thank D. Dearing for use of metabolic cages.

## Author Contributions

Conceived and designed the experiments: SO ARW CHR EDA TC JRE. Performed the experiments: SO ARW LOV LEE. Analyzed the data: SO CHR LOV LEE LAC EDA TC JRE. Contributed reagents/materials/analysis tools: CHR LAC EDA TC. Wrote the paper: SO.

# References

1. Bowen GJ, Ehleringer JR, Chesson LA, Thompson AH, Podlesak DW, et al. (2009) Dietary and physiological controls on the hydrogen and oxygen isotope ratios of hair from mid-20th century indigenous populations. Am J Phys Anthro 139: 494–504.

2. Ehleringer JR, Bowen GJ, Chesson LA, West AG, Podlesak DW, et al. (2008) Hydrogen and oxygen isotope ratios in human hair are related to geography. Proc Natl Acad Sci USA 105: 2788–2793.

3. Bowen GJ, Wassenaar LI, Hobson KA (2005) Global application of stable hydrogen and oxygen isotopes to wildlife forensics. Oecologia 143: 337–348.

4. Hobson KA, Van Wilgenburg SL, Larson K, Wassenaar LI (2009) A feather hydrogen isoscape for Mexico. J Geochem Explor 102: 167–174.

5. Meier-Augenstein W, Fraser I (2008) Forensic isotope analysis leads to identification of a mutilated murder victim. Sci and Just 48: 153–159.

6. Kohn MJ (1996) Predicting animal $\delta^{18}O$: Accounting for diet and physiological adaptation. Geochim Cosmochim Acta 60: 4811–4829.

7. Podlesak D-W, Torregrossa AM, Ehleringer JR, Dearing MD, Passey BH, et al. (2008) Turnover of oxygen and hydrogen isotopes in the body water, $CO_2$, hair, and enamel of a small mammal. Geochim Cosmochim Acta 72: 19–35.

8. Fricke HC, O'Neil JR (1996) Inter- and intra-tooth variation in the oxgyen isotope composition of mammalian tooth enamel phosphate: implications for paleoclimatological and paleobioligical research. Palaeogeo Palaeoclim Palaeoeco 126: 91–99.

9. Dansgaard W (1964) Stable isotopes in precipitation. Tellus 16: 436–467.

10. Bowen GJ, Revenaugh J (2003) Interpolating the isotopic composition of modern meteoric precipitation. Water Resources Res 39: 1299.

11. Gretebeck RJ, Schoeller DA, Socki RA, Davis-Street J, Gibson EK, et al. (1997) Adaptation of the doubly labeled water method for subjects consuming isotopically enriched water. J Appl Physiol 82: 563–570.

12. Longinelli A, Padalino A-P (1980) Oxygen isotopic composition of water from mammal blood: first results. Eur J Mass Spectrom 1: 135–139.

13. Raman A, Schoeller DA, Subar AF, Troiano RP, Schatzkin A, et al. (2004) Water turnover in 458 American adults 40–79 yr of age. Am J Physiol Renal Physiol 286: F394–F401.

14. Chesson LA, Valenzuela LO, O'Grady SP, Cerling TE, Ehleringer JR (2010) Hydrogen and oxygen stable isotope ratios of milk in the USA. J Agric Food Chem 58: 2358–2363.

15. Clementz MT, Holroyd PA, Koch PL (2008) Identifying aquatic habits of herbivorous mammals through stable isotope analysis. Palaios 23: 574–585.

16. Schrier R, Gottschalk C (1997) Diseases of the Kidney. Boston MA: Little, Brown and Company.

17. Lu WT, Juang JH, Hsu B, Huang HS (1998) Effects of high or low dose of streptozocin on pancreatic islets in C57BL/6 and C.B17-SCID mice. Transplant Proc 30(2): 609–610.

18. Polotsky VY, Wilson JA, Haines AS, Scharf MT, Soutiere SE, et al. (2001) The impact of insulin-dependent diabetes on ventilatory control in the mouse. Am J Resp Crit Med 163: 624–632.

19. Rodrigues B, Figueroa D, Mostarda C, Heeren M, Irigoyen MC, et al. (2007) Maximal exercise test is a useful method for physical capacity and oxygen consumption determination in streptozotocin-diabetic rats. Cardiovasc diabetol 6: 38.

20. AMDCC Consortium (2003) in AMDCC Protocols, ed F. B. (The University of Michigan Medical Center). pp 1–3.

21. Bryant JD, Froelich PN (1995) A model of oxygen isotope fractionation in body water of large mammals. Geochim Cosmochim Acta 59: 4523–4537.

22. Luz B, Kolodny Y, Horowitz M (1984) Fractionation of oxygen isotopes between mammalian bone-phosphate and environmental drinking water. Geochim Cosmochim Acta 48: 1689–1693.

23. Schoeller DA, Ravussin E, Schutz Y, Acheson KJ, Baertschi P, Jequier E (1986) Energy expenditure by doubly labeled water: validationin humans and proposed calculation. Am J Physiol Regul Integr Comp Physiol 250: R823–R830.

24. Pflug KP, Schuster KD, Pichotka JP, Forstel H (1979) Fractionation effects of oxygen isotopes in mammals. In: Stable Isotopes: Proceedings of the Third International Conference Klein ER, Klein PD, eds. New York: Academic Press. pp 553–561.

25. Horita J, Wesolowski DJ (1994) Liquid-vapor fractionation of oxygen and hydrogen isotopes of water from the freezing to the critical temperature. Geochim Cosmochim Acta 58: 3425–3437.

26. Cheung MC, et al. (2009) Body surface area prediction in normal, hypermuscular, and obese mice. J Surg Res 153: 326–331.

# 4

# Spatio-Temporal Dynamics of Maize Yield Water Constraints under Climate Change in Spain

Rosana Ferrero[1,4]*, Mauricio Lima[2,3,4], Jose Luis Gonzalez-Andujar[1,3]

**1** Departamento Protección de Cultivos, Instituto de Agricultura Sostenible, Consejo Superior de Investigaciones Científicos (CSIC), Córdoba, Spain, **2** Departamento de Ecología, Pontificia Universidad Católica de Chile, Santiago, Chile, **3** Laboratorio Internacional de Cambio Global, LINCG (CSIC-PUC), Santiago, Chile, **4** Center of Applied Ecology and Sustainability (CAPES), Santiago, Chile

## Abstract

Many studies have analyzed the impact of climate change on crop productivity, but comparing the performance of water management systems has rarely been explored. Because water supply and crop demand in agro-systems may be affected by global climate change in shaping the spatial patterns of agricultural production, we should evaluate how and where irrigation practices are effective in mitigating climate change effects. Here we have constructed simple, general models, based on biological mechanisms and a theoretical framework, which could be useful in explaining and predicting crop productivity dynamics. We have studied maize in irrigated and rain-fed systems at a provincial scale, from 1996 to 2009 in Spain, one of the most prominent "hot-spots" in future climate change projections. Our new approach allowed us to: (1) evaluate new structural properties such as the stability of crop yield dynamics, (2) detect nonlinear responses to climate change (thresholds and discontinuities), challenging the usual linear way of thinking, and (3) examine spatial patterns of yield losses due to water constraints and identify clusters of provinces that have been negatively affected by warming. We have reduced the uncertainty associated with climate change impacts on maize productivity by improving the understanding of the relative contributions of individual factors and providing a better spatial comprehension of the key processes. We have identified water stress and water management systems as being key causes of the yield gap, and detected vulnerable regions where efforts in research and policy should be prioritized in order to increase maize productivity.

**Editor:** Pilar Hernandez, Institute for Sustainable Agriculture (IAS-CSIC), Spain

**Funding:** R. Ferrero gratefully acknowledges receipt of a grant from the Fundación Carolina. J. L. Gonzalez-Andujar and R. Ferrero were supported by FEDER (Fondo Europeo de Desarrollo Regional) and the Spanish Ministry of Economy and Competitiveness funds (AGL2012-33736). R. Ferrero and M. Lima acknowledge financial support from Fondo Basal-CONICYT grant FB-0002. We are grateful to LINCGlobal (Laboratorio Internacional en Cambio Global) for their support. The funders had no role in study design, data collection and analysis, decision to publish, or preparation of the manuscript.

**Competing Interests:** The authors have declared that no competing interests exist.

* E-mail: rferrero@ias.csic.es

## Introduction

Spatio-temporal patterns of agricultural production are clearly influenced both by climate change and agricultural management practices. Recently, many studies have analyzed the impact of climate change on crop productivity [1], but comparing the performance of different crop management systems has rarely been explored (exc. [2]). To be specific, we need to evaluate how and where irrigation practices (e.g. rain-fed versus irrigated) are effective in mitigating the effects of climate change, because water constraints and crop demand in agro-systems could be increased due to climate change [3–7]. Identifying whether there are any differences in the principal bio-physical factors and mechanisms that explain both systems will enable us to improve crop productivity without expanding the cropland area and to diminish the adverse impacts of agriculture for social and ecological systems [8].

We do not know much about crop response to climate change yet, and still less about the differential response between irrigated and rain-fed systems [4]. Increases in agriculture production could potentially come from increases in irrigated crops, because higher yields could be attained with reduced production variability [9].

However, this also depends on soil and management factors that result in spatial patterns of yields [10]. Secondly, irrigation can influence local climate by inducing cooling, but this may depend on the extent of the irrigated area, the level of soil moisture alteration and cloud response to irrigation [11]. Third, average yields in rain-fed systems are commonly 50% or less of yield potential (high yield gap), suggesting ample room for improvement [12] but, again a great spatial variability has been found [13]. Yield gaps could be bigger in cropping systems that experience wider ranges of variation under climate conditions [10]. Fourth, plant population (or density) is known to affect the yield potential at a given location [12] and grain yield stability [14]. However, to our knowledge, there are no previous studies explicitly comparing endogenous processes under different water management systems. Finally, simulation at a broad scale level cannot fully explain the above process, and process-based crop models do not always relate to observed yields [15]. Finer spatial scales and historical data of irrigated versus rain-fed systems could help to compare modelled or simulated yield potentials [12].

Analyzing the sensitivity of irrigated and non-irrigated (rain-fed) crops to past climate changes is crucial to an understanding of the vulnerability of agriculture to climate change in the future, particularly in regions that already suffer from this under present conditions. This paper explores biophysical factors and water management practice constraints to maize (*Zea mays* L.) in Spain. Spatial shifts northwards have been projected for maize, due to the extremely hot, dry summers in south-central Europe [16,17], particularly in Spain [18]. The expected effects of climate change on Spain's agriculture would not be uniform. Mediterranean (arid and semiarid) regions may be particularly sensitive, where a decrease in the general availability of hydric resources and an increase in evaporative demand, especially during summer, will affect irrigation requirements [19]. Namely, it is one of the most prominent "hot-spots" in future climate change projections [20], where a mean reduction of 17% in water resources [21,22] has been predicted. For this drought-prone zone, all climate change scenarios imply the need to significantly increase the contribution of irrigation water. Therefore, identifying and quantifying the links between water management practices and food production is crucial in addressing the intensified conflicts between water scarcity and food safety.

The objective of this paper is to determine how climate variability affects maize production in Spain under irrigated and rain-fed conditions. First, we have analyzed the regulatory structure of maize production dynamics under both water management systems. Second, we have evaluated the mechanisms (in ecological parameters) underlying climate perturbations on maize yields. Third, we have assessed whether the importance of maize production structures (i.e. intrinsic regulation) and climate change perturbations (i.e. exogenous factors) could change according to the type of management (i.e. rain-fed and irrigated) and the geographical location. Fourth, we have estimated the potential yield of each region and water management using the previous models and analyzing the spatial variability of yield losses due to water stress [23]. We have combined information on spatial autocorrelation water stress patterns for maize yields to identify the importance of climate constraints at a regional scale.

## Methods and Materials

### Database

Provincial maize yield levels (*Zea mays*; production per hectare, *kg/ha*) for 1996–2009 were obtained from statistical yearbooks [24]. We studied selected provinces that had both rain-fed and irrigated systems (Fig. 1), and displayed trends in yield fluctuation in Fig. S1. We used Global Historical Climatology Network (GHCND) data on monthly temperature and rainfall (mean, minimum, maximum and extreme; [25]). Various summary statistics of the growing season (July to October) weather were then computed: *EMNT* extreme minimum temperature (°C), *EMXT* extreme maximum temperature (°C), *MMNT* mean minimum temperature (°C), *MMXT* mean maximum temperature (°C), *MNTM* mean temperature (°C), *EMXP* extreme maximum daily precipitation total (l/m2) and *TPCP* total precipitation (l/m2). We also examined carbon dioxide emission ($CO_2$), an important atmospheric gas that contributes to global warming. The annual country-level emissions of $CO_2$ (kt) were taken from the World Bank's World Development Indicators (WDI; [26]).

### Diagnosis and statistical models of yields dynamics

We have analyzed and predicted maize yield responses to the impact of climate change in Spain through the use of models based on the population dynamics theory. Of course this is not a true population in the reproductive sense, but crop systems obey the same rules as all other dynamic systems, both natural and engineered.

First, where necessary, we used sequencing (i.e., splitting the series into two stationary segments) and detrending (i.e., rotating the series around the linear or quadratic trend) to generate a stationary time series. Second, we estimated the logarithmic rate of change of the yield as $R_t = Y_t - Y_{t-1}$ (the same response variable as [1,27]), where $Y_t$ represents the provincial yield in a year $t$(the logarithm of the detrended yield) and $Y_{t-1}$ is the same series with one year of delay (lag 1).

We were able to detect and analyze non-trivial feedback processes by examining their relationship $R_t = f(Y_{t-d})$, where the function $f$ described how the crop yield change rate varied with yield level, and this has been called the *R*-function. We used the partial rate correlation function (or *PRCF*) to estimate the order of the dynamical process and determine how many time lags ($d$) should be included in the model for representing the feedback structure. This function detects the feedback order removing the confounding effect by calculating the partial correlation between $R_t$ and $Y_{t-d}$ with the effects of lower lags removed [28].

We then used the generalized version of the exponential form of the discrete time logistic model [29,30] in terms of the *R*-function to represent pure endogenous models in the function $f$:

$$R_t = r_{\max} - \exp(a Y_{t-d} + c) \tag{1}$$

where $Y_{t-d}$ represents the yield data at time $t-d$ (where $d$ was obtained from *PRCF* function), $r_{\max}$ is a positive constant representing the maximum finite rate of change (and is estimated as the maximum rate of change from the observed data), $c$ is a measure of the ratio between demand and offer of limiting resources and $a$ is the nonlinearity of the curve. The nonlinearity of this model includes a biological realistic property: its net reproductive rate is bounded [29], that is, the performance of any crop must have an upper bound simply because no crop can produce an infinite number of grains that subsequently contribute to the crop yield.

Finally, we used the Royama classification of exogenous effects as a framework to deduce causal mechanisms of the climate change impact on these crop yields in a spatial-temporal study [29]. To include exogenous perturbations, we modelled $r_{\max}$ and $c$ of (1) as linear functions of climate conditions, each of which has an explicit biological interpretation. In this way, we set up mechanistic hypotheses about the exogenous effects of climate on these yields data [29].

If an exogenous factor (i.e. climate or gas emissions) changes $r_{\max}$ and has an additive or independent perturbation effect on crop yield levels, it shifts the *R*–function curve along the *y*-axis ("*vertical*" perturbations):

$$r_{\max}^* = r_{\max} + bZ_{t-d'}$$
$$R_t = r_{\max} - \exp(a Y_{t-d} + c) + bZ_{t-d'} \tag{2}$$

where $Z_{t-d'}$ is the exogenous factor (for lags or $d$ 0 and 1; in logarithm scale). This model produces alterations to both $r_{\max}$ and the carrying capacity (equilibrium point of the population, $R_t = 0$), changing the level of equilibrium and its stability.

If an exogenous factor (i.e. climate or gas emissions) changes $c$ and has a non-additive perturbation effect on crop yield levels, and influences the equilibrium point of the population shifting the *R*-function curve along the *x*-axis ("*lateral*" perturbations):

$$c* = c + bZ_{t-d'}$$
$$R_t = r_{\max} - \exp(a Y_{t-d} + c + bZ_{t-d'}) \tag{3}$$

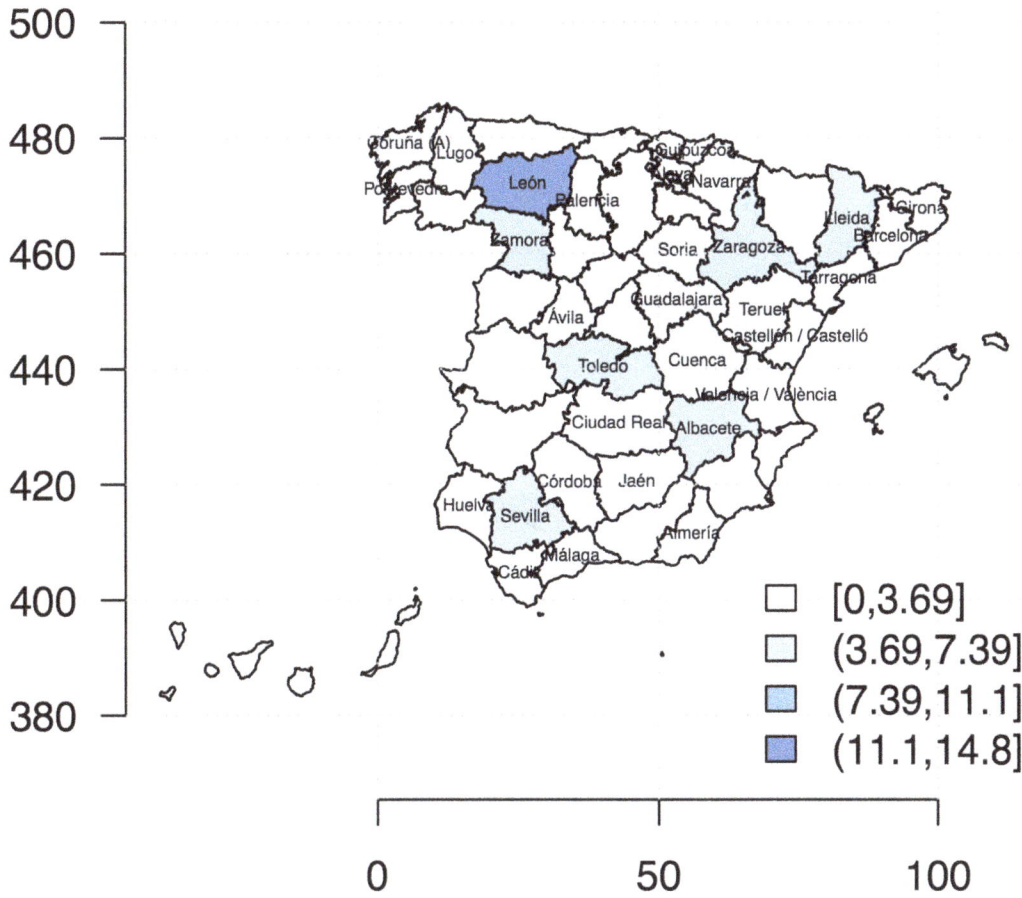

**Figure 1. Definition of study regions (provinces) with percentage of total maize production for 1996–2009.** Only provinces with both irrigated and rain-fed systems were analyzed.

Lateral perturbations do not change the pattern of dynamics around equilibrium because they do not change the slope at the equilibrium.

We fitted Eqs. 1–3 using nonlinear least squares regressions with the *nls* library in the software *R* [31,32]. In particular, the models were fitted by minimizing the Akaike criterion with a correction for finite sample sized ($AIC_c$):

$$AIC_c = (2k - 2\ln L) + \frac{2k(k+1)}{n-k-1}$$

where $k$ is the number of parameters and $L$ is the maximized value of the likelihood function for the model, and $n$ denotes the sample size. Also, we maximized the pseudo $R^2$ measures based on the deviance residual [33]. Models were chosen on the basis of their goodness-of-fit (assessed using root mean square error RMSE and the log-likelihood values), their ability to describe the correct feedback structure, and their appropriateness.

### Yield losses due to suboptimal water availability (YGRw)

We propose a new estimation of the potential yield or equilibrium productivity [34] at the provincial level as the equilibrium value of the models. By solving Eqn. (1–3) for the equilibrium dynamics $Y_t = Y_{t-1} = K$ (when $R_t = 0$), we calculated the maize yield level at equilibrium, sometimes called the carrying capacity ($Mg/ha$). For non-pure endogenous models we made potential yield estimations for each year as the exogenous factor

changed. Then we calculated the percentage of yield losses due to suboptimal water availability ($YGRw$; Eqn. 4; view [35]), which indicated how close the rain-fed yield potential is to the irrigated value for a given site (%).

$$YGR_w = \frac{YP_{IR} - YP_{RF}}{YP_{IR}} \quad (4)$$

We obtained some time-invariant $YGR_w$ values when, in the same province, irrigated and rain-fed $YP$ were estimated from pure endogenous models, so that we calculated the averaged $YGR_w$ for each province, and studied its spatial variability without taking into account the temporal dimension of the data.

We determined whether there was any spatial autocorrelation in $YGR_w$ with the global Moran's $I$ (spatial correlation on average, of an entire map). At this stage, we were not yet trying to determine the causes, although the results could have motivated a hypothesis. We assumed: 1) that there was no spatial patterning due to some underlying but unmodelled factor, and 2) that the assigned spatial weights were those that generated the autocorrelation. Then we tested whether $YGR_w$ was more spatially clustered than by chance. The matrix that represents spatial dependence ($W$) uses a binary indicator of neighbourhood (i.e. the spatial weights, $w_{ij}$, are defined as $w_{ij} = 1$ if the $i$ and $j$ provinces are contiguous neighbours, $w_{ij} = 0$ otherwise, based on rook contiguity; [36]). We used row-standardisation (style $W$) that favours observations with few neighbours. We calculated a non-parametric approach to infer-

## Climate change perturbations in maize yield

**Temperature effects in irrigated maize**

**Temperature effects in rainfed maize**

**Precipitation effects in irrigated maize**

**Precipitation effects in rainfed maize**

**CO2 effects in irrigated maize**

**CO2 effects in rainfed maize**

Negative • Without • Positive

ence on Moran's $I$ using 999 simulations (Monte Carlo permutation test). Also, local indicators of spatial association (or LISA) were calculated to detect "hot spots" where there was a strong autocorrelation, and "cold spots", where there were none. The results were plotted on a Moran scatterplot: the target variable on the $x$-axis, and the (spatially-weighted) sum of neighbouring values on the $y$-axis; these are called spatially lagged values. We identified the high-influence areas.

We analyzed the environmental spatially distributed causes of averaged $YGR_w$ through a Simultaneous Autoregressive Model (SAR; [36]) that considers spatial autocorrelation of residuals:

$$YGR_w = Z^T\beta + e \qquad (5)$$

where, for each province, $YGR_w$ is the percentage of yield losses due to suboptimal water availability, $Z$ is a matrix of averaged climate variables (see Database section; except country-level $CO_2$ emissions), $e = B(Y - Z^T\beta) + \epsilon$ is the error term, and $\epsilon$ represents residual errors (assumed to be independently distributed according to a Normal distribution with zero mean and diagonal covariance matrix $\sigma_e^2$). The error terms are modelled so that they depend on each of the other areas to account for their spatial dependence ($B$ is a matrix that contains the dependence parameters; $B = \lambda W$, where $\lambda$ is a spatial autocorrelation parameter and $W$ is a matrix that represents the spatial dependence explained above). Global Moran's $I$ was computed for the residuals to test if the SAR model accounts for all the spatial autocorrelations in $YGR_w$. For the spatial analysis we used *spdep* library in the software $R$ [37].

## Results

### Regulatory structure and exogenous perturbation models

After sequencing and detrending, all the sites exhibited first-order negative feedback ($PRCF(1)$) as being the most important component of yield growth rate (Figure S2; except for irrigated maize in Vizcaya and rain-fed systems in Tarragona). Major sites showed the highly significant ($p < 0.05$) effect of endogenous processes as determinants of the structure of crop productivity regulation (Table S1).

We evaluated gas emission ($CO_2$) and climate factors (temperature and precipitation; see Database section), as exogenous perturbations of the production curve ($R$-function). Table S1 shows several models that were selected as climate change impacts on maize production for each Spanish region and management system. The stochastic versions of the step-ahead predictions of the models are shown in Fig. S3. As expected, the effects of climate on maize production were not uniform, and depended on the irrigation management system (Figure 2). Maize yields were significantly related to minimum temperatures (possibly night ones) in 11 sites and by maximum temperature in another 5 sites. Generally, there were positive effects of temperature for irrigated systems, except for Almería (for minimum temperature –$EMNT$-) and Ourense (mean temperature –$MNTM$-). However, for rain-fed systems, we detected negative effects of warming on major sites, with the exception of Málaga and Albacete (both for $EMNT$). As expected, precipitation was not important for irrigated systems (except for maximum precipitation –$EMXP$- in Navarra), but it was an important factor in some rain-fed managements. There

were positive effects of precipitation on Teruel and Soria (for a total –$TPCP$- and maximum rainfall), and negative ones in Córdoba ($TPCP$ and $EMXP$) and Zaragoza ($TPCP$) on rain-fed crops. Finally, $CO_2$ emissions negatively affected maize in Lugo (irrigated), Ourense and Soria (both rain-fed), and positively only in Ávila (rain-fed).

Temperature acted mainly as having non-additive (lateral) effects on maize yield dynamics, whereas $CO_2$ emission acted as additive (vertical) effects (Table S1; Figure 3 and S4). Finally, rainfall exerted non-additive effects when it had a negative impact on maize, but when it obtained positive responses the effects were of both types (additive or non-additive; Table S1; Figure S4). For example, Figure 3 shows positive and non-additive (lateral) effects of temperature on rain-fed maize in Albacete and irrigated maize in Sevilla. That is, the increase in temperature had a positive effect on both maize systems, and more so at high yield levels. Figure 3 also indicates a negative and additive (vertical) effect of $CO_2$ emission on Ourense (same strength for all yield levels), and a positive and non-additive (lateral) rainfall effect on rain-fed maize in Soria (more important for high yield levels).

### Relative yield losses due to suboptimal water availability (YGR_w)

We first visualized the spatial relation of $YGR_w$ (Figure 4), where several high $YGR_w$ values were shown in central and southern provinces of Spain. The global Moran's $I$ value ($I = 0.39$) was of an opposite sign and much larger in absolute value than the expectation ($E[I] = -0.034$); this was quite unlikely to be equal to the expectation of no spatial association. The probability of incorrectly rejecting the null hypothesis of no association (type I error) was 0.0021. The Monte Carlo approach also rejects the null hypothesis (the true value for Moran's $I$ is zero; $I_{mc} = 0.406$, $p = 0.005$; Figure S5). The Moran scatterplot (Figure 5; the vector of values and the neighbour list with weights) showed points with a great influence, which are identified by a special symbol and their name. The highest-leverage area is marked on Almería; it has the highest $YGR_w$ (84.56) and a zero weighted spatially-lagged proportion, because it did not have any adjacent areas in the study. Soria and Palencia had low $YGR_w$, and a low spatially-lagged proportion; these are the low-$YGR_w$ neighbourhoods adjacent to low-$YGR_w$ neighbourhoods. They have a great influence on the slope (global Moran's $I$). From Figure 5 it is clear that most of the global Moran's $I$ significance comes from the local Moran's $I$ from high $YGR_w$ in Almería, and low $YGR_w$ associated with low $YGR_w$, in the Soria and Palencia area in the north.

There was clear evidence of local clustering, 6 areas (Ciudad Real, Cuenca, Albacete, Valencia, A Coruña and Pontevedra) showed sufficiently high local Moran's $I$ to reject the null hypothesis with less than a 5% chance of Type I error. These areas were not highlighted in the Moran scatterplot, as they did not greatly influence the global Moran's $I$ but were locally-clustered.

There was a significant spatial correlation in the residuals, because the estimated value of lambda was 0.141 and the $p$-value of the likelihood ratio test 0.0354. Only averaged temperature ($MNTMt$) was significant for the SAR model, suggesting that provinces with higher temperature have larger $YGR_w$ percentages. The model found was: $YGR_w = -64.42 + 5.69 * MNTM_t$ the SAR model, which accounted for the whole spatial autocorrelation

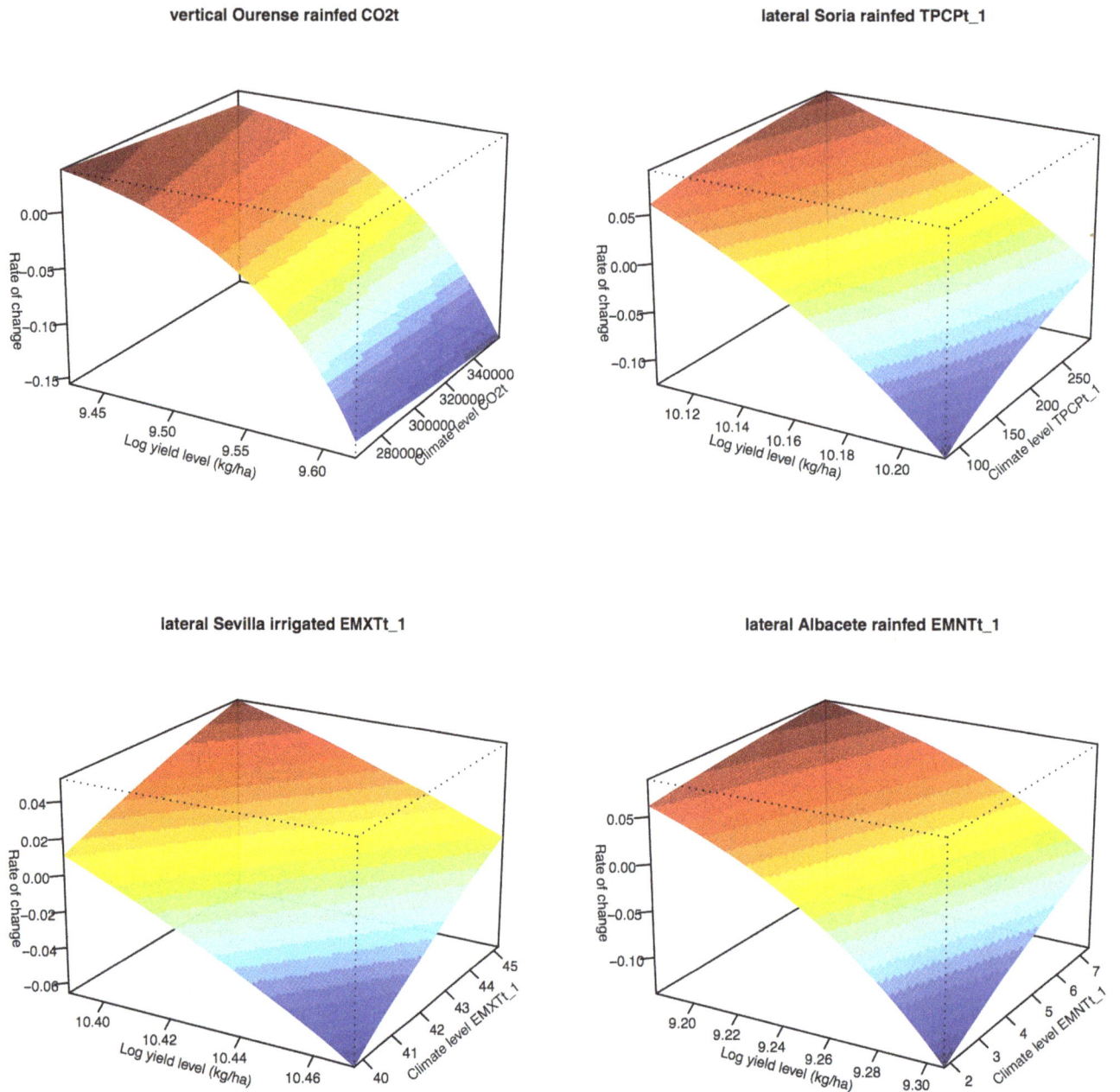

**Figure 3. Yield rate of change against the log observed yield level (with one year of delay) and the exogenous factor that perturbs the productivity function ($R$-function).** Exogenous factors include carbon emissions ($CO2_t$), precipitation ($TPCP_{t\_1}$), and maximum and minimum temperature ($EMXT_{t\_1}$ and $EMNT_{t\_1}$). Additive (vertical) and non-additive (lateral) perturbation effects were detected. Colours indicate the $R$-function value. See Table S1 for description of models and Figure S4 for their graphs.

in $YGR_w$ (global Moran's test for residuals was $I = -0.00811$, $p = 0.422$). Thus, the autocorrelation in the linear model residuals was explained.

## Discussion

In the present study, the impact of climate variability on maize yields in Spanish rain-fed and irrigated systems was investigated for the period 1996–2009. We explored the endogenous structure (regulation) and the exogenous perturbations effects on maize production at a regional scale.

### Regulatory structure: endogenous feedback

We found that maize productivity had a persistently negative effect on crop yields for a one year time delay (first order negative feedback, $PRCF(1)$). Maize productivity was characterized by negative first-order feedback structure in major sites and in both irrigation systems. Namely, there were biomass or density-induced feedback loops in the growth, survival rates, seed germination or grain production rates of individual plants, tending to stabilize their dynamics [38]. In Spain, the seeds produced are used for the next year and, therefore, a year's crop performance could change seed viability and vigour, which also affects the performance of the following crops (changing the demand for resources). Also, a crop

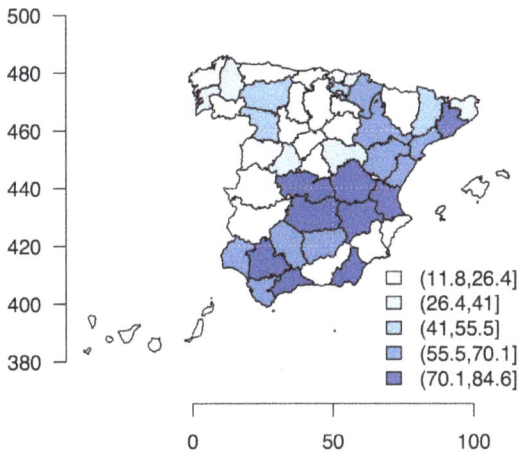

**Figure 4. Relative yield losses due to suboptimal water availability (*YGRw*; %).** The percentage of yield losses due to suboptimal water availability indicates how close rain-fed yield potential are to the irrigated value for a given site.

system could alter habitat conditions; in fact, the frequent practice of crop rotation is a testimony to the importance of negative feedbacks in agricultural systems (i.e. it modifies resource supplies). This produces high-frequency dynamics due to year-to-year endogenous variability in maize yields. Our logistic models appear to capture the essential features of the fluctuations observed, and suggest a mechanistic explanation for the latter. This implies that, to understand the response of maize productivity to climate, we must also know the endogenous feedback structure of the system.

Our models are important to conceptualizing the problem of regulated versus unregulated systems. If a system were to be controlled entirely by an exogenous process (unregulated systems), then the series would perform a random walk and we saw no sign of the generated series becoming stabilized, but it drifted increasingly away from the origin with the passing of time [29]. However, persistence implies regulation (but not necessarily vice versa) and, therefore, the rate of change in a persistent crop productivity system is not statistically independent of the yield level and should be bounded (i.e. regulated systems).

## Climate change effects: exogenous perturbations

In line with previous studies, temperature during the growing season was the most important weather variable influencing maize yields [39]. However, we deciphered the effects of climate on maize productivity providing new interpretations. First, diagnostic analysis suggested that temperature acts mainly as a non-additive (lateral) perturbation in maize productivity. Therefore, the relationship between temperature and maize yields was nonlinear and could not be captured adequately by a linear or quadratic functional relation as in previous studies [40]. Our analysis suggests a biological reason for the nonlinear interaction between climate and maize yield level. Temperature had no direct impacts on yield rate of change (affecting $r_{max}$; additive or vertical effects), but influenced the availability or requirements of some limiting factor or resource (changing $c$; non-additive or lateral effects). There is probably a relationship between extreme heat and plant water stress, increasing water demand and/or soil water content in rain-fed systems, in agreement with the recent results of Lobell *et al.* [41]. This is because, the effects of high temperature are experienced only when the maize yield level is close to equilibrium [29]. This kind of perturbation exerts strong effects on the average

level of yield but few on the intrinsic periodicity induced by endogenous feedback.

Secondly, rain-fed maize yields are negatively affected by temperature increases, but irrigated systems may gain from warming in some regions. As expected, rain-fed crop damage may result from greater water and heat stress during hot growing seasons. However, unexpected positive effects of temperature in irrigated systems are possibly a consequence of heat tolerance, which is consistent with other studies on local adaptation to hot temperatures being able to minimize stress effects [40] or the cooling effect of irrigation [42]. Therefore, we detected some adaptation to heat stress that could mitigate the projected heat-related losses, at least in a few regions with irrigated systems.

Thirdly, climate variability and extreme events are more important than averages. Thus, we detected that minimum temperature was the dominant factor in maize production, in agreement with other recent studies for maize [43–45] and rice [46]. Currently, a new paradigm has been originated: crop yields have declined with a higher minimum or night temperature [46,47] or when there was a marked asymmetry between maxima and minima [48]. One possible explanation includes the facts that the grain-growth rate has increased and that the duration of grain-filling has been shortened as the temperature increased, producing lower crop production (yield levels) [49]. Mohammed & Tarpley [47] proposes a list of the effects of high night temperatures on crop production. Also, our findings are in line with the results of recent research which argue that global minimum temperatures are increasing faster than maximum temperatures, and the need to explore the ecological consequences of this phenomenon [41,50,51]. Therefore, we wish to highlight the importance of considering extreme climate variables in crop production studies, and limiting the use of averages or accumulative climate data which ignore inter-annual variability of climate and extreme events. Our results differ from those of most studies which do not take into account food production structure regulation, and those which use degree-days [40] concepts which assume a cumulative or additive effect of temperature on crop yield and do not adequately account for the effects of extreme temperatures (high or low) either.

In the study period, precipitation was not a major abiotic factor limiting maize yield of cultivated rain-fed crops in Spain. We only detected positive effects of precipitation for irrigated maize in Navarra, Teruel and Soria. Also, growing season rainfall negatively affected rain-fed maize yield in Córdoba and Zaragoza, possibly due to flood and waterlogging problems causing production losses. Again, we agree with Lobell *et al.* [41], who argue that the apparent paradox of the scant effect of precipitation on rain-fed maize yield whereas, on the contrary, there is a water stress effect of temperature, can be solved with the following reflection "*large precipitation changes are required to rival the effect of temperature on water stress, because high temperature affects both water demand and supply*".

As in the study of Long *et al.* 2006 [52], ours study indicates that there was a smaller $CO_2$ effect on maize yield than previously presumed. Impacts of higher $CO_2$ on maize yield were reduced probably because it is a C4 plant, and also because of the national scale of the variable in our study.

## Spatial variability of yield losses due to water stress

We found that the global spatial pattern of yield losses due to water stress is not a random one (Figure 4); there was a high influence in Palencia, Soria (lowest) and Almería (highest). We detected clusters of "cold spots" in northern Spain (A Coruña and Pontevedra) and "hot spots" in central provinces (Ciudad Real,

Cuenca, Albacete, Valencia; Figure 5). Neither cluster greatly influenced the global Moran's $I$ but they were locally-clustered. Moreover, we modelled spatial $YGR_w$ values with climate variables and found that the mean temperature was the highest constraint of maize productivity due to water stress. In conclusion, policy action to decrease the relative yield gap due to water stress on maize productivity has the potential to geographically target high $YGR_w$ areas. Future work will help determine other non-climatic causal relationships between $YGR_w$ and an array of factors that could influence water management practices in maize (e.g. access to water, management technology, soil conditions, etc.).

A recent comparison of simulated and observed yield patterns highlights the value of data in the spatial distribution of yields for understanding the causes of landscape yield variability [10].

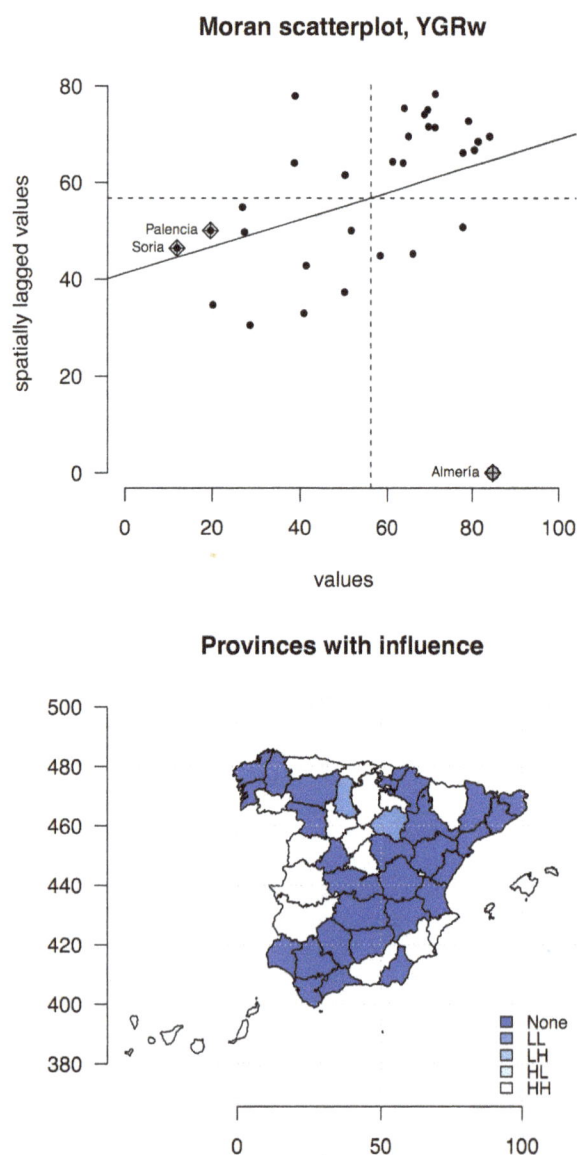

However, to our knowledge, this is the first study explicitly evaluating the spatial pattern of real relative yield gaps due to a water management system and its sensitivity to underlying climate factors. The results demonstrate that spatial patterns of yield loss due to water stress possess substantial information on the relative importance of water management factors for maize productivity.

The need for an analysis to identify and implement adaptation options in agriculture emphasizes the importance of regional scales (federal, provincial, and territorial governments). Global and non-spatial studies can provide only a very partial and potentially misleading insight into the true impact of climate change, where aggregation can indeed conceal vulnerability and climate change costs [53]. However, individual regions (provinces) allow a better analysis of uncertainty and risks, thus providing practical recommendations to farmers.

## Moran scatterplot, YGRw

## Provinces with influence

**Figure 5. Spatial autocorrelation analysis of the relative yield losses due to suboptimal water availability ( YGRw).** Top: Moran scatterplot; bottom: high-influence areas neighbours: no influence (None), high proportion with low proportion neighbours (HL), the reverse (LH), and both high (HH). We define the break between "low" and "high" as the third quartile.

## Conclusions

We identified the same regulation structure for both management systems, i.e. a negative first-order feedback process that tends to stabilize the crop's dynamics. We analyzed the underlying mechanisms of the interaction between climate variation and regulatory structure on maize production. Different climate variables appear to operate differently on maize productivity. We found that the effect of temperature (mainly extreme values) cannot be evaluated independently of crop productivity as in previous studies, because its consequences are experienced only when maize yield level is close to equilibrium (lateral perturbation). We suggest that high maize yield crops are especially vulnerable to weather-related yield variations. These data support the belief that lower yields are more suitable for low-input conditions, because climate might be more severe in crops that interact strongly with productivity [14].

Our results also indicate that it may be important to consider explicitly the irrigation system and spatial variability. Rain-fed agriculture may be at risk as heat waves will be more intense, more frequent and longer (particularly in Seville, Cádiz, Almería, Navarra and Ávila; see Fig. 2). Irrigation seems to allow some tolerance to warming but future levels of water availability would be compromised if water restrictions and irrigation costs increased, as climate change projections indicate. We propose a new framework to estimate yield potential as the equilibrium yield or yield carrying capacity. Climate change is not uniform over Spain and the effectiveness of irrigated and rain-fed management varies with the location, producing different regional vulnerabilities and potential yields. Accordingly, the general strategies for adapting maize productivity to climate change will vary between different zones in Spain.

## Supporting Information

**Figure S1   Time series of maize yield level for rain-fed (red) and irrigated (blue) systems.** Each provinces of Spain were analyzed for 1996–2009.

**Figure S2   Partial rate correlation function ( PRCF).**

**Figure S3   Comparison of observed crop yield levels (points, obs) for the period 1997–2009 with stochastic predictions from models fitted to the data until the year 1996 (broken line, sim) and 95% confidence intervals for forecasts (shaded area, 95PPU).** P-factor is the percent of observations that are within the given uncertainty bounds and R-factor represents the average width of the given uncertainty bounds divided by the standard deviation of the observations. See Table S1 for description of models and variables.

**Figure S4   R-functions: yield rate of change against the log observed yield level (with one year of delay).** Climate factors had vertical (additive) and lateral (non-additive) perturbations on the R-function. Colors indicate the value of the R-function. See Table S1 for description of models and variables.

**Figure S5   A non-parametric approach to inference on Moran's _I_ using 999 simulations (Monte Carlo permutation test).**

**Table S1**   Summary statistics of nonlinear logistic models, 1996–2009. We evaluated pure Endogenous models (E), and additive (or Lateral, L) and non-additive (or Vertical, V) models that also represent the effect of exogenous perturbations. Different crop management systems were analyzed (IR = irrigated and RF = rainfed). %Total percentage of total crop production in Spain, K carrying capacity or potential yield, rmax maximum finite reproductive rate, a non-linearity coefficient, c the ratio between demand and offer of limiting resources, b coefficients for different

exogenous effects, $R^2$ pseudo-coefficient of determination, _logLIK_ log-likelihood, _RMSE_ root-mean-square error and _AICc_ corrected Akaike information criterion. NOTE: \*$p<0.05$, \*\*$p<0.01$, Number of not avaiable data (NA) were indicated by I. _CO2_ carbon dioxide emission (_kt_, country-level emissions), and summary statistics of the growing season weather: _EMNT_ extreme minimum temperature (°C), _EMXT_ extreme maximum temperature (°C), _MMNT_ mean minimum temperature (°C), _MMXT_ mean maximum temperature (°C), _MNTM_ mean temperature (°C), _EMXP_ extreme maximum daily precipitation total (l/m2), _TPCP_ total precipitation (l/m2).

## Acknowledgments

We thank the editor and two anonymous reviewers for their constructive comments, which helped us to improve the manuscript.

## Author Contributions

Conceived and designed the experiments: RF ML JLG-A. Performed the experiments: RF. Analyzed the data: RF. Contributed reagents/materials/analysis tools: RF. Wrote the paper: RF ML JLG-A.

## References

1. Lobell DB, Schlenker W, Costa-Roberts J (2011) Climate trends and global crop production since 1980. Science 333: 616–620.
2. Licker R, Johnston M, Foley JA, Barford C, Kucharik CJ, et al. (2010) Mind the gap: how do climate and agricultural management explain the "yield gap" of croplands around the world? Glob Ecol Biogeogr 19: 769–782.
3. Döll P (2002) Impact of climate change and variability on irrigation requirements: a global perspective. Clim Change 54: 269–293.
4. Turral H, Burke J, Faurès JM, Faurés JM (2011) Climate change, water and food security. Rome Food Agric Organ United Nations: 204.
5. Easterling WE, Aggarwal PK, Batima P, Brander KM, Erda L, et al. (2007) Food, fibre and forest products. In: Parry ML, Canziani OF, Palutikof JP, Linden PJ van der, Hanson CE, editors. Climate Change 2007: Impacts, Adaptation and Vulnerability. Contribution of Working Group II to the Fourth Assessment Report of the Intergovernmental Panel on Climate Change. Cambridge University Press, Cambridge, UK. pp. 273–313.
6. Smith P, Martino D, Cai Z, Gwary D, Janzen H, et al. (2007) Agriculture. In: Metz B, Davidson OR, Bosch PR, Dave R, Meyer LA, editors. Climate Change 2007: Mitigation. Contribution of Working Group III to the Fourth Assessment Report of the Intergovernmental Panel on Climate Change. Cambridge University Press, Cambridge, United Kingdom and New York, NY, USA. pp. 497–540.
7. Medici LO, Reinert F, Carvalho DF, Kozak M, Azevedo RA (2014) What about keeping plants well watered? Environ Exp Bot 99: 38–42.
8. Zou X, Li Y, Gao Q, Wan Y (2011) How water saving irrigation contributes to climate change resilience—a case study of practices in China. Mitig Adapt Strateg Glob Chang 17: 111–132.
9. Mueller ND, Gerber JS, Johnston M, Ray DK, Ramankutty N, et al. (2012) Closing yield gaps through nutrient and water management. Nature 490: 254–257.
10. Lobell DB, Ortiz-Monasterio JI (2006) Regional importance of crop yield constraints: Linking simulation models and geostatistics to interpret spatial patterns. Ecol Modell 196: 173–182.
11. Lobell D, Bala G, Mirin A, Phillips T, Maxwell R, et al. (2009) Regional Differences in the Influence of Irrigation on Climate. J Clim 22: 2248–2255.
12. Lobell DB, Cassman KG, Field CB (2009) Crop Yield Gaps: Their Importance, Magnitudes, and Causes. Annu Rev Environ Resour 34: 179–204.
13. Meng Q, Hou P, Wu L, Chen X, Cui Z, et al. (2013) Understanding production potentials and yield gaps in intensive maize production in China. F Crop Res 143: 91–97.
14. Tokatlidis IS (2013) Adapting maize crop to climate change. Agron Sustain Dev 33: 63–79.
15. Reidsma P, Ewert F, Boogaard H, van Diepen K (2009) Regional crop modelling in Europe: The impact of climatic conditions and farm characteristics on maize yields. Agric Syst 100: 51–60.
16. Olesen JE, Carter TR, Díaz-Ambrona CH, Fronzek S, Heidmann T, et al. (2007) Uncertainties in projected impacts of climate change on European agriculture and terrestrial ecosystems based on scenarios from regional climate models. Clim Change 81: 123–143.
17. Wolf J, Diepen CA (1995) Effects of climate change on grain maize yield potential in the european community. Clim Change 29: 299–331.
18. Iglesias A, Mínguez MI (1995) Prospects for maize production in Spain under climate change. In: Rosenzweig C, editor. Climate Change and Agriculture: Analysis of Potential International Impacts. American Society of Agronomy, Vol. asapecial. pp. 259–273.
19. Moreno JM, Aguiló E, Alonso S, Cobelas MÁ, Anadón R, et al. (2005) A Preliminary General Assessment of the Impacts in Spain Due to the Effects of Climate Change A Preliminary Assessment of the Impacts in Spain due to the Effects of Climate Change. Available: http://www.magrama.gob.es/en/cambio-climatico/temas/impactos-vulnerabilidad-y-adaptacion/plan-nacional-adaptacion-cambio-climatico/eval_impactos_ing.aspx. Accessed 2014 May 8.
20. Rodríguez-Puebla C, Ayuso S, Frías M, García-Casado L (2007) Effects of climate variation on winter cereal production in Spain. Clim Res 34: 223–232.
21. Iglesias A, Garrote L, Quiroga S, Moneo M (2011) A regional comparison of the effects of climate change on agricultural crops in Europe. Clim Change 112: 29–46.
22. Iglesias A, Mougou R, Moneo M, Quiroga S (2010) Towards adaptation of agriculture to climate change in the Mediterranean. Reg Environ Chang 11: 159–166.
23. Neumann K, Verburg PH, Stehfest E, Müller C (2010) The yield gap of global grain production: A spatial analysis. Agric Syst 103: 316–326.
24. MAGRAMA (2012) Ministerio de Agricultura, Alimentación y Medio Ambiente (MAGRAMA). MAGRAMA Stat Databases. Available: http://www.magrama.gob.es/es/estadistica/temas/publicaciones/anuario-de-estadistica/. Accessed 1 April 2012
25. Lawrimore JH, Menne MJ, Gleason BE, Williams CN, Wuertz DB, et al. (2011) An overview of the Global Historical Climatology Network monthly mean temperature data set, version 3. J Geophys Res 116: D19121. Available: http://doi.wiley.com/10.1029/2011JD016187. Accessed 1 April 2012.
26. World Bank, World Development Indicators (WDI) (2012). ESDS Int Univ Manchester. Available: http://data.worldbank.org/data-catalog/world-development-indicators/wdi-2012. Accessed 1 April 2012.
27. Lobell DB, Bänziger M, Magorokosho C, Vivek B (2011) Nonlinear heat effects on African maize as evidenced by historical yield trials. Nat Clim Chang 1: 42–45.
28. Berryman A, Turchin P (2001) Identifying the density-dependent structure underlying ecological time series. Oikos 92: 265–270.
29. Royama T (1992) Analytical population dynamics. London, New York: Chapman & Hall. Springer.
30. Ricker WE (1954) Stock and recruitment. J Fish Res Board Canada 11: 559–623.
31. R Development Core Team (2011) R: A language and environment for statistical computing. R Found Stat Comput Vienna. Available: http://www.r-project.org/. Accessed 2014 May 8.
32. Bates D, Chambers JM (1991) Nonlinear Models. In: Chambers JM, Hastie TJ, editors. Statistical Models in S. Wadsworth & Brooks/Cole, Pacific Grove, California.
33. Cameron AC, Windmeijer FAG (1996) R-Squared Measures for Count Data Regression Models With Applications to Health-Care Utilization. J Bus Econ Stat 14: 209–220.

34. De la Maza M, Lima M, Meserve PL, Gutierrez JR, Jaksic FM (2009) Primary production dynamics and climate variability: ecological consequences in semiarid Chile. Glob Chang Biol: 1116–1126.

35. Liu Z, Yang X, Hubbard K, Lin X (2012) Maize potential yields and yield gaps in the changing climate of Northeast China. Glob Chang Biol 86.

36. Bivand RS, Pebesma E, Gómez-Rubio V (2008) Applied Spatial Data Analysis with R. New York, NY: Springer New York.

37. Bivand R, Altman M, Anselin L, Assunçao R, Berke O, et al. (2013) spdep: Spatial dependence: weighting schemes, statistics and models. R package version 0.5-56.

38. Berryman AA (1999) Principles of population dynamics and their application. Stanley Thrones (Publishers) Limited.

39. Sun BJ, Van Kooten GC (2013) Weather effects on maize yields in northern China. J Agric Sci: 1–11.

40. Butler EE, Huybers P (2012) Adaptation of US maize to temperature variations. Nat Clim Chang 3: 68–72.

41. Lobell DB, Hammer GL, McLean G, Messina C, Roberts MJ, et al. (2013) The critical role of extreme heat for maize production in the United States. Nat Clim Chang 3: 497–501.

42. Lobell DB, Bonfils CJ, Kueppers LM, Snyder MA (2008) Irrigation cooling effect on temperature and heat index extremes. Geophys Res Lett 35: L09705.

43. Muchow RC, Sinclair TR, Bennett JM (1990) Temperature and Solar Radiation Effects on Potential Maize Yield across Locations. 82: 338–343.

44. Chen C, Lei C, Deng A, Qian C, Hoogmoed W, et al. (2011) Will higher minimum temperatures increase corn production in Northeast China? An analysis of historical data over 1965–2008. Agric For Meteorol 151: 1580–1588.

45. Lobell DB (2007) Changes in diurnal temperature range and national cereal yields. Agric For Meteorol 145: 229–238.

46. Peng S, Huang J, Sheehy JE, Laza RC, Visperas RM, et al. (2004) Rice yields decline with higher night temperature from global warming. Proc Natl Acad Sci U S A 101: 9971–9975.

47. Mohammed A, Tarpley L (2007) Effects of High Night Temperature on Crop Physiology and Productivity: Plant Growth Regulators Provide a Management Option. Global Warming Impacts – Case Studies on the Economy, Human Health, and on Urban and Natural Environments. doi: 10.5772/24537.

48. Rosenzweig C, Tubiello FN (1996) Effects of changes in minimum and maximum temperature on wheat yields in the central US A simulation study. Agric For Meteorol 80: 215–230.

49. Muchow RC (1990) Effect of high temperature on grain-growth in field-grown maize. F Crop Res 23: 145–158.

50. Alward RD, Detling J, Milchunas D (1999) Grassland Vegetation Changes and Nocturnal Global Warming. Science (80-) 283: 229–231.

51. Katz RW, Brown BG (1992) Extreme events in a changing climate: Variability is more important than averages. Clim Change 21(3):289–302.

52. Long SP, Ainsworth EA, Leakey ADB, Nösberger J, Ort DR (2006) Food for thought: lower-than-expected crop yield stimulation with rising $CO_2$ concentrations. Science 312: 1918–1921.

53. Bosello F, Carraro C, De Cian E (2009) An Analysis of Adaptation as a Response to Climate Change. SSRN Electron J.

# 5

# Genetic Control of Water Use Efficiency and Leaf Carbon Isotope Discrimination in Sunflower (*Helianthus annuus* L.) Subjected to Two Drought Scenarios

**Afifuddin Latif Adiredjo**[1,2], **Olivier Navaud**[3], **Stephane Muños**[4,5], **Nicolas B. Langlade**[4,5], **Thierry Lamaze**[3], **Philippe Grieu**[1]*

1 Université de Toulouse, INP-ENSAT, UMR 1248 AGIR (INPT-INRA), Castanet-Tolosan, France, 2 Brawijaya University, Faculty of Agriculture, Department of Agronomy, Plant Breeding Laboratory, Malang, Indonesia, 3 Université de Toulouse, UPS-Toulouse III, UMR 5126 CESBIO, Toulouse, France, 4 INRA, Laboratoire des Interactions Plantes-Microorganismes (LIPM), UMR 441, Castanet-Tolosan, France, 5 CNRS, Laboratoire des Interactions Plantes-Microorganismes(LIPM), UMR 2594, Castanet-Tolosan, France

## Abstract

High water use efficiency (WUE) can be achieved by coordination of biomass accumulation and water consumption. WUE is physiologically and genetically linked to carbon isotope discrimination (CID) in leaves of plants. A population of 148 recombinant inbred lines (RILs) of sunflower derived from a cross between XRQ and PSC8 lines was studied to identify quantitative trait loci (QTL) controlling WUE and CID, and to compare QTL associated with these traits in different drought scenarios. We conducted greenhouse experiments in 2011 and 2012 by using 100 balances which provided a daily measurement of water transpired, and we determined WUE, CID, biomass and cumulative water transpired by plants. Wide phenotypic variability, significant genotypic effects, and significant negative correlations between WUE and CID were observed in both experiments. A total of nine QTL controlling WUE and eight controlling CID were identified across the two experiments. A QTL for phenotypic response controlling WUE and CID was also significantly identified. The QTL for WUE were specific to the drought scenarios, whereas the QTL for CID were independent of the drought scenarios and could be found in all the experiments. Our results showed that the stable genomic regions controlling CID were located on the linkage groups 06 and 13 (LG06 and LG13). Three QTL for CID were co-localized with the QTL for WUE, biomass and cumulative water transpired. We found that CID and WUE are highly correlated and have common genetic control. Interestingly, the genetic control of these traits showed an interaction with the environment (between the two drought scenarios and control conditions). Our results open a way for breeding higher WUE by using CID and marker-assisted approaches and therefore help to maintain the stability of sunflower crop production.

**Editor:** Carl J. Bernacchi, University of Illinois, United States of America

**Funding:** The work was supported by a French Government Scholarship (Bourse du gouvernement Français, BGF) and a co-funding by Directorate general of Higher Education, Ministry of Education and Culture, Republic of Indonesia (Beasiswa Luar Negeri, BLN). The funders had no role in study design, data collection and analysis, decision to publish, or preparation of the manuscript.

**Competing Interests:** The authors have declared that no competing interests exist.

* Email: grieu@ensat.fr

## Introduction

Water use efficiency (WUE) as a breeding target can be defined as the ratio of biomass production to water consumption. Breeding for WUE and drought-resistant crop varieties has been a critical area of agricultural research worldwide [1–3]. Substantial efforts have been devoted to identifying and selecting for morphological and physiological traits that increase WUE and yield under rain-fed conditions [2,4–5]. In field conditions, water consumption is usually difficult to determine. Nevertheless, WUE can be represented by measuring leaf carbon isotope discrimination (CID) [6–7]. Because the CID has been demonstrated to be a simple but reliable measure of WUE, the negative correlation between them has been used as an indirect method in selection to improve WUE [8–10]. The principle mechanisms underlying the variation of CID act through variation in the intercellular $CO_2$ concentration $(c_i)$ maintained in leaves [6]. The value of $c_i$ is determined through the coordinated regulation of carboxylation capacity (photosynthesis) and stomatal control of leaf diffusive conductance (transpiration regulation) [6–7].

Genetic variation underlying quantitative traits, such as WUE and CID, which are generally under considerable environmental influence, is governed by quantitative trait loci (QTL) [11–14]. QTL mapping provides a starting point in breeding programs [15–16] and for cloning of the causal mutations by fine mapping.

QTL mapping of WUE is rarely reported. Four QTL associated with WUE have been identified in soybean [17]. The inheritance of WUE has been studied using simple sequence repeat (SSR) markers in alfalfa [18]. In contrast, QTL mapping of CID has been reported by numerous authors. The first QTL identified for CID was reported by Martin and Nienhuis [19]. These authors identified four QTL associated with CID in tomato. Since that time, QTL for CID have been identified across a wide range of species, for example in cotton [20], rice [21], barley [22],

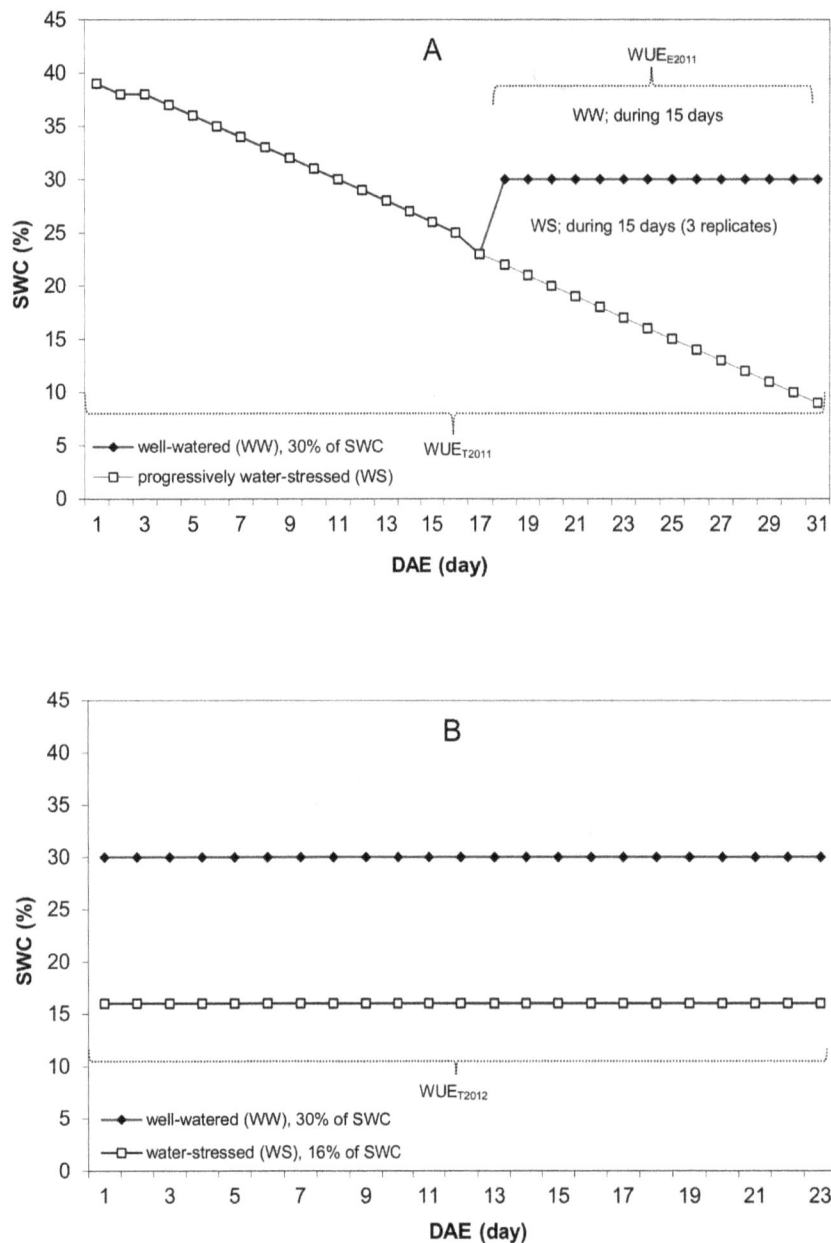

**Figure 1. Principles of the water treatments used in this study.** (A) In experiment 2011, three replicates (each of 150 plants) were subjected to progressive water-stress by water withholding from 1 to 31 DAE. In this experiment a control replicate (150 plants) was watered to maintain non-stressful conditions (SWC = 30%). (B) In Experiment 2012, two replicates (each of 150 plants) were maintained at in stressful conditions SWC = 16% from 1 to 23 DAE whereas two other replicates (each of 150 plants) were irrigated to maintain non-stressful conditions (SWC = 30%). DAE: day after emergence.

*Arabidopsis* [23], and in wheat [24]. However, to our knowledge, QTL of WUE and CID in sunflower have never been reported.

Most of the work identifying QTL of WUE and CID has been done in well-watered conditions, with only one study in a drought situation. There is no report on the QTL identification of WUE and CID of crops subjected to different scenarios of water deficit establishment.

The objectives of the present study are to identify QTL controlling WUE and CID in a population of RILs of sunflower, and to compare QTL associated with these traits in a dual drought scenario: (i) a progressively water-stressed establishment and (ii) a stable water deficit treatment. We are interested in providing new

insights into the genetic architecture of WUE and CID, and in contributing to the potential of sunflower breeding by improved WUE.

## Materials and Methods

### Plant materials

A population of 150 recombinant inbred lines (RILs) was used in two experiments. A population of these RILs was named INEDI and was obtained by single seed descent (self-pollination to at least F8) from a cross between XRQ and PSC8 [25].

## Experiments and trait measurements

Two experiments were conducted in spring 2011 (Exp. 2011) and in spring 2012 (Exp. 2012) under quite similar weather conditions. Plants were grown in a greenhouse at the INRA Auzeville station, Toulouse, France (43°31'46,94" N; 1°29'59,71" E). Greenhouse air temperature was set at $25/18\pm2°C$ (day/night) and relative humidity was about $55-75\pm5\%$.

Three seeds per genotype were sown in a pot (volume: 2 liters) at the beginning of the experiments. The pots contained a mixture of 50% soil (collected from the field), 30% organic matter and 20% sand. These pots were arranged on 100 balances (maximum capacity 30 kg, precision 2 g, model SXS, GRAM, Spain), with six pots per balance (total pot number in greenhouse was 600). Each pot was then covered with a 3 mm layer of polystyrene sheet with a hole in the middle to allow normal plant growth, thus reducing the evaporation of water from the soil surface. Throughout the experiments, the amounts of water in the pots were determined by weighing the pots every day. This weighing recorded the amount of daily water loss, corresponding to the daily transpiration of the plants. For each pot, at the end of the experiment, cumulative daily transpiration was called CWT (the cumulative water transpired). Biomass was separated into leaves and stems at harvest. Total dry aerial biomass (BM) was obtained after drying at 80°C for 48 h. WUE was determined at the end of the experiment, defined as the ratio of BM to CWT. In addition, a dual drought scenario strategy for the two experiments (explained in detail below) was studied.

## Experiment conducted in 2011: scenario of progressive water stress

A randomized complete block design with three replicates was used for the progressive water stress treatments (three replicates $\times$ 150 genotypes = 450 plants; called WS). There was another replicate (150 plants) that was considered as a well-watered treatment, called WW.

At 1 day after emergence (DAE), 17 days after sowing (DAS), all 600 pots were watered to field capacity, corresponding to 39.5% of soil water content (SWC). These 600 pots (WW and WS) were kept without irrigation until 17 DAE (Fig. 1A). In these conditions, stomatal conductance of the plant was still not affected. We calculated that stomatal conductance started to decrease at an average SWC of about 21% (unpublished data).

Starting at 17 DAE, when genotypes reached around 23% of SWC, we irrigated the WW treatment to 30% of SWC and we maintained this SWC by daily irrigation. The WS treatment was kept without irrigation until harvest (during 15 days).

Two determinations of WUE were made. The first was the total water use efficiency, $WUE_{T2011}$, calculated by dividing the BM by the $CWT_{31d}$. $CWT_{31d}$ is the cumulative water transpired during 31 days (from 1 to 31 DAE). The second calculation of WUE was made during the period when the two treatments differed in their soil water content (WW and WS), from 17 to 31 DAE, and called $WUE_{E2011}$ (water use efficiency "estimation"). $WUE_{E2011}$ was calculated by dividing the "estimated biomass" ($BM_E$), by the $CWT_{15d}$, calculated from 17 to 31 DAE. $BM_E = BM - BM_{17}$, where $BM_{17}$ is the biomass estimated at 17 DAE. In addition, the $BM_{17}$ was calculated as follows: $BM_{17} = (LA_{17}/LA_{31}) \times BM$, where $LA_{17}$ and $LA_{31}$ are the leaf areas measured on 17 and 31 DAE, respectively.

## Experiment conducted in 2012: scenario of stable SWC

A randomized complete block design with two treatments and two replicates was performed (300 pots per treatment). Treatments consisted of two levels of stable SWC which was imposed: well-watered (30% of SWC, namely WW) and water-stressed (16% of SWC, namely WS) (Fig. 1B).

At 1 DAE (19 DAS), stable water contents corresponding to 30% of SWC (WW) and 16% of SWC (WS) were maintained for 23 days (Fig. 1B). WUE was calculated by dividing the BM by the $CWT_{23d}$ ($WUE_{T2012}$), where $CWT_{23d}$ is the cumulative water transpired during 23 days (from 1 to 23 DAE).

## Determination of carbon isotope discrimination (CID)

Carbon isotope composition ($\delta$) was calculated relative to the international Pee Dee Belemnite (PDB) standard [26]: $\delta_{plant} = (R_{sa} - R_{sd})/R_{sd} \times 1000$ [‰] where $R_{sa}$ and $R_{sd}$ are the $^{13}C{:}^{12}C$ ratios of the sample and the standard, respectively [27]. Carbon isotope discrimination (CID), a factor related to isotope fractionation by the photosynthetic process relative to the source carbon was then estimated as $CID = (\delta_{air} - \delta_{plant})/(1 + \delta_{plant}/1000)$ where $\delta_{air}$ is the $^{13}C$ composition of atmospheric $CO_2$, which is assumed to be $-8.0‰$ [26]. Before calculating CID, oven-dried leaves of each plant were ground into a homogenous fine powder and 2–3 mg subsamples were weighed and placed into tin capsules (Elemental Microanalysis, UK) to be analyzed using a continuous flow Isotope Ratio Mass Spectrometer (Sercon Ltd., Cheshire, UK) at UC Davis Stable Isotope Facility (California, USA).

## Genetic map construction

A set of 9832 SNPs were used to produce an Infinium HD iSelect BeadChip (Infinium). These SNPs were selected from either genomic re-sequencing or transcriptomic experiments. The gDNA from the INEDI RILs population obtained from the cross between XRQ and PSC8 lines (210 samples) were genotyped with the Infinium array. All genotyping experiments were performed by Integragen (IntegraGen SA, Genopole Campus 1 - Genavenir 8, 5 rue Henri Desbruères, 91000 Evry, France) and the genotypic data were obtained with the Genome Studio software (Illumina) with automatic and manual calling. From the 9832 SNPs, 7094 were technically functional with more than 200 samples having a genotyping data. From this set of 7094 markers, 2576 were polymorphic between XRQ and PSC8 and 2164 did not show distortion of segregation in the population. We used CarthaGène v1.3 [34] to build the genetic maps. We added the genotypic data of markers from a consensus map [35] to the set of the 2164 SNPs to assign them to the appropriate LG to the *group 0.3 8* in CarthaGène. They were ordered using the *lkh 1 -1* function in CarthaGène for each group. The genetic map consisted of 2610 markers located on the 17 LG for a total genetic distance of 1863.1 cM and grouped on 999 different loci. All data will be available through the www.heliagene.org portal.

## Statistical and QTL analysis

The data were first tested for normal distribution with the Kolmogorov-Smirnov test. These data were subjected to analysis of variance (ANOVA) and phenotypic correlation analysis (Pearson's correlation) using the software of statistical package PASW statistics 18 (IBM, New York, USA). Means were compared using a Student-Newman-Keuls (SNK) test ($P<0.05$). The broad sense heritability ($h^2$) was then computed from the estimates of genetic ($\sigma^2 g$) and residual ($\sigma^2 e$) variances derived from the expected mean squares of the analyses of variance as $h^2 = \sigma^2 g / (\sigma^2 g + \sigma^2 e / r)$, where $r$ was the number of replicates.

QTL identification was performed using MCQTL, software for QTL analysis (http://carlit.toulouse.inra.fr/MCQTL/). The MCQTL software package can be used to perform QTL mapping in a multi-cross design. It allows the analysis of the usual populations derived from inbred lines [28]. MCQTL package is comprised of three software applications. The first component, TranslateData reads data from MAPMAKER [29] like files. The second component, ProbaPop computes QTL genotype probabilities given marker information at each chromosome location for each family and stores them in XML formatted files. The last component, Multipop builds the pooled model using the genotype probabilities, computes Fisher tests and estimates the model parameters [28]. The statistical significance of QTLs was assessed using the MCQTL test, which is equal to –log(P-value (F-test)), as described in the MCQTL user guide.

Significant thresholds ($P<0.05$) for QTL detection were calculated for each dataset using 1000 permutations [30] and a genome-wide error rate of 0.01 (Type I error). The corresponding type I error rate at the whole-genome level was calculated as a function of the overall number of markers in the map and the number of markers in each linkage group [31]. In our analysis, the threshold for the Fisher test (–log(P-value (F-test))) was 3.69 for both experiments. This threshold was an average of several thresholds of the traits at a significance level of 5% and was determined after 1000 permutations.

In each experiment, the QTL detection was also performed to identify QTL for the phenotypic response (called "response QTL"), calculated as the difference between two different water treatments (WW and WS). This allowed us to detect chromosome regions having quantitative effects on traits, depending on the environment [32–33].

## Results

### Genotypic variability and phenotypic correlation between water use efficiency (WUE) and carbon isotope discrimination (CID)

In general, a normal distribution was observed for WUE and CID traits across the two experiments and water treatments, except for $WUE_{T2012}$ and CID in Exp. 2012 at WW conditions, the distributions deviate from normality according to the Kolmogorov-Smirnov test (Fig. 2 and 3). As normalizing data through transformation may misrepresent differences among individuals by pulling skewed tails toward the center of the distribution [30], all phenotypic analyses were performed on untransformed data.

Higher mean values for WUE for WS (2.31 to 3.06 g.kg$^{-1}$) than for WW (1.91 to 2.95 g.kg$^{-1}$) (Table S1 and S2) were observed in each experiment. In contrast, higher mean values for CID for WW than for WS were also observed in each experiment. In addition, a similar range of WUE and CID values was observed in both experiments for both WW and WS (for WUE in Exp. 2011 was represented by the $WUE_{E2011}$). In addition, significant genotypic effects were detected for all traits in Exp. 2011 (Table S1), and significant genotypic and SWC effects were detected for all traits in Exp. 2012 (Table 1).

The heritabilities of CID were usually higher than those of WUE in both experiments (CID with $WUE_{T2011}$ or $WUE_{T2012}$), except that the heritability of $WUE_{E2011}$ was higher than that of CID (Table S1 and 1).

Significant negative correlations were observed between WUE and CID in both experiments ($r_p = -0.197$, $P<0.05$; $r_p = -0.409$, $P<0.001$; $r_p = -0.565$, $P<0.001$ for the correlations of $WUE_{T2011}$, $WUE_{E2011}$, $WUE_{T2012}$ with the CID, respectively; Fig. 4, Table S3, S4 and S5). However, when we determined the correlation between WUE and CID for each treatment, we observed a positive correlation between the $WUE_{T2011}$ and CID in Exp. 2011 for WS (Fig. 4 and Table S4). In addition, a significant phenotypic

correlation was observed between Exp. 2011 and 2012 for both WUE and CID (Fig. 5).

### QTL identified for water use efficiency (WUE)

In Exp. 2011, two QTL for $WUE_{T2011}$ were detected for WW and four QTL for $WUE_{E2011}$ were detected for WS (Table 2). For WW, the QTL were located on LG06 and LG11 with the highest likelihood odds ratio (LOD) value at 3 cM (QTL of *WUE11ww.11.1*) (Fig. S1). The marker for the QTL of *WUE11ww.11.1* was identified between the markers of HA005673_395 and HA006174_145 (Fig. 6). For WS, the QTL were located on chromosomes LG03 and LG16 (two QTL for each chromosome) with the highest LOD value at 6 cM, the QTL of *WUEe11ws.16.2*, and the marker of this QTL was HA017124_226. A "response QTL" for WUE (*WUE11diff.06.2*) was collocated with QTL of *WUE11ww.06.1*. In addition, two other "response QTL" were found on LG05 and LG06. The additive effects of the *WUE11ww.06.1* and *WUE11ww.11.1* were $-0.14$ and $0.11$ while the additive effects of the *WUEe11ws.03.1*, *WUEe11ws.03.2*, *WUEe11ws.16.1*, and *WUEe11ws.16.2* were $-0.13$, $0.13$, $0.38$ and $-0.44$, respectively.

In Exp. 2012, two QTL for $WUE_{T2012}$ were detected at WW and one QTL for $WUE_{T2012}$ at WS (Table 3). For WW, the QTL were detected on chromosome LG13 and LG15 with the highest LOD value at 25 cM, the QTL of *WUE12ww.13.1*, and the markers for this QTL was restor (Fig. 6, Fig. S1). For WS, a QTL was detected on chromosome LG09 (QTL of *WUE12ws.09.1*) with the LOD value at 3 cM. The marker for the QTL of *WUE12ws.09.1* was identified between the markers of SSL053 and HA013641_506. In addition, a "response QTL" for WUE (*WUE12diff.13.1*) was co-located with the QTL of *WUE12ww.13.1* and *CID12ww.13.1* (Table 4). The additive effects of *WUE12ws.09.1*, *WUE12ww.13.1* and *WUE12ww.15.1* were $0.20$, $0.04$ and $-0.06$, respectively.

### QTL identified for carbon isotope discrimination (CID)

In Exp. 2011, two QTL for CID were detected at WW and three QTL for CID were detected at WS (Table 2). For WW, the QTL were located on the same chromosomes of LG06 with the highest LOD value at 4.5 cM, QTL of *CID11ww.06.1*, and the marker of this QTL was ORS483 (Fig. 6, Fig. S2). For WS, the QTL were identified on chromosomes LG03, LG06 and LG13 with the highest LOD value at 5.5 cM, the QTL of *CID11ws.03.1*, and the marker of this QTL was HA013974_334. Besides, there was one "response QTL" detected for CID on chromosome LG02 (*CID11diff.02.1*) (Table 4). The additive effects were $-0.15$ and $0.12$ (for QTL of *CID11ww.06.1* and *CID11ww.06.2*) while the additive effects were $-0.13$, $-0.10$, $-0.13$ (for the QTL of *CID11ws.03.1*, *CID11ws.06.1* and *CID11ws.13.1*) (Table 2).

In Exp. 2012, two QTL for CID were detected at WW and one QTL for CID at WS (Table 3). For WW, the QTL were found on chromosomes LG13 and LG15 with the highest LOD value of 8.5 cM, the QTL of *CID12ww.13.1*, and the marker for this QTL was restor (Fig. 6, Fig. S2). For WS, a QTL was found on chromosome LG13 with an LOD value of 2.5 cM; the QTL of *CID12ws.13.1*, and the marker for this QTL was HACG0018_Contig_1_130. The additive effects for *CID12ww.13.1* and *CID12ww.15.1* were $0.20$ and $0.07$, respectively. The additive effect of the QTL of CID at WS (*CID12ws.13.1*) was $0.14$.

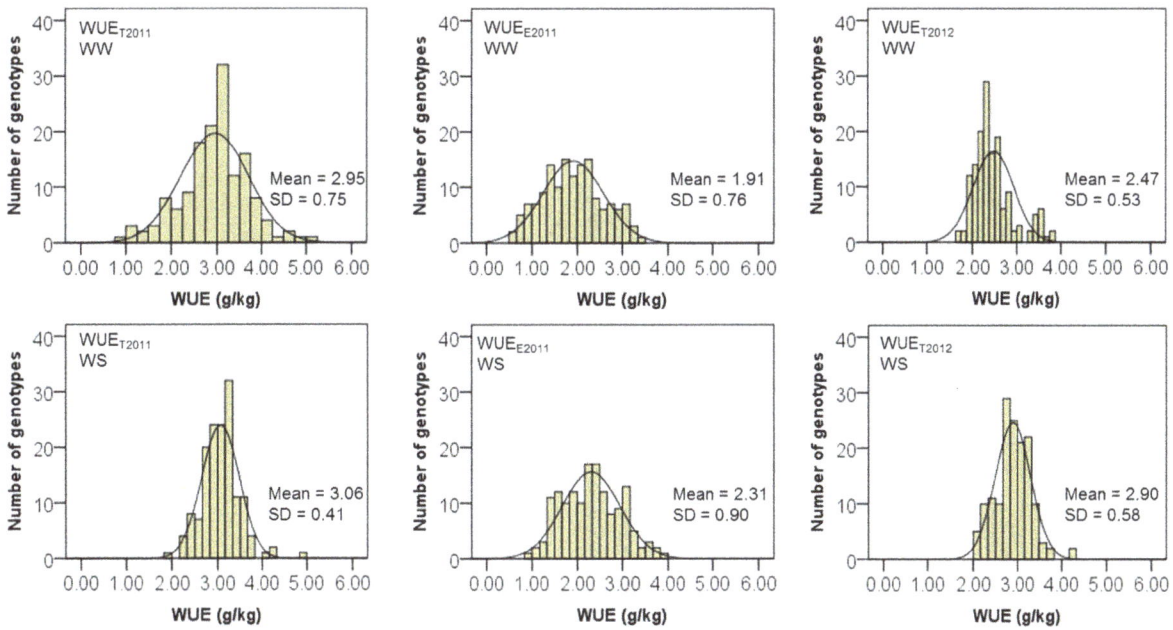

**Figure 2. Frequency distribution for water use efficiency (WUE) in Exp. 2011 and 2012 of 150 recombinant inbred lines (RILs).** $WUE_{T2011}$: total water use efficiency "total" in Exp. 2011; $WUE_{E2011}$: water use efficiency "estimation" in Exp. 2011; $WUE_{T2012}$: water use efficiency "total" in Exp. 2012. WW: well-watered; WS: water-stressed. For $WUE_{T2011}$ and $WUE_{E2011}$ at WW, data represent 150 RILs (n = 150); for $WUE_{T2011}$ and $WUE_{E2011}$ at WS, data represent mean of three replicates of 150 RILs (n = 150); for $WUE_{T2012}$ at WW and WS, data represent mean of two replicates of 150 RILs (n = 150). SD: standard deviation.

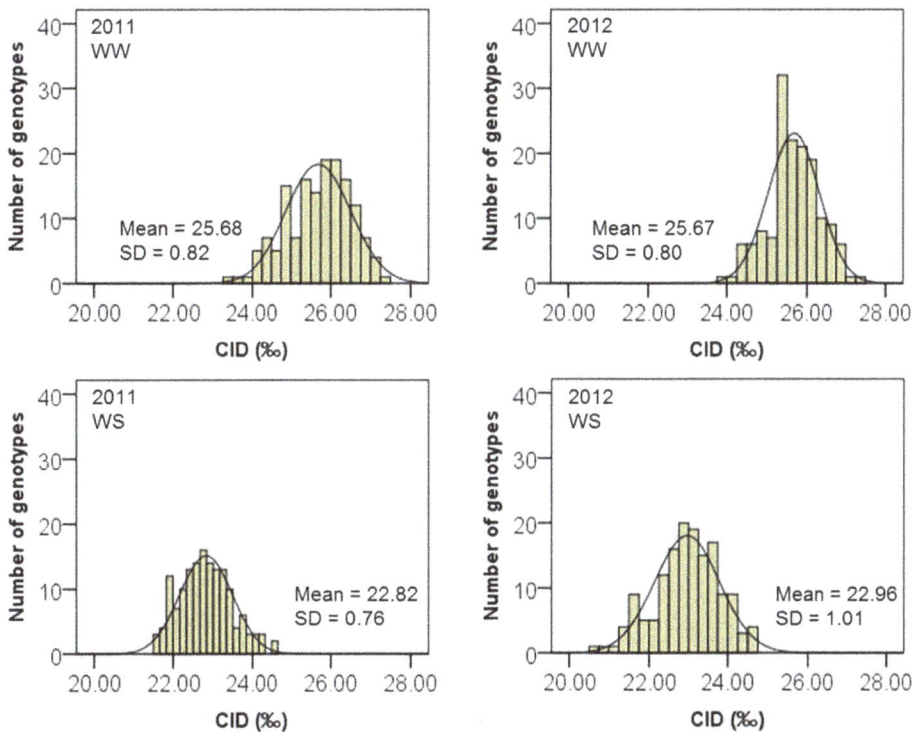

**Figure 3. Frequency distribution for carbon isotope discrimination (CID) in Exp. 2011 and 2012 of 150 recombinant inbred lines (RILs).** WW: well-watered; WS: water-stressed. For CID in Exp. 2011 at WW, data represent 150 RILs (n = 150); for CID in Exp. 2012 at WS, data represent mean of three replicates of 150 RILs (n = 150); for CID in 2012 at WW and WS, data represent mean of two replicates of 150 RILs (n = 150). SD: standard deviation.

**Table 1.** Heritability ($h^2$) and mean square (MS) of analysis of variance (ANOVA) for water use efficiency (WUE), carbon isotope discrimination (CID), biomass (BM) and cumulative water transpired (CWT) for 150 recombinant inbred lines (RILs), two stable soil water contents (SWC) and two replicates in Exp. 2012 (n = 600).

| Trait | $h^2$ | MS | | |
|---|---|---|---|---|
| | | Genotype | Soil water content | Genotype×soil water content |
| $WUE_{T2012}$ | 0.26 | 0.50*** | 28*** | 0.25[ns] |
| CID | 0.41 | 1.68*** | 1100*** | 0.53[ns] |
| BM | 0.36 | 0.51*** | 180*** | 0.29** |
| $CWT_{23d}$ | 0.36 | 40862*** | 31746440*** | 25565*** |

**Significant at $P<0.01$,
***significant at $P<0.001$.
[ns]Not significant.

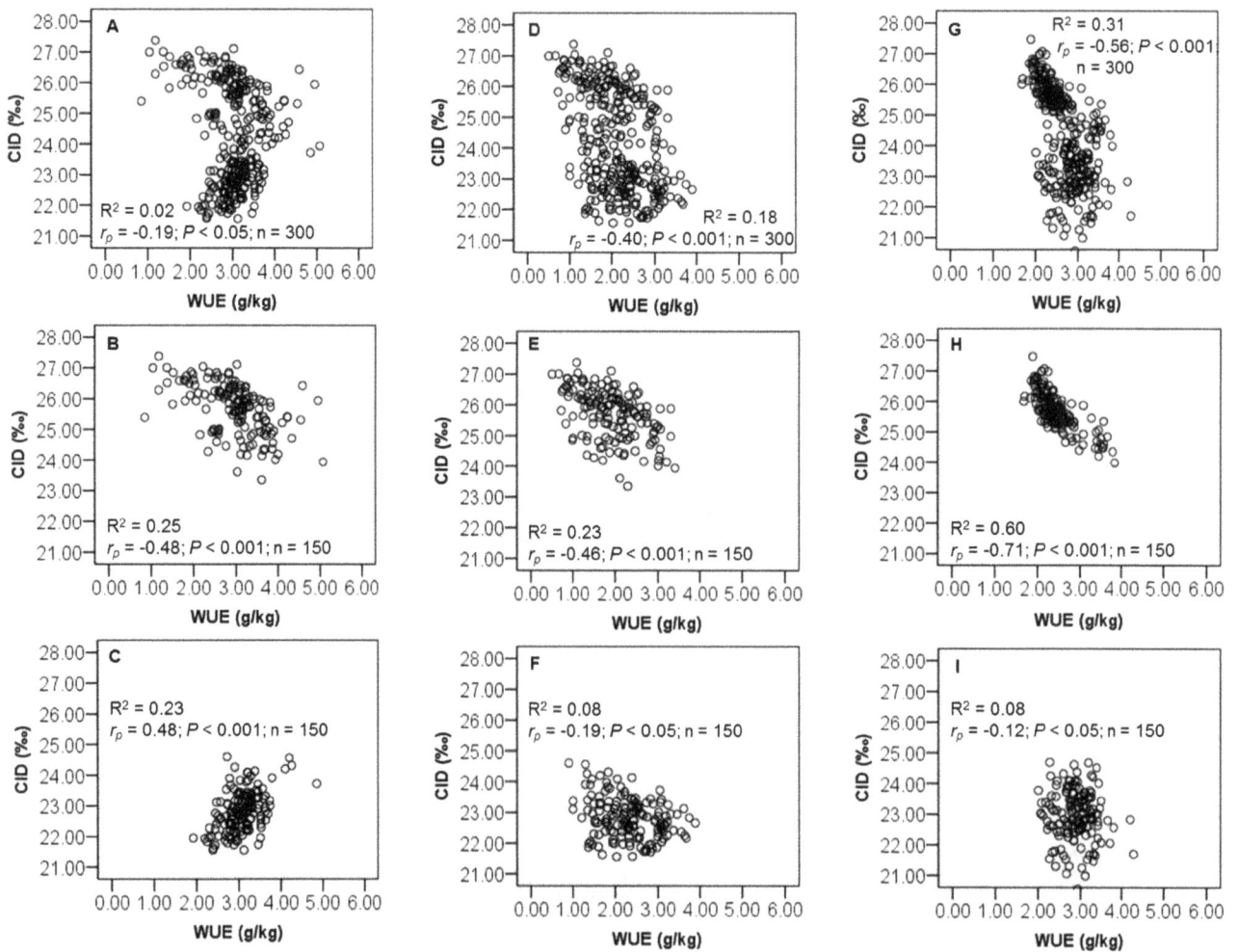

**Figure 4. Relationship between water use efficiency (WUE) and carbon isotope discrimination (CID) of 150 recombinant inbred lines (RILs) in Exp. 2011 and Exp. 2012.** Relationship between (A) $WUE_{T2011}$ and CID in Exp. 2011, (B) $WUE_{T2011}$ and CID at WW in Exp. 2011, (C) $WUE_{T2011}$ and CID at WS in Exp. 2011, (D) $WUE_{E2011}$ and CID in Exp. 2011, (E) $WUE_{E2011}$ and CID at WW in Exp. 2011, (F) $WUE_{E2011}$ and CID at WS in Exp. 2011, (G) $WUE_{T2012}$ and CID in Exp. 2012, (H) $WUE_{T2012}$ and CID at WW in Exp. 2012; (I) $WUE_{T2012}$ and CID at WS in Exp. 2012. Phenotypic correlation ($r_p$) value is provided in each graph.

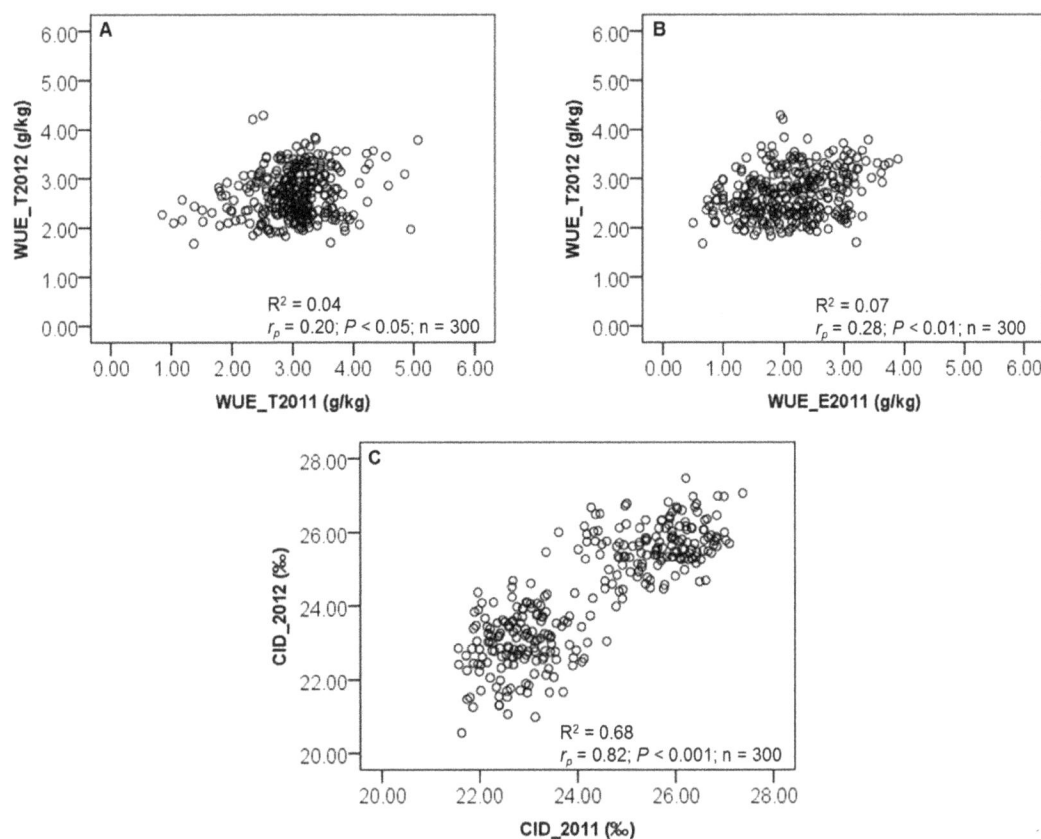

**Figure 5. Relationship between (A, B) WUE and (B) CID values for 150 recombinant inbred lines (RILs) determined in two separate experiments (Exp. 2011 and 2012).** For each trait and experiment, mean of well-watered (WW) and water-stressed (WS) plants were grouped together (n = 300). Phenotypic correlation ($r_p$) value is provided in each graph.

## QTL identified for related traits: biomass (BM) and cumulative water transpired (CWT)

In Exp. 2011, three significant QTL for BM, and one QTL for each of $BM_E$ and $CWT_{31d}$ at WS were identified (Table 2). These QTL were detected on chromosomes LG14, LG15, LG17, LG01 and LG11. There were only two "response QTL" detected for each of $BM_E$ and $CWT_{15d}$. These QTL were detected on the same chromosome, LG06.

In Exp. 2012, seven QTL were identified for BM under both levels of SWC. For $CWT_{23d}$, five significant QTL were detected under both levels of SWC. Further, six "response QTL" for BM and $CWT_{23d}$ were identified on chromosomes LG06, LG09, LG13 and LG15.

## Discussion

### Genetic variation and relationship between WUE and CID

In our experiments, increasing drought lead to an increase in WUE and a decrease in CID. This result was previously reported by Lauteri et al. [36] in sunflower and is well known in other crops, such as durum wheat (*Triticum turgidum* L.) [37], rice (*Oryza sativa* L.) [38] and eucalyptus (*Eucalyptus microtheca*) [39]. In addition, a similar range of values for WUE and CID was observed in the two experiments even though their water stress patterns differed. That was likely because the population had been constructed from parents that had specific responses in non-limited and limited water availability [40–41]. From the phenotypic data,

XRQ exhibited low WUE while PSC8 exhibited high WUE (unpublished data).

CID is highly heritable trait and its heritability is usually higher rather than WUE [7,11]. Nevertheless, in the present study, both of CID and WUE were influenced by environmental variation because the heritability values were below 50% [24]. A previous study [42] has shown that heritabilities for CID, measured on detached sunflower leaves, were above 50% (74–96%), indicating that genetic variance for CID was dominant. However, this result was obtained for plants grown in optimal watering conditions. Consequently, CID appeared dependent on genetic and environmental control. This trait is genetically complex [43], and its expression in leaves and other plant tissues varies with the water supply. In drought conditions, Rebetzke et al. [24] reported that low soil water availability decreases stomatal conductance, which can reduce genetic variance and heritability of CID.

Our work demonstrated the clear relationship between WUE and CID in different water regimes. For each water regime and all genotypes, we observed negative correlations between WUE and CID. These results are in accordance with those of previous work in sunflower [36,42], and with those of numerous authors working on other crops [6,8,44–47]. In one case of progressive water stress, $WUE_{T2011}$ and CID, were positively correlated. This was probably due to the high variability of the soil water content during the progressive drought establishment (SWC was gradually decreased). A similar result was reported on alfalfa genotypes [48]: WUE (mg of dry matter per g $H_2O$) was positively correlated with CID for plants subjected to progressive water stress during 7 days.

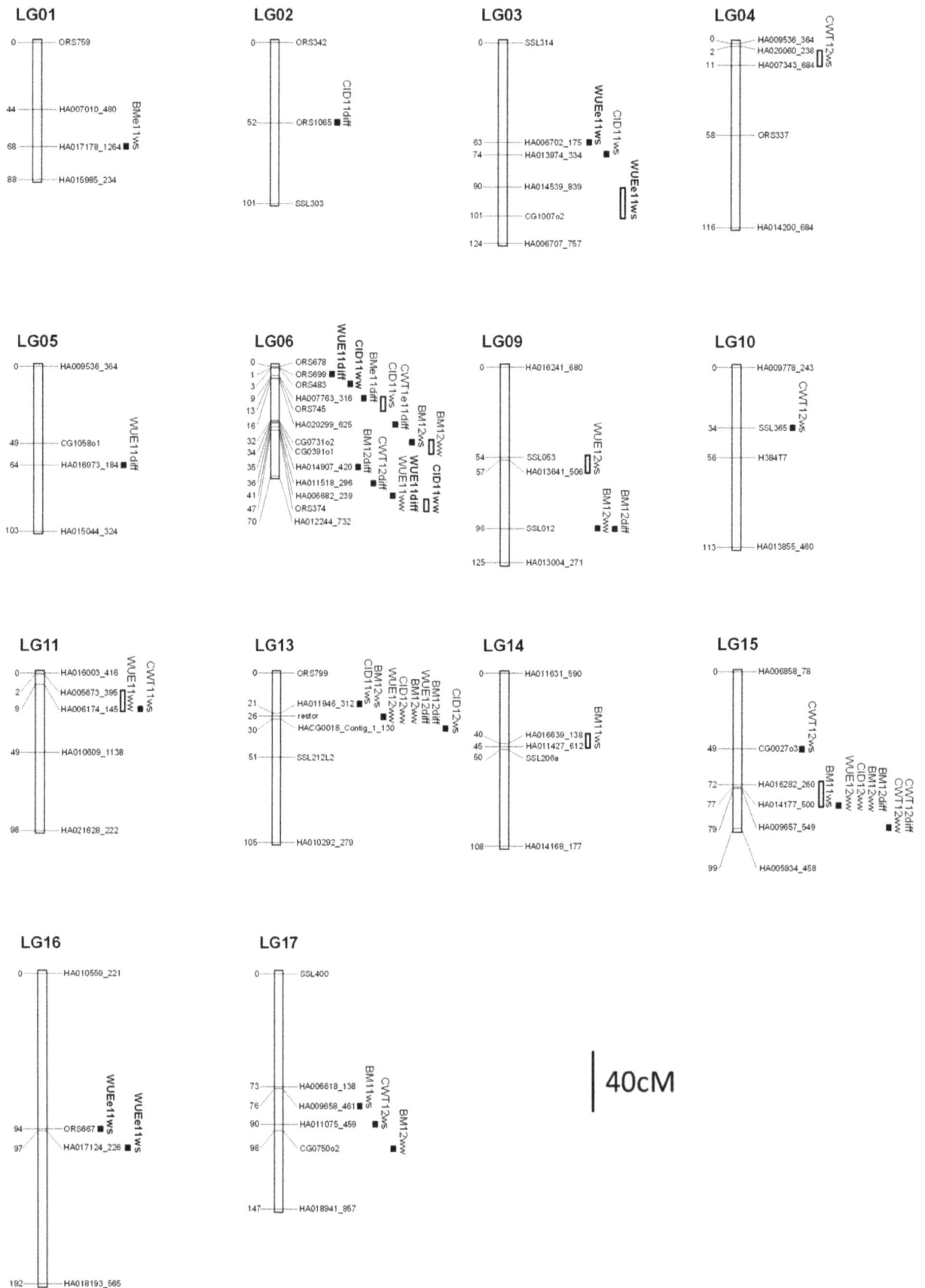

**Figure 6. Genetic locations of QTL for water use efficiency (WUE), carbon isotope discrimination (CID), biomass (BM) and cumulative water transpired (CWT) in the progressive stress experiment (2011) and the stable stress experiment (2012).** Numbers on

the left of linkage groups (LG) indicate the cumulative distance in centimorgan (cM) to the first marker at the top LG. Marker names and QTL are specified to the right of LG. The same QTLs which are found in a LG are shown in bold. Not all these chromosomes contain the complete markers (each chromosome has only been provided by the markers at the top, middle and bottom of LG as well as the markers for identified QTLs). QTL confidence intervals were estimated using the two-LOD confidence region.

In the WW treatment, the high WUE was correlated with high BM and high CWT, while for the WS treatment the high WUE was still correlated with high BM but with low CWT. If increase in WUE is associated with reduced transpiration, such genotypes are often referred to as "conductance type". On the other hand, if increase in WUE is correlated with increased photosynthesis, such genotypes can be categorized as "capacity types" [49–50]. Accordingly, the sunflower genotypes in our study can be categorized as an intermediate between "conductance" and "capacity" type, unlike rice genotypes that have been categorized as "conductance type" [51]. In addition, our results were in agreement with several authors [39,52–53] who have suggested that plants that use water more efficiently by producing greater biomass for a given quantity of water transpired would grow more rapidly, resulting in a positive correlation between WUE and biomass production.

## QTL identified for WUE and CID

Our study is the first to identify QTL for WUE and CID in sunflower subjected to drought. In Exp. 2011, significant regions affecting WUE were identified on four different chromosomes (LG03, LG06, LG11, LG16) in two water treatments and significant regions affecting CID were identified on three different chromosomes (LG03, LG06, LG13) for the same two water treatments. From these QTL, we observed a decrease and an increase of additive effects (XRQ), indicating that genes having both negative and positive effects had been involved in the difference in WUE and CID between the parental lines [54]. In Exp. 2012, the QTL for WUE were detected on three different chromosomes in two water treatments (LG09, LG13 and LG15) and the QTL for CID were identified on two different chromosomes in these two water treatments (LG13 and LG15). All these QTL increased the values of additive effects except the QTL of *WUE12ws.09.1*, indicating that XRQ allele increased the traits. These findings provide an explanation for the underlying genetic basis of the transgressive variation observed in the segregating population. This is in accordance with the argument proposed by Chapman et al. [55] and Vargas et al. [56], namely that a given QTL can have positive or negative additive effects, or none at all, depending on the drought scenario.

The WUE and CID were controlled by several QTL with small genetic additive (XRQ) effects, indicating that WUE and CID were genetically complex traits [2,57]. Reports evaluating genetic analysis for CID in other crops like soybean [58], cotton [59] and rice [54] have identified multiple QTL of smaller effect associated with the trait. However, in the present study, the QTL for WUE and CID explained 42% and 21% of the highest phenotypic variance ($R^2$). These $R^2$ values are higher than those found by previous authors for other crops, for example, rice [10,54], wheat [24] and barley [60–61].

## Expression of QTL for WUE and CID across experiments and water treatments

The locations of QTL might be affected by growth stage [54] and/or environmental change [62–63]. In our results, the QTL for $WUE_{T2011}$ and $WUE_{E2011}$ were found on chromosomes LG03, LG06, LG11 and LG16 (under WW and WS), whereas the QTL for $WUE_{T2012}$ were found on chromosomes LG09, LG13 and LG15 (under WW and WS). These results showed that the expression of QTL for WUE differs with micro-environmental variations. This variation can be explained by the different water regimes in Exp. 2011 and Exp. 2012.

When the same mapping population is phenotyped in different environments, some QTL could be detected in one environment but not in others [63]. Collins et al. [64] noted that QTL can be categorized according to the stability of their effects across environmental conditions. A "constitutive" QTL is consistently detected across most environments, while an "adaptive" QTL is detected only in specific environmental conditions or increases in expression with the level of an environmental factor.

The QTL for CID in Exp. 2011 were detected on chromosomes LG03, LG06 and LG13 (WW and WS), whereas the QTL for CID in Exp. 2012 were detected on chromosomes LG13 and LG15 (under WW and WS). These results indicate that the expression of QTL for CID differs in the two experiments and different water regimes. Despite CID variation is influenced by stomatal conductance and photosynthetic capacity variations [7,37], several QTL of the different water regimes have been detected on the same chromosome [65–67]. This was the case in our study, where the three QTL for CID of the three different water regimes were detected on the same chromosome (LG13). Therefore, the QTL for CID in this study can be considered as a "constitutive" QTL. Additionally, the constitutive QTL for CID was consistent with the result of phenotypic correlation that genotypic ranking for this trait was consistently maintained in the two experiments.

Some QTL for WUE and CID and related traits were located on the same chromosome or on a similar QTL position (co-localization). The QTL for $WUE_{T2012}$ for WW (*WUE12ww.13.1* and *WUE12ww.15.1*) had a similar QTL position (26.20 and 77.10 cM) as the QTL for CID for WW (*CID12ww.13.1*, *CID12ww.15.1*). The QTL for CID (*CID12ws.13.1*) for WS was associated with the QTL for $WUE_{T2012}$ for WW (*WUE12ww.13.1*). This QTL was detected on chromosome LG13 (QTL position: 30.80 cM) near the QTL of *CID12ww.13.1*. The occurrence of QTL associated with different traits at the same locus may be explained by the fact that (i) the QTL are closely linked genetically or (ii) a single locus controls multiple traits and a gene may have pleiotropic effects [54].

We have observed a common genetic basis for WUE and CID in each experiment. Using the same mapping population under different water stress treatments helped us to characterize consistent genomic region (by QTL). Kiani et al. [68] indicated that QTL which was induced only by drought might be associated with mechanism(s) of sunflower drought response and they proposed that the QTL which can reduce trait difference between well-watered and water-stressed conditions should have an effect on drought tolerance because of their contribution to trait stability. Our study in Exp. 2011 showed that the QTL for CID on chromosome LG06 were repeatable across two different water treatments (WW and WS). In Exp. 2012, the QTL for CID on chromosome LG13 have been repeatable across two different water treatments (WW and WS).

All these QTL which are common across different water treatments might be useful for marker-assisted selection (MAS). Identification of QTL influencing several traits could increase the

**Table 2.** Significant quantitative trait loci (QTL) detected for water use efficiency (WUE), carbon isotope discrimination (CID), biomass (BM) and cumulative water transpired (CWT) under under well-watered and progressive water-stressed treatments in Exp. 2011.

| Trait | Treatment | Chromosome | QTL name | QTL position (cM) | Inferior position (cM) | Superior position (cM) | R²[a] (%) | R² global[b] (%) | Additive effect[c] |
|---|---|---|---|---|---|---|---|---|---|
| WUE$_{T2011}$ | WW | LG06 | WUE11ww.06.1 | 41.1 | 0 | 69.5 | 7 | 13 | −0.14 |
| | WW | LG11 | WUE11ww.11.1 | 7.8 | 0 | 17.4 | 9 | 13 | 0.11 |
| CID | WW | LG06 | CID11ww.06.1 | 3 | 0.45 | 14.4 | 12 | 17 | −0.15 |
| | WW | LG06 | CID11ww.06.2 | 47.3 | 23.6 | 60.1 | 8 | 17 | 0.12 |
| WUE$_{E2011}$ | WS | LG03 | WUEe11ws.03.1 | 63.2 | 53.8 | 95.7 | 7 | 21 | −0.13 |
| | WS | LG03 | WUEe11ws.03.2 | 97.7 | 63.9 | 124 | 5 | 21 | 0.13 |
| | WS | LG16 | WUEe11ws.16.1 | 94.1 | 92.7 | 96.1 | 11 | 21 | 0.38 |
| | WS | LG16 | WUEe11ws.16.2 | 97.1 | 96.1 | 99.3 | 15 | 21 | −0.44 |
| CID | WS | LG03 | CID11ws.03.1 | 73.6 | 52 | 76.2 | 15 | 25 | −0.13 |
| | WS | LG06 | CID11ws.06.1 | 11.3 | 0 | 16.6 | 9 | 25 | −0.1 |
| | WS | LG13 | CID11ws.13.1 | 21.2 | 0 | 36.5 | 10 | 25 | −0.13 |
| BM | WS | LG14 | BM11ws.14.1 | 42.4 | 0 | 108 | 5 | 17 | 0.01 |
| | WS | LG15 | BM11ws.15.1 | 76.3 | 0 | 98.9 | 7 | 17 | 0.02 |
| | WS | LG17 | BM11ws.17.1 | 76.1 | 0 | 112 | 6 | 17 | −0.02 |
| BM$_E$ | WS | LG01 | BMe11ws.01.1 | 67.8 | 46.3 | 74.9 | 9 | 9 | 0.02 |
| CWT$_{31d}$ | WS | LG11 | CWTe11ws.11.1 | 9.1 | 0 | 20.5 | 7 | 7 | −4.28 |

WW: well-watered, WS: progressive water-stressed.
[a] Phenotypic variance explained by QTL effect.
[b] Total of phenotypic variances explained by QTL effects.
[c] Additive effect estimated as one-half the difference in homozygotes carrying either allele of parents (XRQ or PSC8). Positive values indicate that XRQ allele increases the trait value, while negative values indicate that PSC8 allele increases the trait value.

**Table 3.** Significant quantitative trait loci (QTL) detected for water use efficiency (WUE), carbon isotope discrimination (CID), biomass (BM) and cumulative water transpired (CWT) under well-watered and water-stressed treatments in Exp. 2012.

| Trait | Treatment | Chromosome | QTL name | QTL position (cM) | Inferior position (cM) | Superior position (cM) | $R^{2a}$ (%) | $R^2$ global[b] (%) | Additive effect[c] |
|---|---|---|---|---|---|---|---|---|---|
| WUE$_{T2012}$ | WW | LG13 | WUE12ww.13.1 | 26.2 | 25.14 | 26.95 | 42 | 45 | 0.2 |
| | WW | LG15 | WUE12ww.15.1 | 77.1 | 13.94 | 90.73 | 6 | 45 | 0.04 |
| CID | WW | LG13 | CID12ww.13.1 | 26.2 | 4.29 | 37.43 | 21 | 26 | 0.2 |
| | WW | LG15 | CID12ww.15.1 | 77.1 | 0 | 98.90 | 6 | 26 | 0.07 |
| BM | WW | LG06 | BM12ww.06.1 | 33.6 | 29.2 | 40.74 | 13 | 40 | 0.1 |
| | WW | LG09 | BM12ww.09.1 | 95.5 | 88.25 | 114.1 | 9 | 40 | 0.08 |
| | WW | LG13 | BM12ww.13.1 | 26.2 | 24.16 | 37.81 | 2 | 40 | 0.17 |
| | WW | LG15 | BM12ww.15.1 | 77.1 | 47.47 | 82.62 | 12 | 40 | 0.1 |
| CWT$_{23d}$ | WW | LG15 | CWT12ww.15.1 | 79.1 | 40.28 | 87.03 | 7 | 7 | 26.07 |
| WUE$_{T2012}$ | WS | LG09 | WUE12ws.09.1 | 55.5 | 33.28 | 83.25 | 9 | 9 | −0.06 |
| CID | WS | LG13 | CID12ws.13.1 | 30.8 | 0 | 62.45 | 7 | 7 | 0.14 |
| BM | WS | LG06 | BM12ws.06.1 | 31.6 | 29.09 | 35.10 | 13 | 26 | 0.02 |
| | WS | LG13 | BM12ws.13.1 | 21.2 | 0 | 29.86 | 9 | 26 | 0.02 |
| | WS | LG17 | BM12ws.17.1 | 98.2 | 68.14 | 111 | 7 | 26 | 0.02 |
| CWT$_{23d}$ | WS | LG04 | CWT12ws.04.1 | 6 | 0 | 21.48 | 10 | 30 | −6.17 |
| | WS | LG10 | CWT12ws.10.1 | 33.6 | 0 | 112.5 | 4 | 30 | 3.23 |
| | WS | LG15 | CWT12ws.15.1 | 49.5 | 28.33 | 80.98 | 10 | 30 | 5.32 |
| | WS | LG17 | CWT12ws.17.1 | 89.8 | 74.7 | 92.29 | 12 | 30 | −6.78 |

WW: well-watered (30% of SWC), WS: water-stressed (16% of SWC).
[a]Phenotypic variance explained by QTL effect.
[b]Total of phenotypic variances explained by QTL effects.
[c]Additive effect estimated as one-half the difference in homozygotes carrying either allele of parents (XRQ or PSC8). Positive values indicate that XRQ allele increases the trait value, while negative values indicate that PSC8 allele increases the trait value.

**Table 4.** Significant "response quantitative trait loci (QTL)" detected for water use efficiency (WUE), carbon isotope discrimination (CID), biomass (BM) and cumulative water transpired (CWT) in Exp. 2011 and Exp. 2012.

| Trait | Experiment | Chromosome | QTL name | QTL position (cM) | Inferior position (cM) | Superior position (cM) | R[2a] (%) | R[2] global[b] (%) | Additive effect[c] |
|---|---|---|---|---|---|---|---|---|---|
| WUE$_{T2011}$ | 2011 | LG05 | WUE11diff.05.1 | 64 | 0 | 103 | 5 | 14 | −0.7 |
| | 2011 | LG06 | WUE11diff.06.1 | 1.3 | 0 | 22.9 | 7 | 14 | −0.1 |
| | 2011 | LG06 | WUE11diff.06.2 | 41.1 | 3 | 69.5 | 6 | 14 | 0.08 |
| CID | 2011 | LG02 | CID11diff.02.1 | 52.1 | 0 | 101 | 7 | 7 | −0.1 |
| BM$_E$ | 2011 | LG06 | BMe11diff.06.1 | 9.3 | 0 | 21.6 | 8 | 8 | −0.1 |
| CWT$_{15d}$ | 2011 | LG06 | CWTe11diff.06.1 | 15.9 | 0 | 22.4 | 9 | 9 | −19 |
| WUE$_{T2012}$ | 2012 | LG13 | WUE12diff.13.1 | 26.2 | 21.9 | 41.6 | 18 | 18 | 0.14 |
| BM | 2012 | LG06 | BM12diff.06.1 | 34.7 | 28.3 | 43 | 10 | 39 | 0.07 |
| | 2012 | LG09 | BM12diff.09.1 | 95.5 | 0.1 | 111 | 10 | 39 | 0.07 |
| | 2012 | LG13 | BM12diff.13.1 | 26.2 | 24.1 | 38.8 | 20 | 39 | 0.14 |
| | 2012 | LG15 | BM12diff.15.1 | 77.1 | 74.1 | 82.2 | 13 | 39 | 0.09 |
| CWT$_{23d}$ | 2012 | LG06 | CWT12diff.06.1 | 35.5 | 24.9 | 43.7 | 10 | 16 | 25.5 |
| | 2012 | LG15 | CWT12diff.15.1 | 79.1 | 0 | 98.9 | 6 | 16 | 19.1 |

[a]Phenotypic variance explained by QTL effect.
[b]Total of phenotypic variances explained by QTL effects.
[c]Additive effect estimated as one-half the difference in homozygotes carrying either allele of parents (XRQ or PSC8). Positive values indicate that XRQ allele increases the trait value, while negative values indicate that PSC8 allele increases the trait value.

efficiency of marker-assisted selection (MAS) and hasten genetic progress [69]. Ribaut et al. [70] noticed that in the design of the best-possible breeding strategy using MAS, additional traits and criteria have to be considered. For each trait of interest, some of the criteria are the number of QTL detected, the percentage of phenotypic variance that they explain, the total percentage of the genome that they represent, and their stability across different environments. Regarding these arguments, our study has shown that CID is the most interesting trait and should be useful for MAS, where three QTL overlapped on chromosome LG06 (CID for WW and WS in Exp. 2011), and three QTL across three different water treatments were co-localized on chromosome LG13 with phenotypic variance ($R^2$) ranges from 7 to 21%. Further, these QTL and other co-localized QTL on chromosomes LG06 and LG13 were identified in the near-centromeric region (inferior to superior position explained from 0 to 60.06 cM, and from 0 to 62.45 cM for LG06 and LG13, respectively), because those chromosomes are classified as a metacentric type [71–72].

## Co-localization of QTL for WUE and CID with related traits

In this study, we also detected QTL for the related traits BM and CWT on the same chromosome of the QTL for WUE and/or CID (for WW and WS). These were observed in Exp. 2012, where two of four QTL for BM for WW (BM12ww.13.1 and BM12ww.15.1) were detected on chromosomes LG13 and LG15, and co-located with the QTL for WUE_T2012 and CID for WW (WUE12ww.13.1, CID12ww.15.1). For WS, the identifications of the QTL for the related traits showed a similar trend. The QTL of BM12ws.13.1 (QTL position: 21.20 cM) was detected on chromosome LG13, as the QTL of CID11ws.13.1, CID12ww.13.1, CID12ws.13.1 and WUE12ww.13.1 have been identified. These indicated the possibility of genetic association of WUE and CID with the accumulation of biomass. Consistent with this, Kiani et al. [68] identified a QTL for total dry matter in water-stressed conditions on chromosome LG13 using another population of sunflower. Interestingly, this QTL overlapped with osmotic adjustment, grain yield, and plant height. Thereby the common genetic basis for WUE, CID, productivity and osmotic adjustment will lead to an improved understanding of drought tolerance genes. In addition, evidence of overlapping QTL of productivity and osmotic adjustment have been observed by several authors [73–75]. However, further study is obviously required to determine the genetic control of osmotic adjustment or hydraulic conductance and their inter-relationships with WUE and CID.

For CWT, the QTL of CWT12ww.15.1 was detected on chromosome LG15 with the QTL position at 79.10 cM near the marker at position of 77.10 cM where the QTL of WUE12ww.15.1 and CID12ww.15.1 have been identified. Not far from these positions, a QTL of CWT12ws.15.1 was also detected (QTL position: 49.5 cM). These indicated out that the cumulative water transpired in WW and WS is genetically and closely related with WUE and CID in non-limited water availability. In addition, the maintenance of biomass accumulation under stable water stress should be considered as an efficiency process between transpiration, biomass accumulation and its partitioning between non-drought and drought conditions [64]. Therefore, the increase in WUE (i.e. the amount of biomass produced per unit of transpired water) might seem to be ideal candidate mechanism for drought-prone environments.

## Identifying the "response QTL" for WUE and CID

In our work, we calculated the "response QTL" to provide new insight into the genetic architecture of WUE and CID, which, unlike a "common" phenotypic trait, is rarely considered in QTL analysis. Water use traits and their response are of primary importance to plant growth and survival. Although we have a growing understanding of the genetic and molecular drivers of water use traits and WUE as well as CID, response QTL of those traits has received relatively little attention.

We detected three QTL of "response QTL" for WUE on chromosomes LG06 and LG13. From these two chromosomes we have also identified the QTL for WUE_T2011 and WUE_T2012 for WW, indicating, at least under the conditions imposed in these experiments, that response QTL was controlled by loci that determine the main trait value under a specific treatment. This was in agreement with Kliebenstein et al. [76–77] who evaluated the response QTL between control and methyl jasmonate (MeJa)-treated plants of *Arabidopsis thaliana*. They reported that significant QTL that influenced response between control and MeJa-treated plants also affected the main trait value in at least one of the two environments, which was called the "allelic sensitivity" model.

In contrast, an independent response QTL, was also observed for several traits, for example the response QTL for WUE_T2011 on chromosome LG15 (WUE11diff.05.1), CID on chromosome LG02 (CIDdiff11.02.1), and CWT_23d on chromosome LG06 (CWT12diff.06.1). This observation was not consistent with Kliebenstein et al. [77], however, it was in agreement with an argument of Schlichting and Pigliucci [78] who suggested the "gene regulation" model must exist, and is not always controlled by loci that are expressed within at least one of the two environments.

As for the prospects for these aspects, characterization of the genes underlying QTL that control the differential WUE and CID regulation might generate a detailed understanding of the molecular and biochemical basis for water use traits in sunflower and how this alters phenotypic response in more complex environments.

## Importance of high WUE or low CID for sunflower breeding: use of the identified markers for MAS

This is the first genetic quantitative analysis and QTL mapping for WUE and CID in sunflower. We investigated two drought scenarios and evaluated genetic variation of sunflower lines to identify genetic control and physiological processes that could explain genotypic differences in the response to drought stress. The present study proved that, in sunflower, selection for CID can be considered in initial screening to improve WUE. However, this merits further investigation in other populations.

Many QTL (particularly for CID) have been reported in the literature. However, very few with large effects have been adequately exploited in crop breeding programs. The majority of the favorable alleles for identified QTL are to be found in journals on library shelves rather than in crop cultivars improved by introgression or selection of these favorable QTL alleles [79]. Nevertheless, Condon et al. [80] reported the release of a new high-yielding wheat variety in droughted environments after a breeding process in which selection for low CID in non-droughted plants led to high WUE.

In conclusion, our results emphasize that the near-centromeric region of chromosomes LG06 and LG13 are a "reliable" region for MAS due to the co-localization of the QTL for CID with several QTL for WUE, BM and CWT. Indeed, the best strategy for using molecular markers should combine selection for QTL involved in the expression of CID.

This paper complements the study of Vincourt et al. [25] and Rengel et al. [81] that exploited the INEDI RIL population in analyzing genetic variation of agronomic and physiological traits, making it possible to establish strategies for a sunflower breeding program and provide a basis for identification of the molecular components of a genotype x environment interaction.

## Supporting Information

**Figure S1   Genetic maps and LOD positions showing the locations of QTLs controlling WUE identified by MCQTL.**

**Figure S2   Genetic maps and LOD positions showing the locations of QTLs controlling CID identified by MCQTL.**

**Table S1   Genotypic variation of water use efficiency (WUE), carbon isotope discrimination (CID), biomass (BM) and cumulative water transpired (CWT) for 150 recombinant inbred lines (RILs) under well-watered (WW) and progressively water-stressed (WS) treatments in Exp. 2011.**

**Table S2   Genotypic variation of water use efficiency (WUE), carbon isotope discrimination (CID), biomass (BM) and cumulative water transpired (CWT) for 150 recombinant inbred lines (RILs) for well-watered (WW) and water-stressed (WS) in Exp. 2012.**

**Table S3   Phenotypic correlations ($r_p$) between water use efficiency (WUE), carbon isotope discrimination (CID), biomass (BM) and cumulative water transpired (CWT) of 150 recombinant inbred lines (RILs) in Exp. 2011 and Exp. 2012.**

**Table S4   Phenotypic correlations ($r_p$) among water use efficiency (WUE), carbon isotope discrimination (CID), biomass (BM) and cumulative water transpired (CWT) of 150 recombinant inbred lines (RILs) under well-watered (WW) and progressive water-stressed (WS) treatments in Exp. 2011.**

**Table S5   Phenotypic correlations ($r_p$) among water use efficiency (WUE), carbon isotope discrimination (CID), biomass (BM) and cumulative water transpired (CWT) of 150 recombinant inbred lines (RILs) under well-watered (WW) and water-stressed (WS) treatments in Exp. 2012.**

## Acknowledgments

We wish to thank team of *Laboratoire des Interactions Plantes-Microorganismes* (LIPM), INRA of Toulouse for providing sunflower materials, genetic maps, and their expert help in QTL analysis. In addition, the authors sincerely thank M. Labarrere for his contribution during the experiments.

Thierry Lamaze and Philippe Grieu are supervisors of Afifuddin Latif Adiredjo.

## Author Contributions

Conceived and designed the experiments: ALA PG TL. Performed the experiments: ALA PG ON. Analyzed the data: ALA PG NL. Contributed reagents/materials/analysis tools: ALA PG NL SM. Wrote the paper: ALA PG. Coordinated research: PG TL.

## References

1. Ehleringer JR, Hall AE, Farquhar GD (1993) Stable isotopes and plant carbon – water relations. San Diego: Academic Press.
2. Richards RA, Rebetzke GJ, Condon AG, van Herwaarden AF (2002) Breeding opportunities for increasing the efficiency of water use and crop yield in temperate cereals. Crop Sci 42: 111–121.
3. Cattivelli L, Rizza F, Badeck FW, Mazzucotelli E, Mastrangelo AM, et al. (2008) Drought tolerance improvement in crop plants: an integrated view from breeding to genomics. Field Crops Res 105: 1–14.
4. Blum A (1996) Crop responses to drought and the interpretation of adaptation. Plant Growth Regul 20: 135–148.
5. Richards RA (1996) Defining selection criteria to improve yield under drought. Plant Growth Regul 20: 157–166.
6. Farquhar GD, Richards RA (1984) Isotopic composition of plant carbon correlates with water use efficiency of wheat genotypes. Aust J Ag Res 11: 539–552.
7. Condon AG, Richards RA, Rebetzke GJ, Farquhar GD (2002) Improving water use efficiency and crop yield. Crop Sci 42: 122–132.
8. Condon AG, Richards RA, Farquhar GD (1993) Relationships between carbon isotope discrimination, water-use efficiency and transpiration efficiency for dryland wheat. Aust J Ag Res 4: 1693–1711.
9. Xu Y, This D, Pausch RC, Vonhof WM, Coburn JR, et al. (2009) Leaf-level water use efficiency determined by carbon isotope discrimination in rice seedlings: genetic variation associated with population structure and QTL mapping. Theor Appl Genet. 118: 1065–1081.
10. This D, Comstock J, Courtois B, Xu Y, Ahmadi N, et al. (2010) Genetic analysis of water use efficiency in rice (*Oryza sativa* L.) at the leaf level. Rice 3: 72–86.
11. Hall AE, Richards RA, Condon AG, Wright GC, Farquhar GD (1994) Carbon isotope discrimination and plant breeding. Plant Breeding Reviews 12: 81–113.
12. Li Z, Pinson SRM, Stansel JW, Park WD (1995) Identification of quantitative trait loci (QTLs) for heading date and plant height in cultivated rice (*Oryza sativa* L.). Theor Appl Genet 91: 374–381.
13. Juenger TE, McKay JK, Hausmann N, Keurentjes JJB, Sen S, et al. (2005) Identification and characterization of QTL underlying whole plant physiology in *Arabidopsis thaliana*: δ¹³C, stomatal conductance and transpiration efficiency. Plant Cell Environ 28: 697–708.
14. Austin DF, Lee M (1996) Genetic resolution and verification of quantitative trait loci for flowering and plant height with recombinant inbred lines of maize. Genome 39: 957–968.
15. Zhang Q (2007) Strategies for developing Green Super Rice. P Natl Acad Sci USA 104: 16402–16409.
16. Chen J, Chang SX, Anyia AO (2011) Gene discovery in cereals through quantitative trait loci and expression analysis in water-use efficiency measured by carbon isotope discrimination. Plant Cell Environ 34: 2009–2023.
17. Mian MAR, Bailey MA, Ashley DA, Wells R, Carter TE, et al. (1996) Molecular markers associated with water use efficiency and leaf ash in soybean. Crop Sci 36: 1252–1257.
18. Julier B, Bernard K, Gibelin C, Huguet T, Lelièvre F (2010) QTL for water use efficiency in alfalfa. In: Huyghe C, ed. Sustainable Use of Genetic Diversity in Forage and Turf Breeding. The Netherlands: Springer.
19. Martin B, Nienhuis J (1989) Restriction fragment length polymorphisms associated with water use efficiency in tomato. Science 243: 1725–1728.
20. Saranga Y, Menz M, Jiang C, Wright RJ, Yakir D, et al. (2001) Genomic dissection of genotype x environment interactions conferring adaptation of cotton to arid conditions. Genome Res 11: 1988–1995.
21. Price AH, Cairns JE, Horton P, Jones HG, Griffiths H (2002) Linking drought-resistance mechanisms to drought avoidance in upland rice using a QTL approach: progress and new opportunities to integrate stomatal and mesophyll responses. J Exp Bot 53: 989–1004.
22. Forster BP, Ellis RP, Moir J, Talam V, Sanguineti MC, et al. (2004) Genotype and phenotype associations with drought tolerance in barley tested in North Africa. Ann Appl Biol 144: 157–168.
23. Hausmann NJ, Juenger TE, Stowe SSK, Dawson TE, Simms EL (2005) Quantitative trait loci affecting δ¹³C and response to differential water availability in *Arabidopsis thaliana*. Evolution 59: 81–96.
24. Rebetzke GJ, Condon AG, Farquhar GD, Appels R, Richards RA (2008) Quantitative trait loci for carbon isotope discrimination are repeatable across environments and wheat mapping populations. Theor Appl Genet 118: 123–137.
25. Vincourt P, As-sadi F, Bordat A, Langlade NB, Gouzy J, et al. (2012) Consensus mapping of major resistance genes and independent QTL for quantitative resistance to sunflower downy mildew. Theor Appl Genet 125: 909–920.
26. Farquhar GD, Ehleringer JR, Hubick KT (1989) Carbon isotope discrimination and photosynthesis. Ann Rev Plant Physiol Plant Mol Biol 40: 503–537.
27. Craig H (1957) Isotopic standards for carbon and oxygen and correction factors for mass spectrometric analysis of carbon dioxide. Geochimi Cosmochim Ac 12: 133–149.

28. Jourjon MF, Jasson S, Marcel J, Ngom B, Mangin B (2005) MCQTL: multi-allelic QTL mapping in multi-cross design. Bioinformatics 21: 128–130.

29. Lincoln SE, Daly MJ, Lander ES (1993) Constructing genetic linkage maps with MAPMAKER/EXP version 3.0. A tutorial and reference manual. Technical Report, 3rd edn, Whitehead Institute for Biomedical Research.

30. Churchill GA, Doerge RW (1994) Empirical threshold values for quantitative trait mapping. Genetics 138: 963–971.

31. Brendel O, Pot D, Plomion C, Rozenberg P, Guehl JM (2002) Genetic parameters and QTL analysis of $\delta^{13}C$ and ring width in maritime pine. Plant, Cell & Environment 25: 945–953.

32. Kleibenstein DJ, Figureuth A, Mitchell-Olds T (2002) Genetic architecture of plastic methyl jasmonate responses in Arabidopsis thaliana. Genetics 161: 1685–1696.

33. Ungerer MC, Halldorsdottir SS, Purugganan MA, Mackay TFC (2003) Genotype–environment interactions at quantitative trait loci affecting inflorescence development in Arabidopsis thaliana. Genetics 165: 353–365.

34. De Givry S, Bouchez M, Chabrier P, Milan D, Schiex T (2005) CarthaGene: multipopulation integrated genetic and radiation hybrid mapping. Bioinformatics 21: 1703–1704.

35. Cadic E, Coque M, Vear F, Besset GB, Pauquet J, et al. (2013) Combined linkage and association mapping of flowering time in Sunflower (Helianthus annuus L.). Theor Appl Genet 126: 1337–1356.

36. Lauteri M, Brugnoli E, Spaccino L (1993) Carbon isotope discrimination in leaf soluble sugars and in whole-plant dry matter in Helianthus annuus L. Grown under different water conditions. In: Ehleringer JR, et al., eds. Stable isotopes and plant carbon – water relations. San Diego. Academic Press, inc. 93–108.

37. Condon AG, Farquhar GD, Richards RA (1990) Genotypic variation in carbon isotope discrimination and transpiration efficiency in wheat. Leaf gas exchange and whole plant studies. Aust J Plant Physiol 17: 9–22.

38. Dingkuhn M, Farquhar GD, SK De D, O'Toole JC, Datta SK (1991) Discrimination of $^{13}C$ among upland rices having different water use efficiencies. Aust J Agri Res 42: 1123–1131.

39. Li C (1999) Carbon isotope composition, water-use efficiency and biomass productivity of Eucalyptus microtheca populations under different water supplies. Plant Soil 214: 165–171.

40. Lauteri M, Scartazza A, Guido MC, Brugnoli E (1997) Genetic variation in photosynthetic capacity, carbon isotope discrimination and mesophyll conductance in provenances of Castanea sativa adapted to different environments. Funct Ecol 11: 675–683.

41. Brendel O, Thiec DL, Saintagne CS, Bodénès C, Kremer A, et al. (2008) Quantitative trait loci controlling water use efficiency and related traits in Quercus robur L. Tree Genet Genomes 4: 263–278.

42. Lambrides CJ, Chapman SC, Shorter R (2004) Genetic variation for carbon isotope discrimination in sunflower: Association with transpiration efficiency and evidence for cytoplasmic inheritance. Crop Sci 44: 1642–1653.

43. Condon AG, Richards RA, Farquhar GD (1992) The effect of variation in soil water availability, vapor pressure deficit and nitrogen nutrition on carbon isotope discrimination in wheat. Aust J Agr Res 43: 935–947.

44. O'Leary MH (1988) Carbon isotopes in photosynthesis. Bioscience 38: 325–336.

45. Rytter RM (2005) Water use efficiency, carbon isotope discrimination and biomass production of two sugar beet varieties under well-watered and dry conditions. J. Agron Crop Sci 191: 426–438.

46. Misra SC, Shinde S, Geerts S, Rao VS, Monneveux P (2010) Can carbon isotope discrimination and ash content predict grain yield and water use efficiency in wheat?. Agr Water Manage 97: 57–65.

47. Rizza F, Ghashghaie J, Meyer S, Matteu L, Mastrangelod AM, et al. (2012) Constitutive differences in water use efficiency between two durum wheat cultivars. Field Crops Res 125: 49–60.

48. Erice G, Louahlia S, Irigoyen JJ, Díaz MS, Alami IT, et al. (2011) Water use efficiency, transpiration and net $CO_2$ exchange of four alfalfa genotypes submitted to progressive drought and subsequent recovery. Environ Exp Bot 72: 123–130.

49. Farquhar GD, Lloyd J (1993) Carbon and oxygen isotope effects in the exchange of carbon dioxide between terrestrial plants and the atmosphere. In: Ehleringer JR, et al., eds. Stable isotopes and plant carbon – water relations. San Diego. Academic Press, inc. 40–70.

50. Scheidegger Y, Saurer M, Bahn M, Seigwolf RTW (2000) Linking stable oxygen and carbon isotopes with stomatal conductance and photosynthetic capacity: A conceptual model. Oecologia 125: 350–357.

51. Impa SM, Nadaradjan S, Boominathan P, Shashidhar G, Bindumadhava H, et al. (2005) Carbon isotope discrimination accurately reflects variability in WUE measured at a whole plant level in rice. Crop Sci 45: 2517–2522.

52. Martin B, Thorstenson YR (1988) Stable carbon isotope composition ($\delta^{13}C$), water use efficiency and biomass productivity of Lycopersicon esculentum, Lycopersicon pennellii, and the F1 hybrid. Plant Physiol 88: 213–217.

53. Wright GC, Hubick KT, Farquhar GD, Rao RCN (1993) Genetic and environmental variation in transpiration efficiency and its correlation with carbon isotope discrimination and specific leaf area in peanut. In: Ehleringer JR, et al., eds. Stable isotopes and plant carbon – water relations. San Diego. Academic Press, inc. 247–267.

54. Laza MR, Kondo M, Ideta O, Barlaan E, Imbe T (2006) Identification of quantitative trait loci for $\delta^{13}C$ and productivity in irrigated lowland rice. Crop Sci 46: 763–773.

55. Chapman S, Cooper M, Podlich D, Hammer G (2003) Evaluating plant breeding strategies by simulating gene action and dryland environment effects. Agron J 95: 99–113.

56. Vargas MF, van Eeuwijk A, Crossa J, Ribaut JM (2006) Mapping QTLs and QTL X environment interaction for CIMMYT maize drought stress program using factorial regression and partial least squares methods. Theor Appl Genet 112: 1009–1023.

57. Ceccarelli S, Acevedo E, Grando S (1991) Breeding for yield stability in unpredictable environments: single traits interaction between traits, and architecture of genotypes. Euphytica 56: 169–185.

58. Specht JE, Chase K, Macrander M, Graef GL, Chung J, et al. (2001) Soybean response to water. Crop Sci 41: 493–509.

59. Saranga Y, Jiang CX, Wright RJ, Yakir D, Paterson AH (2004) Genetic dissection of cotton physiological responses to arid conditions and their interrelationships with productivity. Plant Cell Environ 27: 263–277.

60. Diab A, Merah TB, This D, Ozturk N, Benscher D, et al. (2004) Identification of drought-inducible genes and differentially expressed sequence tags in barley. Theor Appl Genet 109: 1417–1425.

61. Ellis RP, Forster BP, Gordon DC, Handley LL, Keith RP, et al. (2002) Phenotype/genotype associations for yield and salt tolerance in a barley mapping population segregating for two dwarfing genes. J Exp Bot 53: 1163–76.

62. Xu Y, Zhu L, Chen Y, Lu C, Shen L, et al. (1997) Tagging genes for photo-thermo sensitivity in rice using RFLP and microsatellite markers. In: Plant and Animal Genome V, Plant and Animal Genome Conference Organizing Committee, San Diego, California, Poster 149.

63. Xu Y (2002) Global view of QTL: Rice as a model. In Kang MS, ed. Quantitative genetics, genomics and plant breeding. Wallingford, UK. CAB International, 109–134.

64. Collins NC, Tardieu F, Tuberosa R (2008) Quantitative trait loci and crop performance under abiotic stress: where do we stand ?. Plant Physiol 147: 469–486.

65. Morgan JA, LeCain DR, McCaig TN, Quick JS (1993) Gas exchange, carbon isotope discrimination and productivity in winter wheat. Crop Sci 33: 178–186.

66. Reynolds MP, Delgado MI, Gutierrez RM, Larque SA (2000) Photosynthesis of wheat in a warm, irrigated environment I. Genetic diversity and crop productivity. Field Crops Res 66, 37–50.

67. Teulat B, Merah O, Sirault X, Borries C, Waugh R, et al. (2002) QTLs for grain carbon isotope discrimination in field-grown barley. Theor Appl Genet 106: 118–126.

68. Kiani SP, Maury P, Nouri L, Ykhlef N, Grieu P, et al. (2009) QTL analysis of yield-related traits in sunflower under different water treatments. Plant Breeding 128: 363–373.

69. Upadyayula N, da Silva HS, Bohn MO, Rocheford TR (2006) Genetic and QTL analysis of maize tassel and ear inflorescence and architecture. Theor Appl Genet 112: 592–606.

70. Ribaut JM, Jiang C, Gonzalez-de-Leon D, Edmeades GO, Hoisington DA (1997) Identification of quantitative trait loci under drought conditions in tropical maize. 2. Yield components and marker-assisted selection strategies. Theor Appl Genet 94: 887–896.

71. Ceccarelli M, Sarri V, Natali L, Giordani T, Cavallini A, et al. (2007) Characterization of the chromosome complement of Helianthus annuus by in situ hybridization of a tandemly repeated DNA sequence. Genome 50: 429–434.

72. Feng J, Liu LZ, Cai X, Jan CC (2013) Toward a Molecular Cytogenetic Map for Cultivated Sunflower (Helianthus annuus L.) by Landed BAC/BIBAC Clones. G3-Genes Genome Genet 3: 31–40.

73. Teulat B, This D, Khairallah M, Borries C, Ragot C, et al. (1998) Several QTLs involved in osmotic-adjustment trait variation in barley (Hordeum vulgare L.). Theor Appl Genet 96: 688–698.

74. Rachid Al-chaarani G, Gentzbittel L, Huang X, Sarrafi A (2004) Genotypic variation and identification of QTLs for agronomic traits using AFLP and SSR in recombinant inbred lines of sunflower (Helianthus annuus L.). Theor Appl Genet 109: 1353–1360.

75. Saranga Y, Jiang CX, Wright RJ, Yakir D, Paterson AH (2004) Genetic dissection of cotton physiological responses to arid conditions and their inter-relationships with productivity. Plant Cell Environ 27: 263–277.

76. Via SR, Gomulkiewicz R, De jong R, Scheiner SM, Schlicting CD, et al. (1995) Adaptive phenotypic plasticity. Evolution 39: 505–522.

77. Kliebenstein DJ, Figuth A, -Olds TM (2002) Genetic Architecture of Plastic Methyl Jasmonate Responses in Arabidopsis thaliana. Genetics 161: 1685–1696.

78. Schlighting CD, Pigliucci M (1998) Phenotypic evolution: A reactionnorm perspective. Sunderland, MA: Sinauer associates.

79. Hao Z, Liu X, Li X, Xie C, Li M, et al. (2009) Identification of quantitative trait loci for drought tolerance at seedling stage by screening a large number of introgression lines in maize. Plant Breeding 128: 337–341.

80. Condon AG, Richards RA, Rebetzke J, Farquhar GD (2004) Breeding for high water-use efficiency. J Exp Bot 55: 2447–2460.

81. Rengel D, Arribat S, Maury P, Magniette MLM, Hourlier T, et al. (2012) A Gene-Phenotype Network Based on Genetic Variability for Drought Responses Reveals Key Physiological Processes in Controlled and Natural Environments. PLoS ONE 7: e45249.

# 6

# Availability and Temporal Heterogeneity of Water Supply Affect the Vertical Distribution and Mortality of a Belowground Herbivore and Consequently Plant Growth

**Tomonori Tsunoda\*, Naoki Kachi, Jun-Ichirou Suzuki**

Department of Biological Sciences, Tokyo Metropolitan University, Hachioji, Tokyo, Japan

## Abstract

We examined how the volume and temporal heterogeneity of water supply changed the vertical distribution and mortality of a belowground herbivore, and consequently affected plant biomass. *Plantago lanceolata* (Plantaginaceae) seedlings were grown at one per pot under different combinations of water volume (large or small volume) and heterogeneity (homogeneous water conditions, watered every day; heterogeneous conditions, watered every 4 days) in the presence or absence of a larva of the belowground herbivorous insect, *Anomala cuprea* (Coleoptera: Scarabaeidae). The larva was confined in different vertical distributions to top feeding zone (top treatment), middle feeding zone (middle treatment), or bottom feeding zone (bottom treatment); alternatively no larva was introduced (control treatment) or larval movement was not confined (free treatment). Three-way interaction between water volume, heterogeneity, and the herbivore significantly affected plant biomass. With a large water volume, plant biomass was lower in free treatment than in control treatment regardless of heterogeneity. Plant biomass in free treatment was as low as in top treatment. With a small water volume and in free treatment, plant biomass was low (similar to that under top treatment) under homogeneous water conditions but high under heterogeneous ones (similar to that under middle or bottom treatment). Therefore, there was little effect of belowground herbivory on plant growth under heterogeneous water conditions. In other watering regimes, herbivores would be distributed in the shallow soil and reduced root biomass. Herbivore mortality was high with homogeneous application of a large volume or heterogeneous application of a small water volume. Under the large water volume, plant biomass was high in pots in which the herbivore had died. Thus, the combinations of water volume and heterogeneity affected plant growth via the change of a belowground herbivore.

**Editor:** Mari Moora, University of Tartu, Estonia

**Funding:** This research was supported partly by the Japan Society for the Promotion of Science (11J06132 to TT). The funder had no role in study design, data collection and analysis, decision to publish, or preparation to the manuscript. No additional external funding received for this study.

**Competing Interests:** The authors have declared that no competing interests exist.

\* E-mail: ttsunoda@tmu.ac.jp

## Introduction

Water availability and its temporal variability (hereafter, 'water heterogeneity') affect plant biomass growth [1,2]. Plant responses to water frequency vary depending on nutrient availability and soil water content [3–6]. Soil water status also affects soil biota [7], and interacts with soil biota to affect plant growth. Empirical studies are necessary to clarify the interactive effects of water heterogeneity on soil biota and hence on plant growth because of the important influence of soil biota on plant growth and community dynamics [7,8].

In light of the presence of herbivorous insects in the soil, it is important that we elucidate the effects of water availability and heterogeneity on plants [9], because these insects influence the abundance, species diversity, and succession of plants [10–14]. Erb and Lu [15] pointed out that heterogeneity of soil abiotic factors such as moisture and nutrient availability alters the effects of belowground herbivores on plants. The interactions between water availability, heterogeneity, and belowground herbivory are likely to play crucial roles in plant growth.

Water availability and frequency affect the vertical distribution of belowground herbivores, and thus plant growth, because soil insects move vertically in response to changes in soil water status [16,17]. Grubs (Coleoptera: Scarabaeidae) are distributed deep in the soil in response to drought and shallow in the soil in response to irrigation [18]. The carrot-fly larva *Psila rosae* feeds on roots 15 cm below the ground in semi-dry soil, whereas in moist soil it feeds 1 cm from the soil surface [19]. Wireworms (Coleoptera: Elateridae) are distributed deep in the soil in summer but in shallow soil after heavy rain [20]. These changes in the vertical distribution of belowground herbivores affect plant mortality and growth [21]. Therefore, the amount and heterogeneity of water supply that determine soil moisture levels are likely to affect the vertical distribution of belowground herbivores and thus plant growth.

Soil moisture is one of the most important factors affecting the survival and abundance of belowground herbivorous insects [16,17,22]. Soil dryness increases mortality of belowground herbivores [23–26], whereas in moist soil mortality either does not change [24,27,28] or increases [23,29]. These findings suggest that the mortality of belowground herbivores in response to extreme water events has the potential to affect plant growth in various ways.

**Table 1.** Number of replicates, dead plants, and dead belowground herbivores, and the total sample size in biomass analyses of combinations of experimental treatments.

| Water volume | Water heterogeneity | Belowground herbivore | Number of replicates | Number of dead plant | Number of dead belowground herbivore | Sample size in biomass analyses |
|---|---|---|---|---|---|---|
| Small | Homogeneous | Control | 9 | 0 | NA† | 9 |
| | | Bottom | 9 | 0 | 3 | 6 |
| | | Middle | 9 | 0 | 0 | 9 |
| | | Top | 9 | 2 | 1 | 6 |
| | | Free | 9 | 0 | 1 | 8 |
| | Heterogeneous | Control | 9 | 0 | NA† | 9 |
| | | Bottom | 9 | 0 | 4 | 5 |
| | | Middle | 9 | 0 | 2 | 7 |
| | | Top | 9 | 0 | 1 | 8 |
| | | Free | 9 | 0 | 4 | 5 |
| Large | Homogeneous | Control | 9 | 0 | NA† | 9 |
| | | Bottom | 9 | 0 | 2 | 7 |
| | | Middle | 9 | 0 | 6 | 3 |
| | | Top | 9 | 0 | 4 | 5 |
| | | Free | 9 | 2 | 3 | 4 |
| | Heterogeneous | Control | 9 | 0 | NA† | 9 |
| | | Bottom | 9 | 0 | 2 | 7 |
| | | Middle | 9 | 0 | 2 | 7 |
| | | Top | 9 | 1 | 2 | 6 |
| | | Free | 9 | 1 | 3 | 5 |

†NA means not available due to the control treatment.

We conducted a growth experiment to test the hypothesis that changes in the amount and heterogeneity of water supply alter the vertical distribution and mortality of belowground herbivores and thus affect plant growth. Under heterogeneous supply of small amount water, belowground herbivores will become distributed deep in the soil to avoid the dry surface soil and feed on the fine root tips, which will hardly affect plant growth. In contrast, under homogeneous supply of large volume of water, herbivores will become distributed shallow in the soil and detach root connection by grazing, which has impact on plant growth. Soil water status resulting from changes in water supply amount and heterogeneity will also determine the fate of belowground herbivores. Unless belowground herbivores are present, plant growth will be no longer restricted by belowground herbivory.

## Materials and Methods

### Study species

One seedling of *Plantago lanceolata* L. (Plantaginaceae) was grown in each pot, to which we added a third-instar larva, or grub, of the belowground herbivorous insect *Anomala cuprea* Hope (Coleoptera: Scarabaeidae). The short-lived perennial forb, *P. lanceolata*, is cosmopolitan and has a rosette growth form. Seeds of *P. lanceolata* were collected from a population of more than 30 plants on a floodplain of the Tama River in Tokyo (35°38′N, 139°23′E). *P. lanceolata* is not endangered or protected species, and no specific permissions were required for this location to collect seeds of *P. lanceolata*. Larvae of *A. cuprea* feed on various herbaceous species [30,31], including *P. lanceolata* (T.Tsunoda, personal observation).

Grubs were grown from eggs laid in humus by adult *A. cuprea* collected from a floodplain of the Tama River (35°38′N, 139°23′E) in June and July 2012. *A. cuprea* is not endangered or protected species, and no specific permissions were required for this location to collect insects.

### Experimental design

The growth experiment was conducted from September to October 2012 in a plastic film greenhouse under natural sunlight in the experimental garden of Tokyo Metropolitan University (Hachioji, Tokyo; 35°37′N, 139°23′E). The mean annual precipitation in Hachioji has been 1602.3 mm year$^{-1}$ over the past 30 years [32]. Seeds of *P. lanceolata* were sown in a tray of peat moss in a growth chamber (Koitotron PC-02, Koito Industries, Ltd., Kanagawa, Japan) at 25°C. Seven days after sowing, one seedling with cotyledons was transplanted into each plastic pot (20 × 20 × 18 cm deep). Each pot was filled with 4.8 L of a mix of granular red clay and black soil (ratio 1:1 v/v) with 16 g of slow-release fertilizer [Magamp K, 6:40:6:15 (N-P-K-Mg), Hyponex Japan, Osaka, Japan].

The experiment had a three-way factorial randomized block design with nine replications. The factors were water volume with two levels (large, 200 mL of water per day, and small, 100 mL of water per day), water supply frequency with two levels (homogeneous, daily watering, and heterogeneous, watering every 4 days), and belowground herbivore. The certain amount of water for any single watering was determined not to exceed the capacity of the pot. The total volume of water received over the period was the

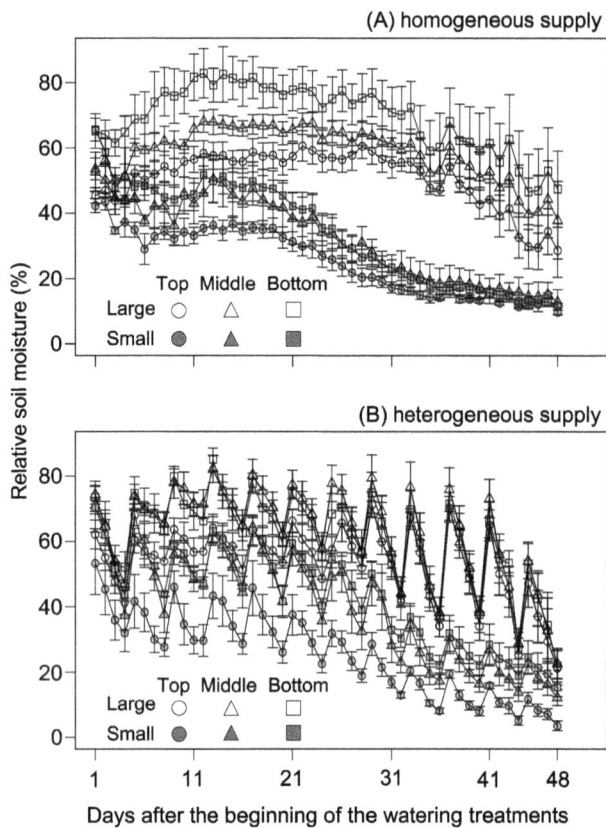

Figure 1. Mean relative soil moisture (± SE). (A) heterogeneous supply treatments; (B) homogeneous supply treatments; Top, middle, bottom: top, middle, and bottom zone treatments, respectively. Large, large water volume supply; Small, small volume supply. The soil moisture content was larger with a large water volume than a small volume. The soil moisture content in the bottom zone was the largest, and that in the top zone the smallest. The temporal variability in soil moisture content under heterogeneous water-supply conditions was larger than that under homogeneous conditions.

cm stainless-steel wire-mesh screens (0.8-mm wire diameter, 5.5-mm mesh) were inserted into the pot soil to divide the soil evenly into three zones in all treatments except the free treatment. In the top, middle, and bottom treatments, one grub was introduced to the relevant zone through a hole (diameter 12 mm) on the side wall of the pot. The hole was closed with plastic film after the grub had been added. In the free treatment in which the grub was able to move freely around the pot, four screens of 20 × 10-cm stainless-steel wire-mesh were inserted, one along each wall of the pot, which assumed perfect mobility of the grub without any other differences from other three treatments. Because the insertion of the wire-mesh did not affect the plant growth in our previous experiment [33], no treatment without wire-mesh was included in this experiment.

For the first week after transplantation of the seedlings, 150 mL water was supplied to all pots every day. The treatments that combined different water volumes with different watering frequencies began at the start of the second week after the transplant. On day 20 after the beginning of the watering treatment, a grub was added to the relevant zone of the pot. In the free treatment, the grub was placed on the soil surface near the centre of the pot and left to burrow underground.

A plant survival was recorded every day after addition of the grub. If the roots became completely detached from the shoot and the leaves wilted, the plant was considered to die. When a plant died, we recorded survival of the grub.

On day 28 after addition of the grub, the plants were harvested and divided into shoots and roots. The shoots and roots were dried at 72°C for 3 days and weighed. At harvest, grub mortalities were also recorded.

## Soil moisture measurement

Soil moisture (water content by volume) was measured with a soil moisture probe (ECH$_2$O, Decagon Devices, Inc., Pullman, Washington, USA) in an additional four replications without grubs of each watering treatment combination and soil zone during the experimental period. Measurements were taken every day, before the watering. Relative soil moisture content was calculated as the difference between the measured value and the minimum value during the experimental period, divided by the range between the maximum and minimum values during the experimental period [34]. The 4-day moving variance of the relative soil moisture content for the current day and the next 3 days was calculated to quantify the variability in water availability in each 4-day watering cycle. The coefficient of variation (CV) and temporal mean of the relative soil moisture content during each watering treatment were calculated. The CV was used as an index of temporal variability in soil moisture content during each watering treatment [34].

same between the homogeneous and heterogeneous supplies for each water volume regime.

There were five types of belowground herbivore treatment: in the top treatment, one grub was placed in the top zone; in the middle treatment, it was placed in the middle zone; and in the bottom treatment, it was placed in the bottom zone. In the control treatment, no grub was added to the pot. In the free treatment, no screens were added to restrict the grub movement. Two 20 × 20

**Table 2.** Effects of water volume (V), water heterogeneity (H), and soil profile (P) on mean relative soil moisture content.

| Source | $\chi^2$ | df | P |
|---|---|---|---|
| V | 38.788 | 1 | <0.001 |
| H | 0.108 | 1 | 0.742 |
| P | 6.048 | 2 | 0.049 |
| V × H | 0.822 | 1 | 0.365 |
| V × P | 0.660 | 2 | 0.719 |
| H × P | 0.551 | 2 | 0.759 |
| V × H × P | 0.599 | 2 | 0.741 |

**Table 3.** Effects of water volume (V), water heterogeneity (H), and soil profile (P) on mean CV of the relative soil moisture.

| Source | $\chi^2$ | df | P |
|---|---|---|---|
| V | 0.333 | 1 | 0.564 |
| H | 6.425 | 1 | 0.011 |
| P | 0.252 | 2 | 0.882 |
| V × H | 0.156 | 1 | 0.693 |
| V × P | 0.230 | 2 | 0.891 |
| H × P | 0.081 | 2 | 0.960 |
| V × H × P | 0.040 | 2 | 0.981 |

## Data analysis

Relative soil moisture content and temporal variability in soil moisture were analyzed by using generalised linear mixed models (GLMMs) under the assumption of a Gaussian error distribution. In this model, the response variable was the relative soil moisture content or the CV; the explanatory variables were water volume, watering frequency, soil zone, and their interactions: the random factors were measuring date and block, with block as nested random effect within date.

Mortalities of plants and grubs were analyzed by using GLMMs with a binomial error distribution and logit-link function. In these models, the response variable was grub or plant mortality; the explanatory variables were water volume, watering frequency, belowground herbivore, and their interactions. The random factor was block.

Plant biomass as the sum of shoot and root biomass of each plant, was analyzed with a GLMM assuming a Gaussian error distribution. In this model, the response variable was plant biomass; the explanatory variables were water volume, watering frequency, belowground herbivore, and their interactions. The random factor was block. For each water volume, plant biomass was analyzed with the model lacking water volume as an explanatory variable.

In the analyses for continuous variables, if the assumption of homogeneity of variance was satisfied with Bartlett's test, then identity-link function was applied. If not, then log-link function was applied. Data from pots in which belowground herbivores died were treated as missing values. Number of samples in the biomass analyses was presented in Table 1.

Because several grubs were found dead at harvest, plant biomass was analyzed with generalised linear models (GLMs) for each water volume to evaluate changes in plant biomass due to

herbivore mortality. In these models, the response variable was plant biomass; the explanatory variables were watering frequency, belowground herbivore, the survival of the belowground herbivore, and their interactions.

All analyses were performed with the statistical software R version 2.15.1 [35]. The lme4 package was used to calculate GLMMs by using maximum likelihood estimation. To determine the effects of the fixed factors (i.e., to calculate the in $P$-values), we used a likelihood ratio test to compare models with and without the variable of interest using a chi-squared test statistic [36]. The data were analyzed by GLMM framework because our data contain binary and continuous variables [37].

## Results

### Soil moisture

The mean relative soil moisture content was larger under a large water volume than a small volume (Figure 1, Table 2). The mean relative soil moisture content in the bottom zone was the largest, and that in the top zone was the smallest (Figure 1; Table 2).

The temporal variability in relative soil moisture content under heterogeneous water-supply conditions was larger than that under homogeneous conditions (Figure 1). The mean CV (%) of the relative soil moisture content was larger under heterogeneous water-supply conditions than under homogeneous conditions (Table 3).

### Plant mortality

Plant mortality changed significantly with differences in the feeding zones of the belowground herbivore (Table 4). No plants died in the middle and bottom treatments, but three died in the top treatment and three in the free treatment. No plants died when a small volume of water was heterogeneously applied (Table 1).

**Table 4.** Effects of water volume (V), water heterogeneity (H), and belowground herbivore (B) on plant mortality.

| Source | $\chi^2$ | df | P |
|---|---|---|---|
| V | 0.753 | 1 | 0.386 |
| H | 0.763 | 1 | 0.382 |
| B | 8.621 | 3 | 0.035 |
| V × H | 6.234 | 3 | 0.101 |
| V × B | 2.156 | 1 | 0.142 |
| H × B | 0.000 | 3 | 1.000 |
| V × H × B | 0.000 | 3 | 1.000 |

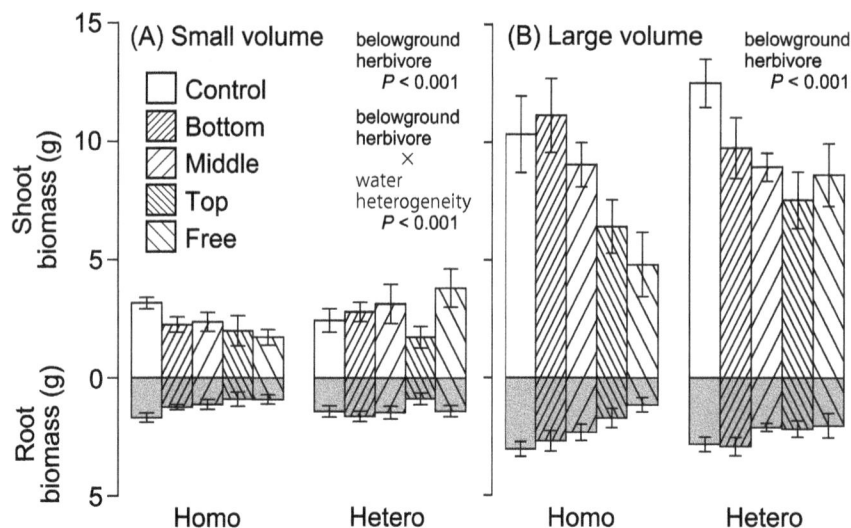

**Figure 2. Mean of the shoot and root biomass (± SE).** (A) Small water volume. (B) Large water volume. Control, no herbivore present; Bottom, Middle, Top; herbivores placed in the bottom, middle, and top zones, respectively, of the pot; Free, herbivore present and free to move in all zones of the pot. Homo, homogeneous water supply; Hetero, heterogeneous water supply. Biomass in the top treatment was the smallest between the middle, bottom, and control treatments in every watering treatment. Biomass in the free treatment was similar to that in the top treatment in every watering treatment, except for the heterogeneously-applied treatment with small volume of water.

Roots of the survived plants reached the bottom zone at the harvest.

### Plant biomass

The three-way interaction between water volume, watering heterogeneity and belowground herbivory, significantly affected mean plant biomass (Figure 2, Table 5A). The herbivory effects on plant biomass differed between the combinations of water-supply volume and heterogeneity: belowground herbivory interacted with heterogeneous conditions and a small water volume (Figure 2A; Table 5B). When a small water volume was supplied, the plant biomass was larger in the control treatment than in the free treatment under homogeneous conditions and smaller than in the free treatment under heterogeneous conditions (Figure 2A). Under both homogeneous and heterogeneous conditions, plant biomass in the top treatment was the smallest between the three zone

**Table 5.** Effects of water volume (V), water heterogeneity (H), and belowground herbivore (B) on plant biomass.

| (A) Three-way analysis | | | |
|---|---|---|---|
| Source | $\chi^2$ | df | P |
| A | 1858.5 | 1 | <0.001 |
| H | 23.367 | 1 | <0.001 |
| B | 512.59 | 4 | <0.001 |
| A × H | 1.084 | 1 | 0.298 |
| A × B | 32.951 | 4 | <0.001 |
| H × B | 31.641 | 4 | <0.001 |
| A × H × B | 27.610 | 4 | <0.001 |
| (B) Two-way analysis at small water volume | | | |
| Source | $\chi^2$ | df | P |
| H | 4.443 | 1 | 0.035 |
| B | 28.933 | 4 | <0.001 |
| H × B | 32.501 | 4 | <0.001 |
| (C) Two-way analysis at large water volume | | | |
| Source | $\chi^2$ | df | P |
| H | 2.096 | 1 | 0.148 |
| B | 20.968 | 4 | <0.001 |
| H × B | 1.304 | 4 | 0.861 |

treatments. Plant biomass in the free treatment was nearly the same as that in the top treatment with a homogeneous water supply and as large as that in the middle or bottom treatment with a heterogeneous water supply.

Under water supply of large volume, plant biomass significantly differed depending on the vertical distribution of a belowground herbivore (Figure 2B; Table 5C). In the treatment with herbivore in the top zone, plant biomass was remarkably small under both water heterogeneity conditions. Plant biomass in the free treatment was the smallest in all treatments with a homogeneous water supply, and plant biomass in the free treatment was slightly larger than in the top treatment with a heterogeneous water supply.

## Mortality of belowground herbivores and its effect on plant biomass

The interaction between water volume and water supply heterogeneity significantly affected mortality of belowground herbivores ($\chi^2 = 5.249$, df $= 1$, $P = 0.022$). With a large water volume, 15 belowground herbivores in the homogeneous treatment died, as did 9 in the heterogeneous treatment (Table 1). With a small water volume, 5 died in the homogeneous treatment and 11 in the heterogeneous one (Table 1). Mortality of belowground herbivores did not differ between the feeding zones ($\chi^2 = 0.895$, df $= 3$, $P = 0.827$).

Mortality of the belowground herbivore significantly affected plant biomass with a large water volume (GLM, $F = 6.793$, df $= 1$, $P = 0.012$) but not with a small water volume (GLM, $F = 0.321$, df $= 1$, $P = 0.573$). Plant biomass in the pots with herbivore mortality was larger than those without mortality.

## Discussion

Plant mortality occurred only in the treatments in which the herbivore was shallow in the soil to sever the aerial shoot from its root system. This plant mortality caused by belowground herbivory is consistent with our previous study [33]. With heterogeneous supply of small volume water, no plant was grazed at the root base and any plants did not die, probably because low moisture in the surface soil was not suitable for grubs to graze there.

The interaction between available volume and heterogeneity of water supply would change the vertical distribution of the grub in the free treatment. With a small volume of water under a heterogeneous supply, the mean plant biomass in the free treatment that was almost equivalent to those in the bottom and middle treatments, suggest the grubs occurred in the bottom and middle zones due to the low soil moisture levels and large moisture variability. Consequently, the herbivory effects on plant growth varied depending on the availability of water supply.

The vertical distribution of the belowground herbivore was consistent with the previous studies in which belowground

herbivores were distributed deep in the soil under dry conditions [18–20]. The belowground herbivores may graze fine roots deep in the soil, which may promotes root turnover and enhances the absorption of resources [38]. Belowground herbivory can thus affect plant growth in either a negative or a positive way, depending on the heterogeneity of the water supply. In contrary, with the large amount of available water regardless of supply patterns, belowground herbivory occurred in shallow soil in the free treatment due to enough soil moisture. Therefore, plant biomass was significantly lower than that in the control treatment.

Mortality of belowground herbivores varied depending on available water, which altered plant growth. When the grubs died with enough volume of available water, plant biomass was large because of the negligible loss of roots. When the available water volume was small, herbivore mortality was higher under a heterogeneous water supply than a homogeneous one, which is consistent with previous studies [23–26]. However, the effects of herbivore mortality on plant biomass were not significant under a small water volume in this experiment because only few herbivores died.

To our knowledge, this is the first experiment to evaluate the simultaneous effects of available water volume, heterogeneity of water supply, and belowground herbivory on plant growth. The results are consistent with our hypothesis: the availability and heterogeneity of water supply changed the vertical distribution and mortality of belowground herbivores, and consequently plant growth. Therefore, the heterogeneity of soil water supply should be considered in root herbivory studies [15].

Severe climatic events attributable to climate change are already having serious consequences for plants and their herbivores [9,39–42]. We observed high mortality and changes in the vertical distribution of belowground herbivores under the most extreme conditions of water supply (i.e. small volume and heterogeneous supply); these consequently affected plant growth. Therefore, changes in the effects of belowground herbivores on plant growth are likely to occur under the severe weather conditions resulting from climate change.

## Acknowledgments

We thank A Takaoka, K Ogata, and S Uchida of Tokyo Metropolitan University for their help with the experiment. We are grateful to Dr. Elly Morrien, an anonymous reviewer, and the academic editor for their constructive comments on our manuscript.

## Author Contributions

Conceived and designed the experiments: TT JIS. Performed the experiments: TT. Analyzed the data: TT. Contributed reagents/materials/analysis tools: TT JIS. Wrote the paper: TT JIS. Provided editorial advice: NK.

## References

1. Novoplansky A, Goldberg DE (2001) Effects of water pulsing on individual performance and competitive hierarchies in plants. Journal of Vegetation Science 12: 199–208. doi:10.2307/3236604.

2. Hagiwara Y, Kachi N, Suzuki J-I (2008) Effects of temporal heterogeneity of watering on size of an annual forb, *Perilla frutescens* (Lamiaceae), depend on soil nutrient levels. Botany 86: 1111–1116. doi:10.1139/B08-064.

3. Maestre FT, Reynolds JF (2007) Amount or pattern? Grassland responses to the heterogeneity and availability of two key resources. Ecology 88: 501–511. doi:10.1890/06-0421.

4. Heisler-White JL, Knapp AK, Kelly EF (2008) Increasing precipitation event size increases aboveground net primary productivity in a semi-arid grassland. Oecologia 158: 129–140. doi:10.1007/s00442-008-1116-9.

5. Knapp AK, Beier C, Briske DD, Classen AT, Luo Y, et al. (2008) Consequences of more extreme precipitation regimes for terrestrial ecosystems. Bioscience 58: 811–821.

6. Hagiwara Y, Kachi N, Suzuki J-I (2010) Effects of temporal heterogeneity of water supply on the growth of *Perilla frutescens* depend on plant density. Annals of Botany 106: 173–181. doi:10.1093/aob/mcq096.

7. Bardgett RD (2005) The Biology of Soil: A Community and Ecosystem Approach. Oxford University Press, Oxford.

8. Bardgett RD, Wardle DA (2010) Aboveground-Belowground Linkages: Biotic Interactions, Ecosystem Processes, and Global Change. Oxford Series in Ecology and Evolution, Oxford University Press, Oxford.

9. Staley JT, Johnson SN (2008) Climate change impacts on root herbivores, pp. 192–213, in SN Johnson and PJ Murray (eds.), Root Feeders – An Ecosystem Perspective. CABI, Wallingford.

10. Brown VK, Gange AC (1992) Secondary plant succession: how is it modified by insect herbivory? Vegetatio 101: 3–13.

11. De Deyn GB, Raaijmakers CE, Zoomer HR, Berg MP, De Ruiter PC, et al. (2003) Soil invertebrate fauna enhances grassland succession and diversity. Nature 422: 711–713.

12. Wardle DA, Bardgett RD, Klironomos JN, Setälä H, van der Putten WH, et al. (2004) Ecological linkages between aboveground and belowground biota. Science 304: 1629–1633.

13. van der Putten WH, Bardgett RD, de Ruiter PC, Hol WHG, Meyer KM, et al. (2009) Empirical and theoretical challenges in aboveground- belowground ecology. Oecologia 161: 1–14. doi 10.1007/s00442-009-1312-2.

14. Stein C, Unsicker SB, Kahmen A, Wagner M, Audorff V, et al. (2010) Impact of invertebrate herbivory in grasslands depends on plant species diversity. Ecology 91: 1639–1650.

15. Erb M, Lu J (2013) Soil abiotic factors influence interactions between belowground herbivores and plant roots. Journal of Experimental Botany 64: 1295–1303. doi:10.1093/jxb/ert007.

16. Villani MG, Wright RJ (1990) Environmental influences on soil macroarthropod behavior in agricultural systems. Annuan Review of Entomology 35: 249–269.

17. Barnett K, Johnson SN (2013) Living in the soil matrix. Advances in Insect Physiology 45: 1–52.

18. Villani MG, Wright RJ (1988) Use of radiography in behavioral studies of turfgrass-infesting scarab grub species (Coleoptera: Scarabaeidae). Bulletin of Entomological Society of America 34: 132–144.

19. Jones OT (1979) Responses of carrot fly larvae, *Psila rosae*, to components of their physical environment. Ecological Entomology 4: 327–334.

20. Lafrance J (1968) The seasonal movements of wireworms (Coleoptera: Elateridae) in relation to soil moisture and temperature in the organic soils of southwestern Quebec. The Canadian Entomologist 100: 801–807.

21. Davidson RL, Roberts RJ (1969) Scarab damage to grass and clover as influenced by depth of feeding. Bulletin of Entomological Research 58: 559–565.

22. Brown VK, Gange AC (1990) Insect herbivory below ground. Advances in Ecological Research 20: 1–58.

23. Campbell RE (1937) Temperature and moisture preferences of wireworms. Ecology 18: 479–489.

24. Moran NA, Whitham TG (1988) Population fluctuations in complex life cycles: An example from *Pemphigus* aphids. Ecology 69: 1214–1218.

25. Briones MJI, Ineson P, Piearce TG (1997) Effects of climate change on soil fauna; responses of enchytraeids, Diptera larvae and tardigrades in a transplant experiment. Applied Soil Ecology 6: 117–134.

26. Riis K, Esbjerg P (1998) Season and soil moisture effect on movement, survival, and distribution of *Cyrtomenus bergi* (Hemiptera: Cydnidae) within the soil profile. Environmental Entomology 27: 1182–1189.

27. Ladd TL, Buriff CR (1979) Japanese beetle: Influence of larval feeding on bluegrass yields at two levels of soil moisture. Journal of Economic Entomology 32: 341–360.

28. Régnière J, Rabb RL, Stinner RE (1981) *Popillia japonica*: Effect of soil moisture and texture on survival and development of eggs and first instar grubs. Environmental Entomology 10: 654–660.

29. Godfrey LD, Yeargan KV (1985) Influence of soil moisture and weed density on clover root curculio *Sitona hispidulus*, larval stress to alfalfa (*Medicago sativa*). Journal of Agricultural Entomology 2: 370–377.

30. Okuno T, Tanaka Y, Kimura Y, Yoneyama S (1978) Diseases and Pests of Flowers and Vegetables in Colour. Hoikusha Publishing CO., LTD. Osaka, Japan (in Japanese).

31. Sakai K, Fujioka M (2007) Atlas of Japanese Scarabaeoidea Vol. 2 Phytophagous group I. Roppon-Ashi Entomological Books. Tokyo, Japan (in Japanese).

32. Japan Meteorological Agency. *AMeDAS, Automated Meteorological Data Acquisition System*, URL http://www.data.jma.go.jp/obd/stats/etrn/index.php (in Japanese).

33. Tsunoda T, Kachi N, Suzuki J-I (2014) Effects of belowground vertical distribution of a herbivore on plant biomass and survival in *Lolium perenne*. Ecological Research 29: 351–355. doi:10.1007/s11284-014-1133-6.

34. James SE, Pärtel M, Wilson SD, Peltzer DA (2003) Temporal heterogeneity of soil moisture in grassland and forest. Journal of Ecology 91: 234–239. doi: 10.1046/j.1365-2745.2003.00758.x.

35. R Development Core Team (2012) *R: A language and environment for statistical computing*. R Foundation for Statistical Computing, Vienna, Austria. ISBN 3-900051-07-0, URL http://www.R-project.org/.

36. Crawley MJ (2007) The R Book. John Wiley and Sons, New York.

37. Bolker BM, Brooks ME, Clark CJ, Geange SW, Poulsen JR, et al. (2008) Generalized linear mixed models: a practical guide for ecology and evolution. Trends in Ecology and Evolution 24: 127–135. doi: 10.1016/j.tree.2008.10.008.

38. Ramsell J, Malloch AJC, Whittaker JB (1993) When grazed by *Tipula paludosa*, *Lolium perenne* is a stronger competitor of *Rumex obtusifolius*. Journal of Ecology 81: 777–786.

39. Staley JT, Hodgson CJ, Mortimer SR, Morecroft MD, Masters GJ, et al. (2007) Effects of summer rainfall manipulations on the abundance and vertical distribution of herbivorous soil macro-invertebrates. European Journal of Soil Biology 43: 189–198. doi:10.1016/j.ejsobi.2007.02.010.

40. Easterling DR, Meehl GA, Parmesan C, Changnon SA, Karl TR, et al. (2000) Climate extremes: observations, modeling, and impacts. Science 289: 2068–2074.

41. Smith MD (2011) An ecological perspective on extreme climatic events: a synthetic definition and framework to guide future research. Journal of Ecology 99: 656–663.

42. Reyer CPO, Leuzinger S, Rammig A, Wolf A, Bartholomeus RP, et al. (2013) A plant's perspective of extremes: terrestrial plant responses to changing climatic variability. Global Change Biology 19: 75–89.

# 7

# The Added Value of Water Footprint Assessment for National Water Policy

**Joep F. Schyns\*, Arjen Y. Hoekstra**

Twente Water Centre, University of Twente, Enschede, The Netherlands

## Abstract

A Water Footprint Assessment is carried out for Morocco, mapping the water footprint of different activities at river basin and monthly scale, distinguishing between surface- and groundwater. The paper aims to demonstrate the added value of detailed analysis of the human water footprint within a country and thorough assessment of the virtual water flows leaving and entering a country for formulating national water policy. Green, blue and grey water footprint estimates and virtual water flows are mainly derived from a previous grid-based ($5 \times 5$ arc minute) global study for the period 1996–2005. These estimates are placed in the context of monthly natural runoff and waste assimilation capacity per river basin derived from Moroccan data sources. The study finds that: (i) evaporation from storage reservoirs is the second largest form of blue water consumption in Morocco, after irrigated crop production; (ii) Morocco's water and land resources are mainly used to produce relatively low-value (in US\$/m$^3$ and US\$/ha) crops such as cereals, olives and almonds; (iii) most of the virtual water export from Morocco relates to the export of products with a relatively low economic water productivity (in US\$/m$^3$); (iv) blue water scarcity on a monthly scale is severe in all river basins and pressure on groundwater resources by abstractions and nitrate pollution is considerable in most basins; (v) the estimated potential water savings by partial relocation of crops to basins where they consume less water and by reducing water footprints of crops down to benchmark levels are significant compared to demand reducing and supply increasing measures considered in Morocco's national water strategy.

**Editor:** Vanesa Magar, Centro de Investigacion Cientifica y Educacion Superior de Ensenada, Mexico

**Funding:** The research was funded by Deltares (http://www.deltares.nl/en). Karen Meijer (Deltares) had a role in study design. Wil van der Krogt (Deltares) had a role in data collection. The writing stage of the manuscript was funded by the Institute for Innovation and Governance Studies (IGS) of the University of Twente (http://www.utwente.nl/igs/). The funders had no role in data analysis, decision to publish, or preparation of the manuscript.

**Competing Interests:** The authors have declared that no competing interests exist.

\* E-mail: j.f.schyns@utwente.nl

## Introduction

Morocco is a semi-arid country in the Mediterranean facing water scarcity and deteriorating water quality. The limited water resources constrain the activities in different sectors of the economy of the country. Agriculture is the largest water consumer and withdrawals for irrigation peak in the dry period of the year, which contributes to low surface runoff and desiccation of streams. Currently, 130 reservoirs are in operation to deal with this mismatch in water demand and natural water supply and to serve for generation of hydroelectricity and flood control [1]. Groundwater resources also play an important role in the socio-economic development of the country, in particular by ensuring the water supply for rural communities [2]. However, a large part of the aquifers is being overexploited and suffer from deteriorating water quality by intrusion of salt water, caused by the overexploitation, and nitrates and pesticides that leach from croplands, caused by excessive use of fertilizers. Surface water downstream of some urban centres is also polluted, due to untreated wastewater discharges.

In 1995, the Moroccan Water Law (no. 10–95) came into force and introduced decentralized integrated water management and rationalisation of water use, including the user-pays and polluter-pays principles. It also dictates the development of national and river basin master plans [3], which are elaborated in accordance with the national water strategy. To cope with water scarcity and pollution, the national water strategy includes action plans to reduce demand, increase supply and preserve and protect water resources [1]. It also proposes legal and institutional reforms for proper implementation and enforcement of these actions. Demand management focuses on improving the efficiency of irrigation and urban supply networks and pricing of water to rationalise its use. Plans to increase supply include the construction of more dams and a large North-South inter-basin water transfer, protection of existing hydraulic infrastructure, desalinization of sea water and reuse of treated wastewater.

Although the national water strategy considers options to reduce water demand in addition to options to increase supply, it does not include the global dimension of water by considering international virtual water trade, nor does it consider whether water resources are efficiently allocated based on physical and economic water productivities of crops (the main water consumers). Analysis of the water footprint of activities in Morocco and the virtual water trade balance of the country therefore might reveal new insights to alleviate water scarcity.

The concept of water footprint was introduced by Hoekstra [4]; this subsequently led to the development of Water Footprint Assessment as a distinct field of research and application [5,6]. The water footprint is an indicator of freshwater use that looks not only at direct water use of a consumer or producer, but also at the indirect water use. As such, it provides a link between human consumption and human appropriation of freshwater systems. Water Footprint Assessment refers to a variety of methods to quantify and map the water footprint of specific processes, products, producers or consumers, to assess the environmental, social and economic sustainability of water footprints at catchment or river basin level and to formulate and assess the effectiveness of strategies to reduce water footprints in prioritized locations. The water footprint of a product is the volume of freshwater used to produce the product, measured over the full supply chain [6]. Three different components of a water footprint are distinguished: green, blue and grey. The green water footprint is the volume of rainwater evaporated or incorporated into the product. Blue water refers to the volume of surface- or groundwater evaporated, incorporated into the product or returned to another catchment or the sea. The grey water footprint relates to pollution and is defined as the volume of freshwater that is required to assimilate the load of pollutants given natural background concentrations and existing ambient water quality standards [6]. The total freshwater volume consumed or polluted within the territory of a nation as a result of activities within the different sectors of the economy is called the water footprint of national production. International trade of products creates 'virtual water flows' leaving and entering a country. The virtual-water export from a nation refers to the water footprint of the products exported. The virtual-water import into a nation refers to the water footprint of the imported products.

Several authors have assessed the water footprint and virtual water trade balance of nations and regions and state the relevance of the tool for well-informed water policy on the national and river basin level [7–10]. In a case study for a Spanish region, Aldaya et al. [10] conclude that water footprint analyses can provide a transparent framework to identify potentially optimal alternatives for efficient water use at the catchment level and that this can be very useful to achieve an efficient allocation of water and economic resources in the region. Chahed et al. [8] state that integration of all water resources at the national scale, including the green water used in rain-fed agriculture and as part of the foodstuffs trade balance, is essential in facing the great challenges of food security in arid countries.

The objective of this study is to explore the added value of analysing the water footprint of activities in Morocco and the virtual water flows from and to Morocco in formulating national water policy. The study includes an assessment of the water footprint of activities in Morocco (at the river basin level on a monthly scale) and the virtual water trade balance of the country and, based on this, response options are formulated to reduce the water footprint within Morocco, alleviate water scarcity and

allocate water resources more efficiently. Results and conclusions from the Water Footprint Assessment are compared with the scope of analysis of, and action plans included in Morocco's national water strategy and river basin plans in order to address the added value of Water Footprint Assessment relative to these existing plans.

The water footprint of Morocco has not been assessed previously on the river basin level on a monthly scale. Morocco has been included in a number of global studies, but these studies did not analyse the spatial and temporal variability of the water footprint within the country [11–13]. Furthermore, this study is the first to include specific estimates of the evaporative losses from the irrigation supply network and from storage reservoirs as part of a comprehensive Water Footprint Assessment. Finally, it is new in providing quantitative estimates of the potential water savings by partial relocation of crop production to regions with lower water consumption per ton of crop by means of an optimization and by reducing water footprints of crops down to benchmark levels.

Several insights and response options emerged from the Water Footprint Assessment, which are currently not considered in the national water strategy of Morocco and the country's river basin plans. Therefore, Water Footprint Assessment is considered to have an added value for formulating national water policy in Morocco.

## Method and Data

### Water Footprint of Morocco's Production

This study follows the terminology and methodology developed by Hoekstra et al. [6]. The water footprint of Morocco's production is estimated at river basin level on a monthly scale for the activities included in Table 1. The river basins are chosen such that they coincide with the action zones of Morocco's river basin agencies (Figure 1A). Due to data limitations, the grey water footprint is analysed on an annual scale and the water footprints of grazing and animal water supply are analysed at national and annual level. The study considers the average climate, production and trade conditions over the period 1996–2005. The water footprints of agriculture, industry and households are obtained from Mekonnen and Hoekstra [13,14], who estimated these parameters globally at a 5 by 5 arc minute spatial resolution. The annual blue water footprint estimates for industries and households by Mekonnen and Hoekstra [13] are distributed throughout the year according to the monthly distribution of public water supply obtained from Ministry EMWE (unpublished data 2013). These distributions are available for the basins Loukkos, Sebou,

**Table 1.** Water footprint estimates included in this study.

| Water footprint of | Components | Period | Source |
| --- | --- | --- | --- |
| Crop production | Green, blue, grey | 1996–2005 | [14] |
| Grazing | Green | 1996–2005 | [13] |
| Animal water supply | Blue | 1996–2005 | [13] |
| Industrial production | Blue, grey | 1996–2005 | [13] |
| Domestic water supply | Blue, grey | 1996–2005 | [13] |
| Storage reservoirs | Blue | - | Own elaboration |
| Irrigation water supply network | Blue | 1996–2005 | Own elaboration |

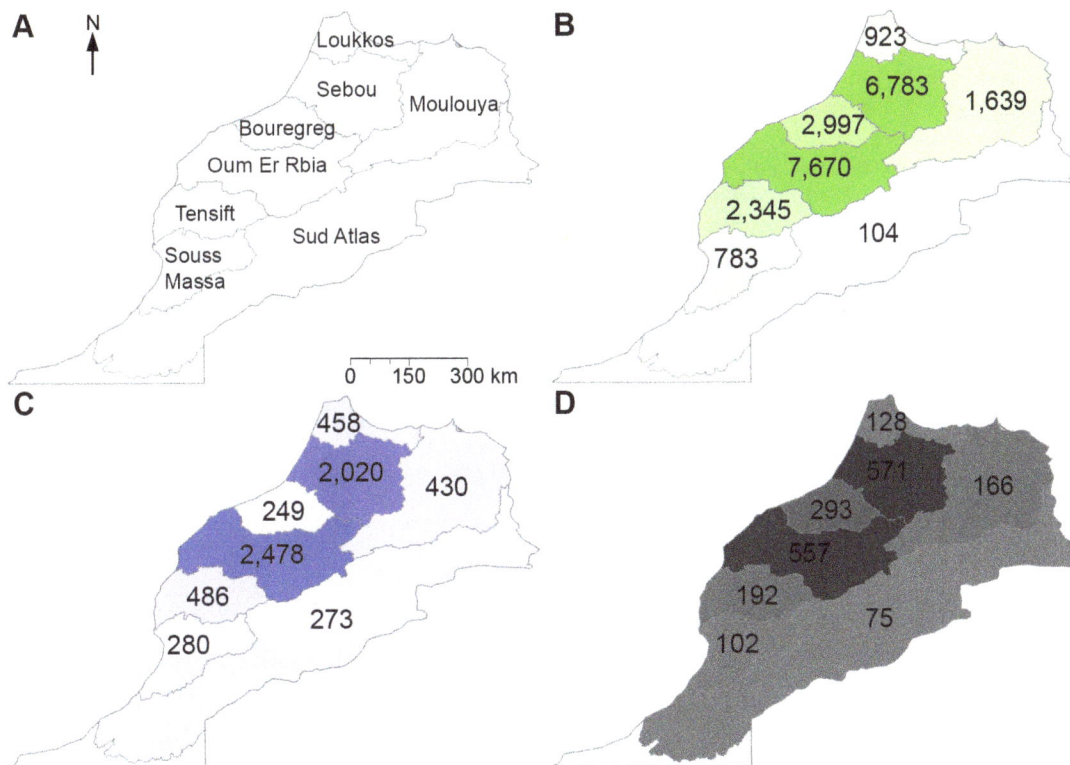

**Figure 1. Water footprint of Morocco's production per river basin.** Period: 1996–2005. Morocco's river basins (A) and total green (B), blue (C) and grey (D) water footprint of Morocco's production per river basin (in Mm³/yr).

Bouregreg and Oum Er Rbia. For the other basins an average of these distributions is taken.

The monthly water footprint of storage reservoirs (in m³/yr) is calculated as the open water evaporation (in m/yr) times the surface area of storage reservoirs (in m²). Data on open water evaporation from the reservoirs in the basins Loukkos, Sebou, Bouregreg and Oum Er Rbia is obtained from Ministry EMWE (unpublished data 2013) and for the other basins from a model simulation with the global hydrological model PCR-GLOBWB carried out by Sperna Weiland et al. [15]. The surface area of reservoirs at upper storage level is derived from Ministry EMWE (unpublished data 2013) and FAO [16]. Since storage levels vary throughout the year (and over the years), and reservoir areas accordingly, this gives an overestimation of the evaporation from reservoirs. To counteract this overestimation, but due to lack of data on monthly storage level and reservoir area, for all months a fraction of the evaporation at upper storage level (43%) is taken as estimate of the water footprint of storage reservoirs. This fraction represents the average reservoir area as fraction of its area at upper storage level, calculated as the average over the reservoirs in the basins Loukkos, Sebou, Bouregreg and Oum Er Rbia for which data on surface area at different reservoir levels is available from Ministry EMWE (unpublished data 2013).

The water footprint of the irrigation supply network refers to the evaporative loss in the network and is estimated based on a factor $K$, which is defined as the ratio of the blue water footprint of the irrigation supply network to the blue surface water footprint of crop production at field level (i.e. crop evapotranspiration of irrigation water stemming from surface water). The blue water footprint of crop production at field level is taken from Mekonnen and Hoekstra [14] and the split to surface water is made according

to the fraction of irrigation water withdrawn from surface water (as opposed to groundwater) per river basin based on data from the associated river basin plans. $K$ is calculated as:

$$K = \left[ \frac{1}{e_a \times e_c} - \frac{1}{e_a} \right] \times f_E$$

in which $e_a$ represents the field application efficiency, $e_c$ the irrigation canal efficiency and $f_E$ the fraction of losses in the irrigation canal network that evaporates (as opposed to percolates). The irrigation efficiencies $e_a$ and $e_c$ are estimated based on data from a local river basin agency and FAO [17]. The value of $f_E$ is assumed at fifty percent. The resultant $K$ for Morocco's irrigated agriculture as a whole is 15%, i.e. the evaporative loss from the irrigation water supply network represents a volume equal to 15% of the blue surface water footprint of crop production at field level on average.

## Water Footprint and Economic Water and Land Productivity of Crops

The water footprint of crops per unit of production (in m³/ton) is calculated by dividing the water footprint per hectare (in m³/ha/yr) by the yield (in ton/ha/yr), for which data are obtained from Mekonnen and Hoekstra [14]. Economic water productivity (in US$/m³) represents the economic value of farm output per unit of water consumed and is calculated as the average producer price for the period 1996–2005 (in US$/ton) obtained from FAO [18] divided by the green plus blue water footprint (in m³/ton). Similarly, economic land productivity (in US$/ha) represents the economic value of farm output per hectare of harvested land and is calculated as the same producer price multiplied by crop yield (in ton/ha), which is also obtained from Mekonnen and Hoekstra [14].

## Virtual Water Flows and Associated Economic Value

Green, blue and grey virtual water flows related to Morocco's import and export of agricultural and industrial commodities for the period 1996–2005 are obtained from Mekonnen and Hoekstra [13], who estimated these flows at a global scale based on trade matrices and water footprints of traded products at the locations of origin. The virtual water export that originates from domestic water resources (another part is re-export) is estimated based on the relative share of the virtual water import to the total water budget:

$$V_{e,dom.res.} = \frac{WF_{national}}{V_i + WF_{national}} \times V_e$$

in which $WF_{national}$ is the water footprint within the nation, $V_i$ the virtual water import and $V_e$ the virtual water export.

The average earning per unit of water exported (in US\$/m$^3$) is calculated by dividing the value of export (in US\$/yr) by virtual water export (in m$^3$/yr). Similarly, the cost per unit of virtual water import is calculated by dividing the import value (in US\$/yr) by virtual water import (in m$^3$/yr). The average economic value of import and export for the period 1996–2005 are derived from the Statistics for International Trade Analysis (SITA) database from the International Trade Centre [19].

## Water Footprint versus Water Availability and Waste Assimilation Capacity

To assess the environmental sustainability of the water footprint within Morocco, the total blue (surface- plus groundwater) water footprint of production is placed in the context of monthly natural runoff and the groundwater footprint in the context of annual groundwater availability. The water needed to assimilate the nitrogen fertilizers that reach the water systems due to leaching is compared with the waste assimilation capacity of aquifers.

The groundwater footprint is calculated by splitting the blue water footprint of crop production, industrial production and domestic water supply according to the fraction withdrawn from groundwater per river basin based on data from the associated river basin plans. Assuming that none of the water abstracted from groundwater for industrial production and domestic water supply returns (clean) to the groundwater in the same period of time, the groundwater footprints of these purposes are increased to equal water withdrawal (as opposed to consumption) by dividing them by the consumptive fractions assumed by Mekonnen and Hoekstra [13]: 5% for industries and 10% for households.

Long-term average monthly natural runoff (1980–2011) for the river basins of Loukkos, Sebou, Bouregreg and Oum Er Rbia is derived from Ministry EMWE (unpublished data 2013). Natural runoff is estimated as the inflow of reservoirs. It is considered undepleted runoff, since large-scale blue water withdrawals come from the reservoirs. For the other basins, long-term average annual natural runoff is derived from the river basin plans for the respective river basins and subsequently distributed over the months according to intra-annual rainfall patterns [20,21] or monthly natural discharge [22]. Due to lack of data, for the Souss Massa basin the same monthly variation is applied as for the adjacent Tensift basin. Groundwater availability is assessed on river basin scale and defined as the recharge by percolation of

rainwater and from rivers, minus the direct evaporation from aquifers. These data are obtained from the river basin plans and from Laouina [23] for the basin of Souss Massa.

Blue water scarcity is defined as the ratio of the total blue water footprint in a catchment over the blue water availability in that catchment [6]. In this study, this ratio is calculated as the total blue water footprint to monthly natural runoff and as the groundwater footprint to annual groundwater availability. Following Hoekstra et al. [24], blue water scarcity values have been classified into four levels of water scarcity. The classification in this study corresponds with their classification, with the note that the current study does not account for environmental flow requirements in the definition of blue water availability, since they are generally not considered in Morocco's river basin plans and local studies on the level of these requirements are lacking. This is compensated for by using stricter threshold values for the different scarcity levels, so that the resultant scheme is equivalent to that of Hoekstra et al. [24]:

- low blue water scarcity (<0.20): the blue water footprint is lower than 20% of natural runoff; river runoff is unmodified or slightly modified.
- moderate blue water scarcity (0.20–0.30): the blue water footprint is between 20 and 30% of natural runoff; runoff is moderately modified.
- significant blue water scarcity (0.30–0.40): the blue water footprint is between 30 and 40% of natural runoff; runoff is significantly modified.
- severe water scarcity (>0.40): the monthly blue water footprint exceeds 40% of natural runoff, so runoff is seriously modified.

The water pollution level is defined as the total grey water footprint in a catchment divided by the waste assimilation capacity [6]. In other words, it shows the fraction of actual runoff that is required to dilute pollutants in order to meet ambient water quality standards. A water pollution level greater than 1 means that ambient water quality standards are violated. The nitrate-related grey water footprint of crop production as computed in this study is assumed to mostly contribute to groundwater pollution and is therefore compared with the waste assimilation capacity of groundwater. As a measure of the latter, we use the actual groundwater availability, calculated as (natural) groundwater availability minus the groundwater footprint.

## Relocation of Crop Production and Reducing Water Footprints of Crops to Benchmark Levels

The potential water savings by changing the pattern of crop production across river basins (which is possible due to spatial differences in crop water use) are quantified by means of an optimization model. The total green plus blue water footprint of twelve main crops in the country (in Mm$^3$/yr) is minimized by changing the spatial pattern of production (in ton/yr) over the river basins under constraints for production demand (in ton/yr) and land availability (in ha/yr). The analysed crops are: almonds, barley, dates, grapes, maize, olives, oranges, sugar beets, sugar cane, mandarins, tomatoes and wheat. Results are compared with a base case, which corresponds with the average green plus blue water footprint of the analysed crops over the period 1996–2005. Land availability is restricted per river basin and taken equal to the average harvested area in the period 1996–2005 obtained from Mekonnen and Hoekstra [14]. Two cases are distinguished: 1) all crops can be relocated; 2) only annual crops (barley, maize, sugar beets, tomatoes and wheat) can be relocated, perennials cannot.

For both cases, the restriction is imposed that the total national production per crop (in ton/yr) should be equal to (or greater than) the total national production of the crop in the base case, which is defined as the average production in the period 1996–2005 obtained from Mekonnen and Hoekstra [14].

Additionally, an assessment is made of the potential water savings by reducing the water footprints of the twelve main crops down to certain benchmark levels. For each basin and crop a benchmark is set in the form of the lowest water consumption (green plus blue) of that crop which is achieved in a comparable river basin in Morocco. In this case, basins are considered comparable when the reference evapotranspiration ($ET_0$ in mm/yr) is in the same order of magnitude (see Table 2). Reference evapotranspiration expresses the evaporating power of the atmosphere at a specific location (and time of the year) and does not consider crop characteristics and soil factors [6]. Differences in soil and development conditions are thus not accounted for.

## Results

### Water Footprint of Morocco's Production

The total water footprint of Morocco's production in the period 1996–2005 was 38.8 $Gm^3$/yr (77% green, 18% blue, 5% grey), see Table 3. Crop production is the largest contributor to this water footprint, accounting for 78% of all green water consumed, 83% of all blue water consumed (evaporative losses in irrigation water supply network included) and 66% of the total volume of polluted water. Evaporative losses from storage reservoirs are estimated at 884 $Mm^3$/yr, which is 13% of the total blue water footprint within Morocco. For most reservoirs, these losses are ultimately linked to irrigated agriculture and in some cases potable water supply.

Largest water footprints (green, blue and grey) are found in the basins Oum Er Rbia and Sebou, the basins containing the main agricultural areas of Morocco (see Figure 1B–D). Together, these two basins account for 63% of the total water footprint of national production. In general, the green water footprint is largest in the rainy period December–May, while the blue water footprint is largest in the period April–September when irrigation water use increases.

In the basins Bouregreg and Loukkos, evaporation from storage reservoirs accounts for 45% and 55% of the total blue water footprint, respectively. Irrigated agriculture is the largest blue water consumer in the other basins, but evaporation from storage reservoirs is also significant in these basins. Main irrigated crops in the Oum Er Rbia basin are maize, wheat, olives and sugar beets,

which together account for 60% of the total irrigation water consumed in the period 1996–2005. In the basin of Sebou, 56% of the blue water footprint of crop production relates to the irrigation of wheat, olives, sugar beets, sugar cane and sunflower seed.

### Water Footprint and Economic Water and Land Productivity of Main Crops

In the period 1996–2005, most green water was consumed by the production of wheat, barley and olives (Figure 2). The largest blue water footprints relate to the production of wheat, olives and maize. For wheat, the number one blue water consuming crop, the blue water footprint was largest in the period March–May and peaked in April.

Water consumption of crops (green plus blue, in $m^3$/ton) varies significantly per river basin due to differences in climatic conditions. In general, water consumption of crops is above country-average in the basins Oum Er Rbia and Tensift and below country-average in the northern basins Bouregreg, Sebou, Loukkos and Moulouya (Figure 3). In the basins Sud Atlas and Souss Massa the picture is not so clear, with some crops having above and others below country-average water footprints (in $m^3$/ton).

The five crops that consumed the most green plus blue water in the period 1996–2005 are the crops with the lowest economic water productivity, ranging from 0.08 US$/$m^3$ for wheat to only 0.02 US$/$m^3$ for almonds (Figure 2). Production of tomatoes yielded 22 times more value per drop than production of wheat. The same five crops also have the lowest economic land productivity, ranging from 375 US$/ha for olives to 112 US$/ha for almonds (Figure 4). The highest value per hectare cultivated was obtained by production of tomatoes.

### Virtual Water Trade Balance of Morocco

Morocco's virtual water trade balance for the period 1996–2005 is shown in Figure 5. Virtual water import exceeds virtual water export, which makes Morocco a net virtual water importer. Only 31% of the virtual water export originates from Morocco's water resources, the other 69% is related to re-export of imported virtual water. By import of products instead of producing them domestically, Morocco saved 27.8 $Gm^3$/yr (75% green, 21% blue and 4% grey) of domestic water in the period 1996–2005, equivalent to 72% of the water footprint within Morocco.

The value of the total virtual water imported in the period 1996–2005 was 12.4 billion US$/yr. Import of industrial products accounted for 83%, import of crop products for 16% and import

**Table 2.** Comparison of river basins based on reference evapotranspiration ($ET_0$ in mm/yr, period: 1961–1990).

| No. | River basin | $ET_0$ (mm/yr) | Considered comparable with no. |
|-----|-------------|----------------|-------------------------------|
| 1 | Sud Atlas | 1,652 | - |
| 2 | Souss Massa | 1,450 | 3 |
| 3 | Moulouya | 1,409 | 2 |
| 4 | Tensift | 1,389 | 5 |
| 5 | Oum Er Rbia | 1,387 | 4 |
| 6 | Sebou | 1,266 | 7,8 |
| 7 | Bouregreg | 1,239 | 6,8 |
| 8 | Loukkos | 1,212 | 6,7 |

Source: $ET_0$ from FAO [31].

**Table 3.** Water footprint of Morocco's production in the period 1996–2005 (in Mm$^3$/yr).

| Water footprint of | Green | Blue | Grey | Total |
|---|---|---|---|---|
| Crop production[a] | 23,245 | 5,097 | 1,378 | 29,719 |
| Grazing[a] | 6,663 | - | - | 6,663 |
| Animal water supply[a] | - | 151 | - | 151 |
| Industrial production[a] | - | 18 | 69 | 88 |
| Domestic water supply[b] | - | 125 | 640 | 765 |
| Storage reservoirs[b] | - | 884 | - | 884 |
| Irrigation water supply network[b] | - | 549 | - | 549 |
| Total water footprint | 29,908 | 6,824 | 2,087 | 38,819 |

Source: [a] [13], [b] Own elaboration.

of animal products for 1%. The average cost of imported commodities per unit of virtual water imported was 0.98 US$/m$^3$. The value of the total virtual water exported in this period was 7.1 billion US$/yr (industrial products: 51%, crop products: 48%, animal products: 1%). The average earning of exported commodities per unit of virtual water exported was 1.66 US$/m$^3$.

The total volume of Morocco's water virtually exported out of the country (i.e. excluding re-export) in the period 1996–2005 is estimated at 1,333 Mm$^3$/yr. This means that about 4% of the water used in Morocco's agricultural and industrial sector is used for making export products. The remainder is used to produce products that are consumed by the inhabitants of Morocco. Virtual export of blue water from Morocco's resources was 435 Mm$^3$/yr, which is to equivalent 3.4% of long-term average natural runoff (13 Gm$^3$/yr).

Most of the virtual water export from Morocco's resources returns relatively little foreign currency per unit of virtual water exported. Export of crop products had the largest share in the virtual water export from Morocco's water resources (1,305 Mm$^3$/yr), returning 0.87 US$/m$^3$ on average. Specific crop products associated with large virtual water export from Moroccan origin are olives, oranges, wheat, sugar beets and mandarins. Out of these products, only export of mandarins (122 Mm$^3$/yr) returned a

value (1.37 US$/m$^3$) larger than the average for crop products (0.87 US$/m$^3$). On the other hand, virtual water export related to Moroccan tomatoes (24 Mm$^3$/yr) yielded 7.13 US$/m$^3$.

## Water Footprint versus Water Availability and Waste Assimilation Capacity

Blue water scarcity manifests itself in specific months of the year (Figure 6; Table 4). The average monthly water scarcity indicates severe water scarcity, more severe than annual (total) water scarcity values suggest. In all basins, the total blue water footprint exceeds natural runoff during a significant period of the year. In the months June, July and August, severe water scarcity occurs in all river basins. Crops with a large blue water footprint in July are: sugar beets in Oum Er Rbia and Sebou; grapes in the basins of Sud Atlas, Souss Massa and Oum Er Rbia; dates in Oum Er Rbia and Sebou; sunflower seed in the Sebou basin; maize in the basin of Oum Er Rbia. Demand for potable water peaks in the months June, July and August due to tourism and evaporation from storage reservoirs is large in these months due to the strong evaporative power of the atmosphere. Annual runoff in the Oum Er Rbia basin is almost completely consumed (inter-basin water transfers not yet considered), which raises the question whether it

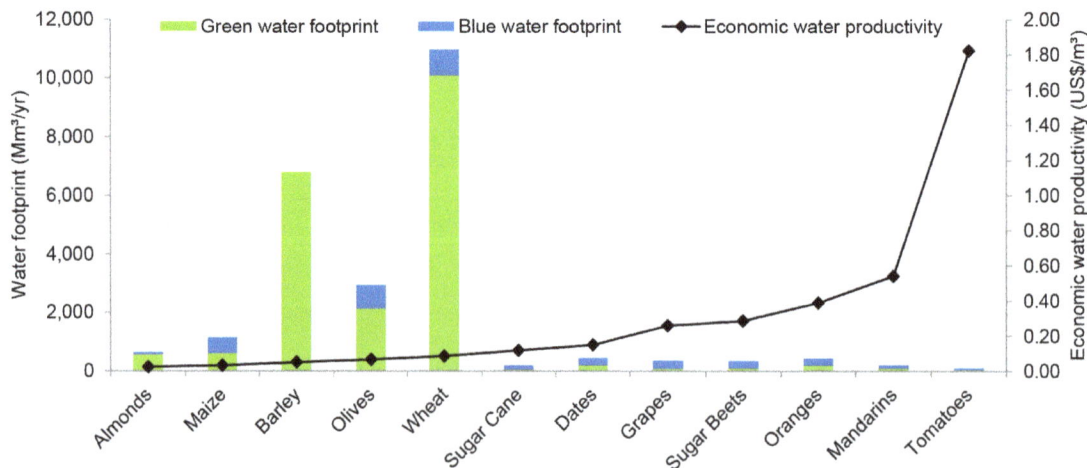

**Figure 2. Economic water productivity and green and blue water footprint of main crops in Morocco.** Period: 1996–2005. Source: Water footprint from Mekonnen and Hoekstra [14], producer prices from FAO [18].

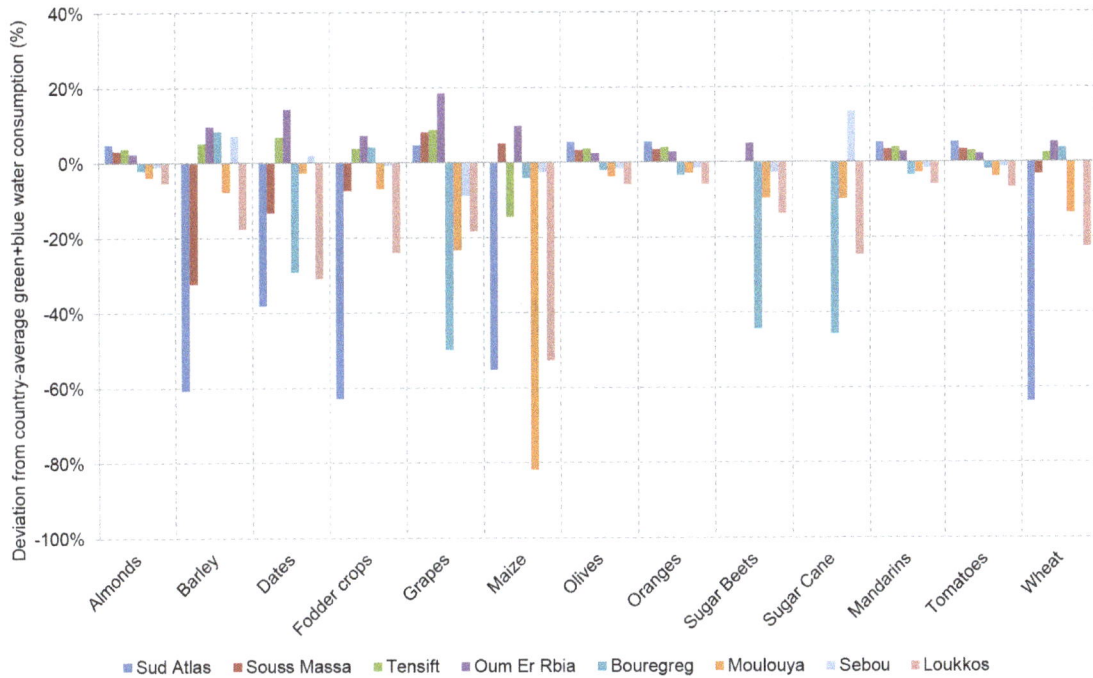

**Figure 3. Variation in green plus blue water consumption (in m³/ton) across river basins.** Period: 1996–2005.

is wise to export water out of this basin to the basins of Bouregreg and Tensift as is common practice.

The total groundwater footprint in Morocco constitutes about half of the country's groundwater availability (Table 5). Groundwater stress is severe in all river basins, except for the basins of Loukkos and Sud Atlas. In the Bouregreg basin, the annual groundwater footprint exceeds annual groundwater availability. As confirmed in the 2012 river basin plan for this basin, most of the aquifers in this basin are indeed overexploited, especially the main aquifers of Berrechid and Chaouia côtière.

In the Bouregreg basin there is no waste assimilation capacity of the groundwater left (because the blue groundwater footprint exceeds groundwater availability), which results in an infinite

water pollution level (Table 6). In the basins of Tensift and Oum Er Rbia, waste assimilation capacity of the groundwater is also exceeded, even by 43 times the natural groundwater availability in the Tensift basin. These findings correspond with figures reported in the river basin plans for these three basins, which indicate severely high nitrate concentrations in the groundwater (at some measurement stations exceeding the maximum permissible limit in drinking water), mainly caused by diffuse nitrate pollution by the irrational use of nitrogen fertilizers, but in the case of the Sahel-Doukkala aquifer in the Oum Er Rbia basin also by the infiltration of untreated domestic wastewater.

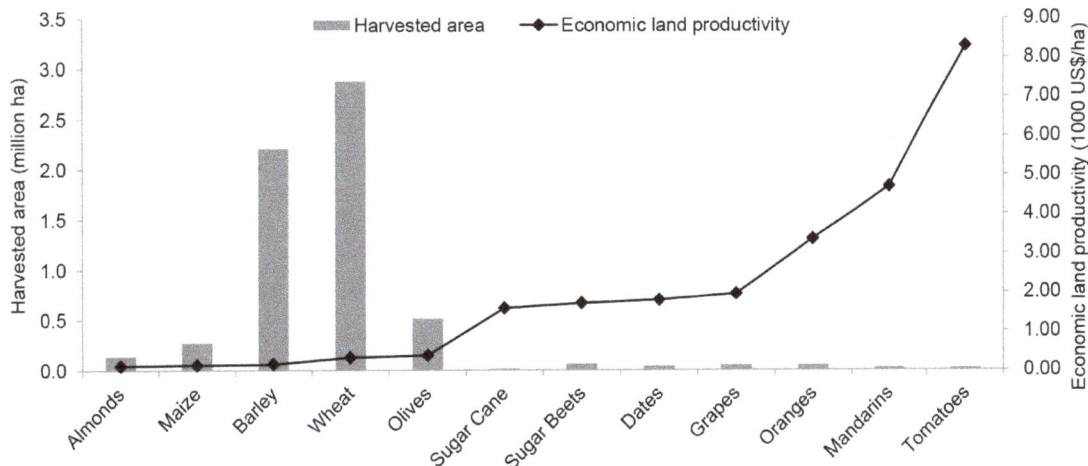

**Figure 4. Economic land productivity and harvested area of main crops in Morocco.** Period: 1996–2005. Source: Harvested area and yield from Mekonnen and Hoekstra [14], producer prices from FAO [18].

**Morocco**

Use of domestic resources in agricultural and industrial sector = 37 Gm³/yr
Natural runoff = 13 Gm³/yr

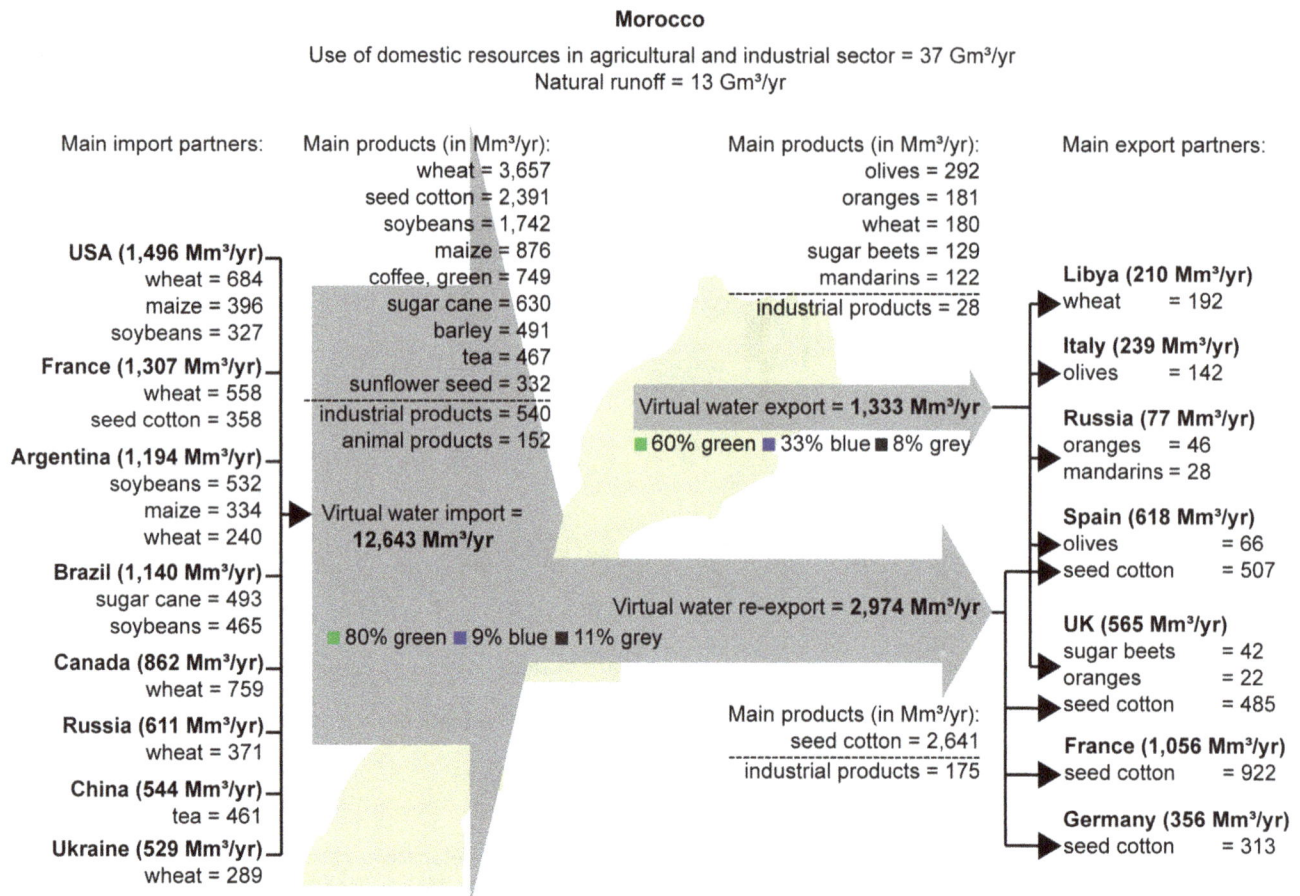

**Figure 5. Morocco's virtual water trade balance related to trade in agricultural and industrial commodities.** Period: 1996–2005. Source: Virtual water import and (total) virtual water export from Mekonnen and Hoekstra [13].

## Reducing the Water Footprint of Crop Production in Morocco

The regional differences in crop water use (Figure 3) provide an opportunity for reduction of the water footprint of crop production in Morocco. Potential water savings (green plus blue) are in the order of 1.9 and 1.2 billion m³ per year when all crops (case A) and when only annual crops (case B) are relocated over the river basins, respectively (Table 7). Blue water savings are 1,276 Mm³/yr in case A and 697 Mm³/yr in case B. These are significant savings when put in the context of Morocco's national water strategy, which includes actions plans to mobilize 1.7 billion m³/yr by 2030 through the construction of 60 large and 1000 small local dams and an additional 0.8 billion m³/yr with the North-South inter-basin water transfer [1].

Largest potential water savings can be obtained by partial relocation of the production of maize and wheat (Table 7), particularly by moving maize production from the Oum Er Rbia basin to the Moulouya basin and wheat production from the Bouregreg basin to the basin of Sebou. Partial relocation of crop production in case A results in decreased water footprints (green plus blue) in all basins, except for the basin of Bouregreg where the water footprint increases (Table 8). In case B, the water footprints in the basins Bouregreg, Sebou and Loukkos increase, particularly due to increased wheat production in these basins, while the water footprints in the other basins decrease. Precipitation in the basins

of Sebou and Loukkos is generally larger than in other parts of Morocco [1].

Reducing the water footprints of crops to benchmark levels leads to a potential green plus blue water saving of 2,768 Mm³/yr, a reduction of 11% (Table 9). Fifty-two per cent of this saving is related to reduced water footprints (i.e. improved water productivities) in the Sebou basin alone. Largest potential water savings are associated with reducing the water footprints of cereals, especially wheat. Blue water savings are estimated at 422 Mm³/yr and are largest in the basins of Sebou and Oum Er Rbia.

## Added Value of Water Footprint Assessment for Morocco's Water Policy

Several insights and response options emerged from the Water Footprint Assessment, which are currently not considered in the national water strategy of Morocco and the country's river basin plans. They include:

(i)    New insights in the water balance of Morocco and the country's main river basins:

● The evaporative losses from storage reservoirs account for a significant part of the blue water footprint within Morocco. This sheds fresh light on the national water strategy that

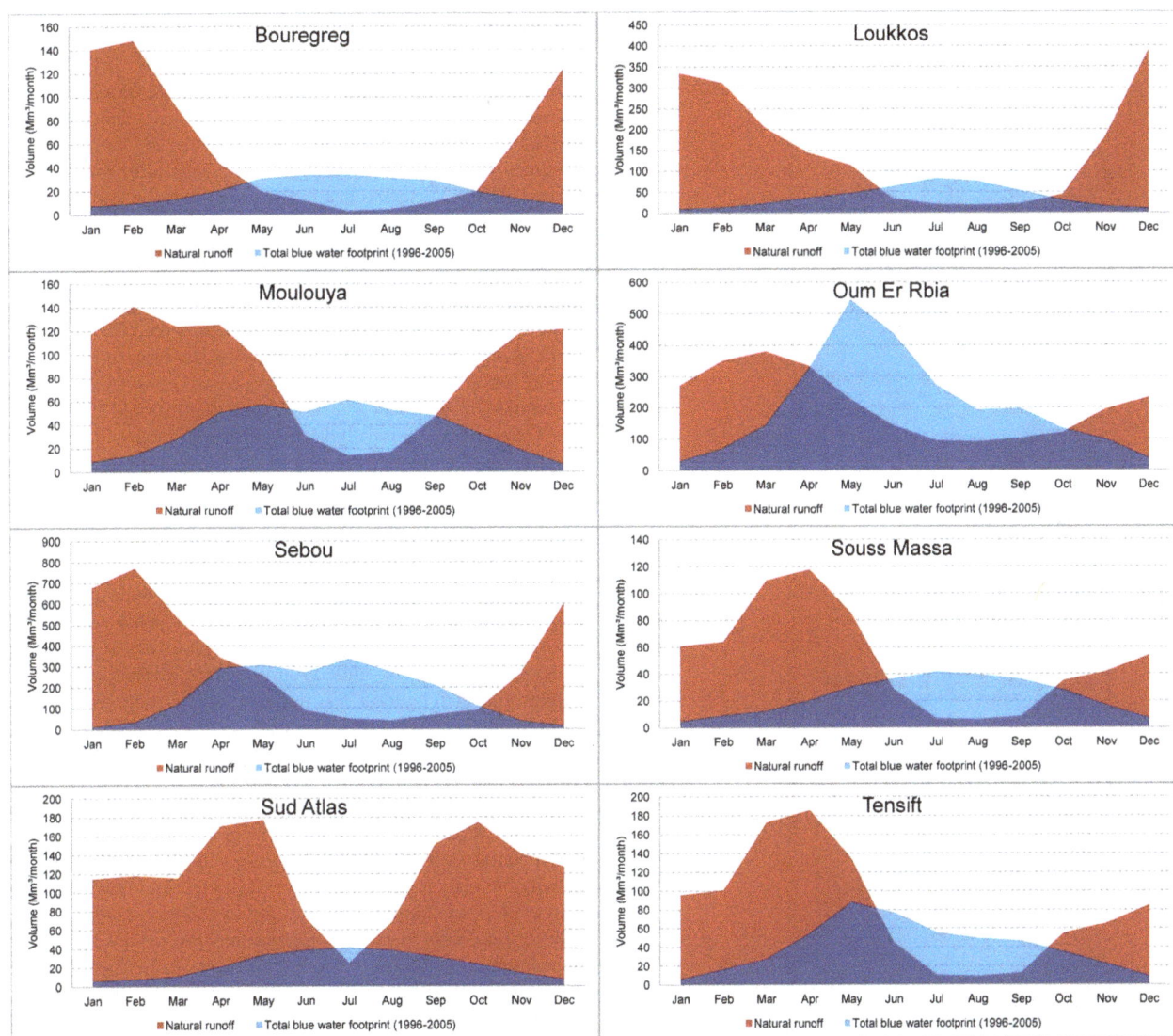

**Figure 6. Total blue water footprint and natural runoff per river basin.** Period of blue water footprint: 1996–2005. Natural runoff is estimated as the long-term average inflow of reservoirs. It is considered undepleted runoff, since large-scale blue water withdrawals come from the reservoirs. The estimates can be considered conservative, because net precipitation in areas downstream of reservoirs is not included.

proposes to build another 60 large and 1000 small dams by 2030.

- Blue water scarcity on a monthly scale is severe and hidden by annual analysis of demand versus supply, which is the common scale of analysis in Morocco's river basin plans.

(ii) New insights in how economically efficient water and land resources are used:

- Analysis of the economic value of crop products per unit of water and land used in the period 1996–2005 indicate that agricultural policy may be better brought in line with water policy by reconsidering which crops to grow.
- It is shown that the export policy in this period was not optimal from a water-economics point of view, which raises the question whether the foreign income generated by export covers the direct and indirect costs of mobilization and (over) exploitation of Morocco's water resources. This might not be

the case considering the costs of the construction and maintenance of the large dams and intra- and inter-basin water transfers in the country and the costs associated with the negative externalities of water (over) consumption, such as the salt-intrusion in Morocco's coastal aquifers.

(iii) New response options to reduce the water footprint of crop production:

- Analysis of the water footprint of the main crops in Morocco and its variation across the river basins offers new ways of looking at reducing water consumption in the agricultural sector. The estimated potential water savings by partial relocation of crops to basins where they consume less water and by reducing water footprints of crops down to benchmark levels are significant compared to demand reducing and supply increasing measures considered in the national water strategy of Morocco.

**Table 4.** Blue water scarcity per river basin.

| River basin | Jan | Feb | Mar | Apr | May | Jun | Jul | Aug | Sep | Oct | Nov | Dec | Tot | Avg |
|---|---|---|---|---|---|---|---|---|---|---|---|---|---|---|
| Bouregreg | 0.05 | 0.06 | 0.14 | 0.47 | 1.57 | 2.89 | 11.3 | 7.30 | 2.78 | 1.01 | 0.19 | 0.06 | 0.37 | 2.32 |
| Loukkos | 0.03 | 0.04 | 0.12 | 0.25 | 0.42 | 1.85 | 4.04 | 4.11 | 2.49 | 0.69 | 0.08 | 0.02 | 0.25 | 1.18 |
| Moulouya | 0.07 | 0.10 | 0.23 | 0.40 | 0.62 | 1.65 | 4.41 | 3.09 | 1.03 | 0.37 | 0.16 | 0.05 | 0.41 | 1.02 |
| Oum Er Rbia | 0.11 | 0.20 | 0.38 | 0.98 | 2.42 | 3.08 | 2.91 | 2.14 | 1.93 | 1.10 | 0.51 | 0.16 | 0.98 | 1.33 |
| Sebou | 0.02 | 0.04 | 0.22 | 0.86 | 1.19 | 3.01 | 6.66 | 6.72 | 3.05 | 1.21 | 0.14 | 0.02 | 0.53 | 1.93 |
| Souss Massa | 0.07 | 0.14 | 0.11 | 0.17 | 0.36 | 1.28 | 6.35 | 6.82 | 4.45 | 0.81 | 0.40 | 0.12 | 0.46 | 1.76 |
| Sud Atlas | 0.05 | 0.07 | 0.09 | 0.12 | 0.19 | 0.54 | 1.67 | 0.56 | 0.21 | 0.14 | 0.10 | 0.06 | 0.19 | 0.32 |
| Tensift | 0.06 | 0.16 | 0.16 | 0.29 | 0.66 | 1.72 | 5.39 | 5.40 | 3.66 | 0.64 | 0.34 | 0.11 | 0.50 | 1.55 |
| Total | 0.05 | 0.09 | 0.22 | 0.56 | 1.03 | 2.23 | 4.15 | 2.98 | 1.55 | 0.66 | 0.22 | 0.06 | 0.52 | 1.15 |

Blue water scarcity is defined as the ratio of the total blue water footprint in a catchment over the natural runoff in that catchment. Classification: low blue water scarcity (<0.20); moderate blue water scarcity (0.20–0.30); significant blue water scarcity (0.30–0.40); severe water scarcity (>0.40).

## Discussion

Morocco's water footprint is mostly green (77%). This underlines the importance of green water resources, also (or especially) in semi-arid countries with a high dependency on blue water, and is in line with other studies showing the dominance of the green over the blue water flow in Africa (and most of the world) [25,26]. The relevance of the green water footprint should not be underestimated. Although rain is free and evaporation happens anyway, green water that is used for one purpose cannot be used for another purpose [27].

Storage reservoir evaporation accounts for a significant share (13%) in the blue water footprint in Morocco. The need for seasonal storage of water is evident given the large mismatch in natural runoff and water demand (Figure 6). However, the large evaporation from reservoirs shows that these should be seen as water consumers, besides their role in water supply. This water footprint can ultimately be linked to the end-purpose of the reservoir, which for most cases in Morocco is primarily serving irrigated agriculture. Therefore, to reduce the need for seasonal storage and hence the water footprint of storage reservoirs, it would be worthwhile to take the timing of crop water demands with respect to natural water availability into account in deciding which crops or crop varieties to grow. Furthermore, local alternatives to the large surface water reservoirs are groundwater dams, which enhance underground water storage in alluvial aquifers and thereby loose less water by evaporation [28].

Our analysis shows that from a strictly water-economics point of view it would be worthwhile to reconsider which crops to grow in Morocco (due to the low value in US$/m$^3$ and US$/ha for some crops compared to others). In practice, the choice of which crops to produce is part of the national strategy regarding food security and of course closely linked to the demand for crops (national and global). Nevertheless, we consider it useful and important to analyse economic water and land productivities (as done in this study) in addition to these considerations. Especially for water-short countries as Morocco it is relevant to evaluate the economic efficiency of water allocation. This also relates to the question whether the foreign income generated by export products, which have a footprint on national resources, outweighs the direct and indirect costs associated with the resource use.

### Uncertainties and Limitations

The water footprint of crop production is largely influenced by the input data used and assumptions made by Mekonnen and Hoekstra [14] and can easily contain an uncertainty of ±20% [14,29,30]. The calculated economic water and land productivities of crops are, apart from the water footprints and yields, dependent on the producer prices. Variations in these prices largely influence the economic water and land productivity of crops. The water footprints of industrial production and domestic water supply are very sensitive to the consumptive fractions applied.

Although figures on water availability are based on data from the river basin plans and the Ministry EMWE (unpublished data 2013), the way they are estimated exactly is often unclear and so is the uncertainty in them. Since natural runoff is estimated as the inflow of reservoirs (thus excluding small-scale local abstractions upstream) and net precipitation in areas downstream of reservoirs is not included, the estimates of natural runoff can be considered conservative.

In general, the river basin plans indicate larger pressure on groundwater resources than suggested in this study. This might be caused by the fact that the river basin plans include more recent withdrawals and because the unit of analysis in this study (river

**Table 5.** Blue water scarcity related to groundwater.

| River basin | Groundwater footprint (Mm³/yr) | Groundwater availability (1996–2005) (Mm³/yr) | Blue water scarcity (−) | Level of water scarcity |
|---|---|---|---|---|
| Bouregreg | 106 | 66 | 1.60 | Severe |
| Tensift | 259 | 262 | 0.99 | Severe |
| Oum Er Rbia | 510 | 667 | 0.77 | Severe |
| Souss Massa | 219 | 349 | 0.63 | Severe |
| Sebou | 689 | 1,502 | 0.46 | Severe |
| Moulouya | 144 | 351 | 0.41 | Severe |
| Loukkos | 93 | 377 | 0.25 | Moderate |
| Sud Atlas | 137 | 697 | 0.20 | Moderate |
| Total | 2,159 | 4,347 | | |

Basins are sorted top-down from highest to lowest scarcity.

basin agency action zone) is larger than the unit used in the river basin plans (individual aquifers), whereby in this study overexploitation of one aquifer might be masked by low exploitation of another. Also local groundwater pollution according to the river basin plans is sometimes worse than the water pollution level estimated here. This could be explained by the fact that the water quality measurements recorded in the basin plans are partly more recent and are measured at specific points, whereas this study considered homogeneous distribution of nitrates in the groundwater.

Given the uncertainties and limitations of the study, the presented water footprint estimates and water scarcity values should be interpreted with care. Nevertheless, the order of magnitude of the estimates in this study gives a good indication to which activities and crops Morocco's water resources are allocated, in which months and basins the water footprints are relatively large or small and where and when this leads to highest water scarcity.

Uncertainties in the estimated potential savings by relocation of crop production and reducing the water footprints of crops to benchmark levels are closely linked to the uncertainties in the estimates of the water footprint of crop production and the results

should be interpreted carefully. However, the order of magnitude of the estimated savings gives a rough indication of the potential of these measures. When considering relocation of crop production it is necessary to assess how the green and blue water footprints of crops manifest themselves on a monthly scale. This study looked at annual water savings, but the associated relocation of crops might well aggravate monthly water scarcity in some river basins. Furthermore, the feasibility and desirability of relocation of crop production are of course largely determined by social and economic factors which are not considered in this study.

## Conclusion

The study finds that: (i) evaporation from storage reservoirs is the second largest form of blue water consumption in Morocco, after irrigated crop production; (ii) Morocco's water and land resources are mainly used to produce relatively low-value (in US$/m³ and US$/ha) crops such as cereals, olives and almonds; (iii) most of the virtual water export from Morocco relates to the export of products with a relatively low economic water productivity (in US$/m³); (iv) blue water scarcity on a monthly scale is severe in all river basins and pressure on groundwater

**Table 6.** Water pollution level related to nitrate-nitrogen in groundwater.

| River basin | Grey water footprint of crop production (1996–2005) (Mm³/yr) | Actual groundwater availability/Waste assimilation capacity (Mm³/yr) | Water pollution level (−) | Waste assimilation capacity exceeded? |
|---|---|---|---|---|
| Bouregreg | 148 | 0 | ∞ | Yes |
| Tensift | 129 | 3 | 43.2 | Yes |
| Oum Er Rbia | 435 | 157 | 2.78 | Yes |
| Sebou | 428 | 813 | 0.53 | No |
| Moulouya | 99 | 207 | 0.48 | No |
| Souss Massa | 51 | 130 | 0.39 | No |
| Loukkos | 63 | 284 | 0.22 | No |
| Sud Atlas | 25 | 560 | 0.04 | No |
| Total | 1,378 | 2,188 | 0.63 | No |

Basins are sorted top-down from highest to lowest pollution level.

**Table 7.** Potential water savings by partial relocation of crop production per crop.

| | Base case green plus blue water footprint (Mm³/yr) | Partial relocation considered for all crops* | | Partial relocation considered for annual crops only** | |
|---|---|---|---|---|---|
| | | Saving (green+blue) (Mm³/yr) | Relative saving (%) | Saving (green+blue) (Mm³/yr) | Relative saving (%) |
| Almonds | 641 | 14 | 2% | 0 | 0% |
| Barley | 6,787 | −116 | −2% | −202 | −3% |
| Dates | 449 | 131 | 29% | 0 | 0% |
| Grapes | 367 | 183 | 50% | 0 | 0% |
| Maize | 1,148 | 939 | 82% | 939 | 82% |
| Olives | 2,951 | 58 | 2% | 0 | 0% |
| Oranges | 440 | 15 | 3% | 0 | 0% |
| Sugar Beets | 353 | 157 | 44% | 157 | 44% |
| Sugar Cane | 200 | 91 | 46% | 0 | 0% |
| Mandarins | 209 | 7 | 3% | 0 | 0% |
| Tomatoes | 99 | 2 | 2% | 2 | 2% |
| Wheat | 10,981 | 413 | 4% | 278 | 3% |
| Total | 24,625 | 1,896 | 8% | 1,174 | 5% |

*All analysed crops are: almonds, barley, dates, grapes, maize, olives, oranges, sugar beets, sugar cane, mandarins, tomatoes and wheat.
**Annual crops are: barley, maize, sugar beets, tomatoes and wheat.

resources by abstractions and nitrate pollution is considerable in most basins; (v) the estimated potential water savings by partial relocation of crops to basins where they consume less water and by reducing water footprints of crops down to benchmark levels are significant compared to demand reducing and supply increasing measures considered in Morocco's national water strategy.

On the basis of these new insights and response options it is concluded that Water Footprint Assessment has an added value for national water policy in Morocco. Water Footprint Assessment forces to look at end-users and -purposes of freshwater, which is key in determining efficient and equitable water allocation within the boundaries of what is environmentally sustainable, both on the river basin and on the national level. This is especially relevant for water-scarce countries such as Morocco. Furthermore, considering the green and grey components of a water footprint provides new perspectives on blue water scarcity, because pressure on blue water resources might be reduced by more efficient use of green water and by less pollution.

**Table 8.** Potential water savings by partial relocation of crop production per river basin.

| | Base case green plus blue water footprint (Mm³/yr) | Partial relocation considered for all crops | | Partial relocation considered for annual crops only** | |
|---|---|---|---|---|---|
| | | Saving (green+blue) (Mm³/yr) | Relative saving (%) | Saving (green+blue) (Mm³/yr) | Relative saving (%) |
| Sud Atlas | 306 | 189 | 62% | 12 | 4% |
| Souss Massa | 903 | 175 | 19% | 14 | 2% |
| Tensift | 2,525 | 388 | 15% | 124 | 5% |
| Oum Er Rbia | 8,498 | 1,229 | 14% | 821 | 10% |
| Bouregreg | 2,813 | −994 | −35% | −95 | −3% |
| Moulouya | 1,737 | 605 | 35% | 412 | 24% |
| Sebou | 6,905 | 154 | 2% | −95 | −1% |
| Loukkos | 939 | 151 | 16% | −19 | −2% |
| Total | 24,625 | 1,896 | 8% | 1,174 | 5% |

*All analysed crops are: almonds, barley, dates, grapes, maize, olives, oranges, sugar beets, sugar cane, mandarins, tomatoes and wheat.
**Annual crops are: barley, maize, sugar beets, tomatoes and wheat.

**Table 9.** Potential water savings by benchmarking water productivities of main crops* (in Mm$^3$/yr).

| | Sud Atlas | Souss Massa | Tensift | Oum Er Rbia | Bouregreg | Moulouya | Sebou | Loukkos | Total |
|---|---|---|---|---|---|---|---|---|---|
| Almonds | 0 | 2 | 1 | 0 | 3 | 0 | 8 | 0 | 14 |
| Barley | 0 | 0 | 0 | 100 | 158 | 222 | 238 | 0 | 717 |
| Dates | 0 | 0 | 0 | 10 | 0 | 4 | 48 | 0 | 63 |
| Grapes | 0 | 20 | 0 | 5 | 0 | 0 | 18 | 4 | 48 |
| Maize | 0 | 13 | 0 | 175 | 32 | 0 | 33 | 0 | 254 |
| Olives | 0 | 9 | 4 | 0 | 10 | 0 | 35 | 0 | 59 |
| Oranges | 0 | 1 | 1 | 0 | 1 | 0 | 6 | 0 | 9 |
| Sugar Beets | 0 | 0 | 0 | 0 | 0 | 0 | 70 | 4 | 73 |
| Sugar Cane | 0 | 0 | 0 | 0 | 0 | 0 | 79 | 10 | 89 |
| Mandarins | 0 | 1 | 0 | 0 | 0 | 0 | 3 | 0 | 4 |
| Tomatoes | 0 | 0 | 0 | 0 | 1 | 0 | 1 | 0 | 3 |
| Wheat | 0 | 14 | 0 | 102 | 417 | 0 | 904 | 0 | 1,436 |
| Total (gn+bl) | 0 | 60 | 6 | 392 | 623 | 226 | 1,444 | 18 | 2,768 |
| Total (blue)** | 0 | 23 | 2 | 113 | 11 | 2 | 258 | 12 | 422 |
| Total (blue) (% of natural runoff) | 0% | 4% | 0% | 4% | 2% | 0% | 7% | 1% | 3% |

*Analysed crops are: almonds, barley, dates, grapes, maize, olives, oranges, sugar beets, sugar cane, mandarins, tomatoes and wheat.
**Assuming that the green/blue water ratio remains the same for all basins and crops.

## Acknowledgments

We like to thank Mesfin Mekonnen (University of Twente, Enschede, Netherlands), Karen Meijer and Wil van der Krogt (Deltares, Delft, Netherlands), Abdelkader Larabi (Mohammadia School of Engineers, Rabat, Morocco) and Siham Laraichi (Ministry of Energy, Mining, Water and Environment, Rabat, Morocco) for their feedback during the various stages of research.

## Author Contributions

Conceived and designed the experiments: JFS AYH. Performed the experiments: JFS. Analyzed the data: JFS AYH. Wrote the paper: JFS AYH.

## References

1. Ministry EMWE (2011) Stratégie Nationale de l'Eau. Available: www.minenv. gov.ma/PDFs/EAU/strategie_eau.pdf, Department of Water, Ministry of Energy, Mining, Water and Environment, Rabat, Morocco. Accessed 2013 January 21.

2. Ministry EMWE (2012) Les eaux souterraines, http://www.water.gov.ma/index.cfm?gen = true&id = 12&ID_PAGE = 42, Department of Water, Ministry of Energy, Mining, Water and Environment, Rabat, Morocco.Accessed 2012 December 8.

3. Official State Gazette (1995) Royal Decree no. 1-95-154 promulgating Law no. 10-95 on water. Available: http://ocid.nacse.org/rewab/docs/Royal_Decree_No_1-95-154_Promulgating_Law_on_Water_EN.pdf. Accessed 2013 January 3.

4. Hoekstra AY, ed. (2003) Virtual water trade: Proceedings of the International Expert Meeting on Virtual Water Trade, Delft, The Netherlands, 12–13 December 2002. Value of Water Research Report Series No. 12, UNESCO-IHE, Delft, The Netherlands. Available: www.waterfootprint.org/Reports/Report12.pdf. Accessed 2013 July 5.

5. Hoekstra AY, Chapagain AK (2008) Globalization of water: Sharing the planet's freshwater resources. Oxford: Blackwell Publishing.

6. Hoekstra AY, Chapagain AK, Aldaya MM, Mekonnen MM (2011) The water footprint assessment manual: Setting the global standard. London: Earthscan.

7. Aldaya MM, Garrido A, Llamas MR, Varelo-Ortega C, Novo P, et al. (2010) Water footprint and virtual water trade in Spain. In: Garrido A, Llamas MR, eds. Water policy in Spain. Leiden: CRC Press. 49–59.

8. Chahed J, Besbes M, Hamdane A (2011) Alleviating water scarcity by optimizing "Green Virtual-Water": the case of Tunisia. In: Hoekstra AY, Aldaya MM, Avril B, eds. Proceedings of the ESF Strategic Workshop on accounting for water scarcity and pollution in the rules of international trade, Amsterdam, 25–26 November 2010. Value of Water Research Report Series No. 54, UNESCO-IHE. Available: http://www.waterfootprint.org/Reports/Report54-Proceedings-ESF-Workshop-Water-Trade.pdf. Accessed 2013 February 4.

9. Hoekstra AY, Mekonnen MM (2012) The water footprint of humanity. Proc Natl Acad Sci U S A 109(9): 3232–3237.

10. Aldaya MM, Martinez-Santos P, Llamas MR (2010) Incorporating the water footprint and virtual water into policy: Reflections from the Mancha Occidental Region, Spain. Water Resources Management 24(5): 941–958.

11. Hoekstra AY, Chapagain AK (2007) Water footprints of nations: water use by people as a function of their consumption pattern. Water Resources Management 21(1): 35–48.

12. Hoekstra AY, Chapagain AK (2007) The water footprints of Morocco and the Netherlands: Global water use as a result of domestic consumption of agricultural commodities. Ecol Econ 64(1): 143–151.

13. Mekonnen MM, Hoekstra AY (2011) National water footprint accounts: the green, blue and grey water footprint of production and consumption. Value of Water Research Report Series No.50, UNESCO-IHE, Delft, The Netherlands. Available: http://www.waterfootprint.org/Reports/Report50-NationalWaterFootprints-Vol1.pdf. Accessed 2012 November 23.

14. Mekonnen MM, Hoekstra AY (2010) The green, blue and grey water footprint of crops and derived crop products. Value of Water Research Report Series No.47, UNESCO-IHE, Delft, The Netherlands. Available: http://www.waterfootprint.org/Reports/Report47-WaterFootprintCrops-Vol1.pdf. Accessed 2012 November 23.

15. Sperna Weiland FC, van Beek LPH, Kwadijk JCJ, Bierkens MFP (2010) The ability of a GCM-forced hydrological model to reproduce global discharge variability. Hydrol Earth Syst Sci 14: 1595–1621. doi:10.5194/hess-14-1595-2010.

16. FAO (2013) AQUASTAT online database. Geo-referenced database of African dams. Food and Agriculture Organization, Rome, Italy. Available: http://www.fao.org/nr/water/aquastat/dams/region/D_Africa.xlsx. Accessed 2013 March 28.

17. FAO (2013) AQUASTAT online database. Country Fact Sheet: Morocco. Food and Agriculture Organization, Rome, Italy. Available: www.fao.org/nr/aquastat/. Accessed 2013 February 22.

18. FAO (2013) FAOSTAT online database. Food and Agriculture Organization, Rome, Italy. Available: http://faostat.fao.org. Accessed 2013 July 6.

19. ITC (2007) SITA version 1996–2005 in SITC [DVD-ROM], International Trade Centre, Geneva.

20. Riad S (2003) Typologie et analyse hydrologique des eaux superficielles à partir de quelques bassins versants représentatifs du Maroc. Unpublished thesis. Available: http://ori-nuxeo.univ-lille1.fr/nuxeo/site/esupversions/e5d351a6-ce4c-4b64-b891-84d85f3d8f02. Accessed 2013 June 19.

21. Tekken V, Kropp JP (2012) Climate-Driven or Human-Induced: Indicating Severe Water Scarcity in the Moulouya River Basin (Morocco). Water 4: 959–982. doi:10.3390/w4040959.

22. JICA MATEE, ABHT (2007) Etude du plan de gestion intégrée des ressources en eau dans la plaine du Haouz, Royaume du Maroc, Rapport intermédiaire. Available: http://eau-tensift.net/fileadmin/user_files/pdf/etudes/JICA_ETUDE_PLAN_DE_GESTION_INTEGREE_RE_HAOUZ.pdf. Accessed 2013 August 12.

23. Laouina A (2001) Compétition irrigation/eau potable en region de stress hydrique: le cas de la region d'Agadir (Maroc). In: Camarda D, Grassini L, ed. Interdependency between agriculture and urbanization: Conflicts on sustainable use of soil and water. Bari: CIHEAM, 2001, 17–31. Available: ressources.ciheam.org/om/pdf/a44/02001585.pdf. Accessed 2013 June 20.

24. Hoekstra AY, Mekonnen MM, Chapagain AK, Mathews RE, Richter BD (2012) Global monthly water scarcity: Blue water footprints versus blue water availability. PLoS ONE 7(2): e32688.

25. Rockstrom J, Falkenmark M, Karlberg L, Hoff H, Rost S, et al. (2009) Future water availability for global food production: The potential of green water for increasing resilience to global change. Water Resour Res 45: W00A12.

26. Schuol J, Abbaspour KC, Yang H, Srinivasan R, Zehnder AJB (2008) Modeling blue and green water availability in Africa. Water Resour Res 44: W07406.

27. Hoekstra AY (2013) The Water Footprint of Modern Consumer Society. London: Routledge.

28. Al-Taiee TM (2012) Groundwater Dams: a Promise Option for Sustainable Development of Water Resources in Arid and Semi-Arid Regions. In: UNESCO. Proceedings of the Second International Conference on integrated water resources management and challenges of the sustainable development, Agadir, 24–26 March 2010. IHP-VII Series on Groundwater No. 4, UNESCO. 35–41.

29. Hoff H, Falkenmark M, Gerten D, Gordon L, Karlberg L, et al. (2010) Greening the global water system. J Hydrol 383(3–4): 177–186.

30. Mekonnen MM, Hoekstra AY (2010) A global and high-resolution assessment of the green, blue and grey water footprint of wheat. Hydrol Earth Syst Sci 14(7): 1259–1276.

31. FAO (2013) Global map of monthly reference evapotranspiration-10 arc minutes. GeoNetwork: grid database. Food and Agriculture Organization, Rome, Italy. Available: www.fao.org/geonetwork/srv/en/resources.get?id = 7416&fname = ref_evap_fao_10min.zip&access = private. Accessed 2013 March 19.

# 8

# An Improved Experimental Method for Simulating Erosion Processes by Concentrated Channel Flow

**Xiao-Yan Chen[1,2]\*, Yu Zhao[1], Bin Mo[1], Hong-Xing Mi[1]**

**1** College of Resources and Environment/Key Laboratory of Eco-environment in Three Gorges Region (Ministry of Education), Southwest University, Chongqing, China,
**2** State Key Laboratory of Soil Erosion and Dryland Farming on the Loess Plateau, Institute of Soil and Water Conservation, CAS and MWR, Yangling, China

## Abstract

Rill erosion is an important process that occurs on hill slopes, including sloped farmland. Laboratory simulations have been vital to understanding rill erosion. Previous experiments obtained sediment yields using rills of various lengths to get the sedimentation process, which disrupted the continuity of the rill erosion process and was time-consuming. In this study, an improved experimental method was used to measure the rill erosion processes by concentrated channel flow. By using this method, a laboratory platform, 12 m long and 3 m wide, was used to construct rills of 0.1 m wide and 12 m long for experiments under five slope gradients (5, 10, 15, 20, and 25 degrees) and three flow rates (2, 4, and 8 L min$^{-1}$). Sediment laden water was simultaneously sampled along the rill at locations 0.5 m, 1 m, 2 m, 3 m, 4 m, 5 m, 6 m, 7 m, 8 m, 10 m, and 12 m from the water inlet to determine the sediment concentration distribution. The rill erosion process measured by the method used in this study and that by previous experimental methods are approximately the same. The experimental data indicated that sediment concentrations increase with slope gradient and flow rate, which highlights the hydraulic impact on rill erosion. Sediment concentration increased rapidly at the initial section of the rill, and the rate of increase in sediment concentration reduced with the rill length. Overall, both experimental methods are feasible and applicable. However, the method proposed in this study is more efficient and easier to operate. This improved method will be useful in related research.

**Editor:** Ben Bond-Lamberty, DOE Pacific Northwest National Laboratory, United States of America

**Funding:** The funding which supported this work is Foundation of State Key Laboratory of Soil Erosion and Dryland Farming on the Loess Plateau under project No. K318009902-1312 (http://english.iswc.cas.cn/). The funders had no role in study design, data collection and analysis, decision to publish, or preparation of the manuscript.

**Competing Interests:** The authors have declared that no competing interests exist.

\* Email: guangguang14@163.com

## Introduction

Soil erosion is a serious environmental problem that threatens agricultural safety and sustainable development due to land degradation [1–5]. As an important component of hill slope soil erosion, rill erosion is especially dangerous on cultivated slopes and upland areas. Eroding rills are formed by concentrated surface runoff and function as sediment source areas and sediment transport vehicles [6–9]. Considering its importance, the mechanism of rill erosion has long been the focus of study.

As early as 1981, Foster et al. [10] differentiated upland soil erosion into inter-rill erosion and rill erosion. They indicated that rill erosion contributed more significantly to sediment production than inter-rill erosion. Since then, many researchers have studied rill morphology, the hydraulics of rill flow, and the rill erosion process. The results indicate that the parameters, such as sediment concentration, soil detachment rate, sediment transport capacity, soil erodibility and critical shear stress, can be used to characterize the rill erosion process. Sediment concentration and soil detachment rate are the most relevant indicators of erosion. Sediment transport capacity expresses the potential sediment entrainment ability of the concentrated flow in rills. Soil erodibility and critical shear stress provide the criteria for determining the occurrence and quantity of rill erosion [11–17].

In recent decades, various research techniques have been applied to study rill erosion. However, information regarding spatially distributed rill erosion is limited. Most existing rill erosion data are spatially integrated, as measured at the rill outlet [18–21]. Spatially integrated rill erosion data do not adequately describe the dynamics of the rill erosion process.

Traditional measurement methods cannot easily quantify the rill erosion process due to the complexity of quantifying erosion among various rill segments. Consequently, new methods that provide dynamic rill erosion process are needed.

Recently, the rare earth element (REE) was utilized to trace the temporal and spatial distribution data of rill erosion [22–24]. The method successfully quantified the rill erosion process. However, the REE method is not economical or efficient, and need special and expensive facilities to conduct the measurement. Lei et al. [13] suggested an experimental method for studying rill erosion process. In their method, the sediment concentrations produced from rills of different lengths are measured. Next, the measured sediment data are integrated to produce a spatially distributed rill erosion process. However, the approach was determined to be time consuming and disruptive to the continuity of the rill erosion process. Aksoy et al. [16,17] performed experimental analysis in a laboratory flume under simulated rainfall by pre-formed rill. They used the experimental data to relate sediment concentration to slope gradient and rainfall intensity.

Rill erosion involves such processes as infiltration of rill, winding of flow path and failure of side walls due to randomized scouring and deposition. These processes are interactive with the most important rill erosion components such as detachment rate, transport capacity, shear stress and sediment concentration along the rill to be computed or estimated.

Along a width-fluctuated rill, the sediment concentrations are much lower than those along a well-defined rill with constant water flow [25]. Naturally developed rills in laboratory experiments involve periodic change of detachment and deposition responsible for rill width fluctuation, widening where deposition occurs and narrowing when scouring takes place [26]. Randomized side wall failure during rill erosion process does cause underestimation of sediment concentration at locality. Still the underestimated sediment load in the water flow is much higher than that after deposition occurrence. Furthermore, water flow of lower sediment concentration causes higher scouring of rill bed to contribute more sediments to compensate the water flow [27,28]:

$$D_r = K_r(\tau - \tau_c)\left(1 - \frac{qc}{T_c}\right) \quad (1)$$

where, $D_r$ (kg m$^{-2}$ s) is the contribution of rill detachment rate to rill sediment load, $K_r$ (s$^{-1}$) is the erodibility of the soil; $\tau$ (kg m$^{-2}$) is the shear stress of flowing water, $\tau_c$(kg m$^{-2}$) is the critical shear strength of soil; $c$ (kg m$^{-1}$) is the sediment concentration and $q$ (m$^2$ s$^{-1}$) is the flow rate of a unit width, $T_c$ (kg s$^{-1}$ m) is the sediment transport capacity of the water flow [28].

Therefore, it seems rational to use well-defined rill for quantitative study of rill erosion.

The objectives of this study were to: 1) develop an improved experimental method for determining the rill erosion process; 2) determine the rill erosion process with experimental data under various hydraulic conditions; 3) assess the sediment concentration and erosion process with previously reported experimental data.

## Methods and Materials

In regular rill erosion, the rill width changes periodically due to the erosion and deposition. And the sediment concentration could be much lower than that in well-defined rill under constant water flow. In order to overcome this problem of rill width fluctuation, well-defined rill was constructed in this experiment.

The experiments were conducted in the State Key Laboratory of Soil Erosion and Dryland Farming on the Loess Plateau, Institute of Soil and Water Conservation, Chinese Academy of Science and Ministry of Water Resources, Yangling, Shanxi Province, China. A platform that was 12 m long and 3 m wide was used as a base to construct a flume that was 12 m long, 0.6 m wide and 0.5 m deep. The flume was sub-divided into six rills that were 0.1 m wide and 12 m long using upright PVC boards to form well-defined rills (Fig. 1). The PVC board surfaces were glued with the experimental soil particles on both sides to create a roughness equivalent to the soil surface so as to minimize the boundary effect on the hydraulic and erosion processes, in case when water flow becomes contact with the boards.

The experimental soil materials were collected from the Ansai Research Station of Soil and Water Conservation on the Chinese Loess Plateau. The soil contains 15.92% clay (<0.005 mm), 63.90% silt (0.005 to 0.05 mm), and 20.18% sand (>0.05 mm), which is classified as a silt-loamy soil. The soil was air dried before being crushed and passed through an 8 mm square sieve.

The bottom 5 cm of the flume was paved with a clay loamy soil (24.9% sand particles, 43.4% silt particles and 31.8% clay

**Figure 1. The experimental setup.** A picture shows the experimental setup.

particles) to achieve a bulk density of approximately 1.5 g cm$^{-3}$ (i.e., an imitation of the plow pan layer). On top of the plow pan layer, the experimental soil was packed in 5 cm thick layers to a total depth of 20 cm, at the bulk density of approximately 1.2 g cm$^{-3}$. The soil near the PVC boards was packed slightly higher than the middle in aim to converge the water flow to the center of the rill and to minimize the boundary effect as much as possible. The prepared rills were saturated with the rain-simulator and allowed to drain for 24 h to ensure an even and homogeneous initial soil moisture profile in an experimental run, and among different experimental runs.

A water tank was used to supply the water flow at the designed discharge rate. An additional specially designed device was used at the rill flow inlet to accelerate the water flow to a velocity level close to the rill flow. Gauze cloth approximately 0.2 m in length was placed at soil surface of the rill inlet to protect the rill surface from being directly scoured by the water flow.

Experimental runs were conducted for five slope gradients (5°, 10°, 15°, 20° and 25°) and three flow rates (2 L min$^{-1}$, 4 L min$^{-1}$, and 8 L min$^{-1}$) with 3 replicates.

Water flow at the designed rate was introduced into the rill at the top end after the flume was adjusted to the designed slope gradient. In Lei et al.'s [13,29] experiments, sediment-laden water samples were collected at outlets of rill of various lengths before the collected data were integrated to produce the sediment process of the entire 8 m rill. Here, after steady water flow in the rill was established, sediment-laden water samples were simultaneously taken along the rill at distances of 0.5 m, 1.0 m, 2.0 m, 3.0 m, 4.0 m, 5.0 m, 6.0 m, 7.0 m, 8.0 m, 10.0 m, and 12.0 m from the water flow inlet, which is an improvement on the previous method. Thus, the sediment delivery process was determined along the rill. Three samples were taken at 1.0-minute interval for each experimental run. Therefore each experiment lasted less than 5 minute after steady flow was established. The average value of the three samples at each location was used to determine the sediment concentration at various rill lengths. The sediment concentrations that were measured at the various locations formed the sediment delivery process along the rill.

## Results and Discussion

### Sediment concentration along the rill

The measured sediment concentrations along the rills under various flow rates and slope gradients are presented in Fig. 2. The experimental data exhibited a well defined trend between the sediment concentration and rill length, which can be described by:

$$S_c = A(1 - e^{-Bx}) \tag{2}$$

where $S_c$ (kg m$^{-3}$) is the sediment concentration, $x$ (m) is the rill length, and $A$ (kg m$^{-3}$) and $B$ (m$^{-1}$) are regression coefficients, with $A$ as the maximum possible sediment concentration and $B$ as the attenuation coefficient to indicate how fast of the reduction in sediment increase.

The experimental data presented in Fig. 2 were regressed with Eq. (2). The regression parameters are listed in Table 1. The regression results indicated that the sediment concentrations for all the experimental runs increased exponentially with the rill length. Polyakov and Nearing [30] and Lei et al. [31] studied the sediment concentrations in rills and concluded that the relationship between sediment concentration and rill length is well-described by Eq. (1) too. The sediment concentration increased rapidly at the initial section of the rill, where the water contains limited amount of sediments. The increase in sediment concentration reduced rapidly along the rill. These sediment concentration distributions

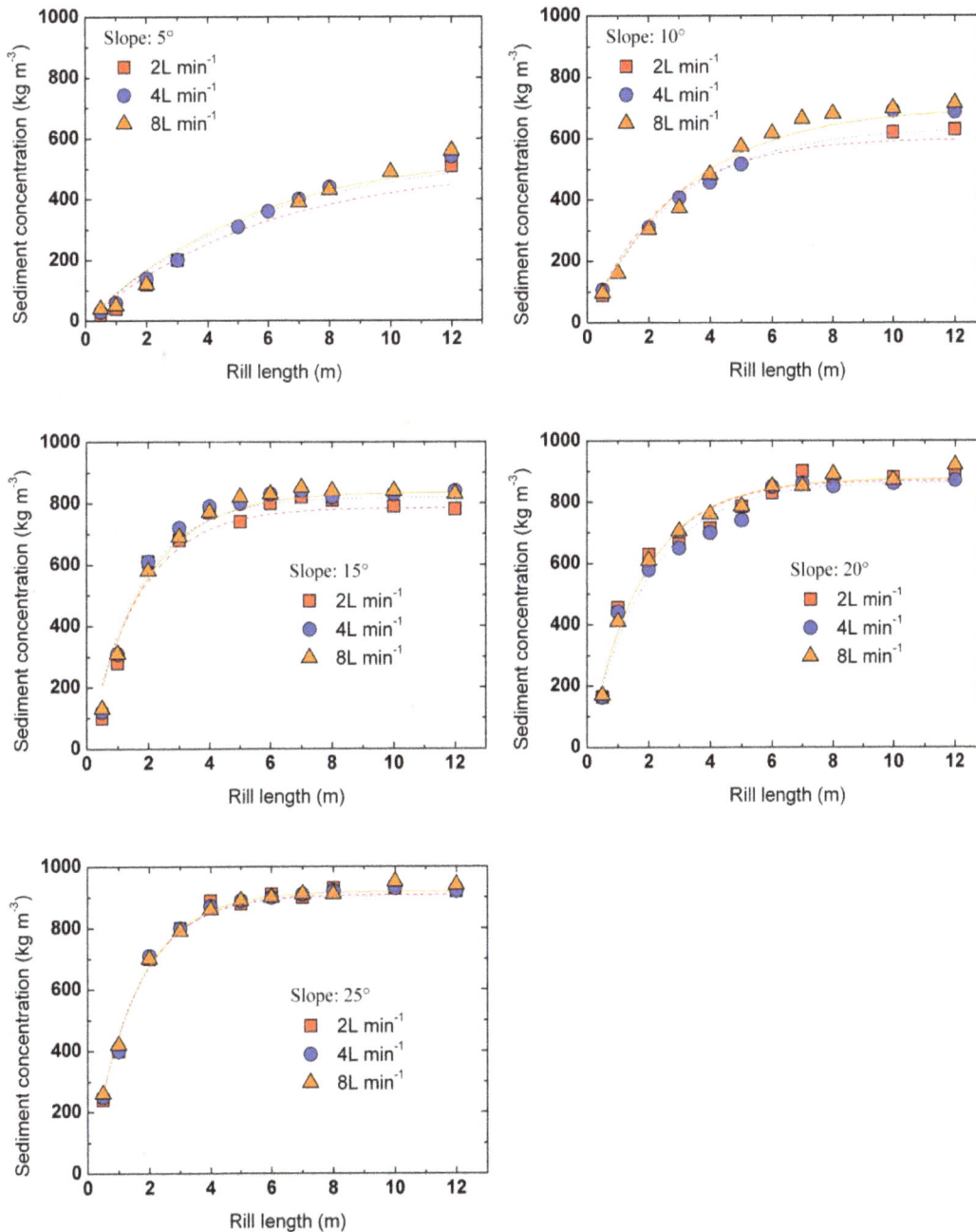

Figure 2. Sediment concentration along the rill at various slope gradients. The variation trend of the measured sediment concentrations along the rill under three flow rates and five slope gradients. The sediment concentration increased exponentially with the rill length for all of the flow rates and slope gradients. The sediment concentration also increased with the slope gradient and flow rate at lower slope gradients (5°, 10° and 15°).

were more evident at steeper slopes. It is possible that clear water (no sediment particles) that is introduced into the rill at the inlet spends limited energy to transport sediments and high energy to detach soil at the upper part of the rill. According to the Water Erosion Prediction Project (WEPP) model [28], clear water has a highest detachment potential. The increase in sediment concentration decreases to zero when the water is saturated with sediments (up to its transport capacity). Almost all the energy of the water flow is used for sediment transportation [32].

In Table 1 all the coefficients of determination ($R^2$) were higher than 0.95, which indicate that Eq. (2) fits well the sediment concentration along the rill. All the values of ($Prob>F$) from the $F$-test were rather low (i.e., less than 8.76 $\mathrm{E}^{-8}$), to indicate that the results were statistically significant ($F$-test with $\alpha = 0.01$).

As indicated in Fig. 2, all the sediment concentrations increased with the rill length and eventually reached the maximum value of $A$. The value of $A$ was the sediment concentration when the water flow was saturated with sediment particles, which represents the sediment concentration at the transport capacity of the water flow. Parameter $B$ was a reduction coefficient that indicates the attenuation of the increase in sediment concentration. From Table 1, the value of $A$ increases with the slope and ranges from 510 to 910. Thus, the sediment transport capacity increased rapidly with the slope gradient. $A$ did not vary significantly with the flow rate under the same slope gradient. The value of $A$ changed slightly between $20°$ and $25°$, which indicates that a critical slope gradient between $20°$ and $25°$ may exist.

In summary, the sediment concentration for sediment saturated flow represents the transport capacity; thus, the sediment transport capacity and the rill length can be estimated.

## Relationship between sediment concentration, slope and flow rate

The sediment concentration increased with the slope and flow rate at lower slope gradients ($5°$, $10°$ and $15°$). This is consistent with the findings of other researchers [33,34]. For rill erosion, the inflow rate and slope gradient are the most influential factors. Steeper slopes and higher flow rates provided greater flow shear

force. These factors enhance soil erosion by either increasing the soil detachment rate or by weakening the protective power of the soil surface to resist erosion [35–37]. Furthermore, the curve defined by Eq. (2) fits well with the sediment concentration under the three flow rates. The sediment concentrations increased with the slope gradient, but they only increased slightly with the flow rate (Fig. 2). For steep slope gradients ($20°$ and $25°$) the sediment curves under three flow rates nearly overlapped at the same slope gradient. The slope may have a greater influence on the sediment concentration than the flow rate. Although the sediment concentrations were approximately the same, the higher flow rate delivered proportionally more sediment (proportional to the flow rate).

To better understand and clarify the relationship between sediment concentration, slope gradient and flow rate, several functions were tried to fit the relationship between the sediment concentration at transport capacity, slopes and flow rates. The following function was found a simple and appropriate fit for the experimental data:

$$S_T = k \cdot S^m \cdot q^n \tag{3}$$

where $S_T$ ($\mathrm{kg\ m^{-3}}$) is the sediment concentration at transport capacity, $S$ is the slope gradient (%), $q$ is the flow rate ($\mathrm{L\ min^{-1}}$) and $m$ and $n$ are regression coefficients. The regression results are listed in Table 2.

Regression analysis indicated that $m$ decreased steadily with slope gradient slope and ranged from 2.45 to 1.69. This indicates that the increase in sediment concentration at transport capacity decreases with the slope gradient. However, $n$, decreased with slope gradient and flow rate, indicating that the increase in sediment concentration declines with the flow rate. All of the coefficients of determination ($R^2$) were higher than 0.84, which indicates that Eq. (3) is able to quantify the sediment concentration at transport capacity of rill erosion. Both the regression coefficients in Eq. (3), $m$ and $n$, were very significant ($F$-test with $\alpha = 0.01$), which illustrates that the slope and flow rate were both critical parameters that influence sediment concentration at transport

**Table 1.** Regression parameters of Eq. (1).

| Slope (°) | flow rate (L min⁻¹) | A(kg m⁻³) | B (m⁻¹) | R² | F | Prob>F |
|-----------|---------------------|-----------|---------|------|----------|-----------|
| 5 | 2 | 510 | 0.17 | 0.95 | 172.74 | 5.09E-08 |
| | 4 | 550 | 0.18 | 0.97 | 1174.60 | 8.76E-08 |
| | 8 | 560 | 0.18 | 0.97 | 593.85 | 2.17E-06 |
| 10 | 2 | 625 | 0.41 | 0.99 | 622.72 | 7.85E-10 |
| | 4 | 635 | 0.35 | 0.96 | 1020.01 | 1.57E-13 |
| | 8 | 675 | 0.31 | 0.98 | 3072.75 | 1.48E-13 |
| 15 | 2 | 785 | 0.60 | 0.95 | 1846.31 | 1.99E-08 |
| | 4 | 820 | 0.60 | 0.97 | 2868.62 | 5.75E-09 |
| | 8 | 835 | 0.56 | 0.98 | 4788.88 | 1.80E-09 |
| 20 | 2 | 870 | 0.56 | 0.96 | 2906.29 | 8.80E-10 |
| | 4 | 865 | 0.52 | 0.96 | 3088.36 | 9.55E-10 |
| | 8 | 875 | 0.56 | 0.98 | 7395.75 | 7.12E-11 |
| 25 | 2 | 910 | 0.68 | 0.99 | 11025.60 | 1.35E-09 |
| | 4 | 905 | 0.69 | 0.99 | 11587.59 | 6.75E-10 |
| | 8 | 910 | 0.68 | 0.99 | 16567.57 | 8.57E-11 |

**Table 2.** Regression coefficients under various slope gradients and flow rates.

| Slope (°) | flow rate (L min⁻¹) | K | m | n | R² | Prob>F |
|---|---|---|---|---|---|---|
| 5 | 2 | 0.83 | 2.45 a | 1.64 a | 0.95 | 4.59E-04 |
| | 4 | 0.89 | 2.23 b | 1.13 b | 0.92 | 4.34E-04 |
| | 8 | 0.87 | 1.99 c | 1.03 bc | 0.95 | 9.51E-04 |
| 10 | 2 | 0.80 | 2.12 bc | 0.96 bc | 0.86 | 7.33E-04 |
| | 4 | 0.74 | 1.94 c | 0.91 c | 0.89 | 3.11E-03 |
| | 8 | 0.78 | 1.87 cd | 0.70 cd | 0.88 | 3.64E-03 |
| 15 | 2 | 0.86 | 1.96 c | 0.61 d | 0.93 | 5.74E-04 |
| | 4 | 0.79 | 1.89 cd | 0.57 de | 0.88 | 1.84E-03 |
| | 8 | 0.83 | 1.76 d | 0.55 de | 0.84 | 1.15E-02 |
| 20 | 2 | 0.82 | 1.87 cd | 0.54 de | 0.95 | 9.73E-04 |
| | 4 | 0.71 | 1.83 cd | 0.49 de | 0.93 | 7.69E-03 |
| | 8 | 0.75 | 1.75 d | 0.38 e | 0.93 | 7.40E-03 |
| 25 | 2 | 0.78 | 1.79 d | 0.41 de | 0.95 | 1.20E-03 |
| | 4 | 0.69 | 1.76 d | 0.39 e | 0.93 | 7.04E-03 |
| | 8 | 0.74 | 1.69 d | 0.38 e | 0.90 | 1.67E-02 |

Footnote: Significant differences between the values are indicated by letters ($p<0.05$).

capacity of rill erosion process. Generally, sediment concentration at transport capacity increased with the slope gradient and flow rate (Table 2). Additionally, the values of $m$ were all greater than $n$; the sediment concentration at the transport capacity increased nearly quadratically ($m \approx 2$) with the slope gradient, but increased less than linearly ($n<1$) with the flow rate. This indicates that the slope is much more important than the flow rate with respect to the sediment transportation [38–40]. Lei et al. [31] suggested that the sediment concentration at transport capacity increased quadratically with slope gradient and exhibited an approximately linear increase with flow rate, which supported our results. Similarly, Aksoy et al. [17] concluded that slope created a linear increase in sediment concentration but the flow rate had practically no effect on sediment concentration.

According to Table 2, the $m$ values of 5°, 10°, 15° and 20° and 25° were significantly different from each other. However, $m$ did not display statistically significant differences between 20° and 25°. Based on the given slope and flow rate, the sediment concentration at the transport capacity increased more significantly with the slope gradients than with flow rates from 5° to 20°. However, when the slope gradient increased from 20° to 25°, the sediment concentration was more stable. The $n$ values revealed similar results: the values at 20° and 25° were obviously different from those at the other slopes, but significant differences did not exist between the two slope gradients. That is to say, the sediment concentration had almost achieved its maximum value (i.e., a certain slope gradient between 20° and 25° was the critical erosion point of the experimental soil).

### Comparison of rill erosion processes measured by various methods

In the experiments by Lei et al. [29], the soil type, designed slope gradients and flow rates were all identical to those of the present study. However, the rill length used by the previous experiment was 8.0 m, whereas the rill length used in the present study was 12.0 m. Furthermore, in the previous study, the sediment concentration process along the 8 m rill was integrated from experimental data of rills of various lengths (i.e., 0.5 m, 1 m,

2 m, etc, up to 8 m), slope gradients and flow rates. In this study, simultaneous and continuous sampling procedures were used. To compare the two data sets, only the data at the upper 8.0 m in this experiment were used (Fig. 3), with the experimental data of discontinuous rill as X-axis and the data of this study as Y-axis. The closer the dots in Fig. 3 are to the 1:1 line, the better the agreement of the two data sets. The data sets indicate that the sediment concentrations from the present experimental method were approximately 1.04, 0.88, 0.83, 0.92, and 0.87 times of the previous experimental data sets at the slope gradients of 5°, 10°, 15°, 20° and 25°, respectively. The coefficients of determination ($R^2$) were all greater than 0.96, which indicates that the two datasets were approximately the same and closely correlated.

The previous experimental technique produced reasonable rill erosion data, but it produced slightly higher sediment concentrations. This could have been caused by random errors in sampling, experiment preparation, etc. The method suggested in this study is more efficient. Simultaneous sampling along the rill is more rational. The relative error produced by these two methods is less than 15%, which suggests that both methods are feasible and applicable. The improved method suggested in this study can measure the rill erosion simultaneously and continuously.

### Conclusions

Given the limits of the method used in previous rill erosion research, an improved and easy to use method of studying rill erosion processes by means of simultaneous sampling of sediment-laden water was suggested. The dynamic changes and distributions of sediment concentrations along the rill were measured for a silty-loam soil over a range of slope gradients and flow rates. The results indicate that sediment concentration increased exponentially with the rill length for all of the flow rates and slope gradients. The sediment concentration also increased with the slope gradient. The data proved that the slope gradient and flow rate are both important parameters for rill erosion, but the slope gradient seemed to be more influential on the sediment concentration. The data computed from the improved method were compared with

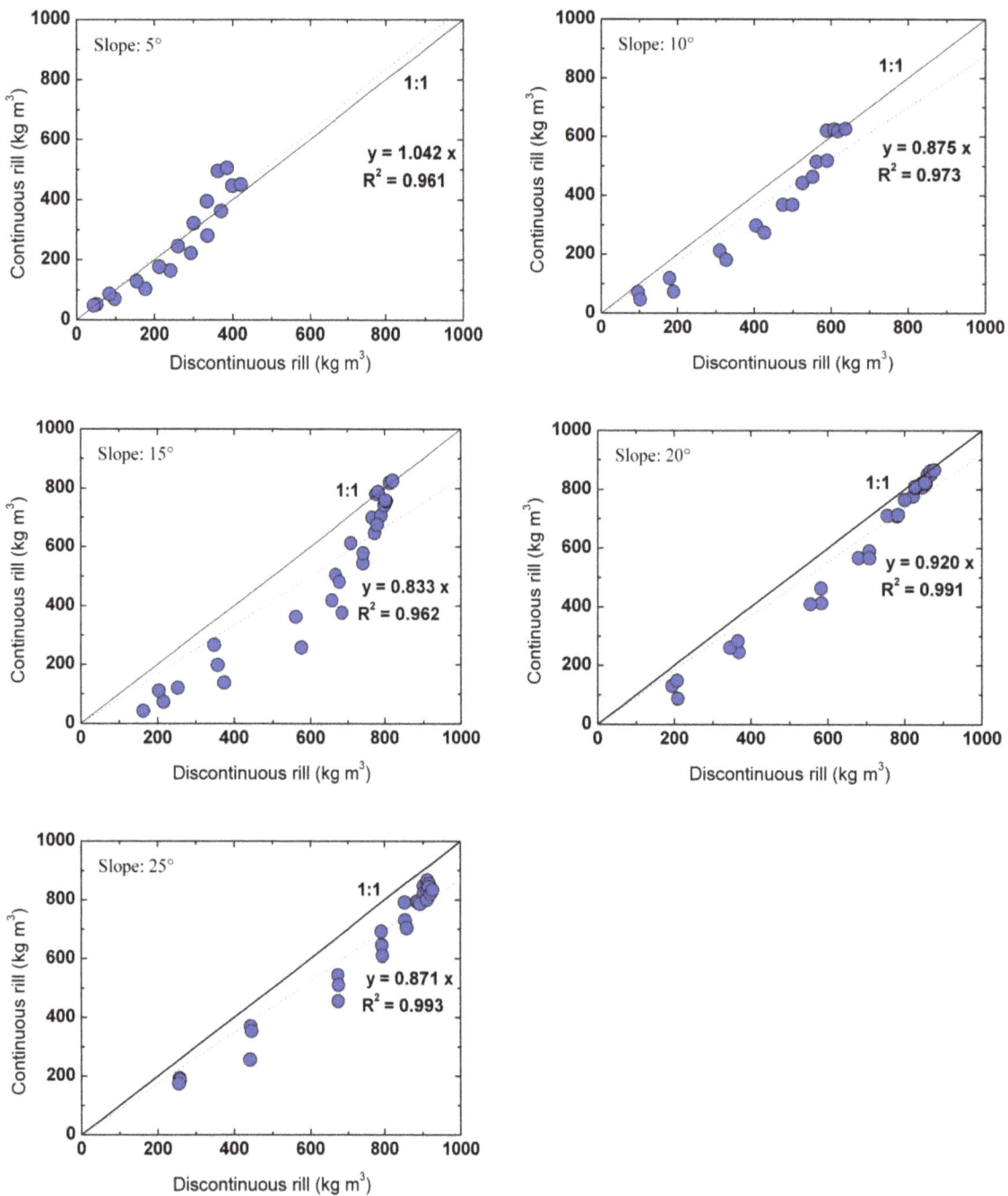

**Figure 3. The sediment concentrations measured from continuous rills in this study and discontinuous rills in Lei et al. (2002).** The sediment concentrations measured from continuous rills in this study and discontinuous rills in Lei et al. (2002) which used a similar methodology. The relative error produce by these two methods is less than 15%, which suggests that both of the methods are feasible.

the data from a previous study, which suggested that this study is a feasible mean to simulate rill erosion process. However, this improved experimental method needs further examination at field scale and more attention should be paid on measures to minimize the boundary effect.

## Acknowledgments

Authors gratefully acknowledge the staff of State Key Laboratory of Soil Erosion and Dry Land Farming on the Loess Plateau for providing guidance of the laboratory apparatus.

## Author Contributions

Conceived and designed the experiments: XYC YZ BM HXM. Performed the experiments: YZ BM HXM. Analyzed the data: YZ HXM BM. Contributed reagents/materials/analysis tools: YZ BM HXM. Wrote the paper: YZ XYC.

# References

1. Wirtz S, Seeger M, Zell A, Wagner C, Wagner JF, et al. (2013) Applicability of Different Hydraulic Parameters to Describe Soil Detachment in Eroding Rills. Plos One 8, doi: 10.1371/journal.pone.0064861

2. Yang DW, Kanae S, Oki T, Koike T, Musiake K (2003) Global potential soil erosion with reference to land use and climate changes. Hydrol Process 17: 2913–2928.

3. Bhattarai R, Dutta D (2007) Estimation of soil erosion and sediment yield using GIS at catchment scale. Water Resour Manag. pp. 1635–1647.

4. Miao CY, Ni JR, Borthwick AGL (2010) Recent changes of water discharge and sediment load in the Yellow River basin, China. Prog Phys Geog 34: 541–561.

5. Miao CY, Duan QY, Yang L, Borthwick AGL (2012) On the Applicability of Temperature and Precipitation Data from CMIP3 for China. Plos One 7.

6. Ellison WD (1949) Protecting the Land against the Raindrops Blast. Sci Mon 68: 241–251.

7. Foster GR, Huggins LF, Meyer LD (1984) A Laboratory Study Of Rill Hydraulics .1. Velocity Relationships. T Asae 27: 790–796.

8. Foster GR, Huggins LF, Meyer LD (1984) A Laboratory Study Of Rill Hydraulics .2. Shear-Stress Relationships. T Asae 27: 797–804.

9. Sun LY, Fang HY, Qi DL, Li JL, Cai QG (2013) A review on rill erosion process and its influencing factors. Chinese Geogr Sci 23: 389–402.

10. Foster GR, Lane LJ, Nowlin JD, Laflen JM, Young RA (1981) Estimating Erosion And Sediment Yield on Field-Sized Areas. T Asae 24: 1253–1263.

11. Tiscarenolopez M, Lopes VL, Stone JJ, Lane LJ (1993) Sensitivity Analysis Of the Wepp Watershed Model for Rangeland Applications .1. Hillslope Processes. T Asae 36: 1659–1672.

12. Gilley JE, Elliot WJ, Laflen JM, Simanton JR (1993) Critical Shear-Stress And Critical Flow-Rates for Initiation Of Rilling. J Hydrol 142: 251–271.

13. Lei TW, Zhang Q, Zhao J, Tang Z (2001) A laboratory study of sediment transport capacity in the dynamic process of rill erosion. T Asae 44: 1537–1542.

14. Gimenez R, Govers G (2002) Flow detachment by concentrated flow on smooth and irregular beds. Soil Sci Soc Am J 66: 1475–1483.

15. Kavvas ML, Yoon J, Chen ZQ, Liang L, Dogrul EC, et al. (2006) Watershed environmental hydrology model: Environmental module and its application to a California watershed. J Hydrol Eng 11: 261–272.

16. Aksoy H, Unal NE, Cokgor S, Gedikli A, Yoon J, et al. (2012) A rainfall simulator for laboratory-scale assessment of rainfall-runoff-sediment transport processes over a two-dimensional flume. Catena 98: 63–72.

17. Aksoy H, Unal NE, Cokgor S, Gedikli A, Yoon J, et al. (2013) Laboratory experiments of sediment transport from bare soil with a rill. Hydrolog Sci J 58: 1505–1518.

18. Zhang XC, Friedrich JM, Nearing MA, Norton LD (2001) Potential use of rare earth oxides as tracers for soil erosion and aggregation studies. Soil Sci Soc Am J 65: 1508–1515.

19. Schuller P, Walling DE, Sepulveda A, Trumper RE, Rouanet JL, et al. (2004) Use of Cs-137 measurements to estimate changes in soil erosion rates associated with changes in soil management practices on cultivated land. Appl Radiat Isotopes 60: 759–766.

20. Li M, Li ZB, Liu PL, Yao WY (2005) Using Cesium-137 technique to study the characteristics of different aspect of soil erosion in the Wind-water Erosion Crisscross Region on Loess Plateau of China. Appl Radiat Isotopes 62: 109–113.

21. Porto P, Walling DE, Callegari G (2013) Using 137Cs and 210Pbex measurements to investigate the sediment budget of a small forested catchment in southern Italy. Hydrol Process 27: 795–806.

22. Lei TW, Zhang QW, Zhao J, Nearing MA (2006) Tracing sediment dynamics and sources in eroding rills with rare earth elements. Eur J Soil Sci 57: 287–294.

23. Li M, Li ZB, Ding WF, Liu PL, Yao WY (2006) Using rare earth element tracers and neutron activation analysis to study rill erosion process. Appl Radiat Isotopes 64: 402–408.

24. Zhu MY, Tan SD, Dang HS, Zhang QF (2011) Rare earth elements tracing the soil erosion processes on slope surface under natural rainfall. J Environ Radioactiv 102: 1078–1084.

25. Lei TW, Nearing MA, Haghighi K, Bralts VF (1998) Rill erosion and morphological evolution: A simulation model. Water Resour Res 34: 3157–3168.

26. Lei TW, Nearing MA (2000) Flume experiments for determining rill hydraulic characteristic erosion and rill patterns. J Hydraul Eng 11: 49–54 (in Chinese).

27. Nearing MA, Norton LD, Bulgakov DA, Larionov GA, West LT, et al. (1997) Hydraulics and erosion in eroding rills. Water Resour Res 33: 865–876.

28. Nearing MA, Foster GR, Lane LJ, Finkner SC (1989) A Process-Based Soil-Erosion Model for Usda-Water Erosion Prediction Project Technology. T Asae 32: 1587–1593.

29. Lei TW, Zhang QW, Zhao J, Xia WS, Pan YH (2002) Soil detachment rates for sediment loaded flow in rills. T Asae 45: 1897–1903.

30. Polyakov VO, Nearing MA (2003) Sediment transport in rill flow under deposition and detachment conditions. Catena 51: 33–43.

31. Lei TW, Zhang QW, Yan LJ (2009) Physically-based rill erosion model. Beijing: Science Press. 229 p.

32. Aksoy H, Kavvas ML (2005) A review of hillslope and watershed scale erosion and sediment transport models. Catena 64: 247–271.

33. Kinnell PIA (2000) The effect of slope length on sediment concentrations associated with side-slope erosion. Soil Sci Soc Am J 64: 1004–1008.

34. Sirjani E, Mahmoodabadi M (2014) Effects of sheet flow rate and slope gradient on sediment load. Arab J Geosci 7: 203–210.

35. Huang C (1998) Sediment regimes under different slope and surface hydrologic conditions. Soil Sci Soc Am J 62: 423–430.

36. Fox DM, Bryan RB (2000) The relationship of soil loss by interrill erosion to slope gradient. Catena 38: 211–222.

37. Miao CY, Ni JR, Borthwick AGL, Yang L (2011) A preliminary estimate of human and natural contributions to the changes in water discharge and sediment load in the Yellow River. Global Planet Change 76: 196–205.

38. Shi ZH, Fang NF, Wu FZ, Wang L, Yue BJ, et al. (2012) Soil erosion processes and sediment sorting associated with transport mechanisms on steep slopes. J Hydrol 454: 123–130.

39. Govers G, Rauws G (1986) Transporting Capacity Of Overland-Flow on Plane And on Irregular Beds. Earth Surf Proc Land 11: 515–524.

40. Zhang GH, Liu YM, Han YF, Zhang XC (2009) Sediment Transport and Soil Detachment on Steep Slopes: II. Sediment Feedback Relationship. Soil Sci Soc Am J 73: 1298–1304.

# 9

# Removing Constraints on the Biomass Production of Freshwater Macroalgae by Manipulating Water Exchange to Manage Nutrient Flux

**Andrew J. Cole\*, Rocky de Nys, Nicholas A. Paul**

MACRO — the Centre for Macroalgal Resources and Biotechnology, and School of Marine and Tropical Biology, James Cook University, Townsville, Queensland, Australia

## Abstract

Freshwater macroalgae represent a largely overlooked group of phototrophic organisms that could play an important role within an industrial ecology context in both utilising waste nutrients and water and supplying biomass for animal feeds and renewable chemicals and fuels. This study used water from the intensive aquaculture of freshwater fish (Barramundi) to examine how the biomass production rate and protein content of the freshwater macroalga *Oedogonium* responds to increasing the flux of nutrients and carbon, by either increasing water exchange rates or through the addition of supplementary nitrogen and $CO_2$. Biomass production rates were highest at low flow rates (0.1–1 vol.day$^{-1}$) using raw pond water. The addition of $CO_2$ to cultures increased biomass production rates by between 2 and 25% with this effect strongest at low water exchange rates. Paradoxically, the addition of nitrogen to cultures decreased productivity, especially at low water exchange rates. The optimal culture of *Oedogonium* occurred at flow rates of between 0.5–1 vol.day$^{-1}$, where uptake rates peaked at 1.09 g.m$^{-2}$.day$^{-1}$ for nitrogen and 0.13 g.m$^{-2}$.day$^{-1}$ for phosphorous. At these flow rates *Oedogonium* biomass had uptake efficiencies of 75.2% for nitrogen and 22.1% for phosphorous. In this study a nitrogen flux of 1.45 g.m$^{-2}$.day$^{-1}$ and a phosphorous flux of 0.6 g.m$^{-2}$.day$^{-1}$ was the minimum required to maintain the growth of *Oedogonium* at 16–17 g DW.m$^{-2}$.day$^{-1}$ and a crude protein content of 25%. A simple model of minimum inputs shows that for every gram of dry weight biomass production (g DW.m$^{-2}$.day$^{-1}$), *Oedogonium* requires 0.09 g.m$^{-2}$.day$^{-1}$ of nitrogen and 0.04 g.m$^{-2}$.day$^{-1}$ of phosphorous to maintain growth without nutrient limitation whilst simultaneously maintaining a high-nutrient uptake rate and efficiency. As such the integrated culture of freshwater macroalgae with aquaculture for the purposes of nutrient recovery is a feasible solution for the bioremediation of wastewater and the supply of a protein resource.

**Editor:** Wei Ning Chen, Nanyang Technological University, Singapore

**Funding:** This research project is supported by the Australian Government through the Australian Renewable Energy Agency (ARENA), and the Advanced Manufacturing Cooperative Research Centre (AMCRC), funded through the Australian Government's Cooperative Research Centre Scheme. The funders had no role in study design, data collection and analysis, decision to publish, or preparation of the manuscript.

**Competing Interests:** The authors have declared that no competing interests exist.

\* Email: andrew.cole3@jcu.edu.au

## Introduction

The production of macroalgal biomass is a developing component of clean technologies for the remediation of wastewater and carbon dioxide within an integrated closed-loop cycle often referred to as industrial ecology [1,2]. The ultimate aim of industrial ecology is to replicate the efficiencies observed in biological systems, where all ecosystem resources are recycled and the waste of one species becomes the food of another [1]. Using this framework, industrial ecology is primarily concerned with shifting industrial processes from linear systems, in which resources move through a system to become waste, to a closed-loop system where wastes become valued as an input for the next production process. A major part of this process is the integration of production systems so that waste products can be easily accessed either as a raw material or as an energy carrier, with the emphasis on processes and practices that reduce greenhouse gas emissions and related environmental impacts of waste streams [3]. On a global scale one of the largest and most consistent sources of industrial waste is nutrients – nitrogen and phosphorous. A large proportion of this waste is created as a by-product of intensive animal agriculture [4]. Globally, nitrogen and phosphorous waste now exceed 138 Tg.y$^{-1}$ and 11 Tg.y$^{-1}$ respectively [5]. These waste nutrients are currently seen as a liability and are either lost to the atmosphere through denitrification or leached into the local environment where nitrogen enrichment and eutrophication problems can occur with global scale impacts [6–12]. However, these excess nutrients need not be relegated to waste but rather could be recycled and utilised as an input resource for the large-scale cultivation of phototrophic organisms, themselves to be recycled again as bioproducts [13–15]. Inorganic nitrogen in particular is the primary limiting nutrient for the production of algae and as such this waste nitrogen could be an ideal resource for the large-scale production of algal biomass [16]. The cultivation and subsequent on-site use of macroalgal biomass at sites with high-nutrient waste streams, such as intensive livestock production, can close the loop between waste production, waste capture and re-use. The on-site use of cultured biomass has the additional benefit of reducing the energy and transportation costs associated with either bringing animal feeds to the farm or transporting algal biomass away.

Generally, the production of macroalgal biomass is highest when cultures are provided with a constant supply of nutrient-rich water [17–20]. In flow-through systems, nutrient supply (or nutrient flux as g nutrients.$m^{-2}$.$day^{-1}$) can be manipulated by either changing the nutrient concentration of the incoming water or by changing the rate of water exchange. This is particularly important for large-scale production facilities which, as a consequence of the large water volumes involved, will only be able to exchange relatively low amounts (e.g. 10–50%) each day. In contrast, the majority of small scale research to date has often used very high water exchange rates (>24 volumes per day) to obtain exceptionally high but commercially unobtainable rates of biomass production [20–22]. A high water exchange rate, in addition to supplying nutrients in excess, also benefits algal production by facilitating the removal of dead algal cells and limits the accumulation of bacteria and fouling organisms which can negatively impact productivity [19,20]. Increasing the rate of water exchange can also benefit biomass production rates by increasing the supply and availability of dissolved inorganic carbon (DIC). Under intensive culture conditions the high density and photosynthetic activity of macroalgae can raise the pH and rapidly deplete the proportion of DIC that is available for photosynthesis [23,24].

The majority of previous work on macroalgal production and the effects of changing nutrient flux have focused on marine species where productivity is positively influenced by increasing the rate of water exchange [18,20]. However, it is currently unclear whether freshwater species conform to this paradigm. It is also unclear how low the rate of water exchange can be before its beneficial aspects are lost and whether a very low water exchange rate, e.g. 10% water exchange per day, supplemented with additional nutrients or any limiting trace elements, will stimulate biomass production as effectively as a higher water exchange rate of low-nutrient water. This includes carbon as a limiting factor in the growth of freshwater macroalgae [25].

The majority of research that has investigated the effect of changing nutrient flux on macroalgal productivity have focused on marine species [17,18,20,21,26], yet paradoxically most major industries with high-nutrient waste streams, such as intensive agriculture, mineral processing, energy production and municipal waste, are located around freshwater, rather than saltwater, environments. Our key target species for the bioremediation of freshwater waste streams are from the genus Oedogonium, a competitively dominant genus of filamentous macroalgae that also has a biochemical composition suitable for a range of biomass applications [25,27,28]. Therefore, this study investigates for the first time how the rate of biomass production of Oedogonium responds to both the rate of water exchange and the supplementary addition of nutrients and carbon when cultured using the wastewater from a freshwater aquaculture farm as a case study for the integration of freshwater macroalgal biomass for bioremediation and bio-products. This study also investigates the effect of these factors on the nitrogen content of freshwater macroalgae with the goal of understanding the relationship between productivity and protein content of Oedogonium in this system.

## Methods

### Ethics statement

Research on algae does not require ethics approval. All algae used in this study originated from stock cultures maintained at the Marine and Aquaculture Research Facility at James Cook University. This research complied with all Australian laws.

This study was undertaken on private land owned by Good Fortune Bay Fisheries LTD. Permission to use this land was granted by the farm manager, Rod Pelling. No other permission was needed to use this land or perform our experiment.

### Study species and site

Oedogonium is a genus of unbranched, uniseriate green algae made up of small cylindrical cells. This genus has a worldwide distribution and is a common component of natural ecosystems where it grows either attached to the substrate or as free floating mats. Oedogonium is a robust and competitively dominant genera that has been identified as a key target group for the bioremediation of freshwater waste streams [27] and as a feedstock biomass for bioenergy applications [25,28,29]. Stock cultures of Oedogonium sp. as described in Lawton et al. [27] (Genebank accession numbers: KC701472), and hereafter referred to as Oedogonium, were maintained at the Marine & Aquaculture Research Facilities Unit (MARFU), at James Cook University (JCU), Townsville (Latitude: 19.33°S; Longitude 146.76°E). The experimental component of this study was conducted at Good Fortune Bay Fisheries Ltd, a 450 tonne per annum freshwater barramundi farm located in Townsville, Australia. Barramundi are cultured in large outdoor ponds, constantly supplied with a supply of clean bore water. Pond effluent water is passively cleaned using reed beds and settlement ponds to reduce suspended solids before being released to the environment. In this experiment, pond effluent water was pumped out of the settlement pond and sand filtered to 150 μm. This water was then used to feed twenty cylindrical tanks (Duraplas AP 1000; 1000 L capacity, 1.19 $m^2$ surface area) to a depth of 70 cm which gave a total culture volume of 853L. Oedogonium was maintained in tumble culture in these tanks through the use of a central aeration ring (45 cm circumference). A 750 μm mesh screen was used to prevent the Oedogonium filaments from exiting the tank with the outgoing water.

### Nutrient flux experiment

This experiment investigated the separate and additive effects of changing water exchange rates and the addition of nitrogen and carbon to cultures on the biomass production and nitrogen content of Oedogonium. Five water renewal rates of 0.1, 0.5, 1, 2.5 and 5 volumes per day were used, with these rates representing realistic water exchange rates that a large-scale algal production facility could use. Exchange rates were controlled using digital flow meters fitted to each tank (Hoselink digital flow meters). These flow meters dispensed an hourly allotment of water into each tank. The elemental composition of the settlement pond water that was used in this experiment was measured at the beginning and end of the experimental period, with elemental composition remaining consistent over this 6 week timeframe (Table 1). To investigate how the concentration of nitrogen and the availability of dissolved inorganic carbon (DIC) influences biomass production, the sand-filtered pond water was split across each water renewal rate into four treatments; pond water (PW), pond water plus nitrogen (PW+N), pond water plus carbon (PW+$CO_2$) and pond water plus nitrogen and carbon (PW+N+$CO_2$). In this study we used a randomised complete block design, which is an appropriate experimental design for longer term studies with uncontrolled variables that differ over time (e.g. temperature, light intensity), and replicates within each block are exposed to the same environmental conditions. This design enabled us to partition the effects of the factors of interest, water exchange rate, carbon addition and nitrogen addition, from any potential environmental effects. In our experiment we had two fixed water treatment factors with two levels of each factor, with and without carbon addition (C) and with and without nitrogen addition (N), and five water exchange (W) rates with five levels; 0.1, 0.5, 1, 2.5 and 5 volumes.day$^{-1}$. In total, we had 20 replicate tanks and each of these tanks was assigned to 1 treatment combination (CxNxW) per replicate block. This experimental set up was replicated over five time periods (five blocks), giving a total of five replicates for each treatment combination.

Additional nitrogen was added to tanks by filling a 150 L conical tank with tap water and enriching this water with 850 g of ammonium chloride. A Kamoer 12 channel dosing pump was then used to supply a pulse addition of 120 mL of this ammonium-enriched water into each treatment tank every hour to provide an additional nitrogen flux of 0.2 mg.L$^{-1}$.hr$^{-1}$. The DIC concentration was increased through the dissolution of $CO_2$ gas into pond water through a speece cone at a rate of 2.5 L.min$^{-1}$. The $CO_2$ supply was controlled via a digital timer and solenoid valve which limited carbon addition to the daylight hours between 8am and 4pm. A 150 mL sample of incoming pond water was collected on days one and three of this experiment to determine the ammonium, nitrite and nitrate concentrations for all treatments, while the total alkalinity was measured from the incoming water to both the PW and PW+$CO_2$ treatments. Water samples were analysed by the Australian Centre for Tropical Freshwater Research at James Cook University. The pH and temperature of each tank was measured between 2–3pm every afternoon. The concentration of DIC and the proportion of each of the three carbon species available ($CO_2$, $HCO_3^-$ or $CO_3^{2-}$) in the culture water were calculated using the software $CO_2$sys [30] based on the total alkalinity of the incoming water and the pH and temperature of each replicate culture. This experiment was conducted over a 6 week period between 6$^{th}$ May and 14$^{th}$ June 2013. During this period the daily water temperature of the tanks ranged between 18.9 and 24.3°C, the mean daily Photosynthetic Active Radiation was 23.2 ($\pm$1.5) mol photons.m$^{-2}$, (range:14.8–31.2 mol.m$^{-2}$) with daily peaks ranging between 1034 and 1914 $\mu$mol photons.m$^{-2}$.s$^{-1}$, and the nitrogen and phosphorous concentration of the pond water ranged between 1.93–2.75 mg.L$^{-1}$ and 0.16–0.53 mg.L$^{-1}$ respectively.

*Oedogonium* was stocked in each tank at a density of 0.35 g.L$^{-1}$ (300 g fresh weight: FW). The algae used to stock these tanks had been continuously cultured for three weeks prior to this experiment using sand filtered pond water at a water renewal rate of 1 tank volume per day (1 vol.day$^{-1}$). The biomass used to stock the tanks had a mean internal nitrogen content of 5.04 ($\pm$0.4) %. To quantify productivity, algal biomass was harvested after 4 days, centrifuged using a domestic washing machine (1000 rpm) to remove excess water, weighed and dried at 60°C for 48 hours before being reweighed to provide dry weight (DW). Algal productivity was calculated using the equation: $P = \{[(B_f - B_I)/FW : DW]/A\}/t$, where $B_f$ and $B_I$ are the final and initial algal biomass, *FW:DW* is the fresh to dry weight ratio, $A$ is the area of culture tanks and $t$ is the number of days in culture. Separate mixed-model ANOVAs were used to test for differences in biomass production rates, internal nitrogen content and maximum daily pH of cultures with water exchange rates (n = 5 treatments) and incoming treatment water (with and without nitrogen; with and without carbon) as our fixed factors, while time (growth trial) was treated as the blocking factor. Residual plots were examined to ensure ANOVA assumptions were met. Tukey's multiple comparison tests were then used to compare the means of treatment groups.

One gram of dried biomass was then sent to OEA Laboratories UK to determine the internal nitrogen content, while the remaining 1 g was sent to the Advanced Analytical Centre at

**Table 1.** Elemental composition of fish pond water.

| Metal | Concentration (mg.L$^{-1}$) |
|---|---|
| Sodium | 271–291 |
| Calcium | 47–59.4 |
| Magnesium | 47–128 |
| Potassium | 5.9–6.0 |
| Strontium | 0.55–0.73 |
| Phosphate | 0.13–0.50 |
| Manganese | 0.05–0.12 |
| Iron | <0.05 |
| Arsenic | <0.01 |
| Selenium | <0.01 |
| Vanadium | <0.02 |
| Zinc | <0.02 |
| Molybdenum | <0.005 |
| Copper | <0.002 |
| Nickel | <0.002 |
| Aluminium | <0.001 |
| Lead | <0.001 |
| Cadmium | <0.0001 |
| Mercury | <0.0001 |

Values are the ranges of two samples taken at the start and end of the experimental period.

JCU to determine the total phosphorous content of the biomass. The internal nitrogen and phosphorous concentrations were then used to calculate the uptake rate and uptake efficiencies of the cultures. Uptake rate was calculated in $g.m^{-2}.day^{-1}$ and was the difference between the initial and final concentrations in the biomass divided by the culture period (4 days) and culture surface area. Uptake efficiency was calculated in % and was the proportion of nitrogen and phosphorous in incoming water that was incorporated into the biomass. Protein content of the *Oedogonium* biomass was calculated by multiplying the nitrogen content by 4.7 following Neveux et al. [28], which is a conservative conversion as it does not include the amino acids Cysteine and Tryptophan [28].

## Results

The average ($\pm$ SE) pH of inflow water was 7.69 ($\pm0.07$) and 6.78 ($\pm0.16$) for PW and PW+$CO_2$ treatment water respectively. The addition of $CO_2$ had only a minor effect on the pH of the low flow treatments with the average daily maximum pH of these treatments ranging between 9.29 and 9.42 (Fig. 1b). The addition of nitrogen to pond water had no significant effect on maximum daily pH values within each tank. However, the pH of cultures was significantly affected by $CO_2$ addition ($P<0.001$) and water exchange rate ($P<0.001$) (Table 2). The pH decreased with both $CO_2$ addition and increasing rates of water exchange. As the rate of water exchange increased the effect of carbon addition on pH became more pronounced with a difference of 0.65 pH units between tanks with ($\sim8.21$) and without ($\sim8.86$) carbon supplementation at a water exchange rate of 5 vol.day$^{-1}$ (Fig. 1a). This interaction between $CO_2$ addition and water exchange was significant ($P = 0.002$) (Fig. 1b).This difference in pH resulted in a 25% increase in the proportion of DIC available to cultures supplemented with $CO_2$ in the lowest water exchange rates and an increase of 70–75% for water exchange rates between 0.5–5 vol.day$^{-1}$ (Fig. 1b).

*Oedogonium* productivity ranged between 3.8 and 23.8 g DW.m$^{-2}$.day$^{-1}$ for all treatments combined, although 76% of these replicate tanks had productivities greater than 12 g DW.m$^{-2}$.day$^{-1}$. Biomass production was significantly influenced by the addition of nitrogen ($P<0.001$) and carbon ($P = 0.002$) and there was a significant interaction between nitrogen addition and the rate of water exchange ($P = <0.001$) (Table 3). This interaction was driven primarily by the negative response of *Oedogonium* to nitrogen supplementation at low flow rates (Fig. 2). At low water exchange rates the addition of nitrogen resulted in the lowest growth rates with mean productivities of 7.8 ($\pm0.9$) and 10.5 ($\pm2.6$) g DW.m$^{-2}$.day$^{-1}$ in the PW+N and PW+N+$CO_2$ treatments, respectively. As the rate of water exchange increased the productivity of the nitrogen addition treatments increased and peaked at a water exchange rate of 2.5 vol.day$^{-1}$. This combination had a mean productivity of 14.8 ($\pm1.7$) g DW.m$^{-2}$.day$^{-1}$ in the PW+N treatment and 16.5 ($\pm1.5$) g DW.m$^{-2}$.day$^{-1}$ and PW+N+$CO_2$ treatment. In contrast, at the lowest water exchange rates *Oedogonium* cultured in raw pond water had the highest productivities. A 10% daily water exchange resulted in a mean productivity of 16.6 ($\pm1.1$) g DW.m$^{-2}$.day$^{-1}$ in pond water and 18.1 ($\pm1.4$) g DW.m$^{-2}$.day$^{-1}$ when $CO_2$ was added to incoming pond water. Increasing water exchange rates in these treatments resulted in a steady 5–15% decline in productivity with the highest water exchange treatment of 5 vol.day$^{-1}$ having mean productivities of 15 ($\pm2.1$) g DW.m$^{-2}$.day$^{-1}$ for PW and 15.9 ($\pm1.1$) g DW.m$^{-2}$.day$^{-1}$ for PW + $CO_2$ treatments. The addition of $CO_2$ to cultures generally had a positive effect on

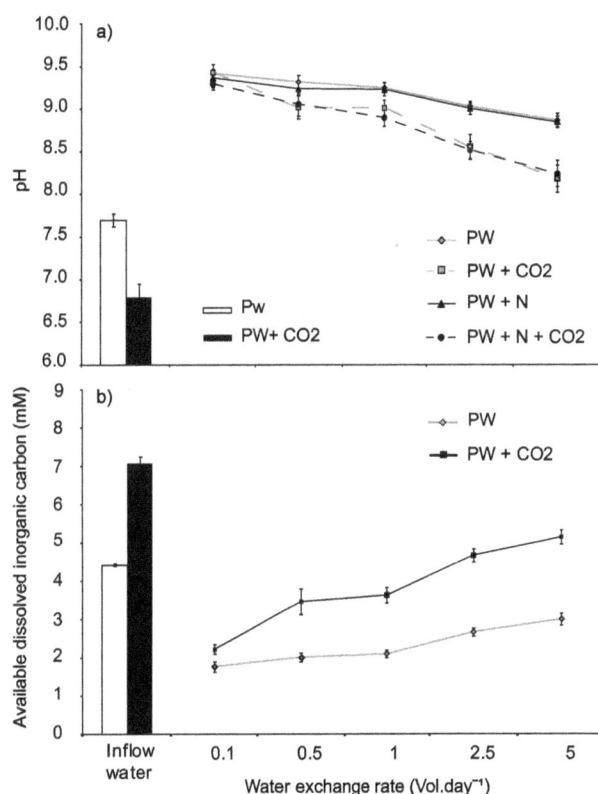

**Figure 1. pH and carbon availability of culture water.** a) Average maximum daily pH and b) total dissolved inorganic carbon available ($CO_2$ and $HCO_3^-$) to *Oedogonium* cultures at five water exchange rates (0.1, 0.5, 1, 2.5 and 5 vol.day$^{-1}$) using fish pond water with and without nitrogen (N) and carbon ($CO_2$) supplementation. In graph b nitrogen treatments have been excluded as nitrogen had no significant effect on carbon availability.

productivity and resulted in up to a 25% increase in productivity at the lowest water exchange rates (Fig. 2).

Settlement pond effluent had a total dissolved inorganic nitrogen content of 2.43 ($\pm1.39$) mg.L$^{-1}$ and a filterable reactive phosphorous concentration of 0.34 ($\pm0.02$) mg.L$^{-1}$. The total nitrogen concentration was made up of 1.06 ($\pm0.16$) mg.L$^{-1}$ as total ammonia nitrogen (TAN), 0.47 ($\pm0.02$) mg.L$^{-1}$ as nitrite-N and 1.35 ($\pm0.18$) mg.L$^{-1}$ as nitrate-N. *Oedogonium* cultured in this water had a total nitrogen uptake rate that ranged between 0.45 ($\pm0.09$) and 1.09 ($\pm0.11$) g.m$^{-2}$.day$^{-1}$ (Fig. 3a) and a mean phosphorous uptake rate ranging between 0.08 ($\pm0.01$) and 0.15 ($\pm0.02$) g.m$^{-2}$.day$^{-1}$ (Fig. 3b), with the lowest uptake rates occurring in the 0.1 vol.day$^{-1}$ treatments for both nutrients. At the asymptote point of uptake rates, which corresponded to a nitrogen and phosphorous flux of 1.5 g.m$^{-2}$.day$^{-1}$ and 0.75 g.m$^{-2}$.day$^{-1}$ respectively, uptake efficiencies of 61 ($\pm3.6$) and 75.2 ($\pm7.5$) % for nitrogen and 15.7 ($\pm2.9$) and 22.1 ($\pm3.5$) % for phosphorous were recorded in the PW and PW+$CO_2$ treatments, respectively. Further increases in nitrogen and phosphorous fluxes did not increase the uptake rate and subsequently decreased the uptake efficiency of nitrogen and phosphorous to lows of 8 and 7% for the highest water exchange treatment which corresponded to a nitrogen and phosphorous flux of 10.1 and 1.7 g.m$^{-2}$.day$^{-1}$ respectively (Fig. 3).

The internal nitrogen content of the *Oedogonium* biomass ranged from 2.5–6.7% across all treatments (Fig. 4). The cultures with the

**Table 2.** Mixed model ANOVA results testing the effect of water exchange rates (0.1, 0.5, 1, 2.5 and 5 vol.day$^{-1}$) and water treatment (Pond water with and without nitrogen and $CO_2$ addition) on the pH of culture water over five replicate trials.

| Source | Effect | DF | MS | F | P |
|---|---|---|---|---|---|
| Water exchange rate (W) | Fixed | 4 | 1.55 | 70 | <0.001 |
| Nitrogen (N) | Fixed | 1 | 0.08 | 3.4 | 0.07 |
| Carbon (C) | Fixed | 1 | 3.32 | 149.6 | <0.001 |
| W X N | Fixed | 4 | 0.04 | 1.6 | 0.18 |
| W x C | Fixed | 4 | 0.1 | 4.6 | 0.002 |
| N X C | Fixed | 4 | 0.01 | 0.5 | 0.48 |
| W x N x C | Fixed | 4 | 0.03 | 1.1 | 0.35 |
| Trial (T) | Random | 4 | 0.04 | 1.6 | 0.18 |
| Error | | 76 | 0.02 | | |

highest productivities, cultured at the lowest water exchange rate, had the lowest internal nitrogen content with a mean nitrogen content of 3.9 ($\pm$0.39) % for PW and 3.54 ($\pm$0.28) % for PW+ $CO_2$ treatments. Increasing the water exchange rate from 0.1 to 0.5 vol.day$^{-1}$ resulted in a 38% and 53% increase in the internal nitrogen content of the *Oedogonium* biomass cultured in the PW and PW+$CO_2$ treatments respectively. Water exchange rates from 0.5– 5 vol.day$^{-1}$ resulted in an internal nitrogen content ranging from 4.94–6.26%, with the mean internal nitrogen content peaking at 5.57 ($\pm$0.11) % and 5.69 ($\pm$0.06) % at a water exchange rate of 2.5 vol.day$^{-1}$for the PW and PW+$CO_2$ treatments respectively (Fig. 3). These nitrogen values translate to a crude protein content, using the 4.7 nitrogen conversion factor, ranging from 16.7 ($\pm$1.3) %, at the lowest water exchange rate, to a high of 26.7 ($\pm$0.3) %, at a water exchange of 2.5 vol.day$^{-1}$when cultured using PW and PW+$CO_2$ water. In the nitrogen addition treatments the internal nitrogen content was relatively stable and ranged between 5.19 and 6.71% with almost two-thirds of the cultures having an internal nitrogen content greater than 5.9%. In these treatments the mean internal nitrogen content peaked at a water exchange rate of 1 vol.day$^{-1}$ with an average nitrogen content of 5.99 ($\pm$0.18) % and 6.13 ($\pm$0.08) % for PW+N and PW+N+$CO_2$ water respectively. The protein content of *Oedogonium* cultured in the

PW+N treatments was largely independent of flow rates and ranged between 26.0 ($\pm$2.7) % and 28.8 ($\pm$0.4) %.

## Discussion

This study has demonstrated that freshwater macroalgae can be cultured using the nutrient wastewater from a freshwater fish farm. We expect that the cultures of *Oedogonium* could be successfully integrated with any industry or process that produces large quantities of similar wastewater. This study has also demonstrated that *Oedogonium* can be successfully cultured in a wide range of nutrient concentrations, is amenable to relatively low water exchange rates and is an extremely robust species that can maintain pure cultures in open systems under field conditions; a problem that severely impacts the scale up of microalgal cultures [31–33]. The biomass production rates in our study were comparable to other studies on freshwater macroalgae, which range between 5–25 g.m$^{-2}$.day$^{-1}$ [15,16,34–37]. These are higher than the 8 g.m$^{-2}$.day$^{-1}$ biomass production rates for microalgae cultured in outdoor ponds using raw industry waste water [38]. Encouragingly, this study demonstrates that there is no need for high flow rates to maintain the production of *Oedogonium* and that a flow rate between 0.5 and 1 vol.day$^{-1}$ is sufficient to maintain high biomass productivity and therefore high uptake efficiency of nitrogen and phosphorous. Additionally the biomass properties of *Oedogonium*, particularly its crude protein content, can be manipulated by controlling the amount of nitrogen the cultures receive, either through changing water renewal rates or through the addition of a concentrated nitrogen waste stream. Culturing *Oedogonium* at flow rates of 0.1 vol.day$^{-1}$, where nitrogen availability is reduced, resulted in a biomass that was highly productive but had a relatively low protein content of 16% (related to a low internal nitrogen content of 3.5%). This biomass is well suited to bioenergy applications, specifically the conversion to liquid fuels through hydrothermal liquefaction [28,39,40]. Likewise the *Oedogonium* biomass produced at higher water exchange rates, 0.5–1 vol.day$^{-1}$, where nitrogen was not limited, resulted in a biomass that had a high protein content (26–28%), which could makes this biomass an ideal feed supplement or replacement option for livestock. Several species of algae have already been successfully incorporated into the diets of livestock, with algae generally having positive effects on animal health, productivity and product quality [41–46].

Worldwide, protein used in the diets of animals exceeds 150 million tonnes annually, with soybean meal accounting for the

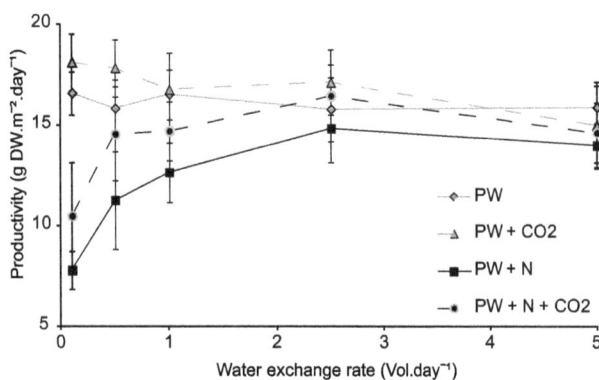

**Figure 2.** *Oedogonium* **productivity.** Biomass production rates of *Oedogonium* cultured at 5 water exchange rates, 0.1, 0.5, 1, 2.5 and 5 vol.day$^{-1}$, using fish pond water (PW) with and without nitrogen (N) and carbon ($CO_2$) supplementation. Values are the means and SE of 5 replicate growth trials.

**Table 3.** Mixed model ANOVA results testing the effect of water exchange rates (0.1, 0.5, 1, 2.5 and 5 vol.day$^{-1}$) and water treatment (Pond water with and without nitrogen and $CO_2$ addition) on the biomass productivity of *Oedogonium* cultured over five replicate trials.

|  | Effect | DF | MS | F | P |
|---|---|---|---|---|---|
| Water exchange rate (W) | Fixed | 4 | 0.04 | 0.28 | 0.60 |
| Nitrogen (N) | Fixed | 1 | 42.26 | 54.72 | <0.001 |
| Carbon (C) | Fixed | 1 | 2.63 | 9.55 | 0.002 |
| W X N | Fixed | 4 | 8.28 | 11.87 | <0.001 |
| W x C | Fixed | 4 | 4.13 | 1.82 | 0.13 |
| N X C | Fixed | 1 | 7.09 | 1.49 | 0.23 |
| W x N x C | Fixed | 4 | 0.06 | 0.10 | 0.98 |
| Trial (T) | Random | 4 | 0.32 | 2.43 | 0.06 |
| Error |  | 76 | 0.02 |  |  |

bulk of this protein [47]. However, the trend of increasing population sizes and the increasing proportion of meat in the diets of developing countries means that in the future unconventional sources of protein will be required to meet global protein needs [47,48]. The crude protein content of *Oedogonium* in this study is lower than that of most species of microalgae (20–60%) [49,50],

but compares well to other species of macroalgae (20–40%) [28,34,51,52]. *Oedogonium* also has a comparable protein content to conventional sources of protein such as whole soybeans at 32–43% [52–55] and is higher than most cereals and grains which range between 8–16% [54,56–59]. Even relatively novel terrestrial sources of crude protein such as Kenaf (*Hibiscus cannabinus*) or *Moringa oleifera* have a leaf protein content ranging between 15–26% [60–62], however when the entire plant biomass is taken into account the total crude protein content is much lower than the 16–28% of *Oedogonium*. In this regard, freshwater macroalgae could be an ideal alternative protein source to feed to livestock, especially if this biomass is cultured using farm waste nutrients and re-used on the same site [34,51,63]. We do not expect the cultured *Oedogonium* biomass to have any animal feed restrictions related to heavy metal accumulation as the water used to culture this biomass is from a freshwater source that is only used for animal production (aquaculture) and has no exposure to anthropogenic contaminants. Further, the same species of *Oedogonium* has previously been cultured in ash dam water (high in contaminants) and the resulting biosolid concentrations of these contaminants were not high compared to seaweeds which are currently used in numerous biomass applications [29,64].

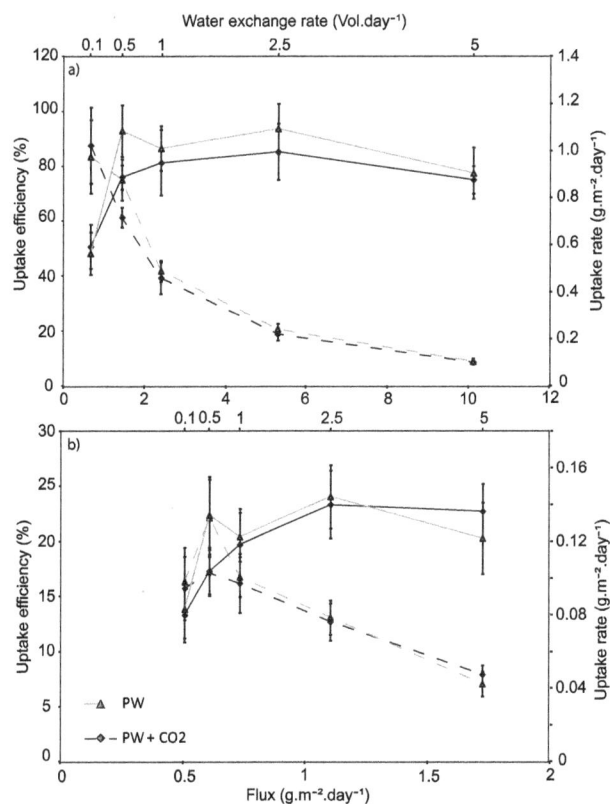

**Figure 3. Nitrogen and phosphorous uptake rates and efficiencies.** Uptake rates (solid lines) and efficiency (dashed lines) of *Oedogonium* when cultured in pond water (grey lines) and pond water +$CO_2$ (black lines) under increasing nutrient flux of a) nitrogen and b) phosphorous. For reference purposes the five water exchange rates that these nutrient fluxes correspond to have been added to the top x-axis.

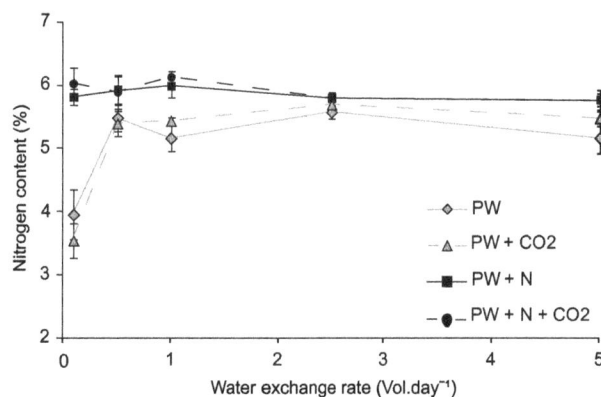

**Figure 4. *Oedogonium* nitrogen content.** Mean (±SE) internal nitrogen content of *Oedogonium* cultured under 5 water exchange rates (0.1, 0.5, 1, 2.5 and 5 vol.day$^{-1}$) using fish pond water (PW) with and without nitrogen (N) and carbon ($CO_2$) supplementation. Note protein content of *Oedogonium* can be calculated by multiplying %N by 4.7 [28].

The ability of *Oedogonium* to adapt and be successfully cultured in a wide range of flow rates makes it particularly suited to larger, industrial-scale cultivation, where large quantities of wastewater can be cleaned whilst simultaneously producing a feedstock biomass that can be tailored to meet different biomass applications. At these larger scales, moving large volumes of water becomes prohibitively expensive. This is one of the limitations of marine macroalgae (seaweed), where relatively high water exchange rates, greater than 12 volumes per day are required to maximise biomass production rates [20,21]. These high water exchange rates provide a constant supply of nitrogen but also prevent carbon limitation as the pH of culture water is kept relatively low [20,23,24]. In contrast, *Oedogonium* had its highest productivity at the lower water exchange rates 0.1–0.5 vol.day$^{-1}$, where nitrogen availability was reduced. At the lowest water exchange rates, the *Oedogonium* biomass was partly utilizing its internal nitrogen reserves to generate new biomass; the internal nitrogen content of these cultures decreased by 22–30% during the culture period. This indicates that processing inorganic nitrogen from the water column is potentially an energetically expensive process relative to using nitrogen that has already been processed and stored within the cell. We expect that this feature can be used to our advantage and that managing cultures with periodic nitrogen limitation followed by pulse additions of nitrogen could result in further increases in biomass production rates. These periodic water renewals would also have the additional benefit of increasing the energy efficiency of the culture system as pumping costs could be further reduced.

While the addition of $CO_2$ to incoming water increased productivity, this increase (between 2 and 25%) was much lower than the 2.5 fold increase in productivity previously identified when *Oedogonium* was cultured in batch conditions with weekly water exchanges [25]. The main difference between these studies is the use of bore water in the current study, which had a total alkalinity ranging between 210–227 mgCO$_3$.L$^{-1}$ compared to 40–60 mgCO$_3$.L$^{-1}$ for the dechlorinated town water which was used in Cole et al. [25]. This higher alkalinity means that for the same pH and temperature, the culture water in the current study had a DIC concentration four times higher than when using dechlorinated townwater. This is an important result as it demonstrates that macroalgae can be successfully cultured at high productivities in areas which do not have access to a source of waste $CO_2$, provided that the alkalinity of the culture water can meet the carbon needs of the algae. This may be a particularly useful approach if *Oedogonium* is used as an industrial tool for carbon capture and offset, whereby each dry weight kilogram of cultured biomass extracts ~400 g of carbon from the water [25].

An unexpected result from this study was the decrease in productivity when pond water was supplemented with nitrogen, especially at low water exchange rates. A consequence of the way nitrogen was added to our tanks is that the ammonium accumulates in the low exchange rate treatments as each hourly addition increases the ammonium concentration until a new equilibrium is reached between the concentration of the ammonium-addition water and the dilution effect of the incoming pond water. In our lowest exchange rates the TAN concentration continually increased over the culture period up to 17 mg.L$^{-1}$ which was four and nine times higher than the tanks receiving a water exchange rate of 0.5 and 1 vol.day$^{-1}$ respectively. The decrease in productivity observed in these cultures is most likely a result of ammonium toxicity. In solution, ammonia occurs in two forms, ammonium ($NH_4^+$) and unionised ammonia ($NH_3$). $NH_3$ is toxic to animals and plants, as it is uncharged, lipid soluble and can traverse biological membranes [65]. The proportion of each

form is primarily dependent upon pH, with the proportion of $NH_3$ increasing rapidly as the pH exceeds 9 [66,67]. At a water temperature of 22°C the proportion of $NH_3$ increases from 12.7% at a pH of 8.5 to 31.5% at pH 9, 59.2% at pH 9.5 and 82.1% at pH 10 [66]. The pH of the low flow treatments (10% exchange) ranged between 9.1–9.36. Subsequently, these low water exchange treatments had a 58% lower productivity than raw pond water, however, as the rate of water exchange increased and the pH of culture water decreased this difference in productivity between treatments converged. The importance of reducing the pH to below 9 can also be seen in our carbon addition treatment, where at exchange rates greater than 0.5 vol.day$^{-1}$ the additional carbon could maintain the pH below 9 and productivity was not significantly impaired. In contrast, in treatments without the addition of $CO_2$ to reduce culture pH, ammonium toxicity decreased productivity until flow rates reached 2.5 volumes per day. A similar effect of ammonium toxicity was observed for duckweed, where productivity decreased linearly as the proportion of $NH_3$ increased and the maximum tolerance level for unionised ammonia was approximately 8 mg $NH_3$-N.L$^{-1}$ [65]. The ability of *Oedogonium* to survive relatively high TAN concentrations with only sublethal effects on growth is a positive result and indicates that *Oedogonium* would be a suitable species to remediate sewage wastewater, which can have TAN concentrations exceeding 20 mg.L$^{-1}$ [38,68], as long as a carbon source is supplied to limit the rise in culture pH and the resultant toxicity of $NH_3$ [65].

## Conclusion

This study has demonstrated that *Oedogonium* can be successfully cultured at low water exchange rates using fish farm discharge water without any need for nutrient or carbon supplementation. We have also demonstrated that the optimal culture of *Oedogonium* occurs at relatively low exchange rates of between 0.5–1 vol.day$^{-1}$ and does not require high concentrations of nitrogen and phosphorous to grow at high productivities. Production rates were generally highest when nutrients were supplied in only slight excess to the theoretical amount required to maintain growth at any given internal nitrogen content. In this study a nitrogen flux of 1.4 g.m$^{-2}$.day$^{-1}$, and a phosphorous flux of 0.6 g.m$^{-2}$.day$^{-1}$ was the minimum needed to maintain *Oedogonium* growth at 16–17 g DW.m$^{-2}$.day$^{-1}$ and a protein content of 25%. A simple model of minimum inputs shows that for every gram of productivity (g DW.m$^{-2}$.day$^{-1}$), *Oedogonium* requires 0.09 g.m$^{-2}$.day$^{-1}$ of nitrogen and 0.04 g.m$^{-2}$.day$^{-1}$ of phosphorous to maintain growth without nutrient limitation whilst simultaneously maintaining a high-nutrient uptake rate and efficiency. In this regard the culture of *Oedogonium* on farms for the purpose of nutrient recovery is an effective bioremediation solution for treating nutrient wastewater while also supplying a useful protein resource to the agricultural industry.

## Acknowledgments

We thank Jonathan Moorhead, Thomas Mannering, Lewis Anderson and Giovanni Del Frari for assistance with experiments and Good Fortune Bay Fisheries Ltd, Kelso for providing space and the use of their wastewater for this experiment.

## Author Contributions

Conceived and designed the experiments: AJC RdN NAP. Performed the experiments: AJC. Analyzed the data: AJC RdN NAP. Contributed reagents/materials/analysis tools: AJC RdN NAP. Wrote the paper: AJC RdN NAP.

# References

1. Frosch RA (1992) Industrial ecology: a philosophical introduction. Proc Natl Acad Sci USA 89: 800–803.

2. Lowe EA, Evans LK (1995) Industrial ecology and industrial ecosystems. J Clean Prod 3: 47–53.

3. Nzihou A, Lifset R (2010) Waste valorization, loop-closing, and industrial ecology. J Ind Ecol 14: 196–199.

4. Tilman D, Cassman KG, Matson PA, Naylor R, Polasky S (2002) Agricultural sustainability and intensive production practices. Nature 418: 671–677.

5. Bouwman L, Goldewijk KK, Van Der Hoek KW, Beusen AHW, Van Vuuren DP, et al. (2011) Exploring global changes in nitrogen and phosphorus cycles in agriculture induced by livestock production over the 1900–2050 period. Proc Natl Acad Sci USA.

6. Liu SM, Zhang J, Chen HT, Zhang GS (2005) Factors influencing nutrient dynamics in the eutrophic Jiaozhou Bay, North China. Prog Oceanogr 66: 66–85.

7. Liu D, Keesing JK, Xing Q, Shi P (2009) World's largest macroalgal bloom caused by expansion of seaweed aquaculture in China. Mar Pollut Bull 58: 888–895.

8. Bianchi TS, DiMarco SF, Cowan Jr JH, Hetland RD, Chapman P, et al. (2010) The science of hypoxia in the Northern Gulf of Mexico: A review. Sci Total Environ 408: 1471–1484.

9. Conley DJ, Paerl HW, Howarth RW, Boesch DF, Seitzinger SP, et al. (2009) Controlling eutrophication: nitrogen and phosphorus. Science 323: 1014–1015.

10. Zhang E, Liu E, Jones R, Langdon P, Yang X, et al. (2010) A 150-year record of recent changes in human activity and eutrophication of Lake Wushan from the middle reach of the Yangze River, China. J Limnol 69: 235–241.

11. Rabalais NN, Turner RE, Gupta BKS, Platon E, Parsons ML (2007) Sediments tell the history of eutrophication and hypoxia in the northern Gulf of Mexico. Ecol Appl 17: S129–S143.

12. Rabalais NN, Turner RE, Díaz RJ, Justić D (2009) Global change and eutrophication of coastal waters. ICES J Mar Sci 66: 1528–1537.

13. Hillman WS, Dudley CJ (1978) The uses of duckweed: The rapid growth, nutritional value, and high biomass productivity of these floating plants suggest their use in water treatment, as feed crops, and in energy-efficient farming. Am Sci 66: 442–451.

14. Pizarro C, Kebede-Westhead E, Mulbry W (2002) Nitrogen and phosphorus removal rates using small algal turfs grown with dairy manure. J Appl Phycol 14: 469–473.

15. Mulbry W, Wilkie A (2001) Growth of benthic freshwater algae on dairy manures. J Appl Phycol 13: 301–306.

16. Mulbry W, Kondrad S, Pizarro C, Kebede-Westhead E (2008) Treatment of dairy manure effluent using freshwater algae: algal productivity and recovery of manure nutrients using pilot-scale algal turf scrubbers. Bioresour Technol 99: 8137–8142.

17. Cohen I, Neori A (1991) Ulva lactuca biofilters for marine fishpond effluents. I. Ammonia uptake kinetics and nitrogen content. Bot Mar 34: 475–482.

18. Neori A, Cohen I, Gordin H (1991) Ulva lactuca biofilters for marine fishpond effluents. II. Growth rate, yield and C: N ratio. Bot Mar 34: 483–490.

19. Demetropoulos CL, Langdon CJ (2004) Enhanced production of Pacific dulse (Palmaria mollis) for co-culture with abalone in a land-based system: effects of seawater exchange, pH, and inorganic carbon concentration. Aquaculture 235: 457–470.

20. Mata L, Schuenhoff A, Santos R (2010) A direct comparison of the performance of the seaweed biofilters, Asparagopsis armata and Ulva rigida. J Appl Phycol 22: 639–644.

21. Neori A, Msuya FE, Shauli L, Schuenhoff A, Kopel F, et al. (2003) A novel three-stage seaweed (Ulva lactuca) biofilter design for integrated mariculture. J Appl Phycol 15: 543–553.

22. Mata L, Gaspar H, Santos R (2012) Carbon/nutrient balance in relation to biomass production and halogenated compound content in the red alga Asparagopsis taxiformis (Bonnemaisoniaceae). J Phycol 48: 248–253.

23. Israel A, Katz S, Dubinsky Z, Merrill J, Friedlander M (1999) Photosynthetic inorganic carbon utilization and growth of Porphyra linearis (Rhodophyta). J Appl Phycol 11: 447–453.

24. Mata L, Silva J, Schuenhoff A, Santos R (2007) Is the tetrasporophyte of Asparagopsis armata (Bonnemaisoniales) limited by inorganic carbon in integrated aquaculture? J Phycol 43: 1252–1258.

25. Cole AJ, Mata L, Paul NA, de Nys R (2013) Using $CO_2$ to enhance carbon capture and biomass applications of freshwater macroalgae. Glob Change Biol Bioenergy: 10.1111/gcbb.12097.

26. Schuenhoff A, Mata L, Santos R (2006) The tetrasporophyte of Asparagopsis armata as a novel seaweed biofilter. Aquaculture 252: 3–11.

27. Lawton RJ, de Nys R, Paul NA (2013) Selecting reliable and robust freshwater macroalgae for biomass applications. PloS one 8: e64168.

28. Neveux N, Magnusson M, Maschmeyer T, de Nys R, Paul NA (2014) Comparing the potential production and value of high-energy liquid fuels and protein from marine and freshwater macroalgae. Glob Change Biol Bioenergy: DOI: 10.1111/gcbb.12171.

29. Roberts DA, de Nys R, Paul NA (2013) The effect of $CO_2$ on algal growth in industrial waste water for bioenergy and bioremediation applications. PloS one 8: e81631.

30. Lewis E, Wallace D, Allison LJ (1998) Program developed for CO2 system calculations: Carbon Dioxide Information Analysis Center, managed by Lockheed Martin Energy Research Corporation for the US Department of Energy.

31. Borowitzka MA (1999) Commercial production of microalgae: ponds, tanks, and fermenters. In: R. Osinga JTJGB, Wijffels RH, editors. Prog Ind Microbiol: Elsevier. pp. 313–321.

32. Greenwell HC, Laurens LML, Shields RJ, Lovitt RW, Flynn KJ (2010) Placing microalgae on the biofuels priority list: a review of the technological challenges. J R Soc Interface 7: 703–726.

33. Scott SA, Davey MP, Dennis JS, Horst I, Howe CJ, et al. (2010) Biodiesel from algae: challenges and prospects. Curr Opin Biotechnol 21: 277–286.

34. Wilkie AC, Mulbry WW (2002) Recovery of dairy manure nutrients by benthic freshwater algae. Bioresour Technol 84: 81–91.

35. Kebede-Westhead E, Pizarro C, Mulbry WW, Wilkie AC (2003) Production and nutrient removal by periphyton grown under different loading rates of anaerobically digested flushed dairy manure. J Phycol 39: 1275–1282.

36. Kebede-Westhead E, Pizarro C, Mulbry WW (2006) Treatment of swine manure effluent using freshwater algae: production, nutrient recovery, and elemental composition of algal biomass at four effluent loading rates. J Appl Phycol 18: 41–46.

37. Craggs RJ, Adey WH, Jessup BK, Oswald WJ (1996) A controlled stream mesocosm for tertiary treatment of sewage. Ecol Eng 6: 149–169.

38. Craggs R, Sutherland D, Campbell H (2012) Hectare-scale demonstration of high rate algal ponds for enhanced wastewater treatment and biofuel production. J Appl Phycol 24: 1–9.

39. Biller P, Ross A (2011) Potential yields and properties of oil from the hydrothermal liquefaction of microalgae with different biochemical content. Bioresour Technol 102: 215–225.

40. Neveux N, Yuen A, Jazrawi C, Magnusson M, Haynes B, et al. (2014) Biocrude yield and productivity from the hydrothermal liquefaction of marine and freshwater green macroalgae. Bioresour Technol 155: 334–341.

41. Papadopoulos G, Goulas C, Apostolaki E, Abril R (2002) Effects of dietary supplements of algae, containing polyunsaturated fatty acids, on milk yield and the composition of milk products in dairy ewes. J Dairy Res 69: 357–365.

42. Singh AP, Avramis CA, Kramer JK, Marangoni AG (2004) Algal meal supplementation of the cows' diet alters the physical properties of milk fat. J Dairy Res 71: 66–73.

43. Gardiner GE, Campbell AJ, O'Doherty JV, Pierce E, Lynch PB, et al. (2008) Effect of Ascophyllum nodosum extract on growth performance, digestibility, carcass characteristics and selected intestinal microflora populations of grower–finisher pigs. Anim Feed Sci Technol 141: 259–273.

44. Dierick N, Ovyn A, De Smet S (2010) In vitro assessment of the effect of intact marine brown macro-algae Ascophyllum nodosum on the gut flora of piglets. Livest Sci 133: 154–156.

45. O'Doherty JV, Dillon S, Figat S, Callan JJ, Sweeney T (2010) The effects of lactose inclusion and seaweed extract derived from Laminaria spp. on performance, digestibility of diet components and microbial populations in newly weaned pigs. Anim Feed Sci Technol 157: 173–180.

46. Craigie JS (2011) Seaweed extract stimuli in plant science and agriculture. J Appl Phycol 23: 371–393.

47. Boland MJ, Rae AN, Vereijken JM, Meuwissen MP, Fischer AR, et al. (2012) The future supply of animal-derived protein for human consumption. Trends Food Sci Technol.

48. Aiking H (2011) Future protein supply. Trends Food Sci Technol 22: 112–120.

49. Becker E (2007) Micro-algae as a source of protein. Biotechnol Adv 25: 207–210.

50. Tokuşoglu Ö, Üunal M (2003) Biomass nutrient profiles of three microalgae: Spirulina platensis, Chlorella vulgaris, and Isochrysis galbana. J Food Sci 68: 1144–1148.

51. Nielsen M, Bruhn A, Rasmussen M, Olesen B, Larsen M, et al. (2011) Cultivation of Ulva lactuca with manure for simultaneous bioremediation and biomass production. J Appl Phycol: 1–10.

52. Angell AR, Mata L, Nys R, Paul NA (2014) Variation in amino acid content and its relationship to nitrogen content and growth rate in Ulva ohnoi (Chlorophyta). J Phycol 50: 216–226.

53. Young VR, Pellett PL (1994) Plant proteins in relation to human protein and amino acid nutrition. Am J Clin Nutrit 59: 1203S–1212S.

54. Price P, Parsons J (1975) Lipids of seven cereal grains. J Am Oil Chem Soc 52: 490–493.

55. Grieshop CM, Fahey GC (2001) Comparison of quality characteristics of soybeans from Brazil, China, and the United States. J Agric Food Chem 49: 2669–2673.

56. Herrera-Saldana R, Huber J, Poore M (1990) Dry matter, crude protein, and starch degradability of five cereal grains. J Dairy Sci 73: 2386–2393.

57. Batajoo KK, Shaver RD (1998) In situ dry matter, crude protein, and starch degradability of selected grains and by-product feeds. Anim Feed Sci Technol 71: 165–176.

58. Mossé J, Baudet J (1983) Crude protein content and amino acid composition of seeds: variability and correlations. Plant Food Hum Nutr 32: 225–245.

59. Ragaee S, Abdel-Aal E-SM, Noaman M (2006) Antioxidant activity and nutrient composition of selected cereals for food use. Food Chem 98: 32–38.

60. Webber III CL (1993) Crude protein and yield components of six kenaf cultivars as affected by crop maturity. Ind Crop Prod 2: 27–31.

61. Mendieta-Araica B, Spörndly R, Reyes-Sánchez N, Spörndly E (2011) Moringa (Moringa oleifera) leaf meal as a source of protein in locally produced concentrates for dairy cows fed low protein diets in tropical areas. Livest Sci 137: 10–17.

62. Aye P, Adegun M (2013) Chemical composition and some functional properties of *Moringa*, *Leucaena* and *Gliricidia* leaf meals. Agr Biol J N Am 4: 71–77.

63. Michalak I, Chojnacka K (2008) The application of macroalga *Pithophora varia* Wille enriched with microelements by biosorption as biological feed supplement for livestock. J Sci Food Agric 88: 1178–1186.

64. Saunders RJ, Paul NA, Hu Y, de Nys R (2012) Sustainable sources of biomass for bioremediation of heavy metals in waste water derived from coal-fired power generation. PloS one 7: e36470.

65. Körner S, Das SK, Veenstra S, Vermaat JE (2001) The effect of pH variation at the ammonium/ammonia equilibrium in wastewater and its toxicity to *Lemna gibba*. Aquat Bot 71: 71–78.

66. Emerson K, Russo RC, Lund RE, Thurston RV (1975) Aqueous ammonia equilibrium calculations: effect of pH and temperature. Can J Fish Aquat Sci 32: 2379–2383.

67. Abeliovich A, Azov Y (1976) Toxicity of ammonia to algae in sewage oxidation ponds. Appl Environ Microbiol 31: 801–806.

68. Wrigley T, Toerien D (1990) Limnological aspects of small sewage ponds. Water Res 24: 83–90.

# 10

# Sensory Ecology of Water Detection by Bats

**Danilo Russo**[1,2]*, **Luca Cistrone**[3], **Gareth Jones**[2]

**1** Laboratorio di Ecologia Applicata, Dipartimento Ar.Bo.Pa.Ve., Facoltà di Agraria, Università degli Studi di Napoli Federico II Portici, Napoli, Italy, **2** School of Biological Sciences, University of Bristol, Bristol, United Kingdom, **3** Forestry and Conservation, Cassino, Italy

## Abstract

Bats face a great risk of dehydration, so sensory mechanisms for water recognition are crucial for their survival. In the laboratory, bats recognized any smooth horizontal surface as water because these provide analogous reflections of echolocation calls. We tested whether bats also approach smooth horizontal surfaces other than water to drink in nature by partly covering watering troughs used by hundreds of bats with a Perspex layer mimicking water. We aimed 1) to confirm that under natural conditions too bats mistake any horizontal smooth surface for water by testing this on large numbers of individuals from a range of species and 2) to assess the occurrence of learning effects. Eleven bat species mistook Perspex for water relying chiefly on echoacoustic information. Using black instead of transparent Perspex did not deter bats from attempting to drink. In *Barbastella barbastellus* no echolocation differences occurred between bats approaching the water and the Perspex surfaces respectively, confirming that bats perceive water and Perspex to be acoustically similar. The drinking attempt rates at the fake surface were often lower than those recorded in the laboratory: bats then either left the site or moved to the control water surface. This suggests that bats modified their behaviour as soon as the lack of drinking reward had overridden the influence of echoacoustic information. Regardless of which of two adjoining surfaces was covered, bats preferentially approached and attempted to drink from the first surface encountered, probably because they followed a common route, involving spatial memory and perhaps social coordination. Overall, although acoustic recognition itself is stereotyped and its importance in the drinking process overwhelming, our findings point at the role of experience in increasing behavioural flexibility under natural conditions.

**Editor:** Brock Fenton, University of Western Ontario, Canada

**Funding:** DR was funded by the Abruzzo, Lazio and Molise National Park (Project: Bats and Forestry, contract n 168 12 April 2010). www.parcoabruzzo.it. The funders had no role in study design, data collection and analysis, decision to publish, or preparation of the manuscript.

**Competing Interests:** The authors have declared that no competing interests exist.

* E-mail: danrusso@unina.it

## Introduction

Because the risk of dehydration is the greatest physiological threat to life on land, drinking water is a fundamental resource for all terrestrial animals [1]. Due to their peculiar morphology and physiology, bats often face the risk of dehydration. Much water is lost through their body surface, especially via the respiratory system and the extensive surfaces of wing membranes [2,3].

Although bats may show physiological adaptations to limit water loss, such as specific qualitative and quantitative chemical composition of the lipid matrix in the epidermis' stratum corneum [4], compensating water loss by drinking is the main mechanism adopted by these mammals to counter dehydration. For this reason even hibernating bats may periodically arouse from torpor to drink [5,6]. The importance of water availability has been emphasised in studies addressing the impact of climate change on bats [7] as well as in those modelling bat distribution patterns [8].

Water also represents a major source of minerals for bats: calcium – reproductive females need to restore the mineral reservoirs mobilized for skeletal development in pups [9,10,11]; or sodium, which is particularly limiting in tropical environments [12]. In some fruit-eating bats, minerals in water are important to counter the effects of secondary plant metabolites largely ingested at times of high energetic demand [13].

Sensory and behavioural adaptations to discover or localise water are therefore subject to strong selective pressure. A groundbreaking study [14] revealed that water recognition is innate and that bats use echoacoustic cues to locate smooth water surfaces. Accordingly, bats perceive any horizontal smooth surface as water because it provides a typical mirror-like reflection of echolocation calls. That study [14] was performed in the laboratory where this recognition process was recorded in 15 species from three families (Rhinolophidae, Vespertilionidae and Miniopteridae), i.e. the phenomenon seems taxonomically widespread among bats. The innate nature of water recognition also appears clear because naïve juveniles show it too [14].

Although it was also suggested [14] that other sensory cues may be integrated for the process of water recognition, at least in a laboratory setting these seemed to be of minor importance compared with echoacoustic cues. Behavioural studies in captivity offer unique chances of effectively controlling the experimental design and the influences of variables potentially affecting the results, but they also involve constraints [15] including the possible influence of an artificial environment, the limited number of individuals tested and the effects of stress on captive subjects. It is thus especially useful to validate the results obtained in captivity with experiments under natural conditions [16].

The main objective of our study was to build on previous laboratory work [14] to confirm its outcome under natural conditions, where we could test a large number of individuals from several species –11, six of which not studied previously [14]. We therefore hypothesise that bats will mistake artificial smooth horizontal surfaces for water in nature, and will attempt to drink from them.

Under natural conditions, where familiar cues and features of the drinking site occur, other sensory information as well as spatial memory may be involved to determine a precise "local" cognitive picture of water, so that any departure from the expected features could be noticed more readily than in a completely unfamiliar setting such as that of a laboratory. Although the echoacoustic recognition of water surface is innate [14], we hypothesise that especially at familiar drinking sites other sensory cues might play an important role. If so, bats should be less likely to be deceived by the artificial layer and more prone to detect and interpret environmental changes based on visual, olfactory, gustative or mechanoreceptorial cues, or spatial memory.

For one model species (*Barbastella barbastellus*) we also compared echolocation sequences emitted by bats approaching real and fake water respectively (an aspect not covered in the previous [14] laboratory study). We hypothesise there will be no difference in echolocation behaviour of bats approaching water and artificial smooth surfaces, providing further evidence that bats approaching fake water do actually try to drink rather than simply attempt to explore the artificial surface.

Besides, our experiment aimed to test the possible influence of learning under natural conditions and in a familiar area, where the location of other easily reachable water sources is known. We hypothesise that an unsuccessful drinking experience such as that determined by replacing real water with an artificial surface mimicking it would, after a few drinking attempts, override the influence of the acoustic water-like cues the latter provided. This would, in turn, prompt the bat to leave the site and move to the closest real water source available.

## Materials and Methods

### Experimental Design

Our experiments were performed in August 2011 at the Abruzzo Lazio and Molise National Park, in the Italian central Apennines, under permit from the Park's authorities according to Law 6 December 1991 n° 394. The study involved no animal capture or handling so the permit only regarded observational work. For our experiment we selected three watering troughs designed to provide water for cattle and used by hundreds of drinking bats every night (D. Russo, *pers. obs.*). All sites were characterized by a similar surrounding habitat dominated by mature beech forest and pastures and were located at 1220–1563 m a.s.l.

For experiment 1, we used two structurally similar, adjoining watering troughs, ca. $6 \times 1.5$ m at each of two sites (sites A and B). In such cases we covered one trough with a 0.5 cm thick transparent Perspex sheet (treatment) whereas the other was left uncovered (control). The sheet was placed immediately above the real water surface. At the third site (site C) we used a single $12 \times 1.5$ m watering trough: in that case we laid a Perspex layer on half of it and left the remaining watering trough surface free. The transparent Perspex did not change the colour of the water surface but introduced potentially significant olfactory and tactile cues (detected by the bats which contacted the artificial surface when attempting to drink from it). We hypothesised that if the latter had no deterring effect on bats, they would show an equal likelihood of

attempting to drink at either surface. To control for possible differences in bat use between the watering troughs, or their sections used as treatments, we repeated the experiment twice at each site, each time covering a different watering trough (or its section, as done for site C).Experiment 2 aimed to test the effect of colour. To test a larger number of bats we covered the watering trough (or a section of it for site C) where in experiment 1 we had recorded a higher drinking activity, this time using a black sheet of Perspex. Even in dim light the colour difference between the surface and the adjoining water was obvious to us and we assumed it to be clear to bats as some vespertilionids can discriminate small differences in brightness even at low light intensities [17]. Then we compared the levels of bat activity between the transparent (recorded in experiment 1) vs. black Perspex layers, assuming that if colour had no deterring effect on bats, they would show an equal likelihood of mistaking either black or transparent Perspex for water.

For both experiments, we preliminarily ensonified both the Perspex layer and water with a natural *Myotis mystacinus* call played back through a Pettersson D1000X detector and an L-400 loudspeaker (Pettersson Elektronik AB, Uppsala); the frequency range for the latter was 10–110 kHz. The speaker's acoustic axis formed an angle of ca. $45°$ with the surface. Qualitative examination of waveforms and spectrograms of the echoes generated by the two surfaces recorded with another D1000X detector placed above the loudspeaker (sampling rate 384000 Hz) suggested there were no detectable difference in structure so we assumed the Perspex surfaces and the water to convey analogous echoacoustic information [14].

### Data Collection and Analysis

Bat activity was filmed continuously with a Sony Handycam HDR – XR520VE (focal distance 5.5–66.0 mm) nightshot videocamera mounted on a 1.8 m tripod positioned at least 2 m away from the watering trough to avoid interference with flying bats. The videocamera was located close to the watering trough's major axis and oriented to include both the covered and uncovered water surfaces. In preliminary tests this setting proved most effective to distinguish the trajectories followed by approaching bats. The scene was illuminated with an additional infrared lamp. Bat echolocation calls were recorded with two Pettersson D1000X bat detectors which continuously sampled in the real-time mode (sampling rate 384000 Hz) and saved recordings onto 4 Gb flashcards. The bat detectors were placed on the edge of each watering trough (or its section) at ca. half of its length and the microphone directed toward its centre. Audio and video recordings were synchronized before starting the experiment so we could associate all filmed bats to their echolocation calls. When black Perspex was used (experiment 2), each minute we also recorded illuminance (in lux) at ground level with a Delta Ohm (Delta Ohm s.r.l., Padua, Italy) photo-radiometer (spectral range 450–760 nm, operational range 0–200,000 lux, resolution $\leq 200$ lux $= 0.1$; $>200$ lux $= 1$). Each recording session started when the first bat approached the drinking site, generally within 20 min after sunset, and lasted 60 min. Temperature was similar across nights (ca. $16°C$) and wind intensity was negligible so these were deemed to exert the same influence on bat activity across different trials.

To record the number of drinking attempts per bat, in the laboratory audio and video recordings were examined synchronously by two operators. In most cases this allowed us to keep track of the movement of all recorded bats and count repeated drinking attempt events. When a bat disappeared from both audio and video recordings it was assumed to have left. In a few such cases we also used the notes made in the field where we attempted to track visually a bat that had disappeared from the video screen to establish whether it had either left or moved to the control watering trough to drink. Although bats that left may have returned later to the site we assume this risk to be negligible at least for those that had drunk successfully. We also ideally divided the area around each watering trough in four quadrants and assigned the bats to one of them according to which quadrants they approached the trough from.

Sound analysis was used to identify bats to species. We used BatSound 4 (Pettersson Elektronik AB, Uppsala) to generate spectrograms with a 512·pt FFT Hamming window, 98% overlap (providing a 975 Hz frequency resolution). For bat identification, one call per sequence was selected at random among those with a good signal-to-noise ratio and measurements were taken according to [18].

For species identification we used simplified versions of the multivariate discriminant functions [18] developed respectively for species emitting frequency modulated calls (FM) and for those whose calls start with a broadband sweep and end with a narrowband tail (FM-QCF). Such functions only covered species representing >1% of bats mistnetted at the experiment sites in summers 2000–2011. The function for species emitting FM-QCF calls included Kuhl's pipistrelle *Pipistrellus kuhlii* (probability of correct identification = 0.98), common pipistrelle *Pipistrellus pipistrellus* (0.98), Savi's bat *Hypsugo savii* (1.00), Leisler's bat *Nyctalus leisleri* (1.00) and Schreiber's bat *Miniopterus schreibersii* (99.1) [Wilk's $\lambda = 0.01783$, $P<0.0001$]. Probabilities of correct identification for species broadcasting FM calls were also high: greater mouse-eared bat *Myotis myotis* (0.81), whiskered bat *Myotis mystacinus* (0.69), Natterer's bat *Myotis nattereri* (0.83), brown long-eared bat *Plecotus auritus* (0.96) and barbastelle bat *B. barbastellus* (0.90) [Wilk's $\lambda = 0.06734$, $P<0.0001$]. Although *M. mystacinus* calls may be easily confused with those of *Myotis daubentonii* [18,19], the latter was never mistnetted at any of the drinking sites used for the experiment during over 10 summers of bat surveys so in our sites the risk of misidentification was ruled out. While *M. mystacinus* is the most abundant bat in the beech forests where the experiments were carried out, *Myotis brandtii* and *Myotis alcathoe* – other possible sources of confusion – are only very rarely encountered there (D. Russo, *pers. obs.*). However, since those species were not covered by the classification function we cannot exclude that we misclassified few bats from those species as *M. mystacinus* but the effects on our analysis are certainly negligible. Other details, including sample sizes used to develop functions, are given in [18]. When possible, after a response was obtained the identification was improved further by looking at diagnostic features of the spectrogram. This was especially useful for *B. barbastellus* whose alternation of call types 1 ad 2 offers unambiguous species recognition [20].

For experiment 1 (transparent Perspex) the variables we tested were respectively the total number of approaches to either surface (Perspex vs. water), the number of individuals performing them, and the number of approaches per subject. We employed a repeated-measure General Linear Model (GLM) ANOVA (factors: site, watering trough covered during the experiment and surface type, i.e. Perspex or water) and entered site as a random factor. We first tested both the variables' main effects and their interactions. We then removed the interactions when not significant. Only final models are presented here. When the residual distribution did not conform to normality according to a Ryan Joiner test, we log-transformed the data to meet the ANOVA assumptions.

For the experiment 2, for each site we first calculated the ratio between the numbers of passes (or bats) associated with either black or transparent Perspex and the corresponding total numbers of passes (or bats), i.e. those recorded over Perspex + those over water, observed during a trial. We used this ratio as a proxy for the number of times bats mistook Perspex for water. Data analysis was done by using two-way GLM ANOVA for repeated measures, entering site as a random factor; treatment was black vs. transparent Perspex.

We restricted further analysis only to trials done with black Perspex. We calculated the same ratios, this time for 5-min intervals and associated them to mean light intensity measured during the sampled interval. We tested whether reduced ambient light would correspond to a higher numbers of passes (or bats) at the covered watering troughs, as predicted if bats are deterred by the colour of the surface approached. Because the distribution of light intensity measurements did not meet the assumptions of a parametric Analysis of Covariance – ANCOVA [21], we applied Quade's non-parametric alternative procedure [22].

## Echolocation Differences during Approach

To determine whether approaching water or Perspex induced any difference in echolocation behaviour, we selected *B. barbastellus* as a model species because as illustrated above we were always fully sure of the identity of recordings.

For 30 approaches (15 for either condition, Perspex or water) we could select sufficient echolocation sequences in situations only differing for surface type, all other variables being equal (i.e. bats approaching the same watering trough from the same direction and under a similar angle of attack). We measured the interpulse interval (IPI, i.e. the time between two consecutive calls) and plotted it vs. time to distinguish between search and approach phases [20] and to identify potential terminal phase events (see "Results"). We used all interpulse intervals corresponding to each phase to determine differences between 1) terminal phase and the remaining approach phase and 2) substrate (water vs. Perspex) by a two-way GLM ANOVA. To avoid pseudo-replication, values were averaged for each sequence and means used for analysis. Data were log-transformed to meet ANOVA assumptions and normality was tested with a Ryan-Joiner test.

We measured duration, starting frequency and terminal frequency from six calls per sequence (three from approach, three from the terminal phase) taken at random from those showing a good signal-to-noise ratio. Duration was measured from oscillograms. We obtained spectrograms with 1024-point FFTs and an FFT Hamming window with a 95% overlap, providing a frequency resolution of 488 Hz. We measured the frequency of maximum energy (FMAXE), the starting (SF) and end (EF) frequencies from power spectra. SF and EF were measured at –25 dB relative to the amplitude of the frequency of maximum energy from the corresponding power spectra [23,24]. In this way we reduced subjectivity and ensured consistency in measurements – e.g. [24]. For search phase calls we refer to call types 1 and 2 following description provided in the scientific literature [18,20,25].

We used mean values obtained from the six calls per sequence to determine differences between phase (approach vs. terminal) and substrate (water vs. Perspex) by a two-way GLM ANOVA.

Statistical tests were carried out with Minitab rel. 13. In all tests, significance was set at $P<0.05$.

## Results

### Experiment 1: Water vs. Transparent Perspex

In the first experiment we recorded 1484 drinking attempts by 299 bats on Perspex and 407 drinking events by 287 bats on water from 11 species (*M. emarginatus, M. myotis, M. mystacinus, M. nattereri,*

*P. auritus*, *B. barbastellus*, *M. schreibersii*, *N. leisleri*, *P. kuhlii*, *P. pipistrellus*, *H. savii*; Figure 1). The most frequently recorded bats were *M. mystacinus* and *B. barbastellus*; some species were only occasionally observed (Figure 1). Only three species (*M. mystacinus*, *B. barbastellus* and *P. auritus*) were recorded at all sites. The maximum number of drinking attempts made by an individual on Perspex largely varied both within and across species, from 1 up to 44 (as seen in one *H. savii*) or 45 (one *M. mystacinus*), but in most cases this was <10. On average, a single bat attempted to drink 5.4±5.6 times from Perspex vs. 1.4±0.6 times from water (ANOVA, $F_{1,49} = 13.35$, $P<0.005$; Figure 1) until it either gave up or moved to drink at the uncovered watering trough. When a bat repeatedly attempted to drink from Perspex, it did so trying again approximately every 2–3 sec, so its behaviour was conspicuous and tracking it from audio and video recordings generally obvious. Bats approaching Perspex actually touched the surface as they did to drink from the uncovered watering trough.

The overall number of approaches (log-transformed values) was only influenced by which watering trough was covered with Perspex during the experiment but did not differ between Perspex and water or across sites (Table 1). The same result was obtained for the number of bats approaching Perspex vs. water. The GLM ANOVA done on the number of drinking attempts per bat carried out on Perspex vs. water provided a different result: a significantly larger number of attempts were made on Perspex than on water (Table 1; Figure 1) but neither which watering trough was covered during the experiment nor the site influenced this variable.

We repeated this analysis only for the species that occurred at all sites during the experiment. For *M. mystacinus* we obtained the same outcome of the analysis carried out on all bats (Table 1).

The total number of *B. barbastellus* approaches was not influenced by surface type, watering trough covered during the trial or site (Table 1). The same results were obtained for the number of *B. barbastellus* bats approaching Perspex vs. water as well as for the number of drinking attempts per bat (Table 1). *P. auritus* made more drinking attempts on Perspex (10.2±11.1) than on water (1.6±2.51) yet the test's significance value was borderline (Table 1). Site, but not watering trough covered had a significant effect on this variable. Both site and watering trough covered influenced significantly the number of bats attempting to drink, but surface type had no effect (Table 1). The number of drinking attempts per bat was significantly higher at watering troughs covered with Perspex (4.3±4.0) than at those (0.6±0.6) left uncovered but neither site nor the watering trough covered influenced this variable (Table 1).

The above analyses illustrated that drinking bats tended to prefer one watering trough over another in all trials. An assessment of the directions followed by bats to reach the watering troughs in the six trials showed that most bats always came from one side (and, in four out of six cases, one quadrant was significantly selected over the others – see Fisher's exact tests in Figure 2) and drank at the first watering trough encountered, so that the first watering trough was disproportionately used over the other regardless of whether it had been covered with Perspex or not (Figure 2).

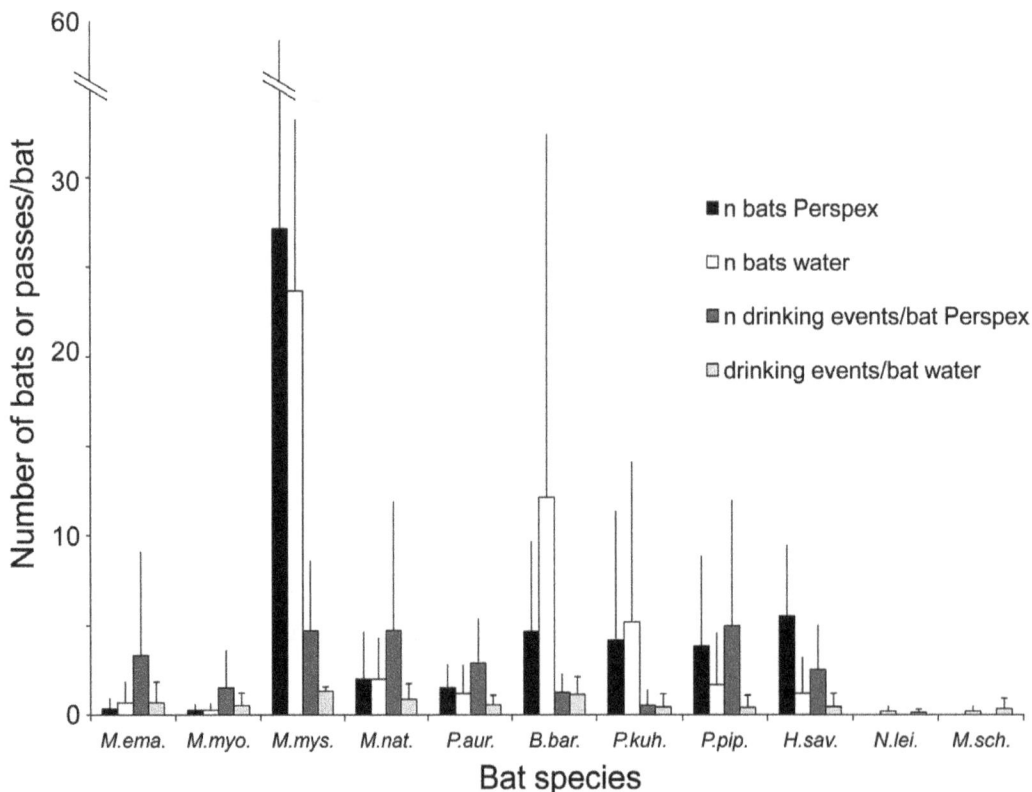

**Figure 1. Mean numbers of bats and mean numbers of drinking attempts per bat on Perspex vs. water categorized by species.** Error bars show the standard deviation. *M. ema.* = *Myotis emarginatus*, *M.myo* = *Myotis myotis*, *M.mys.* = *Myotis mystacinus M.nat.* = *Myotis nattereri*, *P.aur.* = *Plecotus auritus*, *B.bar.* = *Barbastella barbastellus*, *P.kuh.* = *Pipistrellus kuhlii*, *P.pip.* = *Pipistrellus pipistrellus*, *H.sav.* = *Hypsugo savii*, *N.lei.* = *Nyctalus leisleri*, *M.sch.* = *Miniopterus schreibersii*.

**Table 1.** Results of repeated measures General Linear Model ANOVA for the effect of watering trough covered, substrate type – water or Perspex – and site, on: a) the total number of drinking approaches to either surface (Perspex vs. water); b) the number of individual bats performing them; and c) the number of approaches per bat.

| | | d.f. | F | P |
|---|---|---|---|---|
| **All bats** | | | | |
| Overall number of approaches* | Watering trough covered | 1,7 | 22.56 | <0.005 |
| | Substrate | 1,7 | 2.87 | n.s. |
| | Site | 2,7 | 0.39 | n.s. |
| Number of approaching bats | Watering trough covered | 1,7 | 6.83 | <0.05 |
| | Substrate | 1,7 | 0.00 | n.s. |
| | Site | 2,7 | 2.62 | n.s. |
| Number of drinking attempts/bats | Watering trough covered | 1,7 | 2.12 | n.s. |
| | Substrate | 1,7 | 7.16 | <0.05 |
| | Site | 2,7 | 0.70 | n.s. |
| **Myotis mystacinus** | | | | |
| Overall number of approaches* | Watering trough covered | 1,7 | 18.60 | <0.005 |
| | Substrate | 1,7 | 3.69 | n.s. |
| | Site | 2,7 | 1.12 | n.s. |
| Number of approaching bats* | Watering trough covered | 1,7 | 11.29 | <0.05 |
| | Substrate | 1,7 | 0.00 | n.s. |
| | Site | 2,7 | 2.62 | n.s. |
| Number of drinking attempts/bats | Watering trough covered | 1,7 | 1.64 | n.s. |
| | Substrate | 1,7 | 6.16 | <0.05 |
| | Site | 2,7 | 0.70 | n.s. |
| **Barbastella barbastellus** | | | | |
| Overall number of approaches | Watering trough covered | 1,6 | 1.85 | n.s. |
| | Substrate | 1,6 | 0.39 | n.s. |
| | Site | 2,6 | 1.52 | n.s. |
| Number of approaching bats | Watering trough covered | 1,6 | 2.33 | n.s. |
| | Substrate | 1,6 | 1.40 | n.s. |
| | Site | 2,6 | 1.32 | n.s. |
| Number of drinking attempts/bats | Watering trough covered | 1,6 | 2.33 | n.s. |
| | Substrate | 1,6 | 1.40 | n.s. |
| | Site | 2,6 | 1.32 | n.s. |
| **Plecotus auritus** | | | | |
| Overall number of approaches* | Watering trough covered | 1,8 | 2.99 | n.s. |
| | Substrate | 1,8 | 8.01 | 0.047 |
| | Site | 2,8 | 10.28 | <0.05 |
| Number of approaching bats | Watering trough covered | 1,8 | 14.38 | <0.05 |
| | Substrate | 1.8 | 0.01 | n.s. |
| | Site | 2,8 | 9.58 | <0.05 |

**Table 1.** Cont.

| | | d.f. | F | P |
|---|---|---|---|---|
| Number of drinking attempts/bats | Watering trough covered | 1,8 | 0.19 | n.s. |
| | Substrate | 1,8 | 9.03 | <0.05 |
| | Site | 2,8 | 2.31 | n.s. |

Site was entered as a random factor. No interaction was significant so these were removed from final models. (*) = data log-transformed to meet the ANOVA assumptions; d.f. = degree of freedom.

## Experiment 2: Water vs. Black Perspex

Bats equally mistook transparent or black Perspex for water. Neither Perspex type (transparent vs. black Perspex) nor site influenced significantly the ratio between the number of approaches to either surface type and the overall (Perspex + water) number of approaches recorded (transparent vs. black Perspex: $F_{2,5} = 1.27$, n.s.; site: $F_{2,5} = 0.91$, n.s.). An identical result was obtained for the "number of bats approaching Perspex/ overall number of bats" ratio (transparent vs. black Perspex: $F_{2,5} = 14.1$, n.s.; site: $F_{2,5} = 6.03$, n.s.). The analysis at species level was only performed for M. mystacinus because this was the only bat for which sufficient numbers of subjects were recorded at drinking sites (both at covered and uncovered watering troughs) during the experiment done with black Perspex. In this case no significant effect was detected: for the number of approaches ratio, transparent vs. black Perspex: $F_{2,5} = 5.85$, n.s.; site: $F_{2,5} = 3.73$, n.s.; for the number of bats ratio, transparent vs. black Perspex: $F_{2,7} = 3.54$, n.s.; site: $F_{2,5} = 5.52$, n.s.

We then tested whether light or site influenced the number of approaches made by bats on black Perspex divided by the overall approach number (black Perspex + water) within 5-min intervals. During the experiment, mean light intensity (calculated from the 5-min interval values) at the three sites was $1.2 \pm 1.1$ lux, $0.5 \pm 0.3$ lux, and $0.2 \pm 0.1$ lux, lux respectively, whereas the corresponding fraction of the moon illuminated on those nights (data US Naval Observatory, Washington) was 76%, 90%, and 59%. A non-parametric ANCOVA applied to the whole dataset (all species) revealed a positive effect of light ($F_{1,17} = 8.69$, P<0.01) but no effect of site ($F_{2,17} = 2.87$, n.s.); the ratio between the number of bats approaching Perspex and the total (Perspex + water) showed again a positive effect of light ($F_{1,17} = 4.88$, P<0.05) and this time also a site effect ($F_{2,17} = 6.12$, P<0.05). In summary, bats apparently mistook Perspex for water even more frequently when more light was available, i.e. when colour should have been more easily detected.

The same analysis done on the M. mystacinus dataset failed to detect significant differences (number of approaches to black Perspex/overall approach number within 5-min intervals: light, $F_{1,17} = 0.33$, n.s.; site, $F_{2,17} = 2.56$, n.s.; number of bats approaching black Perspex/overall number of bats within 5-min intervals: light, $F_{1,17} = 0.16$, n.s.; site, $F_{2,17} = 3.34$, n.s.).

## Echolocation Differences during Approach to Perspex vs. Water in B. barbastellus

When approaching water or Perspex, B. barbastellus broadcast similar echolocation sequences, only made of calls represented by frequency modulated sweeps. Before initiating the approach (search phase), both call types 1 and 2 were alternated. In the approach phase, calls resembled modified type 2 calls, showing a

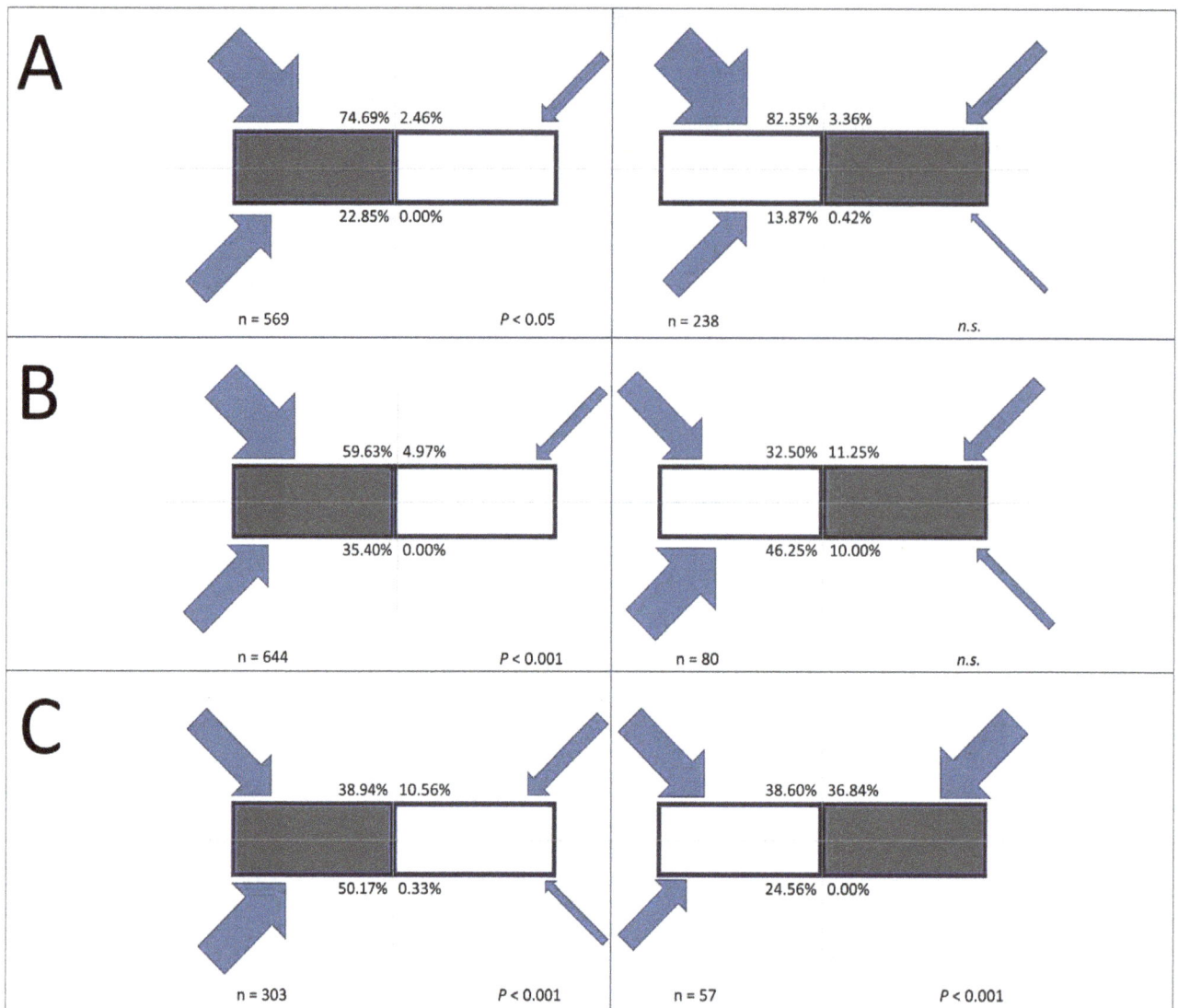

**Figure 2. Schematic bird's eye view of watering troughs manipulated for the experiment at sites A, B and C.** For each site one watering trough was covered with Perspex (grey rectangle), the other was left uncovered (white) so that water was available to bats. Two replicates of the experiment were done at each site, covering a different watering trough each time (left and right respectively). The arrows indicate the general direction (±45°) from which the bats approached the watering trough for drinking. Arrow width is proportional to the numbers of bats approaching from each direction (percent values are also given). Most bats always came from one side (and, in four out of six cases, one quadrant – see P values of Fisher's exact tests in figure) and drank at the first watering trough encountered, so that the latter was disproportionately used over the other regardless of whether it had been covered with Perspex or not.

steeper frequency modulated shape and a broader bandwidth (Table 2). Plotting IPI vs. time we recognized the approach phase characterised by a decreasing IPI trend, and a terminal phase during which IPI did not decrease further (a typical echolocation approach sequence of a drinking *B. barbastellus* is shown in Figure S1). The terminal phase consisted of groups of two or more calls. Within such groups, IPIs measured ca. 10 ms, whereas consecutive groups were spaced out by longer IPIs (ca. 20 ms). For the 30 sequences selected, we obtained mean values/sequence from 483 approach phase (Perspex = 259; water = 224) and 529 terminal phase (Perspex = 238; water = 291) IPIs respectively. IPI did differ between phases but not between substrate types (Table 2). We found significant differences between echolocation calls in the approach and the terminal phase respectively. Compared with the echolocation pulses in the approach phase, those in the terminal

phase had lower FMAXE and EF and a shorter duration, but SF did not differ significantly between phases (Table 2). No significant difference was detected between calls emitted by bats approaching water vs. Perspex, i.e. the two surfaces elicited identical echolocation behaviour.

## Discussion

### Bats Mistake Any Horizontal Smooth Surface for Water in Field Tests

We confirmed that, as in laboratory trials [14], under natural conditions too bats may be deceived by a smooth horizontal surface and exhibit repeated drinking attempts. We are confident that our results reveal a general pattern which we verified for 11 species and many individuals. Echoacoustic cues provided by the

**Table 2.** Descriptive statistics and results of General Linear Model (GLM) Analysis of Variance for echolocation calls of approach and terminal phases recorded from *B. barbastellus* approaching water or Perspex.

| Interpulse interval (ms) | Phase | N | Substrate (SD) | |
|---|---|---|---|---|
| | | | Water | Perspex |
| | Approach | 30 | 34.1 (8.9) | 31.0 (4.5) |
| | Terminal phase | 30 | 13.6 (2.1) | 12.0 (2.2) |
| **GLM ANOVA (*)** | | | | |
| Factor | d.f. | Adj. MS | F | P |
| Phase | 1 | 0.0265 | 325.44 | < 0.001 |
| Substrate | 1 | 2.419 | 3.57 | n.s. |
| Phase x Substrate | 1 | 0.002 | 0.29 | n.s. |
| Error | 56 | 0.007 | | |
| **Duration (ms)** | Phase | N | Substrate (SD) | |
| | | | Water | Perspex |
| | Approach | 30 | 2.3 (0.3) | 1.9 (0.4) |
| | Terminal phase | 30 | 1.2 (1.0) | 1.2 (0.9) |
| **GLM ANOVA** | | | | |
| Factor | d.f. | Adj. MS | F | P |
| Phase | 1 | 11.88 | 122.31 | < 0.001 |
| Substrate | 1 | 0.37 | 3.79 | n.s. |
| Phase x Substrate | 1 | 0.40 | 4.12 | n.s. |
| Error | 56 | | | |
| **FMAXE (kHz)** | Phase | N | Substrate (SD) | |
| | | | Water | Perspex |
| | Approach | 30 | 41.1 (1.4) | 41.3 (2.1) |
| | Terminal phase | 30 | 37.1 (2.7) | 38.0 (3.6) |
| **GLM ANOVA** | | | | |
| Factor | d.f. | Adj. MS | F | P |
| Phase | 1 | 204.98 | 30.76 | < 0.001 |
| Substrate | 1 | 4.76 | 0.71 | n.s. |
| Phase x Substrate | 1 | 1.63 | 0.25 | n.s. |
| Error | 56 | | | |
| **SF (kHz)** | Phase | N | Substrate (SD) | |
| | | | Water | Perspex |
| | Approach | 30 | 48.8 (2.0) | 49.5 (1.3) |
| | Terminal phase | 30 | 49.2 (2.0) | 49.6 (1.6) |
| **GLM ANOVA** | | | | |
| Factor | d.f. | Adj. MS | F | P |
| Phase | 1 | 0.94 | 0.31 | n.s. |
| Substrate | 1 | 5.46 | 1.80 | n.s. |
| Phase x Substrate | 1 | 0.28 | 0.1 | n.s. |
| Error | 56 | | | |
| **EF (kHz)** | Phase | N | Substrate (SD) | |
| | | | Water | Perspex |
| | Approach | 30 | 31.0 (1.7) | 31.1 (2.6) |
| | Terminal phase | 30 | 27.1 (1.2) | 27.4 (1.5) |
| **GLM ANOVA** | | | | |
| Factor | d.f. | Adj. MS | F | P |
| Phase | 1 | 214.70 | 65.30 | < 0.001 |
| Substrate | 1 | 0.58 | 0.18 | n.s. |

**Table 2.** Cont.

| | | | | |
|---|---|---|---|---|
| Phase x Substrate | 1 | 0.28 | 0.09 | n.s. |
| Error | 56 | | | |

(*) = analysis done on log-transformed data. FMAXE = Frequency of Maximum Energy; SF, EF = Starting and Terminal Frequencies taken at -25 dB below the frequency of maximum energy. Interactions between factors are indicated with a 'x' sign. SD = standard deviation, d.f. = degrees of freedom; Adj. MS = adjusted mean squares; n.s. = not significant (P > 0.05).

water-like Perspex surface clearly dominated over all other information. When bats where presented with an equal surface of Perspex and true water, neither the number of total approaches across species nor that for individual species of *M. mystacinus* and *B. barbastellus* differed between the two surface types. *P. auritus* even showed a greater number of approaches to Perspex.

As we showed for *B. barbastellus*, water and Perspex surfaces elicited similar echolocation behaviour, further supporting that bats approaching Perspex tried to drink – as verified by video recordings – rather than to simply explore the surface, i.e. they were unable to distinguish between real water and an artificial smooth surface. In no case did the approaches made by bats represent attempts to capture prey resting on the Perspex layer: the terminal phase we recorded clearly differed from feeding buzzes (fast repetitions of brief calls emitted by bats close to prey; Kalko and Schnitzler 1989) which in foraging *B. barbastellus* show shorter IPIs and are made of two distinct phases, buzz I and buzz II [20,26].

Although we did not analyze them, a terminal phase distinct from a typical feeding buzz was also noticed in the other species recorded (*pers. obs.*) and is probably a generalized pattern used to guide a safe drinking manoeuvre.

In situations where many bats fly together such as a drinking site visited by hundreds to thousands of individuals per night, it can be argued that echo-acoustic detection of obstacles can be confused by the many calls that are broadcast (and the corresponding echoes), so other cues, or spatial memory, could play an important role [27,28]. However, bats have been found to strictly rely on echoacoustic cues even when flying at familiar sites such as roosts, typically in groups of conspecifics [29] and may still effectively avoid objects placed along their route [28].

For water recognition echolocation is so important because of its dual function, i.e. assessing target properties such as density and texture [30], as well as obtaining crucial information to orientate and determine distance to surrounding objects [31]. When approaching water, echolocation is clearly used to recognize the water surface but also to precisely evaluate the distance to water and ensure a safe drinking manoeuvre: any misjudgement would mean to crash into the water and thus expose the bat to heat loss and also increase the risks of predation and injury.

In our experiment, vision had no discernible effect on water recognition, i.e. bats equally mistook transparent or black Perspex for water. Moreover, when the dataset including all species was considered and the effect of light taken into account, bats apparently mistook Perspex for water even more frequently under higher ambient light intensity, contrary to what expected if vision was involved and colour had a deterring role. However, we believe that in this case it is unlikely that the coloured surface was inherently more attractive for bats. We can only speculate that under brighter ambient light vision may have favoured a process of social imitation which has resulted in a concentration of bats over the same watering trough – the one covered with Perspex, where bat activity was more conspicuous because of the repeated drinking attempts occurring there. Visually detecting other drinking bats, along with eavesdropping on their echolocation calls [32] might speed up the process of locating new water sources.

Water colour may be influenced by several factors (some of which may change over time) e.g. water depth, presence of suspended particles, type of substrate, seasonal presence of aquatic plants, colonization by benthic organisms. Colour is also only detectable under sufficient light and thus for a limited time (i.e. around emergence time or on full moon nights). These points would suggest a secondary role of vision for water recognition, all the more in a natural setting.

We are aware that the limited time during which brighter light was available in our test – corresponding to the first half of each trial or less – may have been insufficient to detect any influence of vision, but in designing our experiment we avoided manipulating ambient light levels because we aimed to avoid disrupting natural behaviour patterns.

*Myotis lucifugus* [33] was observed to collide against stationary objects more often in the light, a fact suggesting that bats relied on vision besides echolocation in the presence of visual cues, but their limited visual capabilities at higher illumination levels resulted in a higher collision rate. Small *Myotis* bats such as *M. lucifugus*, and probably *M. mystacinus* which was the dominant species in our samples, have poor visual acuity compared to other bats [34,35] and this may have determined the lack of reaction to visual cues. In laboratory tests it was noticed [14] that rates of drinking attempts at an artificial surface mimicking water dropped when trials took place under lit rather than unlit conditions. However, this was tested on *M. schreibersii*, a species only occasionally present during our trials. For some bat species, vision plays an important role in hunting too [36,37]. The role of colour for water recognition might be more conspicuous in species relying on vision to forage. Although our sample featured one of them, i.e. *P. auritus* [36], the number of drinking attempts we recorded from it in experiment 2 was too limited to detect any effect. Further experiments should explore the existence of interspecific differences in the contribution of vision to water recognition and also test whether surfaces of brighter colours would be more conspicuous to bats.

As far as other sensorial cues (particularly, olfactory or tactile ones) are concerned, our results support the hypothesis that they are negligible as suggested in laboratory studies [14]. We reach this conclusion because if other cues were important for water recognition, the most common reaction of bats to Perspex would have been to immediately refrain from attempting to drink further after the very first approach, which involves a close-range assessment of the substrate.

## A Role for Learning Effects

One of our objectives was to explore whether under natural conditions (where bats may decide to leave to reach another drinking site) an unsuccessful drinking experience as that determined by the artificial surface would override the influence of the acoustic water-like cues the latter provided. We have seen that in several cases the Perspex surface deceived a bat so effectively that, as seen in laboratory trials [14], many consecutive drinking attempts were made before giving up. However, in most cases bats lost motivation after less than 10 attempts and either disappeared from the scene or moved to the uncovered water surface where they drank successfully. Our experimental design was especially suited to test this because uncovered water was available at all sites. In the laboratory experiments [14] bats were presented with a smooth horizontal surface mimicking water and real water was offered only after removing the former. Under such conditions bats showed high rates of drinking attempts at the fake water surface, sometimes 100 or more.

In our experiments, when bats left the watering trough covered with Perspex and moved to the control watering trough they made a marked change in the flight path and the approaching manoeuvre, i.e. they did not appear to randomly sample all potentially available water surfaces until real water was finally located. Based on the unsuccessful drinking attempts, bats must have modified their behaviour as soon as the experience gathered has overridden the influence of echoacoustic information. We do not know whether some of the bats leaving the site may have returned later to try and drink again from Perspex (in this case, the two events would have been recorded separately and attributed to different subjects). Overall, although acoustic recognition itself is stereotyped and its importance in the process overwhelming, our findings suggest that experience plays a role to increase behavioural flexibility. The adaptive value of this flexibility is clear: for example, it allows a bat encountering a frozen water surface or a human-made horizontal artificial surface to save energy by modifying its behaviour to avoid being deceived for too long.

## Effects of Spatial Memory and Social Interactions

On a larger scale, involving site recognition rather than water detection, our experiment pointed at a role for spatial memory. In fact, we found that at all experimental sites one of the two watering troughs received a disproportionately higher number of drinking attempts (as seen from the analysis of the whole dataset and *M. mystacinus*) and was visited by more bats (as found for the whole dataset, *M. mystacinus* and *P. auritus*), regardless of whether it was covered with Perspex or not. It is important to notice that most bats followed the same general route to reach the drinking site, so that the "preferred" drinking watering trough was the one first encountered by approaching bats. This result agrees with the findings presented in a study [38] which explained the adoption of such unidirectional flight paths in terms of cooperation aimed at reducing the risk of collision between bats. Both spatial memory and social interactions may play important roles in determining such preferentially used routes.

## Supporting Information

**Figure S1  Echolocation sequence of a drinking barbastelle bat (*Barbastella barbastellus*).** (a) spectrogram showing the approach and the terminal phases. The red arrow shows the noise produced when the bat makes contact with the water, which was clearly audible in many recordings made over water. (b) Interpulse interval (IPI) plotted vs. time of the same sequence. Note how the terminal phase is made of groups of calls broadcast with a high pulse rate separated by longer IPIs.

## Acknowledgments

We are indebted to the Abruzzo, Lazio and Molise National Park for sponsoring this project and to the Park's scientific service, particularly Cinzia Sulli, for the kind assistance provided. Thanks go to Luciano Bosso and Federica Lacatena for their assistance in the field and to Antonello Migliozzi who helped with figure preparation. We are also grateful to two anonymous reviewers who made especially valuable comments on a previous ms version.

## Author Contributions

Conceived and designed the experiments: DR GJ. Performed the experiments: DR LC. Analyzed the data: DR LC. Contributed reagents/materials/analysis tools: DR. Wrote the paper: DR GJ.

## References

1. Schmidt-Nielsen K (1997) Animal physiology. Adaptation and environment. Cambridge: Cambridge University Press. 607 p.
2. Chew RM, White HE (1960) Evaporative water losses of the pallid bat. J Mammal 41: 452–458.
3. Thomas DW, Cloutier D (1992) Evaporative water loss by hibernating little brown bats, *Myotis lucifugus*. Physiol Zool 65: 443–456.
4. Muñoz-Garcia A, Ro J, Reichard JD, Kunz TH, Williams JB (2012) Cutaneous water loss and lipids of the stratum corneum in two syntopic species of bats. Comp Biochem Physiol A Mol Integr Physiol 161: 208–15.
5. Speakman JR, Racey PA (1989) Hibernal ecology of the pipistrelle bat: energy expenditure, water requirements and mass loss, implications for survival and the function of winter emergence flights. J Anim Ecol 58: 797–813.
6. Boyles JG, Dunbar MB, Whitaker JO, Jr. (2006) Activity following arousal in winter in North American vespertilionid bats. Mamm Rev 36: 267–280.
7. Adams RA, Hayes MA (2008) Water availability and successful lactation by bats as related to climate change in arid regions of western North America. J Anim Ecol 77: 1115–1121.
8. Rainho A, Palmeirim JM (2011) The Importance of Distance to Resources in the Spatial Modelling of Bat Foraging Habitat. PLoS ONE 6(4): e19227. doi:10.1371/journal.pone.0019227.
9. Kwiecinski GG, Krook L, Wimsatt WA (1987) Annual skeletal changes in the little brown bat, *Myotis lucifugus lucifugus*, with particular reference to pregnancy and lactation. Am J Anat 178: 410–420.
10. Barclay RMR (1994) Constraints on reproduction by flying vertebrates-energy and calcium. Am Nat 144: 1021–1031.
11. Bernard RTF, Davison A (1996) Does calcium constrain reproductive activity in insectivorous bats? Some empirical evidence for Schreibers' long-fingered bat (*Miniopterus schreibersii*). S Afr J Zool 31: 218–220.
12. Bravo A, Harmsa KE, Emmons LH (2010) Puddles created by geophagous mammals are potential mineral sources for frugivorous bats (Stenodermatinae) in the Peruvian Amazon. J Trop Ecol 26: 173–184.
13. Voigt CC, Capps KA, Dechmann DKN, Michener RH, Kunz TH (2008) Nutrition or detoxification: why bats visit mineral licks of the Amazonian rainforest. PLoS ONE 3(4): e2011. doi:10.1371/journal.pone.0002011.
14. Greif S, Siemers BM (2010) Innate recognition of water bodies in echolocating bats. Nature Communications, doi: 10.1038/ncomms1110.
15. Siemers BM, Page RA (2009) Behavioral studies of bats in captivity: Methodology, training, and experimental design. In: Kunz TH, Parsons S, editors. Ecological and behavioral methods for the study of bats. 2nd edition, Baltimore: Johns Hopkins University Press. 373–392.
16. Wolff JO (2003) Laboratory studies with rodents: facts or artifacts? Bioscience 53: 421–427.
17. Eklöf J (2003) Vision in echolocating bats. Doctoral thesis Zoology Department, Göteborg University.
18. Russo D, Jones G (2002) Identification of twenty–two bat species (Mammalia: Chiroptera) from Italy by analysis of time-expanded recordings of echolocation calls. J Zool (Lond) 258: 91–103.
19. Vaughan N, Jones G, Harris S (1997) Identification of British bat species by multivariate analysis of echolocation call parameters. Bioacoustics 7: 189–207.
20. Denzinger A, Siemers BM, Schaub A, Schnitzler HU (2001) Echolocation by the barbastelle bat, *Barbastella barbastellus*. J Comp Physiol A 187: 521–528.
21. Huitema BE (1980) The analysis of covariance and its alternatives. New York: Wiley, 445 p.
22. Quade D (1967) Nonparametric Analysis of Covariance by Matching. Biometrics 38: 597–611.
23. Surlykke A, Moss CF (2000) Echolocation behavior of big brown bats, *Eptesicus fuscus*, in the field and the laboratory. J Acoust Soc Am 108: 2419–2429.
24. Siemers BM, Schnitzler HU (2004) Echolocation signals reflect niche differentiation in five sympatric congeneric bat species. Nature 429: 657–661.
25. Goerlitz HR, ter Hofstede HM, Zeale MRK, Jones G, Holderied MW (2010) An aerial-hawking bat uses stealth echolocation to counter moth hearing. Curr Biol 20: 1568–1572.
26. Kalko EKV, Schnitzler H-U (1989) The echolocation and hunting behavior of Daubenton's bat, *Myotis daubentoni*. Behav Ecol Sociobiol 24: 225–238.
27. Griffin DR, Webster FA, Michael CR (1960) The echolocation of flying insects by bats. Anim. Behav 8: 141–154.
28. Goerlitz HR, Genzel D, Wiegrebe L (2012) Bats' avoidance of real and virtual objects: implications for the sonar coding of object size. Behav Processes 89: 61–67.
29. Holland RA, Waters DA (2007) The effect of familiarity on echolocation in the Megachiropteran bat *Rousettus aegyptiacus*. Behaviour 144: 1053–1064.
30. Schmidt S (1988) Evidence for a spectral basis of texture perception in bat sonar. Nature 331: 617–619.
31. Neuweiler G (1988) Foraging ecology and audition in echolocating bats. Trends Ecol Evol 4: 160–166.
32. Fenton MB (2003) Eavesdropping on the echolocation and social calls of bats. Mamm Rev 33: 193–204.
33. Orbach DN, Fenton B (2010) Vision impairs the abilities of bats to avoid colliding with stationary obstacles. PLoS ONE 5(11): e13912. doi:10.1371/journal.pone.0013912
34. Suthers RA (1966) Optomotor responses by echolocating bats. Science 152: 1102–1104.
35. Bell GP, Fenton MB (1986) Visual acuity, sensitivity and binocularity in a gleaning insectivorous bat, *Macrotus californicus* (Chiroptera: Phyllostomidae). Anim Behav 34: 409–414.
36. Eklöf J, Jones G (2003) Use of vision in prey detection by brown long-eared bats, *Plecotus auritus*. Anim Behav 66: 949–953.
37. Rydell J, Eklöf J (2003) Vision complements echolocation in an aerial-hawking bat. Naturwissenschaften 90: 481–483.
38. Adams RA, Simmons JA (2002) Directionality of drinking passes by bats at water holes: is there cooperation? Acta Chiropt 4: 1–5.

# 11

# Responses of Reactive Oxygen Scavenging Enzymes, Proline and Malondialdehyde to Water Deficits among Six Secondary Successional Seral Species in Loess Plateau

**Feng Du[1,2,3]\*, Huijun Shi[1], Xingchang Zhang[1,2,3], Xuexuan Xu[1,2]**

**1** Institute of soil and Water Conservation, Northwest Sci-Tech University of Agriculture and Forestry, Yangling, Shaanxi, China, **2** Institute of soil and Water Conservation, Chinese Academy of Science, Yangling, Shaanxi, China, **3** State key laboratory of soil erosion and dryland farming on Loess Plateau, Yangling, Shaanxi, China

## Abstract

Drought can impact local vegetation dynamics in a long term. In order to predict the possible successional pathway of local community under drought, the responses of some drought resistance indices of six successional seral species in the semi-arid Loss Hilly Region of China were illustrated and compared on three levels of soil water deficits along three growing months (7, 8 and 9). The results showed that: 1) the six species had significant differences in SOD, POD activities and MDA content. The rank correlations between SOD, POD activities and the successional niche positions of the six species were positive, and the correlation between MDA content and the niche positions was negative; 2) activities of SOD, CAT and POD, and content of proline and MDA had significant differences among the three months; 3) there existed significant interactions of SOD, CAT, POD activities and MDA content between months and species. With an exception, no interaction of proline was found. Proline in leaves had a general decline in reproductive month; 4) SOD, CAT, POD activities and proline content had negative correlations with MDA content. Among which, the correlation between SOD activity and MDA content was significant. The results implied that, in arid or semiarid region, the species at later successional stage tend to have strong drought resistance than those at early stage. Anti-drought indices can partially interpret the pathway of community succession in the drought impacted area. SOD activity is more distinct and important on the scope of protecting membrane damage through the scavenging of ROS on exposure to drought.

**Editor:** Silvia Mazzuca, Università della Calabria, Italy

**Funding:** This research was supported jointly by the National Natural Science Foundation of China (41271526 and 41171421), basic scientific fund of Northwest Agriculture and Forestry University (2109021121). The funders had no role in study design, data collection and analysis, decision to publish, or preparation of the manuscript.

**Competing Interests:** The authors have declared that no competing interests exist.

\* E-mail: dufeng@ms.iswc.ac.cn

## Introduction

Drought can impact many ecosystem processes and their function, such as the vegetation succession, biodiversity, productivity and sustainability, etc [1,2]. And drought is a most limiting factor of plant growth and reproduction worldwide [3–6]. Drought can intermittently disturb water cycle of ecosystem, and the carbon and mineral cycle that closely interact with water cycle [6]. These can stunt live beings indirectly through decreasing the resource availability [7]. And also, drought can hurt live beings directly and result in the decrease of their fitness, the dying off of individuals and the dying out of species at local or globe scale [6,8]. In the past, many works had been done on the scope of drought-induced phenomena, processes and mechanismsof plants, anti-drought field practices, and breedings of drought tolerant species or varieties, as well [7,9,10]. Most of these works were dealt with crop plants for the purpose of food, feed and fuel products improvement. Meanwhile, as a fact, other life-support components, as grassland or forest, are more important for the maintenance and conservation of a healthy ecosystem. In recent globe climate change background, the severity and frequency of drought were

reported inclined to some extent [11]. How a local ecosystem responds to that change and what is the possible changing tendency of an ecosystem, e.g., the successional pathway (or trajectory) of rangeland and forest, would make a sense of ecosystem conservation.

Ecosystem' responses to drought are the integrative results of sensitive reactions of live beings and relative sluggish responses of life-support environment. As an important part of ecosystem, the responses of vegetation to drought are the collective reflections of community species. A long term or severe drought may alter or shift the assembly and structure of community and result in degraded succession. The succession is driven by community species' eco-physical traits, such as their fitness, reproductive tragedy and their competitive ability, etc [12,13]. Essentially, these traits root in the comparatively sensitive physical response. All of these are strongly depended on local environment factors, especially the most limiting one, e.g., the soil water in arid or semi-arid area. Two obvious sceneries that drive succession are the difference of eco-physical traits between the early successional species and the later ones [14], and the tolerance difference among co-existing species in stress-impacted area [15].

In Loess Hilly region of China, soil moisture is one of the most limiting factors that determine plant growth and productivity [16]. In most cases, soil moisture is in a water-deficit stress for plant growth in this area. This would be an important driving factor of plant community succession. As co-existing species in communities ordinarily have different drought-resistance process and ability, their responses to local soil moisture deficit and fluctuation were important for community succession pathway. And also, individuals of a species often grow in standing conditions with spatially heterogeneous soil moisture, so their anti-drought abilities have tight link with their abundance in a community. Overall, the anti-drought abilities of co-existing species and the responses of community species to water deficit can differentiate the assembly and structure of a local community, can further shape and ramified the pathway of succession. Hence, the research of anti-drought ability of successional seral species and co-existing species in a community can promote the understanding of the dynamics, causes and pathways of community succession, and help to predict the possible consequences of local environment change under globe climate change context.

Many environmental stresses exert at least part of their effects by causing oxidative damage. Consequences of oxidative stress are species-specifically related with endogenous antioxidant content. Plants' anti-drought involves a series of biological, physical and chemical processes, of which the most critical one is to detoxify radicals. These processes mainly occur in colloid and membrane interface to refrain from membrane hurt for the purpose of normal function. The anti-drought processes and materials were nearly all participated by the enzymatic system. And especially, reactive oxygen scavenging enzymes were triggered primarily when stress occurred and were more important for its function of removing reactive oxygen species (ROS) that is over produced in drought stress [3,4,9,17,18]. The enzymes have Super oxide Dismutase (SOD), Catalase (CAT), and Peroxidase (POD). These enzymes, involving redox status mediation of plant cells, are believed to be the major reliever to stress injuries in causing cellar damage [19]. Besides that, free proline, acts as an important osmoprotectant [20], its content in the leaf under stress conditions is of utmost importance for plant adaptation of environmental stress. The main cellular components susceptible to free radicals are lipids (peroxidation of unsaturated fatty acids in membranes), proteins (denaturation), carbohydrates and nucleic acids. As one of the main products when membrane is attacked by free radical, Malondialdehyde (MDA) will accumulate, and therefore it is often used as a mark of reactive oxygen damage [21].

It was believed that co-existing species' stress tolerant abilities are responsible for the community succession to some extent [22,23]. In the past, we found that soil water of old-fields in Loess Hilly region of China had a decreasing tendency during secondary succession [24,25]. That is to say, species in later succession stage grow in a lower soil water status as compared with those in medium or early succession stages. This research had the following expectations: 1) Later seral species have relative higher levels of reactive oxygen enzymes activities and proline content and low levels of MDA content. 2) The correlations between the activities of reactive oxygen scavenging enzymes (SOD, CAT and POD) and MDA content, and the content of proline and MDA should be negative. As reactive oxygen scavenging enzymes and osmoprotectant can both help to improve the drought resistance ability, while MDA is a mark of plants' hurt under drought.

## Materials and Methods

### 1 Experimental fields

The experiment was conducted in pot cultivation at the drought greenhouse that is affiliated to the State Key Laboratory of Soil Erosion and Dryland Farming on Loess Plateau, China. The greenhouse is functional provided with light, temperature adjustment, air-conditioner and glass rainfall proof.

### 2 Plant materials

The six seral species are *Artemisia scoparia*, *Setaria viridis*, *Artemisia sacrorum*, *Artemisia giraldii*, *Lespedeza dahurica* and *Bothriochloa ischaemum*. According to our previous research, their successional niche positions were $-1.45$, $-1.51$, $0.46$, $1.17$, $1.60$ and $2.00$, respectively [19]. Among of them, *A. scoparia*, *S. viridis* are the dominant species of early stage during secondary succession in semi-arid Loess Hilly region of China. *A. sacrorum* and *A. giraldii* are the dominant species of medium stage, and *L. dahurica* and *B. ischaemum* are later dominant species [24]. The seedlings of the six species were transplanted into plastic pots on May 10, 2012.

### 3 Experimental design

The experiment had species, water deficit and growing stage treatments in completely randomized design. The species treatment included the above six seral species. The water deficit treatment had three deficient or watering levels. The deficient levels had mild, moderate and severe water deficient stresses, equivalent to 80%, 65% and 50% of the field capacity of the pot used soil, respectively. The field capacity of the pot filled soil was measured as 21.5%, hence the watered levels were 17.2%, 14.0% and 10.75%, respectively. The growing stage treatment was represented by sampling months, which had July, August and September. Each treatment was replicated three times.

On May 9 and 10, the experimental pots (with diameter and height) were all filled with 14 kg air-dried local soil in Loess Hilly region. The total nitrogen, phosphorous and potassium of the pot filled soil were measured as 0.075, 0.052 and 1.94%; and the available nitrogen, phosphorous and potassium were 16.07, 2.71 and 109.38 mg.kg$^{-1}$, respectively. The unavailable water content was measured as 4.56% at 15 bar. Ten seedlings of each of the six species for every water deficient levels and sampling months were lifted from local sites of Loess Hilly region. Totally 90 seedlings with largely the same size were lifted on May 11 and 12. On the middle of May, 2011, each plastic pot was transplanted into six vigorous seedlings of one of the six species. In order to assure survivorship, the rest four seedlings were used as substitute plants, which were transplanted into non-experimental pots. The seedlings in both of the two kinds of pots were well watered in the first 15 days of cultivation. The watered level was set at 85% of the field capacity (FC) of the soil, about 18.28%. On June 1st, some withered or weak seedlings were replaced with the stand-by vigorous ones that were previously transplanted in the non-experimental pots. After that, they were cultivated under well-watered condition in the greenhouse until in the middle of June. From June 20, water deficient treatment started and the watering amount of each pot was controlled using weighting method. The weighting and watering of pots were conducted every three days. On July 20, August 15 and September 10, the leaves of the six species of each water stress treatment were sampled and their physiological and biochemical indexes were assayed.

## 4 Measurement of physiological and biochemical indexes and methods

Activities of SOD, CAT and POD, content of proline and MDA were assayed according to the references of Gao [26] (SOD and CAT) and Li [27] (POD). The activity of SOD was measured using the method of photochemical reduction inhibition of nitroblue tetrazolium (NBT). The Cat activity was measured as the decrease of absorbance of $H_2O_2$ at 240 nm per minute. The POD activity was measured using guaiacol oxidation method. The proline content was measured using acid ninhydrin method. The MDA content was measured using the thiobar-bituric acid method.

## 5 Statistical analysis

In this research, five physiological indices as the activities of SOD, CAT and POD, the contents of proline and MDA were measured. Data of the five indices were analyzed using general linear model by Data Processing System [28]. The analysis was three-way univariate ANOVAs with the activities of SOD, CAT and POD, the contents of Pro and MDA as the dependent variables, and months, deficient levels and species as the independent variables. Multivariate comparison was conducted using LSD method when the treatment effects were significant. In order to illustrate the preventive effects of SOD, CAT and POD activities, and proline content on the membrane lipid destruction, the correlations between MDA content and the activities of SOD, CAT and POD, and the correlation between MDA and proline content were analyzed. The correlations between the above indices and successional niche positions of the six species were also analyzed to demonstrate the changing tendency of drought resistant ability along successional succession.

## Results

### 1 Responses of SOD activities of the six species to water deficit

ANOVAs showed that the six seral species had significantly different SOD activities along and among the three growing months of July, August and September (Table 1). Among of them, L. dahurica had highest SOD activities along the three growing months; meanwhile, A. scoparia and A. giraldii had lowest SOD activities (Fig. 1). Except A. giraldii had an exception, the other five species showed a monthly decreased tendency of SOD activities during the three growing season (Fig. 1). A. giraldii had the lowest SOD activities in August, significantly lower than the values in July and September. And also, a significant interaction effect existed between growing months and species, which implied that the six species had different changing tendency of SOD activities along the three months. For example, the multivariate comparison results showed that species of S. viridis, L. dahurica and B. ischaemum had significantly the lowest SOD activities in September. A. sacrorum had the highest SOD activities in July, significantly higher than the activities in August and September. And also, the SOD activities of A. scoparia in July is significantly higher than that in August, but had no significant difference with the values in September.

For the six seral species, no significant differences of SOD activities were found among the three water deficient levels (Table 1 and Fig. 1), although there existed slightly different SOD activities among the three water deficient levels. The severe, moderate and mild deficient levels had averaged SOD activities of 393.95, 388.30 and 378.44 $\mu g.g^{-1}.h^{-1}$, respectively. The measured SOD activities of the six species under moderate and severe

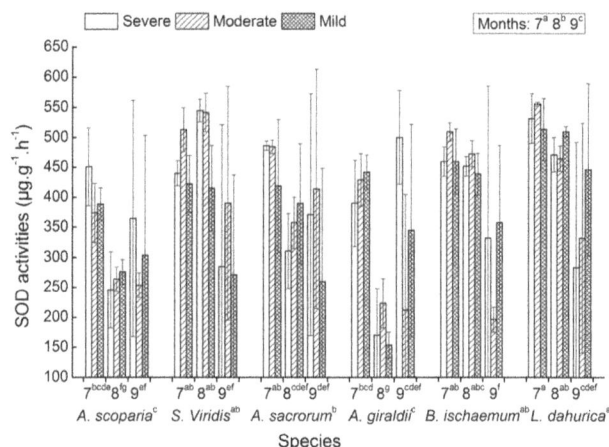

**Figure 1. SOD activities of six successional seral species.** The measurements were conducted on July (7), August (8) and September (9) under three water deficient levels (severe, moderate and mild). Significant multi-comparison results of species and month × species were marked on the second and first label rows of the X-axes, respectively (see associated ANOVAs results of Table 1).

water deficits had slight improvements as compared to the mild one.

## 2 Responses of CAT activities of the six species to water deficit

The six seral species showed a significant decline tendency along the three months' growth (Table 1). In July, August and September, CAT activities of the six species were measured with high, medium and low levels, respectively (Fig. 2). No significant differences of CAT activities were observed among the six seral species and among the three different water deficient levels, but interaction effects of growing month × species, deficient level × species, deficient level × month and month × deficient level × species were all detected significantly. This implied that the response of CAT activities to deficient levels was a growing month related species-specific process, and the process varied with water availability.

The significant interaction between month and species was demonstrated obviously by the significant differences of S. viridis and B. ischaemum among the three growing months (Fig. 2). The differences showed that the CAT activities of S. viridis in September were significantly lower than the activities in July and August, and the CAT activities of B. ischaemum in July were significantly higher than the activities in August and September. The significant interaction between deficient level and species was illustrated by the discrepancy of CAT activities of the six species under the three water deficient levels (Fig. 2). For example, under moderate and mild deficient levels, S. viridis had highest CAT activities than the values of other species; while, under severe deficient level, the CAT activities of S. viridis had no significant differences with other species' values (except of A. sacrorum).

The significant interaction of CAT activities between water deficient level and growing month was illustrated by the multi-comparison discrepancy under the three water deficient levels in July, August and September (Fig. 2). Under mild and moderate deficient levels, the CAT activities of all the six species had its highest values in July; while under severe deficient level, the CAT activities had no significant difference among the three growing months.

**Table 1.** Responses of anti-oxidation, osmoregulation and membrane hurt of six successional seral species to three deficient levels measured on July, August and September.

| Sources | df | SOD | | CAT | | POD | | Pro | | MDA | |
|---|---|---|---|---|---|---|---|---|---|---|---|
| | | F | P | F | P | F | P | F | P | F | P |
| Month | 2 | 19.79 | 1.00E-04 | 13.17 | 1.00E-04 | 12.53 | 1.00E-04 | 6.8 | 3.70E-03 | 11.71 | 1.00E-04 |
| Deficient level | 2 | 0.28 | 0.76 | 2.61 | 7.97E-02 | 0.57 | 0.56 | 4.14E-02 | 0.96 | 0.31 | 0.74 |
| Species | 5 | 6.85 | 1.00E-04 | 1.81 | 1.20E-01 | 6.33 | 1.00E-04 | 0.37 | 0.86 | 33.62 | 1.00E-04 |
| Month × deficient level | 4 | 0.84 | 0.50 | 4.03 | 4.90E-03 | 0.96 | 0.44 | 0.50 | 0.74 | 2.41 | 5.33E-02 |
| Month × species | 10 | 3.84 | 1.00E-04 | 10.67 | 1.00E-04 | 3.00 | 2.20E-03 | 0.28 | 0.98 | 9.07 | 1.00E-04 |
| Deficient level × species | 10 | 1.01 | 0.44 | 3.00 | 2.80E-03 | 1.27 | 0.25 | 0.52 | 0.86 | 2.63 | 6.70E-03 |
| Month × deficient level × species | 20 | 0.94 | 0.55 | 3.23 | 1.00E-04 | 1.44 | 0.12 | 0.29 | 9.97E-01 | 1.77 | 3.36E-02 |

Anti-oxidation traits of the six species were measured as the activities of SOD, CAT and POD enzymes; osmoregulation and membrane hurt were measured as the contents of Pro and MDA, respectively. The table shows the results of the three-way univariate ANOVAs with the activities of SOD, CAT and POD, the contents of Pro and MDA as the dependent variables, and months, deficient levels and species as the independent variables. See Figs. 1 to 5 for means and standard errors. As the leaf samples were limited, some measurements of proline were missed, which lead to the degree freedom of total errors of proline were 83 (Fig. 4). Other items were all 161.

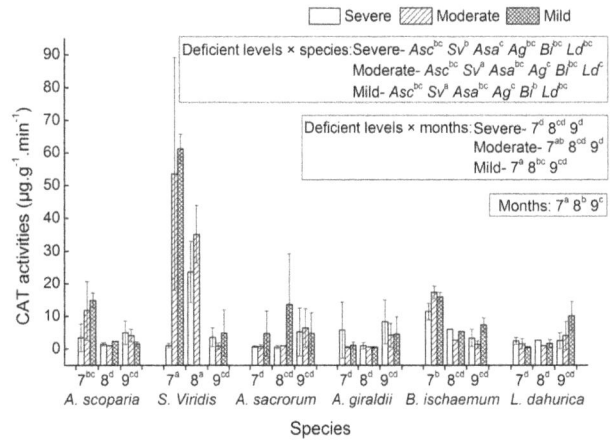

**Figure 2. CAT activities of six successional seral species.** The measurements were conducted on July (7), August (8) and September (9) under three water deficient levels (severe, moderate and mild). Significant multi-comparison results of month × species were marked on the first label row of the X-axes. And significant multi-comparison results of deficient level × species, deficient level × months and months were marked as legends within three blocks, respectively (see associated ANOVAs results of Table 1). In the box of legend zone, species of *A. scoparia, S. viridis, A. sacrorum, A. giraldii, B. ischaemum* and *L. dahurica* were abbreviated as *Asc, Sv, Asa, Ag, Bi* and *Ld*, respectively.

## 3 Responses of POD activities of the six species to water deficit

The POD activities were significantly different among the six seral species and the three growing months (Table 1). *S. viridis, and B. ischaemum* had relative higher POD activities than other species had. And the POD activities showed an increase tendency along the growing season of those three months (Fig. 3). In July, August and September, the POD activities were at low, medium and the highest levels, respectively. No significant differences of POD activities were found among the water deficient levels.

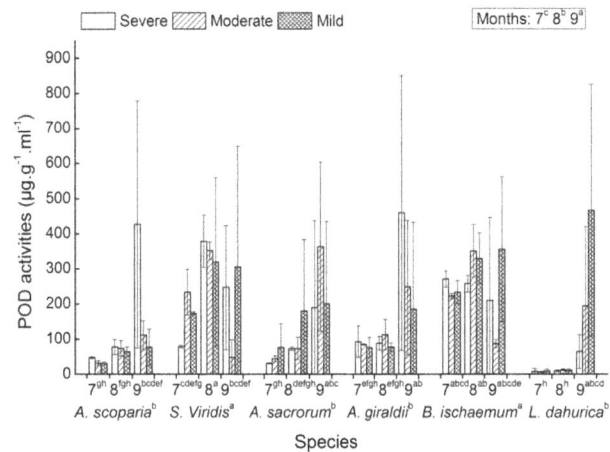

**Figure 3. POD activities of six successional seral species.** The measurements were conducted on July (7), August (8) and September (9) under three water deficient levels (severe, moderate and mild). Significant multi-comparison results of month × species and species were marked on the first and second X-axes' label rows. And significant multi-comparison results of month were marked as a legend within a block (see associated ANOVAs results of Table 1).

The interaction effects between growing months and species were significant. The POD activities of *S. viridis* in August, and *A. sacrorum*, *A. giraldii*, *B. ischaemum* and *L. dahurica* in September were relative high (Table 1 and Fig. 3).

## 4 Responses of proline content of the six species to water deficit

The proline contents of the six species were found significantly different among the three months (Table 1 and Fig. 4). In August, the proline contents were significantly higher than that in July and September. No significant differences were detected among the six species and the three water deficient levels. And also, no significant interaction effects were found.

## 5 Effects of water deficit on the MDA content among the six secondary successional seral species

ANOVAs showed that MDA contents were significantly different among the three growing months and the six seral species. In September, the MDA contents were lower than that in July and August. Among the six species, *A. giraldii* had the highest MDA contents and *L. dahurica* had the lowest contents (Fig. 5). The average MDA contents under severe, moderate and mild deficient levels were 4.17, 4.23 and 3.93 mmol per gram of fresh tissue, respectively. Though MDA contents were a little higher under moderate and severe deficient levels than that under mild deficient level, they were not significantly affected by the water deficient levels (Table 1).

The significant interaction between month and species implied that the six species had their special growing time-related anti-overoxidation characteristics in terms of MDA content (Table 1 and Fig. 5). In July and August, species *A. giraldii* had the highest MDA contents, while in September, the measured MDA contents decreased to a same level with other species (Fig. 5). Though deficient levels showed no significant main effects on the MDA content, there still existed a significant interaction between deficient level and species (Table 1 and Fig. 5). For example, the MDA content of *S. viridis* under severe deficient level was significantly higher than the values under moderate and mild levels. And *A. giraldii* was measured with higher MDA content under moderate than those mild and severe levels (Fig. 5).

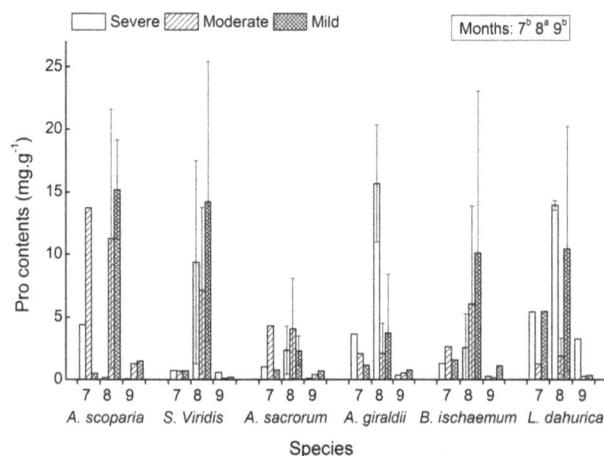

**Figure 4. Proline contents of six successional seral species.** The proline contents were measured on July (7), August (8) and September (9) under three water deficient levels (severe, moderate and mild). Significant multi-comparison results of month were marked as a legend within a block (see associated ANOVAs results of Table 1).

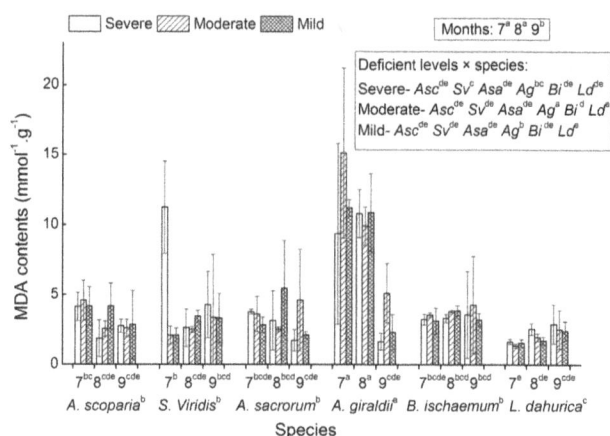

**Figure 5. MDA contents of six successional seral species.** The MDA contents were measured on July (7), August (8) and September (9) under three water deficient levels (severe, moderate and mild). Significant multi-comparison results of month × species and species were marked on the first and second label rows of X-axes. Significant multi-comparison results of month and deficient level × species were marked as legends within two blocks (see associated ANOVAs results of Table 1). In the box of legend zone, species of *A. scoparia*, *S. viridis*, *A. sacrorum*, *A. giraldii*, *B. ischaemum* and *L. dahurica* were abbreviated as *Asc*, *Sv*, *Asa*, *Ag*, *Bi* and *Ld*, respectively.

There also existed a significant interaction among month, deficient level and species (Table 1), which implied that the protection of membrane was a complex growing season and water status related species specific process.

## Discussion

### 1 Reactive oxygen scavenging enzymes system and osmatic adjustment play an important role in the anti-drought process

Normally, there exists a balance in plants between the production of reactive scavenging species (ROS) and quenching activity of antioxidant [29]. But when subjected to environmental adverse, the balance is upset and plants will over produce the active oxygen in excess of the capability of scavenging system, causing oxidative stress and damage. Except for some species that take avoidance or resistance tragedy when exposed to drought, most species have to tolerate drought through physiological or metabolic adaptive ways. Hence, the oxidative stress that ensues from drought is a widespread phenomenon. ROS, are viewed as main toxic cellular metabolites, are highly active molecules that can easily damage membrane and other cellular components. Therefore, on exposure to environmental stress, the scavenging of ROS is a primary process for plants' growth and reproduction. Anti-oxidation mechanisms of the cell include ROS-scavenging system and non-enzymatic antioxidants. Scavenging mechanisms for ROS involve these enzymes: SOD, CAT, POD, ascorbate peroxidase (APX) and glutathione peroxidase (GPX). Among the four major reactive oxygen species [superoxide radical $O_2^-$, hydrogen peroxide $H_2O_2$, hydroxyl radical $\cdot OH$ and singlet oxygen $^1O_2$], hydrogen peroxide $H_2O_2$ and the hydroxyl radical $\cdot OH$ are mostly active toxic and destructive. Superoxide radical $O_2^-$ can be transform into $H_2O_2$ mainly through SOD or ascorbic acid. $H_2O_2$ can mainly be decomposed via the catalysis of CAT and POD.

MDA is one of the products of lipid membrane peroxidation, which can be a mark of stress-caused damage. In our case, the correlations between MDA contents and activities of SOD, CAT and POD were all negative but only the correlation of SOD and MDA was significant. That means SOD acted more effective than ascorbic acid done in the superoxide quenching, at least for the six species. The correlations between MDA contents and the above three enzymes' activities (SOD, CAT and POD) were $-0.32$, $-0.18$ and $-0.058$, with 0.012, 0.16 and 0.66 significant levels, respectively. The insignificant correlations between CAT, POD and MDA implied that antioxidant production does not always significantly result in the enhancement of the anti-oxidative defense. Possible reasons were: 1) Antioxidants act as a cooperative network, employing a series of redox reactions. For example, in our case, $H_2O_2$ is cooperatively scavenged by CAT and POD enzymes. Except for anti-oxidative enzymes, there also exist other non-enzymatic antioxidants that involve the membrane protection to reduce the production of MDA. 2) Other aspects as the compartmentalization of ROS formation and the localization, synthesis and transport of antioxidants would also be the determinants of the effectiveness of ROS scavenging. Some charged ROS molecule as superoxide and hydroxyl radical, they cannot cross biological membrane. 3) The important one is the ability to induce the antioxidant defense is determined by species-specific genome. The plant's response to drought is accompanied by the activation of genes involved in the perception of drought stress and in the transmission of the stress signal.

Except for oxidative stress, plants will also experience or undergo osmotic stress on exposure to drought. With the water loss of tissue, plants will accumulate a range of metabolically benign solutes, collectively known as osmolytes. Osmylytes include proline, betaines, polyols (mannitol, sorbitol, and pinitol) and trehalose, etc. Drought-caused low water potential can accelerate the degradation of structural proteins [30], from which most of osmolytes are synthesized. That is why proline content has negative correlation with relative water content (RWC) of leaves in literatures [31,32].

The accumulation of proline can alleviate the membrane damage and reduce MDA level when plant is tolerating drought stress [33]. The observed correlation between MDA and proline content in our case was not to be an expected negative value. On the contrary it was slightly positive (pairwise Pearson $= 0.039$, $P = 0.77$). In literatures, both positive and negative correlations were reported [34]. This is probably due to: 1) The amino acid of proline is not functionally only for the preventive one of lipid membrane peroxidation. Proline involves several physiological functions and is synthesized if it is in need of these functions when plants experienced stress. These functions include membrane peroxidation prevention related group as osmotic adjustment [35], osmoprotection [36,37], free radical scavenger activity [38], protection of macromolecules from denaturation [39], and other groups that have no direct association with the protection of membrane, as regulation of cytosolic acidity, inhibition of programmed cell death [40] and a source of accumulated carbon and nitrogen during abiotic stress. 2) And there are other ion osmotic adjustments like $K^+$, $Na^+$ and $Cl^-$ in vacuolar, and organic osmolyte like sugar alcohols (e.g. fructans) or ammonium compound (e.g. glycine-betaine) that have the same function. 3) The compartment of proline synthesis and transport may also contribute to the ineffectiveness of membrane protection. 4) The accumulation of proline and their contribution to prevent lipid peroxidation varies among species and phonological stages [30]. For example, in our case, the six species showed different correlations between the measured proline and MDA contents.

The correlations of A. scoparia, S. viridis, A. sacrorum, A. giraldii, B. ischaemum and L. dahurica were 0.36, $-0.047$, 0.14, 0.35, $-0.12$ and $-0.16$. The accumulation of proline mainly mediated by water potential, which is an integrative eco-physical process that is controlled by the water balance between root water absorbability and leaf water withhold. For those species with strong water absorbing and well water retention, the accumulation of proline will not be activated only when their water potential decreases dramatically. Besides that, as proline was associated with morphogenesis, it will migrate to reproductive tissues at flowering and productive period [41,42]. Also, some authors have suggested the presence of proline protein carriers (AtProT1 and AtProT2) capable of removing the proline synthesized in the leaf and root to the reproductive organs [43]. The six seral species are in flowering and productive phase in september. Perhaps this is why the measured proline contents in leaves were relative low in September (Figure 4).

## 2 Differences of anti-drought traits of the six seral species

The activity improvement of reactive oxygen scavenging enzymes and the accumulation of osmolytes are common in many plant species in response to drought stress. Oxidative stress induced by drought often lead to the lipid membrane peroxidation. Consequently, SOD, CAT, POD, Pro and MDA are often used as important indices for evaluating the cellular redox balance and the anti-oxidative abilities. High values of SOD, CAT and POD activities and proline content, and low MDA content are the reflection of a species with high anti-drought ability. According to the responses of the five indices to water deficits (Table 1, Fig. 1–5), the drought resistant abilities of the six species can be ordinated and the adaptive responsive pattern to drought can be illustrated at large. The significant differences of SOD, POD and MDA activities among the six species plus their mean value have the following ranks of drought resistance: SOD ($\mu g.g^{-1}.h^{-1}$) – L. dahurica (456.51), S. viridis (424.80), B. ischaemum (408.70), A. sacrorum (388.12), A. scoparia (324.12) and A. giraldii (318.57); POD ($\mu g.g^{-1}.ml^{-1}$) – B. ischaemum (257.55), S. viridis (237.37), A. giraldii (158.27), A. sacrorum (136.99), A. scoparia (104.41) and L. dahurica (87.28); MDA ($mmol.g^{-1}$) – L. dahurica (2.07), A. scoparia (3.30), A. sacrorum (3.32), B. ischaemum (3.56), S. viridis (3.88) and A. giraldii (8.50). Except L. dahurica, whose ranking position of POD had obvious inconsistency with SOD and MDA, the ranks of drought resistance deduced from the three items of SOD, POD activities and MDA content are roughly coincided. The ranks can interpret partially at which successional niche position will the six seral species dominate or get their highest abundances during secondary succession in Loess Hilly Plateau. The rank correlations between the deduced ranks and successional niche positions were 0.14, 0.086 and $-0.14$ (spearman, pairwise), respectively. This implied that species at later successional stage tend to have high SOD and Cat activity than those at early stage, and species at later stage have relative good anti-peroxidation ability than species at early stage have.

The SOD and CAT activities had a decreasing tendency (Fig. 1 and 2), whereas POD activity tend to incline at later growth period (Fig. 3). These suggest that the quenching of ROS is a kind of growth period associated synergistic function. The significant interaction between growing months and species implied that the change of anti-oxidative enzymes is species-specific phenological seasonal cycle (Table 1 and Fig. 1–3). Normally, plants have evolved seasonal cycle of physiological activity, which are phenological season related and succeed from long term's acclimation to local habitats. The anti-oxidative enzymes and osmolytes, as a part of habitats-depended physiological activity, will be naturally associated with metabolic activity strength and growth rate. The high activity of SOD and CAT in July may be ascribed to the high levels of metabolism and photosynthesis when light intensity and temperature reach their climax. The high levels of photosynthesis and metabolism provide ample substrate and energy for the synthesis of the two enzymes to avoid photoinhibitory. And vice versa, the quick growth and high levels of photosynthesis and metabolism need the smooth quenching of ROS when local habitats are in good conditions for photosynthesis and growth.

The impacts of water deficits on the five indices were not found significant as expected. This probably due to two reasons: the first one is the oxidative stress load from the seasonal light intensity or phonological cycle may overweight the load by water deficits; the second one is the watering or deficit levels is not enough to differentiate the responses of all the six species, although species responded variedly in terms of CAT activity and MDA content (Table 1 and Fig. 2 and 5).

## 3 The implication of stress tolerant traits of co-existing species to succession and ecosystem conservation

Finding the connections between co-existing species' physiological responses and the downstream ecological events is vital to understand the vegetation dynamics when environment is altered. The eco-physical traits and the relative competitive ability of co-existing species are two determinants that are responsible for the species assemblages and community succession. Especially, the adaptability of co-existing species to the most limiting environmental factor is more important in the determining their fitness and abundance in a community, which will impel community succession and determine the successional tendency or pathway. In the future, the accelerating globe change of climate and frequent local extreme climate events might shift species' temperature, water and nutrition niche. With these backgrounds, non-native species from adjacent areas may cross frontiers and become new elements of the biota, which will promote biological invasion [44]. Plants are the main active participants of mineral, water and carbon cycle through their absorbing, fixation and release. The potential vegetation change caused by climate change may influence the pattern and strength of local chemical elements cycle in the long term.

## Author Contributions

Conceived and designed the experiments: FD. Performed the experiments: HJS XCZ XXX. Analyzed the data: FD HJS. Contributed reagents/materials/analysis tools: FD XCZ. Wrote the paper: FD HJS.

## References

1. Ledger ME, Edwards FK, Brown LE, Milner AM, Woodward G (2011) Impact of simulated drought on ecosystem biomass production: an experimental test in stream mesocosms. Glob Change Biol 17: 2288–2297.
2. Anderson-Teixeira KJ, Miller AD, Mohan JE, Hudiburg TW, Duval BD, et al. (2013) Altered dynamics of forest recovery under a changing climate. Glob Change Biol: 1–21.
3. Shao HB, Chu LY, Wu G, Zhang JH, Lu ZH, et al. (2007) Changes of some anti-oxidative physiological indices under soil water deficits among 10 wheat (Triticum aestivum L.) genotypes at tillering stage. Colloids Surf B Biointerfaces 54: 143–149.
4. Shao HB, Liang ZS, Shao MA, Wang BC (2005) Changes of anti-oxidative enzymes and membrane peroxidation for soil water deficits among 10 wheat genotypes at seedling stage. Colloids Surf B Biointerfaces 42: 107–113.
5. Ludlow MM, Muchow RC (1990) A Critical Evaluation of Traits for Improving Crop Yields in Water-Limited Environments. Adv Agron 43: 107–153.
6. Van der Molen MK, Dolman AJ, Ciais P, Eglin T, Gobron N, et al. (2011) Drought and ecosystem carbon cycling. Agric For Meteorol 151: 765–773.
7. Elizamar CDS, Manoel BDA, André DDAN, Carlos DDSJ (2013) Drought and Its Consequences to Plants – From Individual to Ecosystem. Responses Org Water Stress.
8. Ostle NJ, Smith P, Fisher R, Ian Woodward F, Fisher JB, et al. (2009) Integrating plant–soil interactions into global carbon cycle models. J Ecol 97: 851–863.
9. Ali A, Alqurainy F (2006) Activities of antioxidants in plants under environmental stress. The Lutein – prevention and treatment for age-related diseases. Trivandrum, Inida: Transworld Research Network. 187–256. Available: http://www.cabdirect.org/abstracts/20063204010.html;jsessionid=72FCA1820C86D7F9536027037824D3AB?freeview=true&gitCommit=4.13.29. Accessed 2013 April 20.
10. Farooq M, Wahid A, Kobayashi N, Fujita D, Basra SMA (2009) Plant Drought Stress: Effects, Mechanisms and Management. In: Lichtfouse E, Navarrete M, Debaeke P, Véronique S, Alberola C, editors. Sustainable Agriculture. Springer Netherlands. 153–188.
11. De Carvalho MHC (2008) Drought stress and reactive oxygen species: production, scavenging and signaling. Plant Signal Behav 3: 156–165.
12. Diaz S, Hodgson JG, Thompson K, Cabido M, Cornelissen JHC, et al. (2004) The plant traits that drive ecosystems: Evidence from three continents. J Veg Sci 15: 295–304.
13. Shipley B (2010) From plant traits to vegetation structure: chance and selection in the assembly of ecological communities. Cambridge University Press. 275
14. Bazzaz FA (1979) The physiological ecology of plant succession. Annu Rev Ecol Syst 10: 351–371.
15. Connell JH, Slatyer PO (1977) Mechanisms of succession in natural communities and their role in community stability and organization. He Am Nat 111: 1119–1144.
16. Ning T, Guo Z, Guo M, Han B (2013) Soil water resources use limit in the loess plateau of China. Agricultural Sciences 04: 100–105.
17. Wu BG, Glenn WT (1985) The positivecorrelations between the activity of superoxide dismutase and dehydration tolerance in wheat seedlings. Acta Botanica Sinica 27: 152–160.
18. Tan Y, Liang Z, Shao H, Du F (2006) Effect of water deficits on the activity of anti-oxidative enzymes and osmoregulation among three different genotypes of Radix Astragali at seeding stage. Colloids Surf B Biointerfaces 49: 60–65.
19. Ruth G (2002) Oxidative Stress and Acclimation Mechanisms in Plants. Arabidopsis Book 49: 1–19.
20. Shabala S, Mackay A (2011) Ion transport in Halophytes. In: Turkan I, editor. Advances in botanical research: Plant Responses to Drought and Salinity StressDevelopments in a Post-Genomic Era. USA: Elsevier Academic Press, Vol. 57. p. 165, 505.
21. Jiang MY, Jing JH, Wang ST (1991) Water stress and membrane-lipid peroxidation in plants. Acta Univ Agric. Boreali-occidentalis 19: 88–94.
22. Picket STA, Collins SL, Armesto JJ (1987) Models, mechanisms and pathways of succession. Bot Rev 53: 335–371.
23. MacDougall AS, Turkington ROY (2004) Relative importance of suppression-based and tolerance-based competition in an invaded oak savanna. J Ecol 92: 422–434.
24. Du F, Shao HB, Shan L, Liang ZS, Shao MA (2007) Secondary succession and its effects on soil moisture and nutrition in abandoned old-fields of hilly region of Loess Plateau, China. Colloids Surf B Biointerfaces 58: 278–285.
25. Du F, Liang ZS, Shan L, Tan Y (2005) Effects of old-fileds successional revegetation on soil moisture in Hilly Loess Region of Northern Shaanx. Journal of Natural Resources. 20: 33–37.
26. Gao JF (2006) Guide to botanical physiological experiment. Higher Education Press 152 p.
27. Li HS (2006) Theory and techniques of botanical biochemical experiments Higher Education Press 2006. 298 p.
28. Tang QY, Feng MG (2007) DPS Data processing system: Experimental design, statistical analysis, and data mining. Beijing: Science Press.528 p.
29. Smirnoff N (1993) The role of active oxygen in the response of plants to water deficit and desiccation. New Phytol 125: 27–58.
30. Lemos JM, Vendruscolo ECG, Schuster I, dos Santos MF (2011) Physiological and Biochemical Responses of Wheat Subjected to Water Deficit Stress at Different Phenological Stages of Development. J Agric Sci Technol B 1: 22–30.
31. Chutipaijit S, Cha-Um S, Sompornpailin K (2009) Differential accumulations of proline and flavonoids in indica rice varieties against salinity. Pak J Bot 41: 2497–2506.
32. Claussen W (2005) Proline as a measure of stress in tomato plants. Plant Sci 168: 241–248. doi:10.1016/j.plantsci.2004.07.039
33. Ozden M, Demirel U, Kahraman A (2009) Effects of proline on antioxidant system in leaves of grapevine (Vitis vinifera L.) exposed to oxidative stress by $H_2O_2$. Sci Hortic 119: 163–168. doi:10.1016/j.scienta.2008.07.031
34. Chutipaijit S, Cha-um S, Sompornpailin K (2011) High contents of proline and anthocyanin increase protective response to salinity in Oryza sativa L. spp. indica. Aust J Crop Sci 5: 1191–1198.
35. Ashraf M, Foolad MR (2007) Roles of glycine betaine and proline in improving plant abiotic stress resistance. Environ Exp Bot 59: 206–216.
36. Kishor PK, Hong Z, Miao GH, Hu CAA, Verma DPS (1995) Overexpression of [delta]-pyrroline-5-carboxylate synthetase increases proline production and confers osmotolerance in transgenic plants. Plant Physiol 108: 1387–1394.

37. Okuma E, Soeda K, Fukuda M, Tada M, Murata Y (2002) Negative correlation between the ratio of $K^+$ to $Na^+$ and proline accumulation in tobacco suspension cells. Soil Sci Plant Nutr 48: 753–757.

38. Shao H, Chu L, Shao M, Jaleel CA, Hong-mei M (2008) Higher plant antioxidants and redox signaling under environmental stresses. C R Biol 331: 433–441.

39. Okuma E, Murakami Y, Shimoishi Y, Tada M, Murata Y (2004) Effects of exogenous application of proline and betaine on the growth of tobacco cultured cells under saline conditions. Soil Sci Plant Nutr 50: 1301–1305.

40. Sivakumar P, Sharmila P, Pardha Saradhi P (2000) Proline Alleviates Salt-Stress-Induced Enhancement in Ribulose-1,5-Bisphosphate Oxygenase Activity. Biochem Biophys Res Commun 279: 512–515.

41. Nanjo T, Kobayashi M, Yoshiba Y, Sanada Y, Wada K, et al. (1999) Biological functions of proline in morphogenesis and osmotolerance revealed in antisense transgenic Arabidopsis thaliana. Plant J 18: 185–193.

42. Igarashi Y, Yoshiba Y, Takeshita T, Nomura S, Otomo J, et al. (2000) Molecular cloning and characterization of a cDNA encoding proline transporter in rice. Plant Cell Physiol 41: 750–756.

43. Rentsch D, Hirner B, Schmelzer E, Frommer WB (1996) Salt stress-induced proline transporters and salt stress-repressed broad specificity amino acid permeases identified by suppression of a yeast amino acid permease-targeting mutant. Plant Cell Online 8: 1437–1446.

44. Walther G-R, Post E, Convey P, Menzel A, Parmesan C, et al. (2002) Ecological responses to recent climate change. Nature 416: 389–395.

# 12

# Perceptional and Socio-Demographic Factors Associated with Household Drinking Water Management Strategies in Rural Puerto Rico

**Meha Jain**[1]*, **Yili Lim**[1], **Javier A. Arce-Nazario**[2], **María Uriarte**[1]

**1** Department of Ecology, Evolution and Environmental Biology, Columbia University, New York, New York, United States of America, **2** Department of Biology, University of Puerto Rico in Cayey, Cayey, Puerto Rico, United States of America

## Abstract

Identifying which factors influence household water management can help policy makers target interventions to improve drinking water quality for communities that may not receive adequate water quality at the tap. We assessed which perceptional and socio-demographic factors are associated with household drinking water management strategies in rural Puerto Rico. Specifically, we examined which factors were associated with household decisions to boil or filter tap water before drinking, or to obtain drinking water from multiple sources. We find that households differ in their management strategies depending on the institution that distributes water (i.e. government PRASA vs community-managed non-PRASA), perceptions of institutional efficacy, and perceptions of water quality. Specifically, households in PRASA communities are more likely to boil and filter their tap water due to perceptions of low water quality. Households in non-PRASA communities are more likely to procure water from multiple sources due to perceptions of institutional inefficacy. Based on informal discussions with community members, we suggest that water quality may be improved if PRASA systems improve the taste and odor of tap water, possibly by allowing for dechlorination prior to distribution, and if non-PRASA systems reduce the turbidity of water at the tap, possibly by increasing the degree of chlorination and filtering prior to distribution. Future studies should examine objective water quality standards to identify whether current management strategies are effective at improving water quality prior to consumption.

**Editor:** Joan Muela Ribera, Universitat Rovira i Virgili, Spain

**Funding:** Funding was provided by awards to MU from National Science Foundation DEB 0620910 and the Crosscutting Initiatives program, Earth Institute, Columbia University. This study was partially supported by Award Number P20MD006144 from the National Center on Minority Health and Health Disparities. The content is solely the responsibility of the authors and does not necessarily represent the official views of the National Center on Minority Health and Health Disparities or the National Institutes of Health. This work was also supported by the National Science Foundation Graduate Research Fellowship under Grant No. 11-44155 awarded to MJ. The funders had no role in study design, data collection and analysis, decision to publish, or preparation of the manuscript.

**Competing Interests:** The authors have declared that no competing interests exist.

* E-mail: mj2415@columbia.edu

## Introduction

Over 700 million people across the globe do not have access to clean drinking water, leading to high levels of chronic waterborne illnesses [1–3]. This is particularly problematic in rural communities that do not receive adequately treated water from government facilities and may not have access to appropriate technologies to treat water locally [4,5]. Scientists and policymakers have long considered the best ways to improve access to potable water, yet identifying the most effective ways to manage drinking water is difficult given that it is typically managed by multiple public and private agencies [6–8]. Drinking water is often extracted and treated at different spatial scales (e.g. regional, watershed, and household level), resulting in management by various stakeholders that act at each of these scales (e.g. governmental, private, and household sectors; [9,10]. Given the complexity of drinking water management, policy makers and agencies (e.g. World Health Organization) over the past decade have increasingly recognized the importance of household water management, particularly in regions where government and community water treatment facilities are ineffective [11,12].

Households play an important role in determining the water quality experienced by individuals, as households are the last point of management prior to consumption [4,12].

To target the most successful interventions, it is important to understand the socio-cultural context of current household water management decisions [13]; by understanding how households manage their drinking water and why, policymakers can more effectively target intervention strategies to improve water quality prior to consumption. Though most households in a given community face the same water quality at the tap, some may treat their water prior to consumption while others may not [14,15]. This variation in household water management is influenced by a variety of factors, including knowledge of water treatment practices prior to distribution, perceptions of water quality at the tap, and socio-demographic characteristics of the decision-maker [14,16,17]. For example, previous studies have found that households are more likely to treat their tap water when they believe that government or community treatment facilities are ineffective [18,19], or when they believe that water quality is low at the tap [15]. While previous studies have examined the importance of these factors individually, few studies have considered these multiple drivers within the same analysis. Doing so is important because it identifies which factors are the most influential for household decision-making. This knowledge can then be used to identify and target interventions that are in line with current household perceptions, which has been shown to result in a greater rate of intervention uptake and success [20].

Our study assesses which factors most strongly influence household water management decisions, specifically whether households filter or boil their tap water prior to consumption or whether they obtain drinking water from multiple sources, in rural Puerto Rico. It is important to understand household water management in this region because previous studies have suggested that broader water management institutions do not always provide adequate water quality at the tap, particularly in rural, mountainous regions that are far from government treatment facilities [21]. There are two broad categories of institutions that manage drinking water for the island's four million people: government-managed Puerto Rico Aqueduct and Sewer Authority (PRASA) systems (which serve approximately 3.8 million people), and private and community non-PRASA systems (which serve approximately 400 communities, or up to 250,000 people), which are found primarily in mountainous regions that are too far to be connected to PRASA treatment facilities [21,22]. While the non-PRASA category encompasses a range of management strategies, given decentralized management where each community typically develops their own management plan, it is widely believed that non-PRASA communities in general are exposed to low water quality at the tap due to ineffective management of water prior to distribution. The Puerto Rico Department of Health (PRDOH) considers non-PRASA systems to be a health threat since they typically do not comply with federal water quality standards [23]. This is because about fifty percent of non-PRASA systems obtain water from surface sources and there is little or no monitoring of water quality in these communities [24]. Previous studies estimate that 30% of non-PRASA systems lack any water treatment infrastructure [22], and water is not treated consistently even when water infrastructure exists [22,25]. PRASA systems on the other hand typically filter and chlorinate water at treatment facilities before distribution and provide water quality assessments required by the U.S. Federal Potable Water Standards. Despite centralized management, PRASA systems are often plagued by water shortages and high rates of sediment loading and turbidity, which can result in non-compliances with the US Environmental Protection Agency (EPA) water quality standards [26]. This is because many filtration plants, particularly in mountainous regions, are not equipped to handle water filtration during periods of heavy rainfall [23], which is especially problematic given Puerto Rico's high frequency of tropical storms [27].

Given the possibility of inadequate water treatment by non-PRASA and PRASA facilities, some households have developed management strategies that are thought to improve drinking water quality prior to consumption. These strategies include filtering or boiling tap water or obtaining water from alternate sources like private wells and local markets. In this study, we assessed which perceptional factors that have been postulated to be important in previous literature are most associated with households that undertake water management strategies in rural, mountainous Puerto Rico [15,18,19]. Specifically, we predict the following in order of importance:

(1) households will have different management techniques depending on whether water is provided by government

(PRASA) or community (non-PRASA) institutions likely due to differences in water quality at the tap;

(2) households that have problems with institutional water management prior to distribution are more likely to treat water;

(3) households are more likely to treat water if they perceive that water from the tap is of low quality;

(4) households that have less knowledge about how their water is treated prior to distribution are more likely to treat their water.

We quantify the relative importance of these various factors for household decision-making to better guide future water quality assessments and interventions in rural Puerto Rico. While our results are specific to Puerto Rico, we argue that our methodology can also be implemented in other regions to better understand the drivers of household water management and more effectively target interventions to those households vulnerable to low water quality.

## Methods

### Study site

Data were collected in eight different community sectors within the Cayey Mountain range in Puerto Rico from June to August of 2009. Our study focused on communities in this region because they are thought to be at high risk for low water quality given that they are rural and found in mountainous terrain, which makes them difficult to connect to PRASA treatment facilities. We specifically focused on villages found in Cayey and Patillas municipalities (Figure 1), which contain a large number of non-PRASA communities. Both municipalities are similar in socio-economic and development status. The median household income was $10,923 in Cayey and $9,375 in Patillas in 2000, which were lower than the island average of $13,189 [28]. We selected PRASA and non-PRASA communities that were adjacent to one another in each of the two municipalities. This was possible when we interviewed communities at the boundary where PRASA systems stopped serving communities with piped government water. This paired sampling design reduced possible confounding effects from socio-economic and geographic factors and allowed us to better assess whether households make different decisions based on if PRASA or non-PRASA institutions manage their water. Initial communities (n = 2) were selected based on where our field team had previous experience and knew PRASA and non-PRASA communities were adjacent to one other. We then used a snowball technique and visited additional communities (n = 6) that were suggested to us by the initial community contact [29]. While the communities that we selected for sampling were not entirely selected at random given this snowball technique, we believe that they are representative of the broader region given that each of our four pairs of PRASA and non-PRASA communities were spread across a wide geographic area in the Cayey mountain range (up to 15 km between our four sites).

### Data collection

We surveyed 218 respondents across the eight community sectors considered in our study. Each community sector ranged in size from 50 to 200 households, but to ensure comparability we selected adjacent PRASA and non-PRASA communities that were approximately the same size. We aimed to interview 20 to 30 households in each community, and selected survey households at random distributed equally throughout each community. A summary of the number of survey respondents in PRASA verus

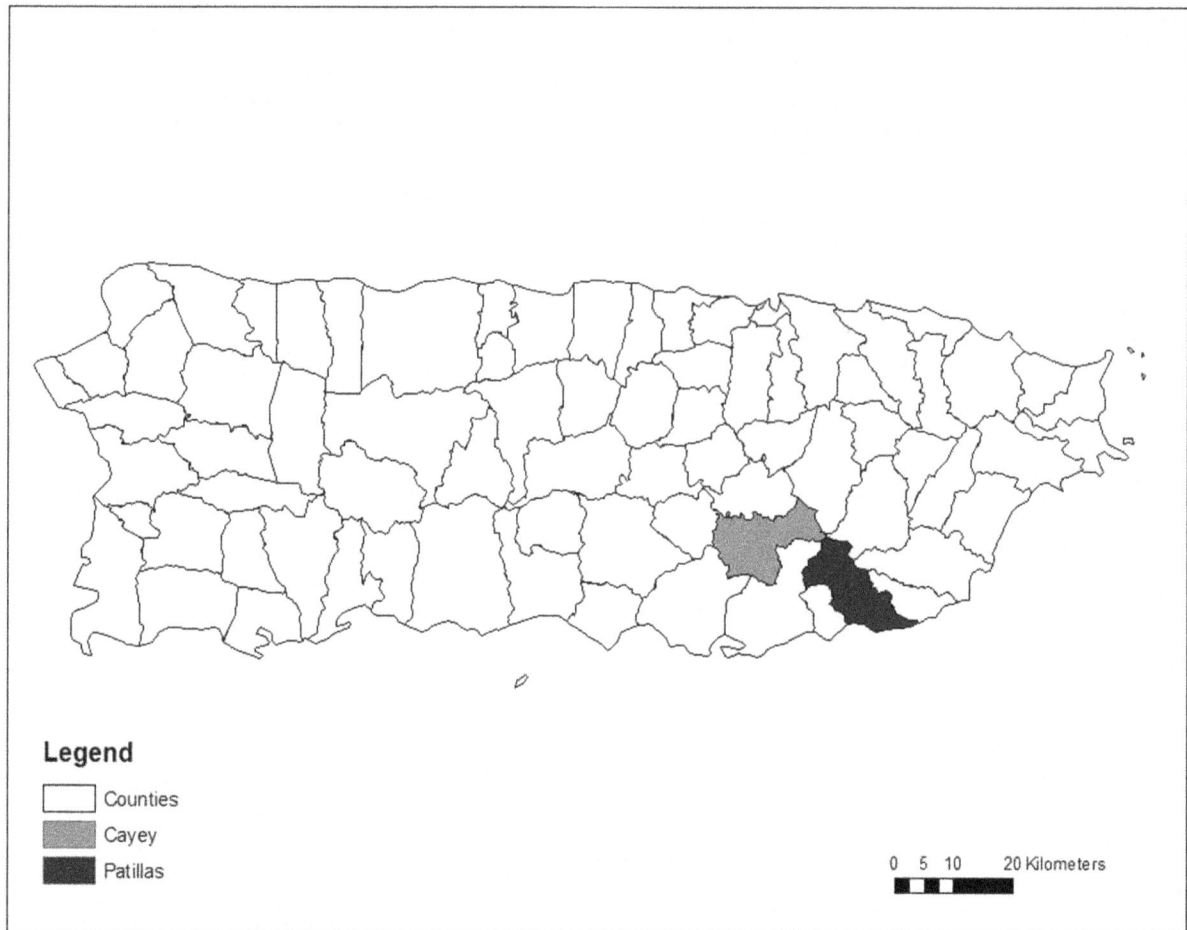

**Figure 1. Map of Study Region in Puerto Rico.** Municipalities where surveys were conducted are highlighted in gray. We did not list specific communities that we visited to keep the communities we surveyed anonymous.

non-PRASA communities is given in the supplementary information (Table S1). We then spoke to the household member who answered the door and identified which member of the household was in charge of household water management decisions. If that family member was home, we then conducted the oral structured survey with that family member. If the family member in charge of water management decisions was not at home we skipped that household and did not include it in our survey sample.

**Ethics statement.** Surveys were approved by the Columbia University Institutional Review Board under protocol number IRB-AAAE0079 and informed consent was written. Surveys were conducted in Spanish by local research assistants. We asked all respondents if we could audio record their interviews in order to keep a record of responses and to assist in confirming written responses and only did so if the interviewee gave permission. Our survey instrument contained questions related to whether households undertake any drinking water management prior to consumption, the respondent's perceptions of institutional water management and water quality at the tap, and socio-demographic information for the respondent. Details about each question are listed below, and all data collected were self-reported.

We asked respondents how they managed their drinking water sources prior to consumption, which serves as the dependent variable in our analyses. We grouped responses into two different types of strategies that households may undertake to cope with

inadequate water quality. One coping strategy is to increase *the number of drinking water sources* used in the household. Households may diversify sources of drinking water by purchasing bottled water or obtaining drinking water from a personal well. The second coping strategy considered in this study is if households *treat tap water before drinking*. If households believe that their tap water is of inadequate quality, they may filter or boil it before drinking.

We also collected data on the following variables that have been suggested to be important for household water management decisions in previous studies. These variables serve as covariates in our statistical models and we discuss specific data that were collected for each variable of interest. As outlined in the introduction, we believe that management institution type, problems with institutional water management, perceptions of water quality, and knowledge of water treatment will influence household decisions to manage drinking water.

**Management institution.** We considered *the type of institution that manages water* (i.e. *PRASA or non-PRASA*) as a fixed effect because the way that specific institutions manage water may influence household decision-making. This may occur if institutions influence the behavior of households via uniform rules and norms [30]. Institutions may also affect household decision-making if they expose all households in a given community to the same quality of resource. Previous studies have shown that mismanagement of water treatment by institutions may negatively impact water quality experienced by all households within the distribution system [14].

**Problems with institutional management.** As a broad measure of whether households believe that institutions effectively manage water, which has been shown to be important in the previous literature [18,19], we asked households whether they *have problems with how their water is managed* by PRASA or non-PRASA operators. We predict that respondents who have more problems with institutional management are more likely to treat tap water since they may believe that their water was inadequately treated before distribution.

**Perceptions of water quality at the tap.** Even though all households in a given community are exposed to the same water quality at the tap, varying perceptions may lead to heterogeneous behavior among decision-makers. Previous studies have shown that perceptions of water quality are strong drivers of household water management decisions [15]. To assess water quality perceptions, we asked respondents to rank the *quality of their tap water* on a scale of 1 to 4, where 1 equals poor water quality and 4 equals excellent water quality. We predict that households that believe they have poor water quality are more likely to develop coping strategies.

**Knowledge of institutional management.** Given that previous studies have suggested that increased knowledge of institutional management practices influences individual decision-making [31], we asked respondents whether they *knew how their water was treated* before it is piped to their homes. We predict that households that have less knowledge of how their water was treated by management institutions are more likely to treat water given that they may not trust that their water was treated prior to distribution. Previous studies have suggested a link between increased knowledge, transparency, and trust [32,33].

**Socio-economic and demographic variables.** Various socio-economic and demographic factors, such as income, age, and gender of the decision-maker, can influence household decisions [34,35]. We considered the age and gender of the respondent as controls in our analysis, but did not include income in our final models because only half of our interviewees responded to this question. Income data were collected as self-reported annual income for the household in $10,000 US increments (e.g. $10,000–$20,000, $20,000–$30,000, etc.). However, to test whether income may be important for water management decisions in our region, we ran our statistical models on the subset of data with income. We found that the income variable was never significant (p>0.05), suggesting that it is not a significant driver of water management decisions in this region. Furthermore, since we are interested in quantifying the relative importance of various perceptional and socio-demographic factors for decision-making, excluding income from the analysis should not impact our results; instead, it would at most reduce the amount of variance explained by our models.

## Statistical analyses

We conducted three sets of analyses to identify how water management and the drivers of water treatment decisions varied across households in our study. First, we used ANOVA to compare institution types for our two dependent variables of interest: the number of water sources and water treatment. We also compared the distribution of our covariates between institution type using ANOVA analyses. These simple comparisons illustrate whether there were significant differences in coping strategies, perceptions, and socio-demographic factors between households in PRASA and non-PRASA communities.

In a second set of analyses, we used separate logistic regressions to assess the effects of all covariates (Table 1) on the two response variables of interest. To assess whether these covariates have different effects on household decision-making in PRASA and non-PRASA communities, we included interactions between management institution (i.e. PRASA, non-PRASA) and the other covariates. To avoid parameter tradeoffs and clarify interpretation of the results, we dropped covariates that had a correlation >0.4. Based on this criterion, we dropped gender from our analysis. We then conducted stepwise variable selection using $AIC_c$ to select the best model [36]. To facilitate the interpretation of effect magnitudes among covariates, all continuous predictors were standardized by subtracting their mean and dividing by twice their standard deviation [37]. Goodness of fit was calculated using the universal goodness of fit le Cessie and Houwelingen test [38] in the Design package (Version 2.3-0) in R Project Software (R Statistical computing 2012, Version 2.14.1 was used for all analyses).

Finally, to assess the relative importance of each variable, we dropped each variable one at a time from the best logistic regression model and compared the $AIC_c$ from the resulting model with the $AIC_c$ from the best model. Variables that contributed most to model fit, and therefore were the most important in our analysis, had the largest change in $AIC_c$ between the best model and the model with the variable in question dropped [39].

## Results

### ANOVA results

Several variables differed between PRASA and non-PRASA households (Table 2). Considering water management strategies, non-PRASA households were significantly more likely to obtain water from multiple sources, whereas PRASA households were significantly more likely to treat their tap water before drinking. This simple analysis suggests that households in PRASA and non-PRASA communities mitigate perceived low water quality in different ways. Considering perceptional variables, Non-PRASA households were significantly more likely to know how their institutions managed drinking water prior to distribution and non-PRASA households were also more likely to report higher water quality than PRASA households (Table 2).

### Logistic regression models

The most important predictor of household decisions to obtain water from multiple sources was the institution that manages water (e.g. PRASA vs non-PRASA; Table 3, Figure 2A). Respondents in non-PRASA communities were more likely to obtain water from multiple sources than those from PRASA communities. Using the le Cessie and Houwelingen goodness of fit test, there is not a significant difference between observed and predicted values from the model suggesting good model fit (z = 0.78, sd = 0.19, p = 0.44).

The best predictors of household decisions to treat tap water before drinking were the institution that manages water, perceptions of water quality, and the interaction between the institution that manages water and problems with institutional management (Table 3, Figure 2B). PRASA households were significantly more likely to treat their water before drinking than non-PRASA households. Households that reported lower water quality were also more likely to treat their tap water, regardless of water management institution. Finally, the significant interaction between the institution that manages water and whether a household reported problems with institutional management suggests that non-PRASA households that had problems with institutional management were more likely to treat tap water before drinking than PRASA households. Le Cessie and

**Table 1.** Description and hypothesized relationship for each of the variables considered in our statistical models.

| Variable | Variable Code | Description | Hypothesis |
|---|---|---|---|
| Number of Water Sources | Num Source | Number of drinking water sources (0= one source, 1= multiple sources) | Dependent Variable |
| Treat Water | Treat Water | Whether a household filtered or boiled tap water before drinking (0= No, 1= Yes) | Dependent Variable |
| Institution Type | Water System | Which water system the household receives water from (i.e. PRASA =0, Non-PRASA =1) | + |
| Knowledge of Treatment | Treatment Knowledge | Identified if individual had knowledge of how institution (PRASA or Non-PRASA) treated water before it arrives at the tap (i.e. No =0, Yes =1) | - |
| Reported Problems with Institutional Management | Problems | Whether the respondent reported problems with the way institutions manage water (i.e. No =0, Yes =1) | + |
| Perceptions of Water Quality | Water Quality | Self-reported quality of drinking water from the tap (i.e. poor =1, fair =2, good =3, excellent =4) | - |
| Demographic Data | Age | Age | Control |
| Gender | Gender | Gender (0= Male, 1= Female) | Control |

Variable, coding method, description, and the hypothesized relationship with the likelihood of adopting coping strategies for all covariates considered in both statistical models. A positive relationship indicates that the variable would lead to increased coping, as defined by a higher likelihood of treating water and obtaining water from multiple sources.

Houwelingen goodness of fit test indicated a good fit between predicted and observed data (z = − 1.31, sd = 0.14, p = 0.19).

### Variable importance

To understand the relative importance of each covariate considered in our logistic models (Table 1), we conducted a full model logistic regression and assessed the importance of each factor based on its contribution to model fit as measured by the change in AIC$_c$ when that variable was dropped from the full model. In the model that predicted which households were more likely to obtain water from multiple sources, we found that the institution that manages water contributed most to model fit (Figure 3A). This suggests that whether households were from PRASA or non-PRASA communities was the most important variable for predicting whether households obtain water from multiple sources. The remainder of the variables in the model contributed little to model fit.

For the model that identified whether households treat or do not treat water, the institution that manages water was also the best predictor (Figure 3B). This suggests that whether households are

from PRASA or non-PRASA communities was the most important variable to explain whether households treat or do not treat their water. Perceptions of water quality also contributed significantly to model fit (Figure 3B) suggesting that this variable is also important.

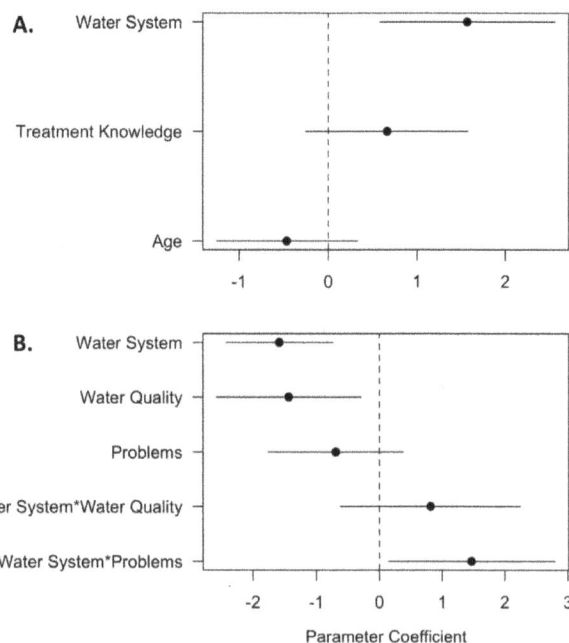

**Table 2.** Comparison of each variable considered in our statistical models by institution type (PRASA vs non-PRASA).

| | Mean value by Institution | | ANOVA results | | |
|---|---|---|---|---|---|
| Variable | PRASA | Non-PRASA | d, f | F | P |
| Number of Water Sources | 0.06 | 0.26 | 1, 187 | 14.28 | <0.001* |
| Treat Water | 0.71 | 0.42 | 1, 187 | 16.26 | <0.001* |
| Treatment Knowledge | 0.49 | 0.73 | 1, 187 | 11.74 | <0.001* |
| Problems | 0.38 | 0.49 | 1, 187 | 2.47 | 0.12 |
| Water Quality | 2.41 | 2.98 | 1, 187 | 21.69 | <0.001* |
| Age | 53.20 | 50.25 | 1, 187 | 1.35 | 0.25 |

Mean value by institution (i.e. PRASA, Non-PRASA) and ANOVA results (degrees of freedom, F-statistic, p-value) are reported for each variable. * indicates p<0.05.

**Figure 2. Parameter Estimate Plot of All Variables Considered in the Two Models that Predict Household Water Management Strategies.** Standard errors are plotted as black lines. The variable is significant if standard error bars do not cross the zero axis. For the number of water sources (A), institution type is significant (p<0.005). For whether households treat water (B), institution type (p<0.001), perceptions of water quality (p<0.05), and the interaction between institution type and if households have a problem with institutional management (p<0.05) are significant.

**Table 3.** Results for each statistical model predicting which factors are associated with household water management strategies.

| Response Variable | Covariates considered in logit model | Parameter Coefficient (Standard Error) | p value | N | GOF (p value) |
|---|---|---|---|---|---|
| Number of Water Sources | Water System | 1.57 (0.52) | <0.005* | 189 | 0.44 |
| Number of Water Sources | Treatment Knowledge | 0.66 (0.48) | 0.17 | 189 | 0.44 |
| Number of Water Sources | Age | −0.47 (0.42) | 0.27 | 189 | 0.44 |
| Treat Water | Water System | −1.58 (0.44) | <0.001* | 189 | 0.19 |
| Treat Water | Water Quality | −1.43 (0.60) | 0.02* | 189 | 0.19 |
| Treat Water | Problems | −0.69 (0.56) | 0.23 | 189 | 0.19 |
| Treat Water | Water System*Water Quality | 0.81 (0.75) | 0.28 | 189 | 0.19 |
| Treat Water | Water System*Problems | 1.47 (0.70) | 0.04* | 189 | 0.19 |

Variables considered, parameter coefficients with standard error, p values, sample size, and goodness of fit for both of the full models including interaction terms. The first model predicts whether households obtain water from one or more sources, and the second model predicts whether households treat or do not treat their water. Significance of at least 5% is highlighted with a *.

## Discussion

Policy-makers and agencies have increasingly recognized the importance of household water management for potable water provisioning given that households are the last point of management prior to consumption [13]. By understanding which factors most influence household water management, policy makers can better identify and target intervention strategies that improve access to clean drinking water. In this study, we examined household water management in rural Puerto Rico. It is important to understand household level management in these communities given that both government PRASA and community non-PRASA water treatment may be ineffective at providing clean water at the

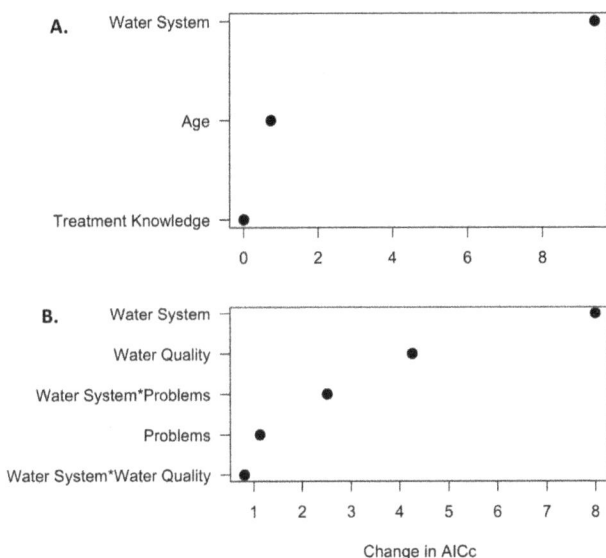

**Figure 3. Importance of Each Covariate for Model Fit in the Two Models that Predict Household Water Management Strategies.** Change in AIC$_c$ for each of the covariates considered in the full logit model for the number of drinking water sources (A) and whether households treat or do not treat water (B). Larger changes in AIC$_c$ values suggest that the variable contributed more to overall model fit. In both analyses (A and B), the institutional variable Water System (i.e. PRASA, non-PRASA) is the variable that contributes most to overall model fit. In the analysis of whether households treat water (B), water quality perceptions were also an important variable.

tap. Specifically, we analyzed (1) whether households obtained water from multiple sources or filtered or boiled tap water before drinking, and (2) which perceptional and socio-demographic factors were most associated with these management decisions. Our analysis suggests that three of our four initial predictions are correct: households manage water differently based on whether they are in PRASA or non-PRASA communities, households are more likely to treat water if they have problems with institutional management, and households are more likely to treat water if they believe that their tap water is of low quality (Figure 2). The fourth factor we predicted to be important in our analysis, whether households had knowledge of how water was treated prior to distribution, was not significant in our analyses.

The institution that manages water (i.e. PRASA vs non-PRASA) was the strongest driver of household drinking water management (Figure 3). PRASA households were more likely to filter or boil their water before drinking, whereas non-PRASA households were more likely to obtain water from multiple sources (Figure 2A). Differences in management strategies between PRASA and non-PRASA communities may be due to differences in perceptions of low water quality, possibly because of differences in water quality at the tap [14]. In PRASA communities, our informal discussions with community members indicate that perceptions of low water quality are due to the bad taste and odor of tap water, which community members attribute to over-chlorination. PRASA treatment facilities typically add chlorine to water prior to distribution, which has been associated with a reduction in bacteria such as *Escherichia Coli* (E. Coli) [40,41]. However, based on our informal interviews with community members across our survey area, it is possible that PRASA systems are over-chlorinating water in this region; these anecdotal claims are bolstered by objective water quality measures collected by the government for the *barrios* (sub-districts) considered in our study, which show periods when chlorine levels are higher than those recommended by the EPA (> 4.0 ppm, Fig. S1) [24,42,43]. Thus, in PRASA communities, families may filter or boil their tap water in order to improve the smell and taste of water prior to consumption. In non-PRASA communities, discussions with community members suggest that perceptions of low water quality are due to turbidity, which community members attribute to the lack of treatment by non-PRASA institutions. Based on discussions with community members and the operators of non-PRASA systems, it appears as if water was not regularly treated (e.g. via chlorine addition or filters) in storage tanks prior to distribution, which resulted in increased water turbidity at the tap. Households mitigated this perceived low water quality by obtaining water from other sources, like store-bought bottled water or filtered water from friends and relatives in PRASA communities.

Second, households are more likely to treat water if they believe that their water was ineffectively managed by treatment facilities prior to distribution. This corroborates previous studies that show that households and communities increase water management efforts if they believe that government or private agencies ineffectively manage water prior to distribution [18,19]. This result is only significant for non-PRASA communities (Figure 2B), suggesting that perceptions of institutional effectiveness drive decisions to treat water only in non-PRASA households. Institutional perceptions may play a stronger role in non-PRASA relative to PRASA communities because institutional management of drinking water is decentralized; given decentralized management, households in non-PRASA communities often play a stronger role in community-level water management than do households in PRASA communities, where water management is centralized within government agencies. Informal discussions with non-PRASA community members support this interpretation: non-PRASA households state that they feel a strong connection to water management institutions due to increased knowledge of treatment practices (Table 2) and the ability to participate in water management by speaking with local water operators or attending community meetings.

Finally, we found that perceptions of water quality were significant predictors of whether households were more likely to treat their water via filtering and boiling (Figure 2B). These results corroborate previous studies that find that households are more likely to manage their water if they perceive that their tap water is of low quality [15]. It is important to note that we only examined water quality perceptions and not objective water quality metrics at the household level, and it is unclear how well these two measures correlate with one another. If these two measures are not related, this could lead to water management decisions that result in low drinking water quality. For example, households may perceive that their water is of good quality, resulting in no treatment at the tap, when in reality objective water quality measures show that water treatment is required prior to consumption. Future studies should measure objective water quality standards in this region both before and after household treatment of drinking water to determine whether households are accurately perceiving low water quality and treating water effectively.

Based on the three main findings outlined above, we have several recommendations to improve water quality management in this region. First, we argue that both PRASA and non-PRASA institutions would likely improve water quality if they took household perceptions into account and understood how households manage water after it is distributed to the tap. Specifically, PRASA systems may improve water quality if they take steps to improve the taste and odor of tap water. If this low water quality is caused by over-chlorination as many people in PRASA communities believe, these systems should reduce the amount of chlorine used or let chlorinated water sit in storage tanks to allow for dechlorination prior to distribution while controlling for environmental variables that may increase chlorination byproducts [44]. Non-PRASA systems, on the other hand, may benefit by reducing the amount of turbidity at the tap, possibly by filtering water prior to distribution; this, and chlorination, may reduce perceived low water quality at the household scale. Second, objective water quality assessments should be coupled with these household level survey results to focus intervention strategies on the most

vulnerable populations, particularly those households that have low water quality but do not treat their water or that treat their water ineffectively. For example, PRASA households perceive low water quality due to bad taste and odor possibly caused by over-chlorination, however, one of the main strategies to mitigate this problem is filtering tap water. Yet to dechlorinate water, expensive active carbon filters are required [45] and these filters were typically not used in this region, suggesting that household strategies to filter water may be ineffective at reducing chlorine content. Finally, given that perceptions of institutional effectiveness appear to influence household management decisions, particularly in non-PRASA communities, we argue that these agencies should strengthen perceptions of institutional effectiveness by increasing the involvement of local community members in water management decisions. If community members have an increased say in how water is managed prior to distribution, it is likely that there will be improved water management given that household-level concerns about water quality are more likely to be addressed [46,47].

It is important to note that this study examined household perceptions of water quality and management, and it is possible that these perceptions are inaccurate when compared to objective measures. For example, most PRASA households believed that the bad taste and odor of tap water were caused by over-chlorination at treatment plants prior to distribution, but it is possible that the bad taste and odor were caused by other factors, like the addition of air or exposure to old pipes during the distribution process [48,49]. Future work should quantify objective water quality and assess whether current management strategies are effective at improving water quality prior to consumption. Second, we conducted our analyses based on survey data collected for over 200 people who live in the Cayey Mountain range. It is possible that our results would differ if we increased the scope of this study, particularly to other regions in Puerto Rico that may have different management strategies in PRASA and non-PRASA systems. Future studies should conduct similar perceptional studies across the island to better identify how universal the findings of this study are. Finally, it is important to note that we used the broad category of non-PRASA to encompass a wide range of institutions. Given that non-PRASA management is decentralized and individual communities are making water management decisions, it is possible that each non-PRASA system managed water slightly differently prior to distribution. We argue, however, that the coarse institutional categorization of non-PRASA is important particularly for policy given that the government uses this coarse categorization in water quality and compliance monitoring [23]. Future work should examine the heterogeneity in water management across non-PRASA systems to identify whether certain management strategies result in different outcomes for water quality and management at the household scale.

In conclusion, this study highlights the importance of social surveys and decision-making analyses to better identify how households currently manage drinking water and which factors influence household management decisions. Our results suggest that both community-level properties, like the type of institution that manages water prior to distribution, and household-level factors, like water quality perceptions, are important for predicting household-level water management behavior. By understanding household perceptions of both water quality and treatment of water prior to distribution, policy-makers can better identify and target intervention strategies that are tailored to current household decision-making. This is important given that previous studies have suggested that policies have a higher chance of uptake and success if they are created considering the local context [20].

## Supporting Information

**Figure S1   Free chlorine levels in ppm in PRASA and non-PRASA communities across our survey area.** Data for PRASA communities were obtained from government databases collected at the barrio level, and data for non-PRASA communities were collected by our field team across several of our study communities of interest. These data suggest that free chlorine levels are typically lower in non-PRASA communities than PRASA communities, and several PRASA measurements have free chlorine levels higher than those recommended by the EPA (4.0 ppm, dotted horizontal line). This suggests that there may be over-chlorination in some PRASA communities.

**Table S1   Number of interviewees in Non-PRASA and PRASA communities in our two study municipalities.** We do not provide specific names of the communities or sectors surveyed in order to keep anonymity of our participants.

## Acknowledgments

We would like to thank our field assistants Natalia Rodriguez and Yazmin Rivera, who conducted the household interviews and helped enter data during the summer of 2009, and Derek Berezdivin, who also helped enter data in 2010.

## Author Contributions

Conceived and designed the experiments: MJ YL MU. Performed the experiments: MJ YL. Analyzed the data: MJ. Wrote the paper: MJ MU JA YL.

## References

1. Teunis P, Medema GJ, Kruidenier L, Havelaar AH (1997) Assessment of the risk of infection by Cryptosporidium or Giardia in drinking water from a surface water source. Water Research.
2. Hellard ME, Sinclair MI, Forbes AB, Fairley CK (2001) A randomized, blinded, controlled trial investigating the gastrointestinal health effects of drinking water quality. Environ Health Perspect 109: 773–778.
3. Prüss A, Kay D, Fewtrell L, Bartram J (2002) Estimating the burden of disease from water, sanitation, and hygiene at a global level. Environ Health Perspect 110: 537–542.
4. Trevett AF, Carter RC, Tyrrel SF (2004) Water quality deterioration: A study of household drinking water quality in rural Honduras. International Journal of Environmental Health Research 14: 273–283.
5. Hunter PR, Toro GIR, Minnigh HA (2010) Impact on diarrhoeal illness of a community educational intervention to improve drinking water quality in rural communities in Puerto Rico. Bmc Public Health 10,
6. Cash DW, Adger WN, Berkes F, Garden P, Lebel L, et al. (2006) Scale and cross-scale dynamics: Governance and information in a multilevel world. Ecology and Society 11.
7. Berkes F (2006) From community-based resource management to complex systems: The scale issue and marine commons. Ecology and Society 11.
8. Sarker A, Ross H, Shrestha KK (2008) A common-pool resource approach for water quality management: An Australian case study. Ecological Economics 68: 461–471.
9. Lebel L, Garden P, Imamura M (2005) The politics of scale, position, and place in the governance of water resources in the Mekong region. Ecology and Society 10.
10. Saravanan VS (2008) A systems approach to unravel complex water management institutions. Ecological Complexity 5: 202–215.
11. Mintz E, Bartram J, Lochery P, Wegelin M (2001) Not Just a Drop in the Bucket: Expanding Access to Point-of-Use Water Treatment Systems. American Journal of Public Health 91: 1565–1570.
12. Clasen TF, Cairncross S (2004) Editorial: Household water management: refining the dominant paradigm. Tropical Medicine & International Health 9: 187–191.
13. Sobsey MD (2002) Managing water in the home: accelerated health gains from improved water supply. World Health Organization.
14. Gartin M, Crona B, Wutich A, Westerhoff P (2010) Urban Ethnohydrology: Cultural Knowledge of Water Quality and Water Management in a Desert City. Ecology and Society 15: 36.
15. Hu Z, Morton LW, Mahler RL (2011) Bottled Water: United States Consumers and Their Perceptions of Water Quality. Int J Env Res Pub He 8: 565–578.
16. Fielding KS, Russell S, Spinks A, Mankad A (2012) Determinants of household water conservation: The role of demographic, infrastructure, behavior, and psychosocial variables. Water Resour Res 48.
17. Sabau G, Haghiri M (2008) Household willingness-to-engage in water quality projects in western Newfoundland and Labrador: a demand-side management approach. Water and Environment Journal 22: 168–176.
18. Zérah M (2000) Water, unreliable supply in Delhi. Manohar Publishers.
19. Katuwal H, Bohara AK (2011) Coping with poor water supplies: empirical evidence from Kathmandu, Nepal. J Water Health 9: 143–158.
20. Jehu-Appiah C, Aryeetey G, Agyepong I, Spaan E, Baltussen R (2012) Household perceptions and their implications for enrolment in the National Health Insurance Scheme in Ghana. Health Policy and Planning 27: 222–233.
21. Molina-Rivera W (1998) Estimated water use in Puerto Rico, 1995. Washington D.C.: US. Geological Survey Open-File Report.
22. Guerrero-Preston R, Norat J, Rodriguez M, Santiago L, Suarez E (2008) Determinants of compliance with drinking water standards in rural Puerto Rico between 1996 and 2000: a multilevel approach. P R Health Sci J 27: 229–235.
23. Quinones F (2005) PRASA has ample water supplies. Water Industry News.
24. Environmental Protection Agency (2010) Puerto Rico Aqueduct and Sewer Authority (PRASA) Pollutant Discharge Settlement. Environmental Protection Agency.
25. Toro GR, Minnigh HA (2004) Regulation and Financing of Potable Water Systems in Puerto Rico: A Study in Failure in Governance. AWRA Dunde: Scotland.
26. de Cardenas SC (2011) Does private management lead to improvement of water services? Lessons learned form the experience of Bolivia and Puerto Rico. University of Iowa.
27. Boose E, Serrano M, Foster D (2004) Landscape and regional impacts of hurricanes in Puerto Rico. Ecol Monogr 74: 335–352.
28. US Bureau of the Census (2000) Census of population: social and economic characteristics. Washington D.C. USA: Department of Commerce, Economics, and Statistics Administration.
29. Biernacki P, Waldorf D (1981) Snowball Sampling: Problems and Techniques of Chain Referral Sampling. Sociological Methods & Research 10: 141–163.
30. North DC (1991) Institutions. The Journal of Economic Perspectives 5: 97–112.
31. Makutsa P, Nzaku K, Ogutu P, Barasa P (2001) Challenges in implementing a point-of-use water quality intervention in rural Kenya. American Journal of Public Health 91: 1571–1573.
32. Peters RG, Covello VT, McCallum DB (1997) The determinants of trust and credibility in environmental risk communication: an empirical study. Risk Analysis 17: 43–54.
33. Palanski ME, Kahai SS, Yammarino FJ (2011) Team Virtues and Performance: An Examination of Transparency, Behavioral Integrity, and Trust. J Bus Ethics 99: 201–216.
34. Jorgensen B, Graymore M, O'Toole K (2009) Household water use behavior: An integrated model. Journal of Environmental Management 91: 227–236.
35. Below TB, Mutabazi KD, Kirschke D, Franke C, Sieber S, et al. (2012) Can farmers' adaptation to climate change be explained by socio-economic household-level variables? Global Environmental Change 22: 223–235.
36. Hurvich C, Tsai C (1989) Regression and Time-Series Model Selection in Small Samples. Biometrika 76: 297–307.
37. Gelman A, Hill J (2007) Data Analysis Using Regression And Multilevel/ Hierarchical Models. Cambridge Univ Pr.
38. le Cessie S, van Houwelingen J (1991) A Goodness-of-Fit Test for Binary Regression-Models, Based on Smoothing Methods. Biometrics 47: 1267–1282.
39. Burnham KP, Anderson DR (2002) Model Selection and Multimodel Inference. 2nd ed. New York: Springer.
40. Payment P, Trudel M, Plante R (1985) Elimination of viruses and indicator bacteria at each step of treatment during preparation of drinking water at seven water treatment plants. Appl Environ Microbiol 49: 1418–1428.
41. Arnold BF, Colford JM (2007) Treating water with chlorine at point-of-use to improve water quality and reduce child diarrhea in developing countries: a systematic review and meta-analysis. Am J Trop Med Hyg 76: 354–364.
42. Autoridad de Acueductos y Alcantarillados (2001-2010) Water Quality Reports.
43. Puerto Rico Department of Public Health (1996-2009) Water Quality Reports.
44. Chowdhury S, Champagne P, McLellan PJ (2009) Science of the Total Environment. Science of the Total Environment, The 407: 4189–4206.
45. Worley JL (2000) Evaluation of Dechlorinating Agents and Disposable Containers for Odor Testing of Drinking Water. Virginia Polytechnic Institute and State University.
46. Olsson P, Folke C, Berkes F (2004) Adaptive Comanagement for Building Resilience in Social-Ecological Systems. Environmental Management 34: 75–90.
47. Larson AM, Soto F (2008) Decentralization of Natural Resource Governance Regimes. Annu Rev Env Resour 33: 213–239.
48. Sangodoyin AY (1993) Water quality in pipe distribution systems. Environmental Management and Health.
49. Young WF, Horth H, Crane R, Odgen T, Arnott M (1996) Taste and Odour Threshold Concetrations of Potential Potable Water Contaminants. Water Research 30: 331–340.

# 13

# Shift in the Microbial Ecology of a Hospital Hot Water System following the Introduction of an On-Site Monochloramine Disinfection System

**Julianne L. Baron**[1,2]**, Amit Vikram**[3]**, Scott Duda**[2]**, Janet E. Stout**[2,3]**, Kyle Bibby**[3,4]*

**1** Department of Infectious Diseases and Microbiology, University of Pittsburgh, Graduate School of Public Health, Pittsburgh, Pennsylvania, United States of America, **2** Special Pathogens Laboratory, Pittsburgh, Pennsylvania, United States of America, **3** Department of Civil and Environmental Engineering, University of Pittsburgh, Swanson School of Engineering, Pittsburgh, Pennsylvania, United States of America, **4** Department of Computational and Systems Biology, University of Pittsburgh Medical School, Pittsburgh, Pennsylvania, United States of America

## Abstract

Drinking water distribution systems, including premise plumbing, contain a diverse microbiological community that may include opportunistic pathogens. On-site supplemental disinfection systems have been proposed as a control method for opportunistic pathogens in premise plumbing. The majority of on-site disinfection systems to date have been installed in hospitals due to the high concentration of opportunistic pathogen susceptible occupants. The installation of on-site supplemental disinfection systems in hospitals allows for evaluation of the impact of on-site disinfection systems on drinking water system microbial ecology prior to widespread application. This study evaluated the impact of supplemental monochloramine on the microbial ecology of a hospital's hot water system. Samples were taken three months and immediately prior to monochloramine treatment and monthly for the first six months of treatment, and all samples were subjected to high throughput Illumina 16S rRNA region sequencing. The microbial community composition of monochloramine treated samples was dramatically different than the baseline months. There was an immediate shift towards decreased relative abundance of Betaproteobacteria, and increased relative abundance of Firmicutes, Alphaproteobacteria, Gammaproteobacteria, Cyanobacteria and Actinobacteria. Following treatment, microbial populations grouped by sampling location rather than sampling time. Over the course of treatment the relative abundance of certain genera containing opportunistic pathogens and genera containing denitrifying bacteria increased. The results demonstrate the driving influence of supplemental disinfection on premise plumbing microbial ecology and suggest the value of further investigation into the overall effects of premise plumbing disinfection strategies on microbial ecology and not solely specific target microorganisms.

**Editor:** Stefan Bereswill, Charité-University Medicine Berlin, Germany

**Funding:** This project was funded by a grant from the Alfred P. Sloan Foundation (grant B2013-12). The funders had no role in study design, data collection and analysis, decision to publish, or preparation of the manuscript.

**Competing Interests:** The authors have declared that no competing interests exist.

* Email: BibbyKJ@pitt.edu

## Introduction

Drinking water distribution systems, including premise plumbing, contain a diverse microbiological population [1]. Once new pipes have been added to an existing system, microbial colonization begins rapidly, with microbial communities being established in as little as one year [2]. For the purposes of this study, the 'microbial community' is defined as planktonic microbes within the hospital hot water system during the study period. The microbial ecology of drinking water distribution systems varies widely, depending upon system parameters such as disinfection scheme [3], hydraulic parameters [4], location in the system, age of the system [5], and pipe materials [6]. Microbes are capable of corroding pipes within distribution systems, possibly releasing harmful chemicals such as lead [7–9]. It is largely believed that within a drinking water distribution system, the disinfection scheme is one of the primary factors controlling the abundance and make-up of microbes [3,6,10]. Additionally, the effectiveness of disinfection in removing pathogens from drinking water is

mediated by the microbial ecology of the drinking water system [1]. However, the impact of on-site disinfection on premise plumbing microbial ecology is not well understood, motivating the current study.

The complex microbial ecology of premise plumbing systems can serve as a reservoir for opportunistic pathogens, such as *Legionella* spp., non-tuberculous Mycobacteria, *Pseudomonas* spp., *Acinetobacter* spp., *Stenotrophomonas* spp., *Brevundimonas* spp., *Sphingomonas* spp., and *Chryseobacterium* spp. [11–13]. Biofilms and amoeba within the water system can protect opportunistic pathogens from disinfection [1,14–16], and may even allow their regrowth and increase in pathogenicity [17–19]. As an example of the utility of microbial ecology-based approaches, a recent landmark microbial ecology-based study showed that biofilms in showerheads are actually enriched in opportunistic pathogens, creating the potential for an aerosol route of infection [20]. Additionally, antibiotic resistance genes have been detected in the biofilms of drinking water distribution systems [21,22]. Each of these points highlight the necessity for a greater understanding of premise plumbing microbial ecology.

Premise plumbing systems have an approximately ten-times greater microbial load than full-scale drinking water distribution systems, due to many factors including greater water stagnation and surface area to volume ratio [23,24]. Premise plumbing systems of hospitals are of particular concern, as hospitals may contain immunocompromised patients [25], who may not be protected by current drinking water monitoring standards [26], and who would be more susceptible to infections caused by opportunistic pathogens. To date, the majority of on-site disinfection systems have been installed in hospitals, creating a valuable testing ground to observe the impact of on-site disinfection systems on premise plumbing microbial ecology prior to more widespread application.

In addition to use in on-site systems, monochloramine as a secondary disinfectant has been advocated in the US as an effective method to reduce the production of disinfection-by-products [27,28] and control biofilm growth within water distribution systems [29]. While monochloramine is able to penetrate biofilms better than alternative disinfectants, this may not result in a reduction in biofilm growth [8]. Additionally, chloramine treatment requires the addition of an excess of ammonia, which may cause increased growth by ammonia-oxidizing bacteria [28], such as members of the genera *Nitrospira* spp. and *Nitrosomonas* spp. [30]. Bacterial nitrification is known to increase the degradation rate of monochloramine [31], thereby reducing the expected longevity and effectiveness of chloramine. Denitrifying bacteria have previously been identified in chlorami-nated drinking water systems [32]; however, this topic has not been fully explored in the literature.

The effectiveness of chloramination in removing opportunistic pathogens in premise plumbing remains unclear [27]. On-site monochloramine addition has been proposed as a disinfection strategy for the control of *Legionella* [33–36], but long-term studies have not yet been conducted [33,34]. Recently, a culture-based study of monochloramine on-site disinfection in a hospital's hot water system for the purpose of *Legionella* control demonstrated a significant reduction in *L. pneumophila* and no change in nitrate or nitrite levels [37]. Observed discrepancies in system performance are potentially due to differing microbial ecologies or water chemistries of the systems tested. A more holistic view of system microbial ecology, such as presented in this study, may allow more efficient application of supplemental disinfection.

Despite the obvious importance of the microbial ecology of drinking water systems in modulating disinfectant effectiveness and as a reservoir for opportunistic pathogens, there is a notable lack of studies detailing the shift in microbial diversity and composition in response to on-site disinfection. The objective of this study was to determine the effects of on-site monochloramine disinfection on the microbial ecology of a hospital hot water system. Both the microbial ecology of hot water systems and the response of premise plumbing microbial ecology to on-site disinfection are not currently well described in the literature. This study utilizes 216 samples taken from 27 sites and pooled into five composites for two time points prior to and six time points following the addition of on-site monochloramine addition. Samples were analyzed utilizing Illumina DNA sequencing of the microbial community 16S rRNA region and results demonstrate a dynamic shift of the microbial ecology of a hospital's hot water system in response to monochloramine addition.

## Materials and Methods

### Hospital setting

For these activities no specific permissions were required for these locations. This study took place in a 495-bed tertiary care hospital complex in Pittsburgh, PA. The building has 12 floors and receives chlorinated, municipal cold water. The hospital's hot water system was treated with the Sanikill monochloramine injection system (Sanipur, Lombardo, Flero, Italy). Monochlor-amine was dosed to a target concentration between 1.5 and 3.0 ppm as $Cl_2$. Details regarding monochloramine dosing and water chemistry are included in Text S1.

### Sample collection and processing

Hot water was collected from 27 sites throughout the hospital at two time points before monochloramine injection (three months and immediately prior) and monthly for the first six months of monochloramine application. Water samples were collected from a variety of locations throughout the hospital (Table 1). Samples were taken from hot water tanks, the hot water return line, faucets in the intensive care units, rehabilitation suites including both automatic and standard faucets, and other patient rooms on the upper floors. The faucets in the intensive care units are located on the third, fourth, and fifth floors. The faucets in the rehabilitation suites are located on floors six and seven and represent both electronic sensor (automatic) faucets and standard faucets. The final grouping of sites was from short-term use patient rooms located on floors eight, nine, ten, eleven, and twelve. At each site, hot water was flushed for one minute prior to sample collection into sterile HDPE bottles with enough sodium thiosulfate to neutralize 20 ppm chlorine (Microtech Scientific, Orange, CA). For hot water tank sampling, the drain valve was opened, allowed to flush for one minute, then sampled into sterile HDPE bottles as described above. Following sampling, 100 mL of sample water was filtered through a 0.2 µm, 47 mm, polycarbonate filter membrane (Whatman, Florham Park, NJ), placed into 10 mL of the original water sample, and vortexed vigorously for 10 seconds as described in methods ISO Standards 11731:1998 and 11731:2004 for *Legionella* isolation. Five mL of each concentrated sample was frozen at −80°C until DNA extraction.

### DNA extraction, PCR, and Sequencing

Frozen water samples were thawed and pooled as described in Table 1. The 27 samples were divided into five pools including the hot water tanks and hot water return line (HWT), floors 3–5 (the intensive care units, F3), floors 6 and 7 automatic faucets (the rehabilitation suites' automatic faucets, F6A), floors 6 and 7 standard faucets (the rehabilitation suites' standard faucets, F6S), and floors 8–12 (the short-term use patient rooms, F8). These samples were then filtered through 0.2 µm, 47 mm, Supor 200 Polyethersulfone membranes (Pall Corporation), housed in sterile Nalgene filter funnels (Thermo Scientific; Fisher). Filter membranes were subjected to DNA extraction using the RapidWater DNA Isolation Kit (MO-BIO Laboratories) as described by the manufacturer. PCR was performed in quadruplicate using 16S rRNA region primers 515F and 806R including sequencing and barcoding adapters as previously described [38]. These primers amplify an approximately 300 base pair region of the rRNA region spanning variable regions 3 and 4. The specificity of this primer set is considered to be well optimized and 'nearly universal' [39]; analysis of these primers against the 97% Greengenes 13.5 OTU database demonstrated a specificity of 99.9% and 98.3% for the 515f and 806r primers, respectively. Dreamtaq Mastermix (Thermo Scientific) was used and PCR product was checked on

**Table 1.** Sample pool description, abbreviation, and number of pooled sites.

| Sample Description | Sample Abbreviation | Number of Pooled Sites |
|---|---|---|
| Outlets of Hot Water Tanks and Hot Water Return Line | HWT | 3 |
| Floors 3–5 Patient Room Faucets | F3 | 4 |
| Floors 6 & 7 Patient Room Automatic Faucets | F6A | 7 |
| Floors 6 & 7 Patient Room Standard Faucets and Showers | F6S | 7 |
| Floors 8–12 Patient Room Faucets | F8 | 6 |
| Technical Replicates of Floors 8–12 Patient Room Faucets | F8rep | 6 |

Hot water was collected after a one-minute flush from the following locations throughout the hospital.

a 1% agarose gel. An independent negative control was run for each sample and primer set and all negative controls were negative for PCR amplification. PCR products were pooled and purified using the UltraClean PCR Clean-Up Kit (MO-BIO Laboratories). Each sample then underwent additional cleaning with the Agencourt AMPure XP PCR purification kit (Beckman Coulter) and quantified using the QuBit 2.0 Fluorometer (Invitrogen). Following quantification, 0.1 picomoles of each sample PCR product were pooled. The sample pool underwent two additional clean up steps with a 1.5:1 ratio of Agencourt AMPure XP beads followed by a 1.2:1 bead ratio (Beckman Coulter) to eliminate primer dimers. Samples were sequenced on an in-house Illumina MiSeq sequencing platform as previously described [38].

### Data analysis

Data was analyzed within the MacQIIME (http://www.wernerlab.org/software/macqiime) implementation of QIIME 1.7.0 [40]. Sequences were parsed based upon sample-specific barcodes and trimmed to a minimum quality score of 20. Operational taxonomic units (OTUs) at 97% were then picked against the Greengenes 13.5 database using UCLUST [41] for taxonomic assignment. Following assignment, 7,000 successfully assigned sequences from each sample were chosen at random to allow for even downstream analyses and even cross-sample comparison. Observed OTUs were defined as observed species whereas unassigned sequences were removed from subsequent analyses (closed reference OTU picking). Alpha-diversity evenness was calculated using the 'equitability' metric within QIIME. Beta diversity analyses were conducted by UNIFRAC analysis [42]. OTUs were also open-reference picked, where unassigned sequences are placed in the taxa "other" and therefore not removed. Discussion and results from this open-reference OTU picking analysis is included in Text S1. Open-reference OTU picking did not result in a shift in any fundamental conclusions with the exception of the increase in the genus *Stenotrophomonas* spp. following monochloramine addition; closed-reference OTU picking is presented for higher-quality taxonomic assignment. Morisita-Horn indices were calculated as previously described [43,44]. Sequences are available under MG RAST accession numbers 4552832.3 to 4552878.3.

### Results

### Sequence Data

Sequencing reads were split by sample-specific barcodes, trimmed to a minimum quality score of 20, and placed into OTUs at 97% through comparison with the Greengenes 13.5 coreset. For each sample, 7,000 sequences with assigned taxonomy were selected to allow for even comparison across samples. Two

types of OTU picking were done for this study: closed reference (sequences were compared to a reference set of sequences for OTU clustering, sequences not matching one of these pre-defined sequences were discarded) and open reference (sequences were compared to each other for OTU picking, sequences not mapping to the reference database were grouped as 'other') in Text S1.

### Alpha Diversity

Alpha diversity (number of observed OTUs) of samples treated with monochloramine was significantly higher than samples from the baseline months (Figure 1). Prior to treatment, the average number of observed OTUs at 97% similarity was $151.2\pm39.7$, whereas during treatment the average number of observed OTUs was $225.2\pm61.2$ ($p<0.001$) (Figure 1). This shift was not associated with a statistically significant loss of sample evenness (Figure S1). The same statistical trends in alpha diversity were observed for open-reference picked OTUs (Figure S2).

### Beta Diversity

Beta diversity (sample interrelatedness) was analyzed using weighted UNIFRAC [42]. The principal coordinate analysis (PCoA) plot from this analysis is shown in Figure 2. Samples from the first two months prior to treatment cluster together whereas those following disinfection tend to cluster by sample site more strongly than sample time (Figure 2). The same trend was observed for open-reference picked OTUs (Figure S3).

### Taxonomic Comparison

Figure 3 shows the phyla-level taxonomy for each of the sample pools. Phyla <1.3% relative abundance are listed as 'minor phyla'. Prior to treatment, samples from all locations were similarly

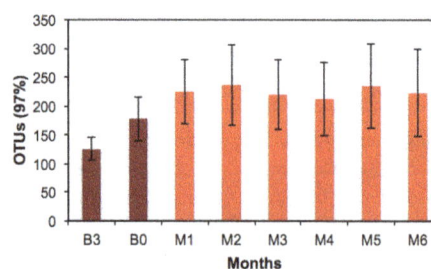

**Figure 1. Comparison of the number of OTUs (97% similarity) for each month.** Bars represent standard deviation. Each sample pool was normalized to 7,000 sequences. Samples from B3 and B0 represent those taken three months and immediately prior to monochloramine treatment, respectively. Samples from M1, M2, M3, M4, M5, and M6 were taken monthly during the first six months of treatment.

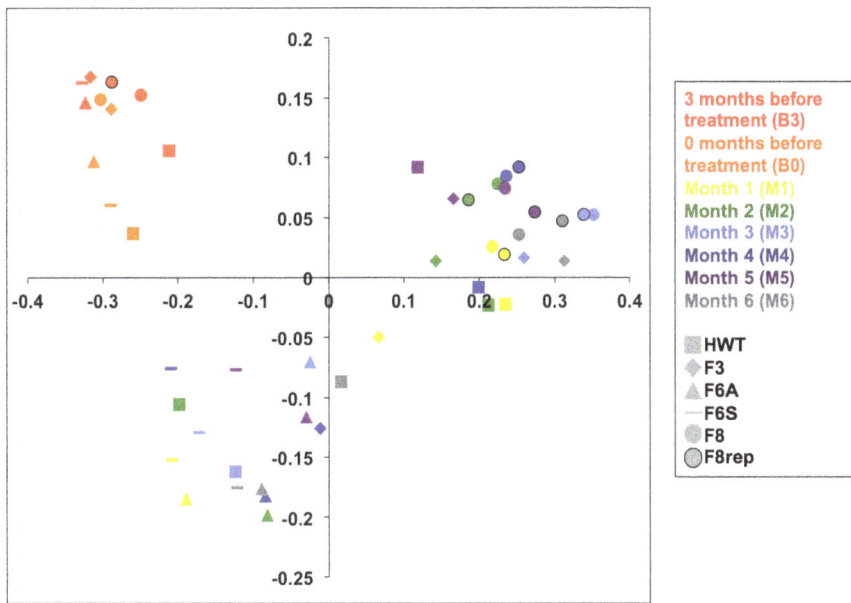

**Figure 2. PCoA analysis of samples pools.** Samples that cluster more closely together share a greater similarity in microbial community structure. Colors represent months sampled whereas shapes represent sample pool. Samples from B3 and B0 represent those taken three months and immediately prior to monochloramine treatment, respectively. Samples from M1, M2, M3, M4, M5, and M6 were taken monthly during the first six months of treatment.

structured, predominantly comprised of Betaproteobacteria, with lesser quantities of Firmicutes, Bacteroidetes, Alphaproteobacteria, and Gammaproteobacteria (Figure 3 Panels A–E). Following initiation of treatment (M1) there was a shift away from the predominance of Betaproteobacteria and towards a greater relative abundance of Firmicutes, Alphaproteobacteria, Gammaproteobacteria, and minor fractions of Cyanobacteria and Actinobacteria (Figure 3 Panels A–E). The same taxonomy trends were observed for open-reference picked data (Figure S4 Panels A–E).

The samples from the hot water tank (HWT) from pre-treatment months (B3 and B0) were approximately 60% Betaproteobacteria with approximately 35% Firmicutes, Bacteroidetes, Alphaproteobacteria, and Gammaproteobacteria in aggregate (Figure 3 Panel A). Following treatment the relative abundance of Betaproteobacteria was reduced to approximately 20% and Firmicutes, Alphaproteobacteria, and Gammaproteobacteria subsequently increased to comprise an average of 78% of the total relative abundance (Figure 3 Panel A).

The microbial community profile of samples from the lower floors of the hospital (intensive care units, F3) was slightly different than those of the hot water tank samples but a similar trend was observed (Figure 3 Panel B). Over 65% of pre-treatment samples were Betaproteobacteria with Firmicutes, Bacteroidetes, Alphaproteobacteria, and Gammaproteobacteria accounting for a combined 20% of community relative abundance (Figure 3 Panel B). Following treatment the amount of Betaproteobacteria and Bacteroidetes decreased to an average of 23% relative abundance, while the relative abundance of Firmicutes and Alphaproteobacteria increased sharply to approximately 68% (Figure 3 Panel B).

In spite of being from the same rooms, the taxonomic composition of samples from F6A and F6S differed after treatment (Figure 3 Panels C and D). Prior to treatment both the automatic (F6A) and standard (F6S) faucets in the rehabilitation suites contained 65–80% Betaproteobacteria, with Bacteroidetes, Alphaproteobacteria, Gammaproteobacteria, and Cyanobacteria

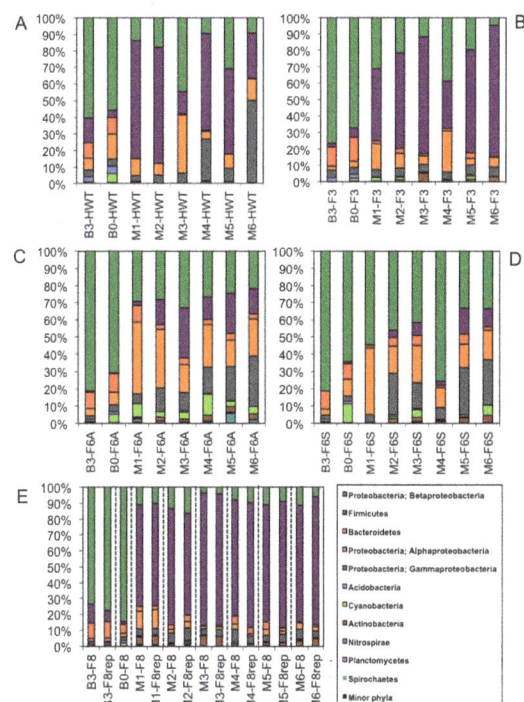

**Figure 3. Taxonomic assignments of sequences from HWT (hot water tank samples) (Panel A), F3 (floors 3–5) (Panel B), F6A (floors 6 and 7 automatic faucets) (Panel C), F6S (floors 6 and 7 standard faucets) (Panel D), F8 (floors 8–12) and F8rep (replicate barcoded PCRs of samples from floors 8–12) (Panel E).** Samples from B3 and B0 represent those taken three months and immediately prior to monochloramine treatment, respectively. Samples from M1, M2, M3, M4, M5, and M6 were taken monthly during the first six months of treatment. Black lines in Panel E separate pairs of replicates.

accounting for the other 20–35% of relative abundance (Figure 3 Panels C and D). However, after monochloramine application, the automatic faucets (F6A) underwent a 50% reduction in the total relative abundance of Betaproteobacteria and became enriched in Firmicutes, Alphaproteobacteria, Gammaproteobacteria, Actinobacteria, and Spirochaetes (Figure 3 Panel C). The standard faucets (F6S) lost only 26% of Betaproteobacteria, but also saw an increase in members of the Firmicutes, Alphaproteobacteria, Gammaprotobacteria, and Actinobacteria phyla from an average relative abundance of 10% before treatment to 46% after monochloramine addition (Figure 3 Panel D).

Prior to treatment, the microbial community in samples from the upper floors of the hospital (short-term use patient rooms, F8) resembled most of the other baseline samples with over 70% Betaproteobacteria and approximately 20% of Firmicutes, Bacteroidetes, Alphaproteobacteria, Gammaproteobacteria, Acidobacteria, and Cyanobacteria (Figure 3 Panel E). Following monochloramine treatment, the relative abundance of Betaproteobacteria was reduced from approximately 70% to 10% and replaced by Firmicutes, which increased from 7% of the relative abundance in the baseline months to 74% after treatment (Figure 3 Panel E). There was only a slight increase, from 2% to 9% relative abundance, in the amount of Gammaproteobacteria and Actinobacteria present (Figure 3 Panel E).

## Sample Replicates

Separately amplified and barcoded technical replicates of sample pool F8 for 7 of the 8 sample pools were also sequenced to verify technical reproducibility. There is no replicate for month B0. UNIFRAC analysis demonstrated that the replicates from each month cluster very closely (Figure 2). All of the samples from F8 in samples M1–M6 and their replicates (circles and outlined circles) clustered together in the upper-right hand quadrant (Figure 2). Morisita-Horn analyses of replicates demonstrate high levels of community similarity, ranging from 0.990 (M2) to 0.9998 (M3). These results further validate the technical reproducibility of the methodology (Figure 3 Panel E) [43,44]. The open-reference picked UNIFRAC analysis and taxonomy also show replicates to have similar profiles to their original samples (Figure S3 and S4 Panel E). Morisita-Horn analyses of these samples showed similarly high levels of community similarity ranging from 0.991 (M2) to 0.9992 (M1).

## Genera Containing Opportunistic Pathogens

Sequence data was further analyzed to observe the change in genera containing opportunistic pathogens of interest during treatment. Genera analyzed were: *Legionella* spp., *Pseudomonas* spp., *Acinetobacter* spp., and *Stenotrophomonas* spp. (Gammaproteobacteria group); *Brevundimonas* spp. and *Sphingomonas* spp. (Alphaproteobacteria group); *Chryseobacterium* spp. (Bacteroidetes group); and *Mycobacterium* spp. (Actinobacteria group). These genera are of special interest as some to all of the species contained within them are pathogens; however, the nature of short-read 16S rRNA region sequence analysis is such that species-level pathogens cannot be definitively identified. Trends demonstrated by this analysis could be used to direct future analyses targeting opportunistically pathogenic organisms more specifically. Analysis of the relative abundance of each of these organism groups over time shows a statistically significant increase in relative abundance for *Acinetobacter* (p = 0.0054), *Mycobacterium* (p = 0.0017), *Pseudomonas* (p = 0.031) and *Sphingomonas* (p = 0.034) as treatment progressed (Figure 4). *Brevundimonas*, *Chryseobacterium*, Legionellaceae, and *Stenotrophomonas* did not demonstrate a statistically significant increase in relative abundance following treatment (Figure 4).

The open-reference picked data demonstrated an increase in the same opportunistic pathogen containing genera as the closed-reference picked data, *Acinetobacter* (p = 0.004), *Mycobacterium* (p = 0.002), *Pseudomonas* (p = 0.015), and *Sphingomonas* (p = 0.025), but also showed a significant increase in the genera *Stenotrophomonas* (p = 0.03) (Figure S5).

## Nitrification and Denitrification

Additionally, we investigated the shift in relative abundance of representative genera associated with nitrification and denitrification (Figure 5). There was no statistically significant difference in the relative abundance of the potential nitrifiers *Nitrospira* and Nitrosomonadaceae, before (mean = 0.0015 ± 0.0018) and after treatment (mean = 0.0005 ± 0.0011) (p = 0.175). Other nitrifier-containing genera such as *Nitrosococcus*, *Nitrobacter*, *Nitrospina*, or *Nitrococcus*, were not identified in any samples. The total relative abundance of genera containing denitrifiers (*Thiobacillus*, *Micrococcus*, and *Paracoccus*) underwent a statistically significant increase before (mean = 0.00005 ± 0.000074) and after treatment with monochloramine (mean = 0.0029 ± 0.0029) (p = 0.026). The denitrifier-containing genera *Rhizobiales* and *Rhodanobacter* were not identified in any samples. The same trends were observed in open-reference picked data (Figure S6).

## Discussion

Our study objective was to examine the shift in the microbial ecology of a hospital hot water system associated with the introduction of on-site monochloramine addition. To evaluate the shift in microbial community structure we sampled 27 sites in a hospital and pooled samples into 5 groups for 8 sample time points. Sites were pooled based on their location and use in the hospital and faucet type (automatic versus standard). This study took place during the first U.S. trial of the Sanikill on-site monochloramine generation system (Sanipur, Brescia, Italy) [45–47]. These samples were subjected to DNA extraction, 16S rRNA region barcoded PCR, and Illumina sequencing to analyze the response of the microbial ecology to the addition of monochloramine.

The microbial population shift in response to monochloramine addition was immediate. The number of OTUs observed (alpha diversity) significantly increased following monochloramine treatment (Figure 1). It is possible that the overall loss of dominance of initially abundant microbial groups (e.g. Betaproteobacteria) allowed for a greater number of other bacterial species to grow, or for selected individuals to die off, thereby increasing the alpha diversity. Samples from different sites taken before monochloramine treatment were comprised of similar microbial populations and samples taken after treatment were distinct from samples taken in the baseline months (Figures 2 and 3, Figures S3 and S4). Interestingly, it appears that following monochloramine treatment the location of sampling matters more in sample similarity (beta diversity) than does the month they were taken (Figure 2, Figure S3). Microbial communities from the lower floors' intensive care units (F3) and the upper floors' short term patient rooms (F8) were more similar than to the floors 6 and 7's rehabilitation suites (F6A and F6S) automatic and standard faucet samples. These sites were located in single patient rooms in rehabilitation units and may experience as much use as some locations on the lower and upper floors, which include the trauma burn unit, the intensive care unit (ICU), the neonatal ICU, and the cardiovascular ICU. The HWT samples from earlier months of treatment closely resembled floors 6 and 7 (F6A and F6S) whereas the HWT microbial ecology from

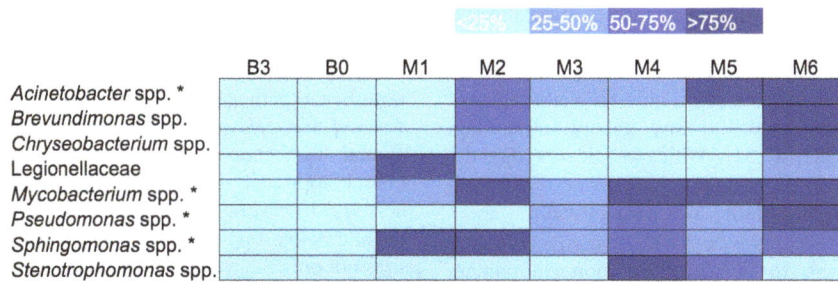

**Figure 4. Relative abundance of different genera of opportunistic waterborne pathogens.** Samples color coded into four groupings calculated by 25% of the maximum relative abundance for each organism. Months with the least relative abundance are lightest in color, whereas months with the highest relative abundance are darkest. *denotes a statistically significant increase in the relative abundance of this organism following treatment.

the later months was more related to the lower (F3) and upper floors (F8).

We investigated the possible differences in microbial ecology between automatic and standard faucets as it has been previously demonstrated that opportunistic pathogens, including *Legionella* [48] and *Pseudomonas aeruginosa* [49], are detected more frequently and in greater concentrations in automatic faucets. It has been suggested that the reason for the differences between automatic and standard faucets could be due to water flow, temperature, and structural issues. Automatic faucets may have diluted monochloramine concentrations due to low flow and poor flushing [48,49] and automatic faucets also contain mixing valves, which are made of materials such as rubber, polyvinylchloride, and plastic, which more easily support the growth of biofilms [48,49]. Potentially due to these biofilms, the increased colonization can persist even following disinfection with chlorine dioxide [48]. We observed a differential reduction in the relative abundance of Betaproteobacteria in standard and automatic faucets following treatment. The automatic faucets lost 50% of their relative abundance of Betaproteobacteria whereas the standard faucets only saw an average 26% reduction.

There was an overall shift towards less relative abundance of Betaproteobacteria, and more relative abundance of Firmicutes, Alphaproteobacteria, Gammaproteobacteria, Cyanobacteria and

Actinobacteria after monochloramine treatment. A previous microbial ecology study of a simulated drinking water distribution system treated with monochloramine demonstrated a different trend, with an increase in specific genera within the Actinobacteria, Betaproteobacteria, and Gammaproteobacteria phyla [3]. The dissimilarity of these studies may be due to the fact that the latter occurred in a cold water system whereas our study was in a hot water supply.

Several waterborne pathogen-containing genera were examined for changes in relative abundance due to monochloramine treatment. The relative abundance of a few of the waterborne pathogen-containing genera examined, including *Acinetobacter*, *Mycobacterium*, *Pseudomonas*, and *Sphingomonas*, showed an increase after monochloramine treatment. Other studies have described an increase in some of these organisms including *Legionella*, *Mycobacterium*, and *Pseudomonas* in chloraminated water [3,6] as well as biofilms treated with monochloramine [50]. Feazel et al. previously demonstrated that *Mycobacterium* spp. can be enriched in showerhead biofilms compared to the source water [20]. An increased relative abundance of *Mycobacterium* spp. due to monochloramine treatment is of concern, specifically if this increase in relative abundance is due to the presence of more viable mycobacterial cells. These microorganisms may pose a specific threat of aerosol exposure to immunocompromised

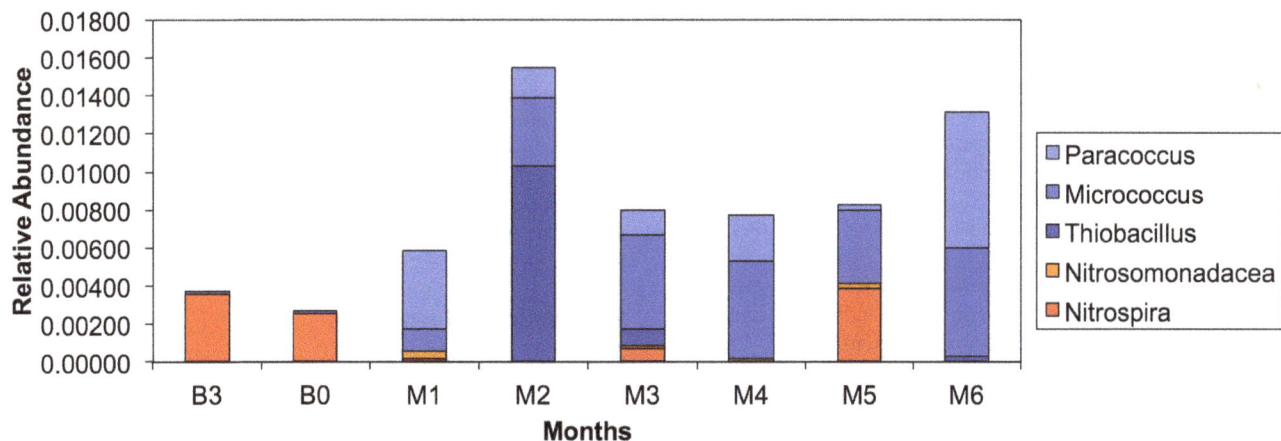

**Figure 5. Relative abundance of genera containing nitrifying (*Nitrospira* and Nitrosomonadacea) and denitrifying bacteria (*Thiobacillus, Micrococcus,* and *Paracoccus*).** No other genera associated with nitrification (*Nitrosococcus, Nitrobacter, Nitrospina,* or *Nitrococcus,*) or denitrification (*Rhizobiales* and *Rhodanobacter*) were found in any of our samples. The x-axis represents sampling months with months B3 and B0 being before monochloramine treatment and months M1–M6 representing the first six months of treatment. The y-axis represents the relative abundance.

patients who reside in buildings with an increased abundance of these organisms in hot water [20]. Interestingly, a recent study demonstrated that while the concentration of live bacteria is reduced after monochloramine treatment, only the viable microbial community structure is altered and genera containing opportunistic pathogens persist [51]. While we did not directly quantify microorganisms in the samples collected or verify that microorganisms detected were viable, our parallel culture-based study observed a statistically significant reduction in culturable total bacteria and *Legionella* species following monochloramine treatment (Table S1) [45–47,52].

Previous studies have found an increase in nitrification in chloraminated systems, which effectively decreased monochloramine concentration [6,31]. This chemical decay led to higher levels of *Legionella*, *Mycobacterium* spp., and *P. aeruginosa* at earlier water ages than in chlorinated simulated distribution systems [6]. A change in potentially nitrifying bacteria following monochloramine addition was not observed in the culture-based portion of this study [45–47], consistent with our molecular observations. Concentrations of nitrate and nitrite remained fairly stable throughout the study months, with the exception of a spike in nitrate levels in M6 (Table S1) [52]. We observed a statistically significant increase in the relative abundance of genera associated with denitrification in monochloramine treated samples. A previous study found a high absolute abundance, up to 200,000 cfu/mL, of potentially denitrifying bacteria in a chloraminated system even after regular flushing [32]. The highest relative abundance of bacterial genera associated with denitrification occurred during M6 when there was a spike in nitrate concentrations (Table S1) [52]. However, in months 1 and 2 there was also a large relative abundance of these bacteria present with fairly low nitrate concentrations, suggesting that some other factor might be important in their relative abundance. We do not believe that these trends were due to seasonality in our study as microbiological data were largely consistent across the study period. However, the possibility for seasonal effects cannot be excluded.

A notable increase in the relative abundance of the genus *Alicyclobacillus* spp. (Firmicutes phylum) was observed following monochloramine treatment, from an average of 4.1±4.5% of the microbial population prior to treatment to an average of 40.9±27.1% following treatment (p<0.001). This genera is comprised primarily of spore-formers that are of concern in food spoilage [53], and has previously been detected in drinking water [54]. The high relative abundance of *Alicyclobacillus* spp. suggests a potentially dominant role in chloraminated hot water system microbial ecology worthy of future investigation.

The incidence of reported Legionnaires' disease cases increased threefold from 2000 to 2009 [55]. This fact, coupled with an increasingly elderly and immunocompromised population [55], has lead to an increased concern about *Legionella* and other opportunistic waterborne pathogens. Additionally, the American Society of Heating, Refrigerating, and Air-Conditioning Engineers (ASHRAE) has recently proposed Standard 188P for the prevention of legionellosis associated with premise plumbing systems [56]. This standard serves to reduce the risk of *Legionella* infections through a risk management approach [56]. For these reasons, on-site disinfection has become progressively important to protect patients in hospitals and long-term care facilities from waterborne opportunistic pathogens. An increased understanding of the influence of on-site disinfection on premise plumbing microbial ecology is necessary to maximize effectiveness and to limit undesired side effects.

This study demonstrates that there exists the potential for unwanted consequences of supplemental disinfectant addition for the removal of *Legionella* such as the potential enrichment of other waterborne pathogens, including *Acinetobacter*, *Mycobacterium*, *Pseudomonas*, and *Sphingomonas*. Understanding the impact of supplemental disinfection on water system microbial ecology, through a holistic approach, is necessary to maximize disinfectant effectiveness and to ensure that supplemental disinfectant does not select for alternative opportunistic pathogens. A recent review emphasizes not only the role of disinfectants but also other system factors that may impact microbial ecology such as temperature, pipe material, organic carbon, presence of automatic faucets, and point-of-use filtration [24]. The authors suggest a probiotic approach to opportunistic pathogen control which would either add microbes that can outcompete these pathogens, remove key species, or using engineering controls to favor benign organisms that are antagonistic to opportunistic pathogens [24]. This systematic, probiotic, approach to premise plumbing opportunistic pathogen management is an inventive concept for dealing with the diverse microbial ecology of these systems, but requires a greater understanding of the drivers of premise plumbing microbial ecology, such as provided by this study.

In conclusion, we observed a shift in the microbial ecology of a hospital's hot water system treated with on-site chloramination. This shift occurred immediately following monochloramine treatment. Prior to treatment, the bacterial ecology of all samples was dominated by Betaproteobacteria; following treatment, members of Firmicutes and Alphaproteobacteria dominated. Differences in community composition were seen in different locations within the hospital as well as between automatic and standard faucets. This suggests that water from different locations and outlet types should be sampled to get a more thorough picture of the microbiota of a system. There was an increase in the relative abundance of several genera containing opportunistic waterborne pathogens following the onset of monochloramine treatment, including *Acinetobacter*, *Mycobacterium*, *Pseudomonas*, and *Sphingomonas* and genera associated with denitrification. The benefits and risks of each supplemental disinfection strategy should be evaluated before implementation in any building, especially in hospitals, long term care facilities, and other buildings housing immunocompromised patients. This work demonstrates the effects of a supplemental monochloramine disinfection system on the microbial ecology of premise plumbing biofilms. Given the importance of premise plumbing microbial ecology on opportunistic pathogen presence and persistence, understanding the driving influence of supplemental disinfectants on microbial ecology is a crucial component of any effort to rid premise plumbing systems of opportunistic pathogens. As additional facilities turn to on-site water disinfection strategies, more long-term studies on the effects of disinfectants on microbial ecology in premise plumbing are needed as well as those evaluating a probiotic approach to opportunistic pathogen eradication.

## Supporting Information

**Figure S1 Sample evenness for closed-reference OTU picking.** No statistically significant different was observed for samples taken prior to or following monochloramine addition.

**Figure S2 Alpha diversity for open-reference OTU picking.** A statistically significant difference was observed for samples taken prior to or following monochloramine addition (p = 0.046).

**Figure S3  Beta diversity for open-reference OTU picking.** Samples from before monochloramine treatment clustered together whereas following treatment samples clustered by location more so than month of treatment.

**Figure S4  Taxonomic assignment of sequences from HWT (hot water tank samples) (Panel A), F3 (floors 3–5) (Panel B), F6A (floors 6 and 7 automatic faucets) (Panel C), F6S (floors 6 and 7 standard faucets) (Panel D), F8 (floors 8–12) and F8rep (replicate barcoded PCRs of samples from floors 8–12) (Panel E) for open-reference OTU picking.**

**Figure S5  Relative abundance of waterborne pathogen containing genera for open-reference OTU picking.** A statistically significant increase in *Acinetobacter* spp., *Mycobacterium* spp., *Pseudomonas* spp., *Sphingomonas* spp., and *Stenotrophomonas* spp. was observed following treatment.

**Figure S6  Relative abundance genera containing nitrifying (*Nitrospira* and Nitrosomonadacea) and denitrifying bacteria (*Thiobacillus*, *Micrococcus*, and *Paracoc-* cus) for open-reference OTU picking.** No other genera containing nitrifying bacteria (*Nitrosococcus*, *Nitrobacter*, *Nitrospina*, or *Nitrococcus*,) or denitrifying bacteria (*Rhizobiales* and *Rhodanobacter*) were found in our samples.

**Table S1  Physicochemical data obtained during the study.**

**Text S1  Supplementary Information.** Water chemistry and monochloramine dosing methods, description of minor phyla observed, and open-reference OTU picking results.

## Acknowledgments

We would like to thank the staff of Special Pathogens Laboratory for their assistance in sample collection.

## Author Contributions

Conceived and designed the experiments: JLB JES KB. Performed the experiments: JLB AV SD. Analyzed the data: JLB JES KB. Wrote the paper: JLB KB.

## References

1. Berry D, Xi C, Raskin L (2006) Microbial ecology of drinking water distribution systems. Curr Opin Biotechnol 17: 297–302.
2. Martiny AC, Jorgensen TM, Albrechtsen HJ, Arvin E, Molin S (2003) Long-term succession of structure and diversity of a biofilm formed in a model drinking water distribution system. Appl Environ Microbiol 69: 6899–6907.
3. Gomez-Alvarez V, Revetta RP, Santo Domingo JW (2012) Metagenomic analyses of drinking water receiving different disinfection treatments. Appl Environ Microbiol 78: 6095–6102.
4. Douterelo I, Sharpe RL, Boxall JB (2013) Influence of hydraulic regimes on bacterial community structure and composition in an experimental drinking water distribution system. Water Res 47: 503–516.
5. Henne K, Kahlisch L, Brettar I, Hofle MG (2012) Analysis of structure and composition of bacterial core communities in mature drinking water biofilms and bulk water of a citywide network in Germany. Appl Environ Microbiol 78: 3530–3538.
6. Wang H, Masters S, Edwards MA, Falkinham JO, Pruden A (2014) Effect of Disinfectant, Water Age, and Pipe Materials on Bacterial and Eukaryotic Community Structure in Drinking Water Biofilm. Environ Sci Technol 48: 1426–1435.
7. White C, Tancos M, Lytle DA (2011) Microbial community profile of a lead service line removed from a drinking water distribution system. Appl Environ Microbiol 77: 5557–5561.
8. Zhang Y, Griffin A, Rahman M, Camper A, Baribeau H, et al. (2009) Lead contamination of potable water due to nitrification. Environ Sci Technol 43: 1890–1895.
9. Zhang Y, Triantafyllidou S, Edwards M (2008) Effect of nitrification and GAC filtration on copper and lead leaching in home plumbing systems. Journal of Environmental Engineering-Asce 134: 521–530.
10. Mathieu L, Bouteleux C, Fass S, Angel E, Block JC (2009) Reversible shift in the alpha-, beta- and gamma-proteobacteria populations of drinking water biofilms during discontinuous chlorination. Water Res 43: 3375–3386.
11. Squier C, Yu VL, Stout JE (2000) Waterborne Nosocomial Infections. Curr Infect Dis Rep 2: 490–496.
12. Perola O, Nousiainen T, Suomalainen S, Aukee S, Karkkainen UM, et al. (2002) Recurrent Sphingomonas paucimobilis-bacteraemia associated with a multi-bacterial water-borne epidemic among neutropenic patients. J Hosp Infect 50: 196–201.
13. Mondello P, Ferrari L, Carnevale G (2006) Nosocomial Brevundimonas vesicularis meningitis. Infez Med 14: 235–237.
14. Buse HY, Ashbolt NJ (2011) Differential growth of Legionella pneumophila strains within a range of amoebae at various temperatures associated with in-premise plumbing. Lett Appl Microbiol 53: 217–224.
15. Emtiazi F, Schwartz T, Marten SM, Krolla-Sidenstein P, Obst U (2004) Investigation of natural biofilms formed during the production of drinking water from surface water embankment filtration. Water Res 38: 1197–1206.
16. Berry D, Horn M, Xi C, Raskin L (2010) Mycobacterium avium Infections of Acanthamoeba Strains: Host Strain Variability, Grazing-Acquired Infections, and Altered Dynamics of Inactivation with Monochloramine. Appl Environ Microbiol 76: 6685–6688.
17. Swanson MS, Hammer BK (2000) Legionella pneumophila pathogesesis: a fateful journey from amoebae to macrophages. Annu Rev Microbiol 54: 567–613.
18. Lau HY, Ashbolt NJ (2009) The role of biofilms and protozoa in Legionella pathogenesis: implications for drinking water. J Appl Microbiol 107: 368–378.
19. van der Wielen PW, van der Kooij D (2013) Nontuberculous mycobacteria, fungi, and opportunistic pathogens in unchlorinated drinking water in The Netherlands. Appl Environ Microbiol 79: 825–834.
20. Feazel LM, Baumgartner LK, Peterson KL, Frank DN, Harris JK, et al. (2009) Opportunistic pathogens enriched in showerhead biofilms. Proc Natl Acad Sci U S A 106: 16393–16399.
21. Schwartz T, Kohnen W, Jansen B, Obst U (2003) Detection of antibiotic-resistant bacteria and their resistance genes in wastewater, surface water, and drinking water biofilms. FEMS Microbiol Ecol 43: 325–335.
22. Shi P, Jia S, Zhang XX, Zhang T, Cheng S, et al. (2013) Metagenomic insights into chlorination effects on microbial antibiotic resistance in drinking water. Water Res 47: 111–120.
23. NRC (2006) Drinking Water Distribution Systems: Assessing and Reducing Risks.
24. Wang H, Edwards MA, Falkinham JO, Pruden A (2013) Probiotic Approach to Pathogen Control in Premise Plumbing Systems? A Review. Environmental Science & Technology 47: 10117–10128.
25. Williams MM, Armbruster CR, Arduino MJ (2013) Plumbing of hospital premises is a reservoir for opportunistically pathogenic microorganisms: a review. Biofouling 29: 147–162.
26. Williams MM, Braun-Howland EB (2003) Growth of Escherichia coli in model distribution system biofilms exposed to hypochlorous acid or monochloramine. Appl Environ Microbiol 69: 5463–5471.
27. Wang H, Edwards M, Falkinham JO 3rd, Pruden A (2012) Molecular survey of the occurrence of Legionella spp., Mycobacterium spp., Pseudomonas aeruginosa, and amoeba hosts in two chloraminated drinking water distribution systems. Appl Environ Microbiol 78: 6285–6294.
28. Regan JM, Harrington GW, Noguera DR (2002) Ammonia- and nitrite-oxidizing bacterial communities in a pilot-scale chloraminated drinking water distribution system. Appl Environ Microbiol 68: 73–81.
29. LeChevallier MW, Cawthon CD, Lee RG (1988) Inactivation of biofilm bacteria. Appl Environ Microbiol 54: 2492–2499.
30. Hoefel D, Monis PT, Grooby WL, Andrews S, Saint CP (2005) Culture-independent techniques for rapid detection of bacteria associated with loss of chloramine residual in a drinking water system. Appl Environ Microbiol 71: 6479–6488.
31. Zhang Y, Edwards M (2009) Accelerated chloramine decay and microbial growth by nitrification in premise plumbing. Journal American Water Works Association 101: 51.
32. Nguyen C, Elfland C, Edwards M (2012) Impact of advanced water conservation features and new copper pipe on rapid chloramine decay and microbial regrowth. Water Res 46: 611–621.
33. Lin YE, Stout JE, Yu VL (2011) Controlling Legionella in hospital drinking water: an evidence-based review of disinfection methods. Infect Control Hosp Epidemiol 32: 166–173.

34. Stout JE, Goetz AM, Yu VL (2011) Hospital Epidemiology and Infection Control; Mayhall CG, editor: Lippincott Williams, & Wilkins.

35. Flannery B, Gelling LB, Vugia DJ, Weintraub JM, Salerno JJ, et al. (2006) Reducing Legionella colonization in water systems with monochloramine. Emerg Infect Dis 12: 588–596.

36. Pryor M, Springthorpe S, Riffard S, Brooks T, Huo Y, et al. (2004) Investigation of opportunistic pathogens in municipal drinking water under different supply and treatment regimes. Water Sci Technol 50: 83–90.

37. Marchesi I, Cencetti S, Marchegiano P, Frezza G, Borella P, et al. (2012) Control of Legionella contamination in a hospital water distribution system by monochloramine. American Journal of Infection Control 40: 279–281.

38. Caporaso JG, Lauber CL, Walters WA, Berg-Lyons D, Huntley J, et al. (2012) Ultra-high-throughput microbial community analysis on the Illumina HiSeq and MiSeq platforms. ISME J 6: 1621–1624.

39. Walters WA, Caporaso JG, Lauber CL, Berg-Lyons D, Fierer N, et al. (2011) PrimerProspector: de novo design and taxonomic analysis of barcoded polymerase chain reaction primers. Bioinformatics 27: 1159–1161.

40. Caporaso JG, Kuczynski J, Stombaugh J, Bittinger K, Bushman FD, et al. (2010) QIIME allows analysis of high-throughput community sequencing data. Nat Methods 7: 335–336.

41. Edgar RC (2010) Search and clustering orders of magnitude faster than BLAST. Bioinformatics.

42. Lozupone C, Knight R (2005) UniFrac: a new phylogenetic method for comparing microbial communities. Appl Environ Microbiol 71: 8228–8235.

43. Morisita M (1959) Measuring of the dispersion of individuals and analysis of the distributional patterns. Memoirs of the Faculty of Science, Kyushu University, Series E (Biology) 2.

44. Horn HS (1966) Measurement of "overlap" in comparative ecological studies. The American Naturalist 100: 419–424.

45. Stout JE, Duda S, Kandiah S, Hannigan J, Yassin M, et al. (2012) Evaluation of a new monochloramine generation system for controlling Legionella in building hot water systems. Association of Water Technologies Annual Convention and Exposition.

46. Kandiah S, Yassin MH, Hariri R, Ferrelli J, Fabrizio M, et al. (2012) Control of Legionella contamination with monochloramine disinfection in a large urban hospital hot water system. Association for Professionals in Infection Control and Epidemiology Annual Conference.

47. Duda S, Kandiah S, Stout JE, Baron JL, Yassin MH, et al. (2013) Monochloramine disinfection of a hospital water system for preventing hospital-acquired Legionnaires' disease: lessons learned from a 1.5 year study. The 8th International Conference on Legionella.

48. Sydnor ER, Bova G, Gimburg A, Cosgrove SE, Perl TM, et al. (2012) Electronic-eye faucets: Legionella species contamination in healthcare settings. Infect Control Hosp Epidemiol 33: 235–240.

49. Yapicioglu H, Gokmen TG, Yildizdas D, Koksal F, Ozlu F, et al. (2012) Pseudomonas aeruginosa infections due to electronic faucets in a neonatal intensive care unit. J Paediatr Child Health 48: 430–434.

50. Revetta RP, Gomez-Alvarez V, Gerke TL, Curioso C, Santo Domingo JW, et al. (2013) Establishment and early succession of bacterial communities in monochloramine-treated drinking water biofilms. FEMS Microbiol Ecol.

51. Chiao TH, Clancy TM, Pinto A, Xi C, Raskin L (2014) Differential resistance of drinking water bacterial populations to monochloramine disinfection. Environ Sci Technol 48: 4038–4047.

52. Duda S, Kandiah S, Stout JE, Baron JL, Yassin MH, et al. (2014) Evaluation of a new monochloramine generation system for controlling Legionella in building hot water systems. Submitted for publication.

53. Jensen N, Whitfield FB (2003) Role of Alicyclobacillus acidoterrestris in the development of a disinfectant taint in shelf-stable fruit juice. Letters in Applied Microbiology 36: 9–14.

54. Revetta RP, Pemberton A, Lamendella R, Iker B, Santo Domingo JW (2010) Identification of bacterial populations in drinking water using 16S rRNA-based sequence analyses. Water Research 44: 1353–1360.

55. Centers for Disease Control and Prevention (2011) Legionellosis–United States, 2000–2009. Morbidity and Mortality Weekly Report: 1083–1086.

56. BSR/ASHRAE (2011) Proposed New Standard 188P, Prevention of Legionellosis Associated with Building Water Systems. Atlanta, GA: American Society of Heating, Refrigerating, and Air-Conditioning Engineers, Inc.

# 14

# Removal Efficiency of Radioactive Cesium and Iodine Ions by a Flow-Type Apparatus Designed for Electrochemically Reduced Water Production

**Takeki Hamasaki, Noboru Nakamichi, Kiichiro Teruya, Sanetaka Shirahata\***

Department of Bioscience and Biotechnology, Faculty of Agriculture, Kyushu University, Higashi-ku, Fukuoka, Japan

## Abstract

The Fukushima Daiichi Nuclear Power Plant accident on March 11, 2011 attracted people's attention, with anxiety over possible radiation hazards. Immediate and long-term concerns are around protection from external and internal exposure by the liberated radionuclides. In particular, residents living in the affected regions are most concerned about ingesting contaminated foodstuffs, including drinking water. Efficient removal of radionuclides from rainwater and drinking water has been reported using several pot-type filtration devices. A currently used flow-type test apparatus is expected to simultaneously provide radionuclide elimination prior to ingestion and protection from internal exposure by accidental ingestion of radionuclides through the use of a micro-carbon carboxymethyl cartridge unit and an electrochemically reduced water production unit, respectively. However, the removability of radionuclides from contaminated tap water has not been tested to date. Thus, the current research was undertaken to assess the capability of the apparatus to remove radionuclides from artificially contaminated tap water. The results presented here demonstrate that the apparatus can reduce radioactivity levels to below the detection limit in applied tap water containing either 300 Bq/kg of $^{137}$Cs or 150 Bq/kg of $^{125}$I. The apparatus had a removal efficiency of over 90% for all concentration ranges of radio–cesium and –iodine tested. The results showing efficient radionuclide removability, together with previous studies on molecular hydrogen and platinum nanoparticles as reactive oxygen species scavengers, strongly suggest that the test apparatus has the potential to offer maximum safety against radionuclide-contaminated foodstuffs, including drinking water.

**Editor:** Vishal Shah, Dowling College, United States of America

**Funding:** All experiments were performed using Kyushu university's finance (Trust Accounts No. JAKF650803). The funder had no role in study design, data collection and analysis, decision to publish, or preparation of the manuscript.

**Competing Interests:** The authors have declared that no competing interests exist.

\* Email: sirahata@grt.kyushu-u.ac.jp

## Introduction

The Great East Japan Earthquake of magnitude 9 struck the northeastern coast of Japan on March 11, 2011. The earthquake caused a catastrophic tsunami, with the wave height of nearly 40.5 m, which caused failures in the nuclear reactor cooling system in the Fukushima Daiichi Nuclear Power Plant (FDNPP) [1,2]. Soon after, these failures triggered hydrogen explosions in the nuclear reactors, discharging radioactive steam and liberating various radionuclides into the air over several days [2,3]. Following the incident, natural factors such as wind flow, air streams, and rainfall caused dispersion and precipitation of various levels of radionuclides on land surfaces and vegetation in the Tohoku and Kanto regions [4–8]. Radionuclides were also detected in Fukuoka, 1,000 km away from the FDNPP [9], indicating the wide spread of the radioactive plume over Japan. Urgent action to cope with the situation involves decontamination of terrestrial and aquatic radioactivity sources, including drinking water. Incineration of contaminated materials such as plants, wood bark, garbage, and house wreckage is one choice for disposition, although it leaves cesium-enriched ash. An entire system for safe incineration, removal of ash radioactivity and safe disposal has been reported, with promising results [10]. Numerous conventional methods using ion exchange, various membrane processes, coagulation and co-precipitation and other technologies for eliminating radionuclides from radioactive wastewaters have been reported to be effective [11,12]. Numerous approaches have been shown to remove radionuclides from contaminated water, including a mixture of activated carbon and/or zeolite-based media [13–15], co-precipitation with zinc hexacyanoferrate (II) followed by precipitation [16], sorption of radionuclides with biomaterials such as diatomite [17], Prussian blue immobilized diatomite or alginate/calcium beads or magnetic nanoparticles [18–20], arca shell [21], sulphuric acid-modified persimmon waste [22], nickel (II) hexacyanoferrate (III) functionalized walnut shell [23], mesoporous silica monoliths conjugated with dibenzo-18-crown-6 ether [24], and cobalt ferrocyanide impregnated anion exchange beads [25]. Additionally, a layered chalcogenide with a $CdI_2$ crystal structure for adsorbing several cations has been explored [26].

Although these technologies are encouraging for removal of various levels of radionuclides and further improvements are expected to arise in the future, securing safe drinking water is also of prime importance. Rainwater samples collected in Fukushima in early April, 2011 have been reported to contain $^{131}$I (1470±26.5 Bq/L), $^{134}$Cs (100±25.3 Bq/L) and $^{137}$Cs (129±9.47 Bq/L) [27]. The fallout contaminates surface waters, including lakes and rivers, which are the main sources for preparing tap water to supply the residents in these regions. As a result, drinking water prepared from several water purification plants was reported to be contaminated. Subsidiary methods to reinforce conventional water purification systems have been reported to eliminate radioactivity from contaminated water sources. The efficacies of the coagulation-flocculation-sedimentation method in water purification plants, with removal efficiencies of 17% and 56% for $^{131}$I and $^{134}$Cs, respectively [28,29], and radionuclide absorption by algal strains for environmental remediation [30,31] have been assessed. Another significant point to consider is the contamination of drinking water via distribution system such as pipes, storage tanks, water pumps and heaters, which may be persistent contaminating sources. A recent review concluded that cesium appears to be removed by flushing water pipes with a low pH solution containing sodium or magnesium as ion competitors [32]. However, further assessment will be required before applying this approach to the vast areas of regional contamination. Approximately one month later, the radioactivity levels had decreased to below the limit values in the water purification plants [33]. Whereas even after 2 years, total Cs radioactivities above the limit values are reported in some foodstuffs, such as Chinese mushrooms, rice, soybean, adzuki-bean and several fish obtained from the areas surrounding the FDNPP [34,35]. Moreover, low levels of radioactive Cs species are still detected in the drinking water of many cities around FDNPP [36]. These results imply that the fallout still remains on land surfaces and nearby mountain areas and that rainfall wash down is a highly probable contaminant of tap water sources [7,8]. Precautions to avoid consumption of such foodstuffs, including drinking water, have been taken by measuring radioactivity levels prior to distribution. Nevertheless, following the accident, the concentrations of $^{131}$I in the tap water distributed by these purification plants were 210 Bq/L in Tokyo, 189 Bq/L in Ibaraki, and 220 Bq/L in Chiba, all of which exceeded the upper limit of $^{131}$I concentration set as 100 Bq/L for infants under 1 year of age by the Ministry of Health, Labour and Welfare, 1947 [3,37]. Therefore, it is highly desirable to have terminal security systems that can achieve the removal of even lower levels of radioactive contaminants in tap water because, for example, radiocesium accumulates in the body. However, only limited studies examining removability of radionuclides from household water purifiers are available to date. Several domestic pot-type water purifiers have been suggested as a possible final security treatment to eliminate contaminated radionuclides in tap water [27,38]. Although most of these pot-type water purifiers are efficacious, with varying degrees of radionuclide removal from contaminated water, they are useless against the biological effects exerted by unconscious ingestion of radionuclides via drinking water and/or foodstuffs.

Ionizing radiation emitted by ingested radionuclides causes water radiolysis by acting on the water molecules, which comprise approximately 80% of body weight [39]. Water radiolysis yields a variety of reactive oxygen species (ROS) including hydrogen peroxide ($H_2O_2$), the hydroxyl radical ($^{\bullet}OH$), superoxide anion radicals ($^{\bullet}O_2^-$), and other molecular species [40]. These free radicals cause extensive oxidative damage to biologically critical macromolecules such as DNA, RNA, proteins and lipids [41–45]. Such damage eventually induces cellular apoptosis or carcinogenic transformation [46,47]. Therefore, an ideal apparatus should have the potential to provide both the elimination of radionuclides prior to ingestion and protection from detrimental ROS effects generated by the accidentally and/or unconsciously internalized radionuclides.

Considering these requirements, an apparatus designed to produce electrochemically reduced water (ERW) could be thought to fulfill such demands because it contains two functional units; an electrolysis unit for molecular hydrogen enrichment, and a micro-carbon carboxymethyl (CM) cartridge unit for removing various impurities. ERW produced from tap water by this apparatus contains as much as 0.587 ppm dissolved hydrogen (Table 1, [48]). Dissolved molecular hydrogen has been shown to exert a radioprotective effect in both *in vitro* and *in vivo* studies [49,53]. These compelling results strongly support the suggestion that molecular hydrogen dissolved in ERW could function as a radioprotective agent in the body. Moreover, ERW was shown to contain platinum nanoparticles (Pt NPs) at up to 2.5 ppb as an ROS scavenger, liberated from Pt-electrodes during electrolysis [39,54].

As for the second requirement, a micro-carbon CM cartridge unit composed of a nonwoven-fabric filter, several types of activated carbon and an ion-exchange material was present in the current test apparatus to remove particulate matters, microorganisms and 13 designated impurities [55]. However, this micro-carbon CM cartridge has not been assessed for its ability to remove radionuclides from contaminated tap water. Therefore, the present research was aimed at evaluating whether the test apparatus as a whole is capable of removing radionuclides from contaminated tap water.

## Materials and Methods

### Chemicals

Cesium chloride (CsCl) and potassium iodide (KI) were purchased from Wako Pure Chemical Industries (Osaka, Japan).

### Radioisotopes

$^{137}$CsCl [0.2021 MBq/g] and Na$^{125}$I [12.950 TBq/g] were purchased from Japan Radioisotope Association (JRIA, Tokyo, Japan). We used $^{125}$I because Kyushu University Radioisotope Center has an approval to use this radionuclide. Tap water distributed by the Fukuoka City Waterworks Bureau, Fukuoka, Japan was used in all experiments except ultrapure water (Milli Q water, Merck Millipore, Tokyo, Japan) for the preparation of standard solutions for inductively coupled plasma-mass spectrometry (ICP-MS) analysis.

### Electrochemically reduced water (ERW)-producing apparatus

A water flow-type apparatus, Trim Ion NEO, was provided by Nihon Trim Co. Ltd., Osaka, Japan as the test apparatus. This test apparatus is composed of two units, a micro-carbon CM cartridge unit (Fig. 1B) and an electrolysis unit (Fig. 1C). Tap water flows into the cartridge unit, where tap water passes through the nonwoven-fabric filter to remove macroparticles, and pre-cleaned water flows into mixed layers of activated charcoal powders and cationic ion-exchange material to remove most of the impurities, including dissolved lead and 13 other elements that must be removed. The remaining contaminants, such as microorganisms and iron rust particles larger than 0.1 μm in size, are also eliminated by the cartridge (Fig. 1B). The micro-carbon CM cartridge unit is certified to withstand filtration of at least 12 tons of tap water per year or 35 liters per day for 1 year. In the present study, we used a new cartridge unit for each experiment. Purified tap water flows into the electrolysis unit, which is composed of five platinum (Pt)-coated electrode plates, separated by semi-permeable membranes and the water is electrolyzed while passing through the gaps between the electrodes (Fig. 1C). Platinum-

**Table 1.** Characteristics of the sample waters.

| | Tap Water | Filtered Water | ERW | | | |
| | | | Lv 1 | Lv 2 | Lv 3 | Lv 4 |
|---|---|---|---|---|---|---|
| pH | 7.6±0.0 | 7.6±0.0 | 8.0±0.0 | 8.5±0.0 | 9.1±0.0 | 9.4±0.1 |
| ORP (mV) | 555.3±15.5 | 550.0±20.1 | 140.0±5.0 | 110.0±7.5 | −673.3±2.5 | −688.0±9.5 |
| EC (ms/m) | 49.3±0.1 | 49.5±0.1 | 49.7±0.1 | 49.7±0.1 | 49.0±0.2 | 48.1±0.2 |
| DH (ppb, µg/l) | N.D. | N.D. | 70.0±19.3 | 163.3±18.0 | 321.7±47.5 | 587.0±44.6 |
| DO (ppm, g/l) | 7.5±0.0 | 7.5±0.0 | 7.5±0.1 | 7.1±0.1 | 6.6±0.2 | 6.1±0.3 |

Filtered water: tap water was passed through the micro carbon cartridge without electrolysis. Lv 1: electrochemically reduced water (ERW) generated by electrolyzing the filtered water at level 1 with constant electric current at 50 volts (V) upper limit voltage and a flow rate of 1.8–2.0 l/min. Likewise, other ERWs were produced using identical conditions, except selecting the Lv 2 to Lv 4 switch. ORP: oxidation-reduction potential. EC: electrical conductivity. DH: dissolved hydrogen. DO: dissolved oxygen. Measurements were conducted at ambient temperatures. N.D.: Not Detected.

coated titanium electrodes are certified for at least 1,400 hours use without a marked deterioration with respect to the efficacy of water electrolysis, suggesting that the loss of a small amount of Pt nanoparticles from the surface of the electrode will not significantly affect the electrolysis efficacy of the device used here. Electrolyzed tap water near the cathode typically exhibits a high pH, low dissolved oxygen, high negative redox potential and a high concentration of dissolved hydrogen (0.4–0.9 ppm) (Table 1, [48]). Water produced in this manner, with the above characteristics, is designated as ERW. The test apparatus is designed to produce five types of water; four types of ERW (Levels 1–4) electrolyzed with a constant electric current for each level (0.8 to 4.2 A) at a maximum of 50 volts and one type of filtered water without electrolysis (Table 1). ERW is produced near the cathode, as indicated by the thick right-facing arrows in Fig. 1c, and positively charged radioactive Cs ions will be attracted to the cathode side during electrolysis, resulting in an increased concentration of $Cs^+$ ions in ERW, dependent upon the current intensity. Conversely, negatively charged I ions will be attracted to the anode side, resulting in a decreased concentration of I ions in ERW. The electrolysis currents were increased in the order of levels 1 to 4, where Level 4 represents the strongest current, reflecting the highest dissolved hydrogen (DH) and the lowest oxidation-reduction potential (ORP) (Table 1). When the radioactivity of ERW at level 4 is measured as being lower than the background level, then one can conclude that the radioactivity of ERW at levels 1 to 3 is lower than the background level. ERW at levels 1 to 3 is usually used for drinking and at level 4 is used for cooking. We have included Table 1 to aid the readers understanding of the four types of ERW.

### Preparation of non-radioactive sample water (CsCl, KI)

Tap water was used as a control. CsCl solutions of 20 liters each with concentrations of 20 and 2,000 ppb were prepared using tap water. Likewise, KI solutions with concentrations of 100 and 4,000 ppb were prepared. These solutions are designated as sample waters. The test system was arranged by placing an adjustable speed pump between the sample waters and the test apparatus to mimic tap water pressure, connected to the inlet of the test apparatus, as shown in Fig. 1A. The water flow rate was set to 1.8–2.0 L/min by adjusting the pump speed throughout the entire experiment. In the experiment, 1–2 liters of tap water was used to wash and equilibrate the system each time the sample concentrations were changed. Fifteen milliliters of filtered, ERW and relevant control waters were collected for ICP-MS analysis. The

removal efficiency was calculated according to a previously described equation [38], shown in Tables 2 and 3.

### ICP-MS analysis of Cs and I elements in ERWs

Sample waters were passed through the apparatus, and collected filtered waters were quantitated using ICP-MS (Agilent 7500c, Agilent Technologies Co. Ltd., Santa Clara, CA, USA) in the Radioisotope Center at Kyushu University.

### Preparation of radioactive sample water ($^{137}$CsCl and Na$^{125}$I)

Stock solution of $^{137}$CsCl was diluted with 20 liters of tap water to prepare concentrations of 15,000, 3,000, 300, and 30 Bq/Kg. Likewise, Na$^{125}$I stock solution was diluted with 20 liters of tap water to prepare concentrations of 15,000, 1,500, and 150 Bq/Kg. All other experimental conditions, such as water flow rate, system equilibration, the electrolysis conditions of the apparatus were carried out as closely as possible to those used for the non-radioisotope experiments, except that 10 ml of each of the sample waters were collected for radioactivity counting.

### Radioactivity counting of $^{137}$Cs and $^{125}$I in sample waters

Radioactive sample waters were passed through the apparatus, and collected waters were quantitated using a gamma counter (AccuFLEX γ ARC-7001, Hitachi Aloka Medical, Ltd., Tokyo, Japan) in the Center of Advanced Instrumental Analysis at Kyushu University. To evaluate the effect of the electrolysis step on radionuclide removal, filtered waters were electrolyzed by a constant current (4.2 A) at level 4 and radioactivities of ERW were quantitated as above.

### Statistical analysis

All experiments were performed in triplicate. Data are expressed as means ± SD for each experiment.

## Results

### Analysis of Cs and I elements in the filtered water

Prior to radioisotope experiments, CsCl and KI solutions were prepared as described in the Materials and Methods section and their removability was tested. The background Cs concentration in tap water was similar to that for the filtered water (Fig. 2A, column 0 ppb). When 20 and 2,000 ppb CsCl solutions were used, the measured values of the filtered water indicate that the test apparatus had a higher removability (87.4%) for the 20 ppb CsCl

A

B                                                                          C

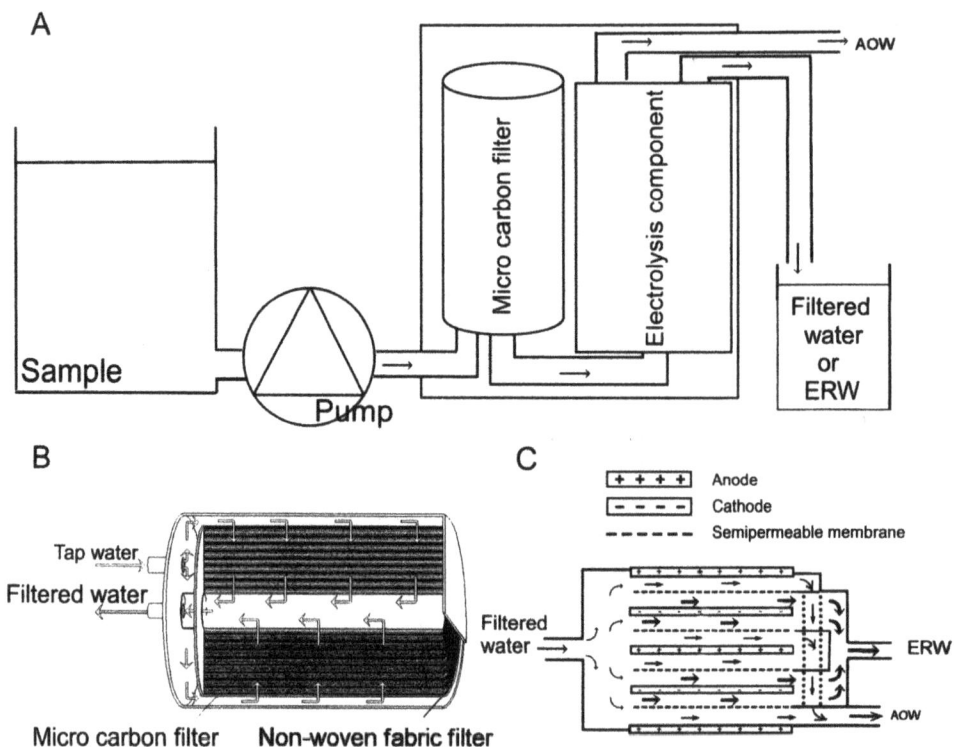

**Figure 1. Schematic of the flow-type electrolysis apparatus.** The test apparatus is composed of two units, a micro-carbon CM cartridge (B) and an electrolysis unit (C). The overall water flow and equipment set up is shown in (A). Sample water is connected to an adjustable speed pump to maintain a flow rate of 1.8–2.0 l/min and expelled to the inlet of the electrolysis unit (A). Tap water passes through the nonwoven-fabric filter, the mixed layers of activated charcoal powders and cationic ion-exchange material to make filtered water (B). Filtered water then flows into the electrolysis unit composed of platinum-coated 5 electrode plates separated by semi-permeable membranes (C). Filtered water will be electrolyzed at levels 1, 2, 3 and 4 at a maximum of 50 volts while passing through the gaps between the electrodes.

solution than for the 2,000 ppb CsCl solution (58.2%) (Fig. 2A, Table 2). Similar experiments using KI solutions were carried out and the results are shown in Fig. 2B. The background I concentrations in tap water and that for the filtered water were similar (Fig. 2B, column 0 ppb). Removal efficiency after filtration for 100 ppb and 4,000 ppb KI solutions were 91.7% and 84.6%, respectively (Table 3). These results demonstrate that the micro-carbon CM cartridge is capable of removing Cs and I ions at all concentration ranges tested (Fig. 2).

## Removal efficiency of $^{137}$CsCl and Na$^{125}$I in the filtered water

Because the test apparatus removed Cs and I ions efficiently, assays were extended to examine the removability of $^{137}$CsCl and Na$^{125}$I. The natural background counts in tap water and filtered water were below the detection limit of the gamma counter (Fig. 3A and B, column 0). Tap water containing 30 (0.0067 ppb as Cs ions), 300 (0.0642 ppb as Cs ions), 3,000 (0.636 ppb as Cs ions) and 15,000 (3.16 ppb as Cs ions) Bq/kg of $^{137}$CsCl as controls showed the expected radioactive counts (Fig. 3A and 3B, white bar

**Table 2.** Removal efficiencies (%) for Cs ion and $^{137}$Cs.

| Measured (Loaded) amounts | | Removal efficiency (%) |
|---|---|---|
| as Cs ion (ppb) | as $^{137}$Cs (Bq/kg) | |
| 1976.47 (2000) | 0 | 58.2 |
| 20.55 (20) | 0 | 87.4 |
| #3.1600 | 16212.0 (15000) | 96.9 |
| #0.6360 | 3262.0 (3000) | 96.9 |
| #0.0642 | 329.0 (300) | 99.2* |
| #0.0067 | 34.9 (30) | 92.5* |

Removal efficiency (%) = (1−[A]/[B])×100 according to [38]. [A], [B]: concentrations of Cs and $^{137}$Cs after and before filtration. Each solution was filtered only, without electrolysis. *: [A] values were below the detection limit. *: [A] values used to calculate removal efficiency were below the detection limit. #: equivalent ppb values calculated from the radioactivities loaded. Values within parentheses were prepared and loaded amounts or radioactivities of cesium.

**Table 3.** Removal efficiencies (%) for I and $^{125}$I ions.

| Measured (Loaded) amounts | | Removal efficiency (%) |
|---|---|---|
| as I ion (ppb) | as $^{125}$I (Bq/kg) | |
| 3891.0 (4000) | 0 | 84.6 |
| 130.0 (100) | 0 | 91.7 |
| #0.0000197 | 14993.0 (15000) | 99.4 |
| #0.00000351 | 1788.0 (1500) | 99.3 |
| #0.000000196 | 146.3 (150) | 99.5* |

Removal efficiency (%) = (1−[A]/[B])×100 according to [38]. [A], [B]: concentrations of I and $^{125}$I solutions after and before filtration. Each solution was filtered only, without electrolysis. *: [A] values used to calculate removal efficiency were below the detection limit. #: equivalent ppb values calculated from the radioactivities loaded. Values within parentheses were prepared and loaded amounts or radioactivities of iodine.

at each concentration) with a high correlation coefficient (Fig. 3C, $R^2 = 0.999$). Control waters were then passed through the micro-carbon CM cartridge and the filtrate radioactivities were measured (Fig. 3A and B). It was found that the radioactivities of the filtered water for $^{137}$CsCl were reduced significantly (Fig. 3A and 3B) and removal efficiency was 96.9%, even after loading 15,000 Bq/kg of $^{137}$CsCl (Table 2).

To evaluate Na$^{125}$I removability, we prepared Na$^{125}$I containing sample waters as described above. The natural background count in tap water and filtered water exhibited values below the detection limit (Fig. 4A and B, column 0). Tap water containing 150 (0.000196 ppt as I ions), 1,500 (0.00351 ppt as I ions) and 15,000 (0.0197 ppt as I ions) Bq/kg of Na$^{125}$I as controls showed expected radioactive counts (Fig. 4A and 4B, white bar at each concentration) with a high correlation coefficient (Fig. 4C, $R^2 = 0.999$). Radioactive control tap waters were passed through the micro-carbon CM cartridge, reducing the filtrate radioactivities significantly (Fig. 4A and B), with a removal efficiency of over 99% (Table 3). Thus, the micro-carbon CM cartridge was demonstrated to efficiently remove radioactivities up to 15,000 Bq/kg of Na$^{125}$I.

### Effect of electrolysis on the removal efficiency of $^{137}$Cs and $^{125}$I

In parallel with the preceding experiments, we evaluated the effects of the electrolysis step in terms of efficiencies for $^{137}$Cs and $^{125}$I removal from the filtered radioactive water. Filtered water was electrolyzed at the highest current level of 4. In this experiment, we selected 300 Bq/kg of $^{137}$Cs water, which loaded 30 times more radioactivity than the upper limit value of 10 Bq/kg for drinking water set by the government [56]. Under these conditions, the radioactivity in ERW remained below the detection limit (Fig. 5A). Similarly, we evaluated the removability of $^{125}$I by the highest electrolysis level of 4. In this case, we selected 150 Bq/kg of $^{125}$I, which is a loading of 1.5 times more radioactivity than the upper limit of 100 Bq/L of $^{131}$I concentration for infants under 1 year of age set by the Ministry of Health, Labour and Welfare, 1947 [37]. The radioactive iodine level in ERW remained below the detection limit (Fig. 5B). Therefore, the results indicate that the cartridge substantially removed $^{137}$Cs and $^{125}$I from tap water prior to the electrolysis step, thereby assuring undetectable levels of radioactivity in ERW produced at the highest current level of 4, which has the highest attraction for $^{137}$Cs$^+$, and thus the results hold true for ERWs produced with the current levels 1 to 3.

### Discussion

The FDNPP accident liberated various radionuclides, including $^{131}$I, $^{132}$I, $^{134}$Cs, and $^{137}$Cs [57]. Amongst these radionuclides, $^{131}$I can enter the body through inhalation and by ingesting contaminated foodstuffs including drinking water, which then rapidly concentrates in the thyroid gland, where β-radiation exposure takes place. As its half-life is 8 days, radioactivity levels are expected to be reduced substantially over several months. Therefore, an obvious precaution is not to ingest $^{131}$I-contaminated or doubtful foodstuffs including drinking water. Water supply law in Japan limits the lowest chlorine concentration in tap water outlet at 0.1 mg/L [58]. Dissolved $^{131}$I is reported to form various species in tap water such as the radioactive iodide ion ($^{131}$I$^-$), hypoiodous acid (HO$^{131}$I), the iodate ion ($^{131}$IO$_3^-$), iodine molecules ($^{131}$IO$_2$) and organic $^{131}$I. $^{131}$I$^-$ reacts with chlorine and is transformed mainly into HOI at neutral pH. HOI is further transformed into IO$_3^-$ by reacting with chlorine [29], and as a result, almost all iodine is converted to the iodate ion (IO$_3^-$) in tap water due to the oxidation by chlorine [59]. It is reported that $^{131}$I$^-$ removal is increased by water containing 0.1–0.5 mg/L chlorine, with lower concentrations of powdered activated charcoal [29]. However, granular and powdered activated carbons were reported to remove $^{131}$I at about 30–40% efficiency. Additionally, it has been reported that $^{125}$I$^-$ and $^{125}$I$_3^-$ were prepared from $^{125}$I and used to test the removability of these species by a granular type charcoal, which resulted in a small amount of adsorption [60]. These results may partly explain the inefficient removability by activated charcoal reported by others, through selective adsorption of iodate and iodine [38,60,61]. Activated carbon was shown to remove iodide (I$^-$) more efficiently than iodate (IO$_3^-$) [27]. Therefore, it appears that combinations between the types of activated carbon/charcoal and iodine species affect overall removability. In the present experiments, we used tap water distributed by the Waterworks Bureau of the City of Fukuoka, expected to contain at least 0.1 mg/L chlorine. Thus, $^{125}$I is mostly, if not completely, converted to iodate ions (IO$_3^-$) by chlorine in the tap water. In the present results, KI and $^{125}$I were efficiently removed from tap waters by the micro-carbon CM cartridge, suggesting that iodide and iodate ions were removed. The micro-carbon CM cartridge is composed of a nonwoven-fabric filter and activated carbons consisting of a coconut shell activated carbon powder, a coconut shell activated carbon conjugated with a silver compound for antimicrobial effect, and an amorphous titanosilicate-based inorganic compound (BASF Co, Germany) molded with a fibrous binder for shaping. This cartridge was used in the present test apparatus to remove

**A**

**B**

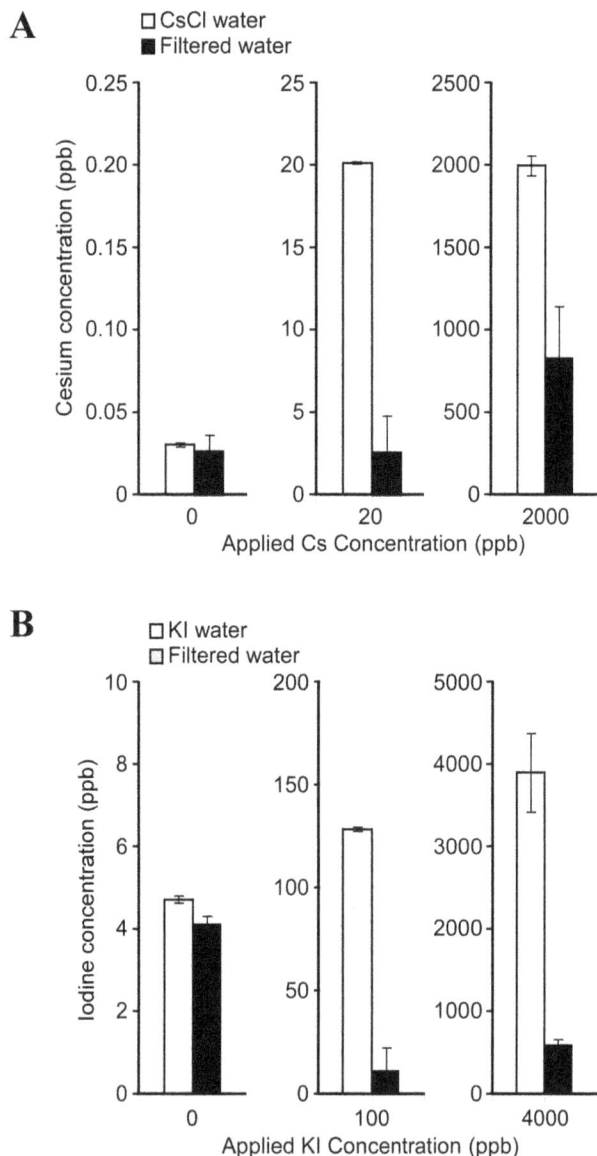

**Figure 2. Measurement of Cs and I elements in filtered waters.** CsCl solutions at concentrations of 0, 20 and 2,000 ppb were passed through the test apparatus. Collected filtered waters were used to measure Cs concentration by ICP-MS (A). KI solutions at concentrations 0, 100 and 4,000 ppb were passed through the test apparatus. Collected filtered waters as in (A) were used to measure I concentration by ICP-MS (B). White bar: Tap water, gray bar: Filtered water. Experiments were carried out in triplicate.

particulate matters, microorganisms, and for qualified removability of 13 designated impurities, tested according to the standard method set by JIS S 3201, 2004 (Domestic Water Purifier Quality Test) [55]. It is worth mentioning that the test apparatus effectively removed I and $^{125}$I (applicable to $Cs^+$ and $^{137}$Cs), even though water was supplied to the apparatus through a pump simulating tap water outlet pressure to attain 1.8–2.0 L/min flow rate, which markedly reduced the contact time of water with the activated carbon surfaces and ion-exchangers compared with those in pot-type water purifiers. It has been reported that the above-mentioned molded activated carbons can replace the hollow fiber membrane filter that is commonly used in other water purifiers to eliminate materials larger than 0.1 µm in size [55]. Incidentally,

hollow fiber membranes do not contribute to the elimination of iodate ($IO_3^-$) ions because their radius is 0.326 nm, even when their radius is increased several fold in water [61]. Additionally, the ineffectiveness of removing $^{131}$I by boiling tap water has been reported [3].

Cesium is an alkaline earth metal that exists as a monovalent cation form ($Cs^+$) in water and in soils [27]. We found that $Cs^+$ could be efficiently removed by the micro-carbon CM cartridge tested here. The mechanism for the removal of $Cs^+$ remains to be investigated. The $Cs^+$ removal efficiency by the apparatus was 87.4% at 20 ppb, which is comparable to that of several pot-type water purifiers that have efficiencies of around 90% for tap water containing 40–50 µg/l (ppb) cesium chloride [38]. A removal efficiency of 58.2% for $Cs^+$ appears to be low at the highest concentration (1976.5 ppb) loading. This lower removal efficiency could be explained by the amount of $Cs^+$ getting close to system over loading because this amount is 625.3 times more $Cs^+$ ion loading than the 3.16 ppb $Cs^+$ ion calculated from the highest radioactivity (16,212 Bq/kg) loading where 96.9% removability was attained (Table 2). Therefore, the apparatus could remove $^{137}$Cs with above 96% efficiency for less than a 3.16 ppb CsCl loading and the removal efficiency is higher than that reported for two commercialized pot-type water purifiers, composed of activated charcoal and an ion exchanger, or activated charcoal, ceramics and a hollow fiber membrane, with 84.2–91.5% efficiencies for rain water samples [27]. Another set of experiments using commercialized four pot-type purifiers made of materials similar to those above assessed iodine and cesium removability, with efficiencies of approximately 85% and 75–90%, respectively [38]. Others also tested Cs removability using a spongiform adsorbent made of Prussian blue caged within the diatomite cavities and carbon nanotubes, by contacting for 10 hours with low levels of $^{137}$Cs, yielding a 99.93% removal efficiency [18]. The present test apparatus showed a removal efficiency of over 96% for Cs and I, which is competitive with or better than previously reported removal efficiencies ranging from 75% to 99.93%. It is emphasized here that the advantages of the test apparatus are that it has long been used for domestic use, is easy to operate, provides a sufficient amount of purified water instantaneously (max. 5 l/min.) and offers an established system for proper disposal and/or recycling of used cartridges. Following the FDNPP accident, tap water contamination monitoring revealed that the maximum of the sum of $^{134}$Cs and $^{137}$Cs was 180.5 Bq/kg on March 2011 in Tamura, Fukushima Prefecture. It was also reported that a sum of $^{134}$Cs and $^{137}$Cs less than 32 Bq/kg was sporadically detected in tap water during 22 days of monitoring after the accident [29]. The water purification plants take precautions not to distribute contaminated water through constant monitoring to meet the latest upper limit value, set by the government as 10 Bq/kg for drinking water, effective from April 1, 2012 [56]. In reality, the detection of greater than 10 Bq/kg radioactivities in tap water in general public is most likely to be the result of accidental and sporadic contamination events. In any case, the test apparatus was demonstrated to decontaminate radiocesium levels to below the detection limit, even when tap water was contaminated by up to 300 Bq/kg radiocesium. When loading 300 Bq/kg of $^{137}$Cs to the cartridge, the removability obtained was a conditional value of 99.2% and leaving the remaining radioactivity to be below the upper limit value of 10 Bq/kg set by the Government. This indicates that the filtered water right before entering into the electrolysis unit still contains a trace amount of $^{137}$Cs and the following electrolysis step may produce $^{137}$Cs enriched ERW. The test apparatus is a powerful electrolysis device yet finely tuned to produce various levels of dissolved hydrogen electric current

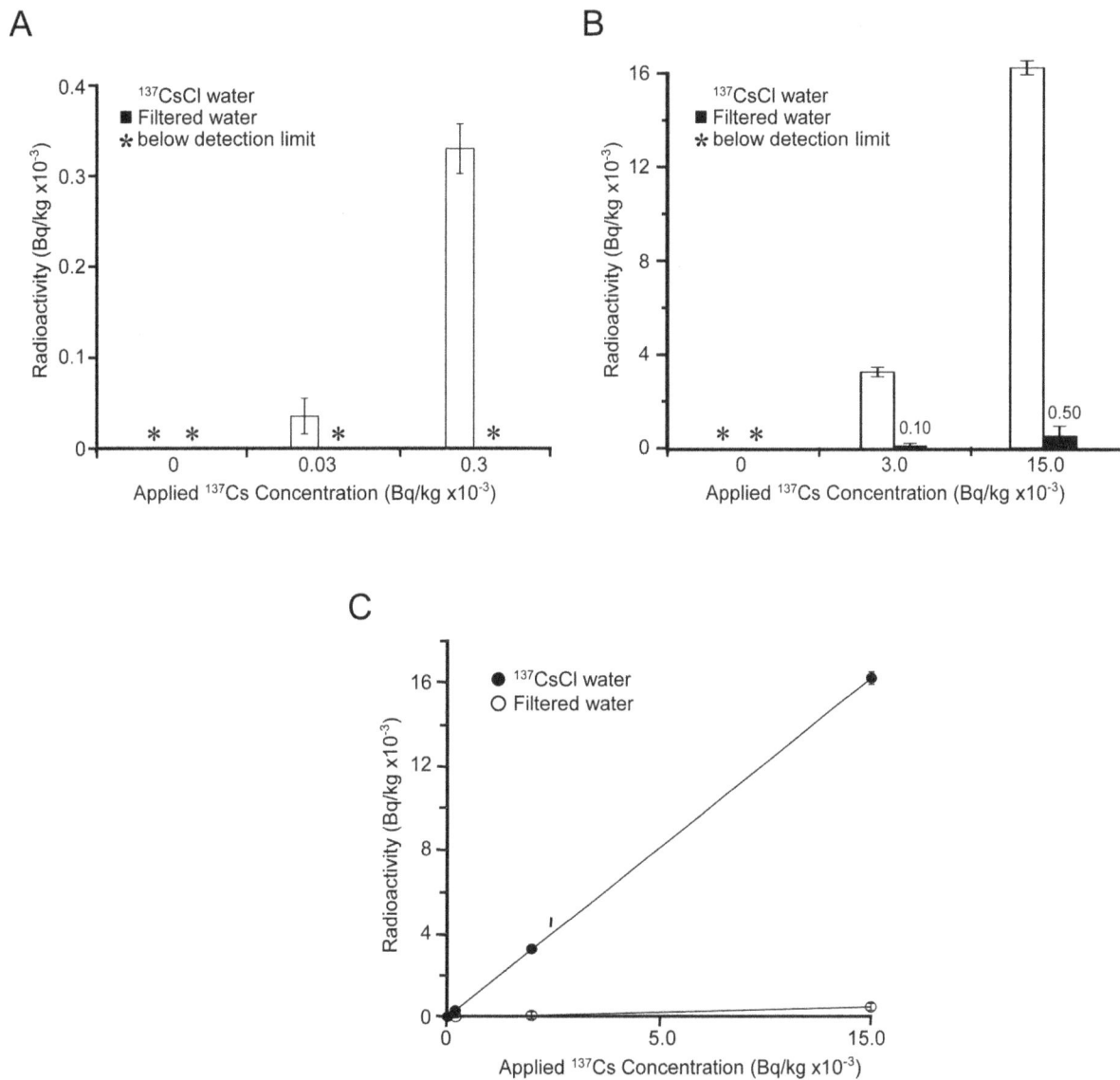

**Figure 3. Measurement of $^{137}$Cs in sample waters.** $^{137}$CsCl solutions at concentrations of 0, 0.03, 0.3, 3.0 and 15.0 KBq/kg were passed through the test apparatus. Collected filtered waters were used to measure $^{137}$Cs counts by an AccuFLEX $\gamma$ ARC-7001 gamma counter (A and B). White bar: $^{137}$CsCl solutions before filtration, gray bar: $^{137}$CsCl solutions after filtration. Radioactivities before and after filtration were evaluated by linear-regression analysis (C). $\bullet$: $^{137}$CsCl solutions before filtration, $\bigcirc$: $^{137}$CsCl solutions after filtration. Experiments were carried out in triplicate.

dependently (Table 1). Under these conditions, we could not definitely exclude a slight possibility that the electrolysis step contributes to $^{137}$Cs enrichment in ERW. Thus, the only way to clarify such uncertainty was to conduct the experiments as shown in Fig. 5. Moreover, we judged that it is not sufficient enough by just showing the removability of the cartridge filter unit alone and extrapolating the results for evaluating the entire flow-type system. To this end, we decided to measure the radioactivity in ERW, which allows evaluating cartridge unit and electrolysis unit simultaneously. Therefore, the evaluation of the filtering unit in combination with the electrolysis unit as the complete flow-type system was necessary. Our concern for negatively charged I ion was less intense compared to Cs$^+$ ion due to higher removability by the cartridge unit and attracted to the anode side. Nevertheless, we confirmed the removability of the flow-type system experimentally to provide the data set with Cs$^+$ data.

It is commonly regarded that tap water prepared from lakes and rivers contains varying amounts of organic and inorganic materials. In the present study, we considered these to have a significant impact on removability by the test system because such materials are most likely to compete with the very small amounts of $^{137}$Cs and $^{125}$I ions present. Only experiments using low levels of radionuclides will answer the question of whether such interactions between the constituents and added radionuclides may affect removability by this apparatus. Another reason to use lower levels of radionuclides is that even a small amount of $^{137}$Cs dissolved in water is difficult to remove [11,20] and accumulates in the body, causing prolonged exposure. Moreover, the fact is that low levels of radioactive Cs species currently contaminate drinking water in many cities around FDNPP [36]. This may be partly attributed to the limited removability of solubilized cesium by the conventional coagulation-sedimentation process [11,29]. It is therefore extremely important, for the residents of affected regions,

A

B

C

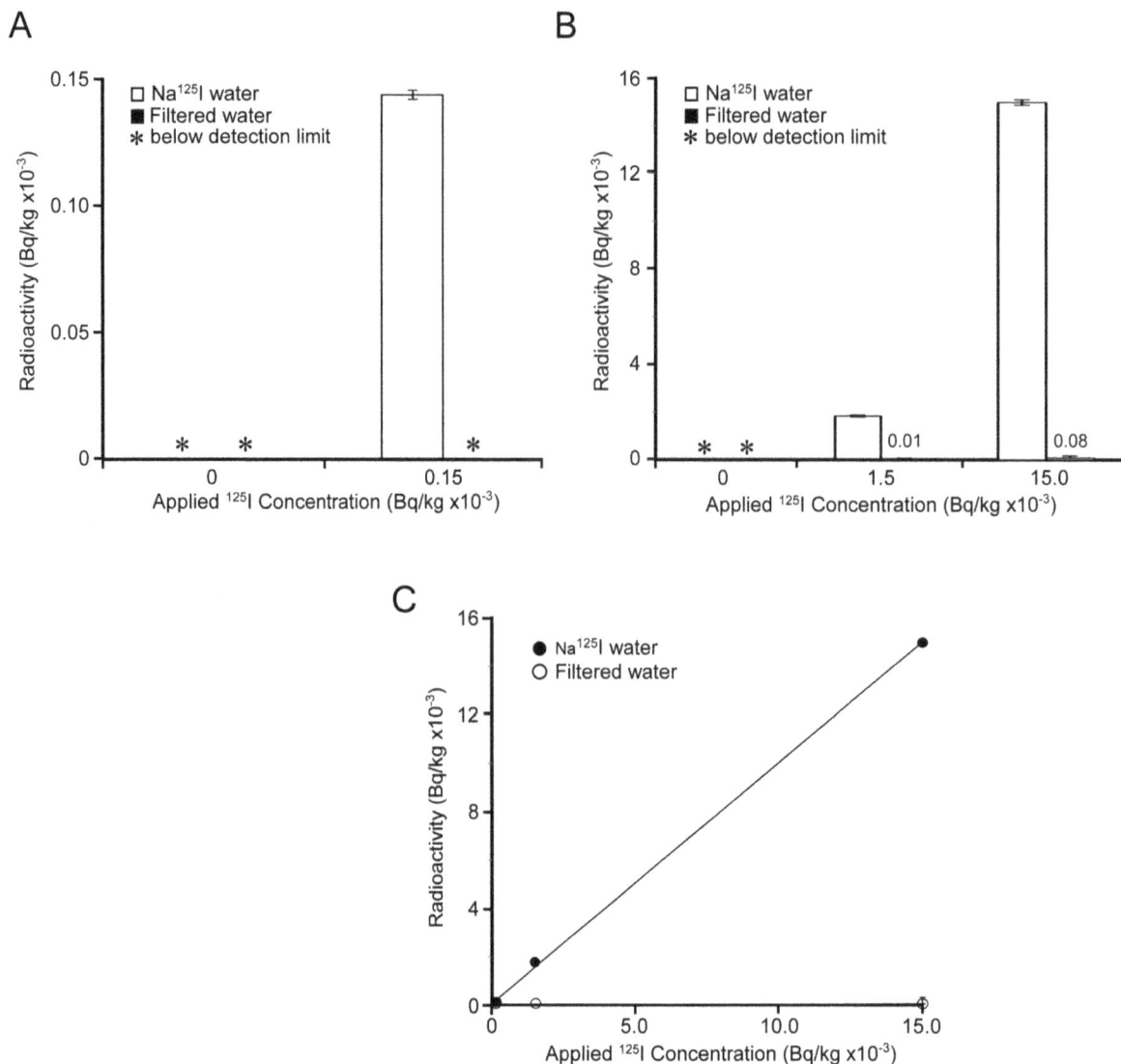

**Figure 4. Measurement of $^{125}$I elements in sample waters.** Na$^{125}$I solutions at concentrations of 0, 0.15, 1.5 and 15.0 KBq/kg were passed through the test apparatus. Collected filtered waters were used to measure $^{125}$I counts by an AccuFLEX $\gamma$ ARC-7001 gamma counter (A and B). White bar: Na$^{125}$I solutions before filtration, gray bar: Na$^{125}$I solutions after filtration. Radioactivities before and after filtration were evaluated by linear-regression analysis (C). ●: Na$^{125}$I solutions before filtration, ○: Na$^{125}$I solutions after filtration. Experiments were carried out in triplicate.

to find a way to remove even small amounts of nuclear contaminants from drinking water.

Another concern related to radiocesium is its longer half-life and a characteristic of ready transfer to the human diet through plants [62]. Precipitated Cs$^+$ binds to clay minerals rather tightly [27], and depth distribution studies reveal that approximately 80% of total radiocesium is retained in the upper 2.0 cm of tested soil samples [63]. Another study estimated that $^{137}$Cs could reach a depth of only 18 cm after 300 yr [37]. These characteristics of surface area retention of radiocesium in addition to its long physical half-life ($^{134}$Cs, $T_{1/2} = 2.06$ yr; $^{137}$Cs, $T_{1/2} = 30.17$ yr) could be a secondary contamination source for vegetation via roots. Uptake of radiocesium from the root is thought to occur via the potassium transport system and is distributed rapidly within the plants [62]. Indeed, many agricultural products are reported to be contaminated by radiocesium and are their marketing is restricted [64,65]. Ingestion of radiocesium-contaminated food-stuffs will expose the gastrointestinal tract and be absorbed into

tissues and organs in the body. Gastrointestinal, reproductive and hematopoietic systems are sensitive to ionizing radiation due to their high turnover rate [57,66]. As an example, the degeneration of small intestinal mucosa cells is caused by free radicals produced from the interactions of radiation energy with intracellular water molecules [66]. Water radiolysis generates a variety of ROS that cause extensive oxidative damages to biologically critical macro-molecules, leading to cell death [40,42–45]. Therefore, providing a method to counter radiation hazards caused by accidentally ingested radioactive waters and foodstuffs will be a great contribution to human health.

ERW is regarded as beneficial to health because of its ROS scavenging ability [39]. ERW produced from tap water by this apparatus could contain as much as 0.587 ppm of dissolved hydrogen (Table 1, [48]). This hydrogen concentration in ERW is relatively high for a flow-type electrolysis apparatus when compared with the concentration of 1.6 ppm hydrogen in 100% hydrogen-saturated water [53]. Such dissolved molecular hydro-

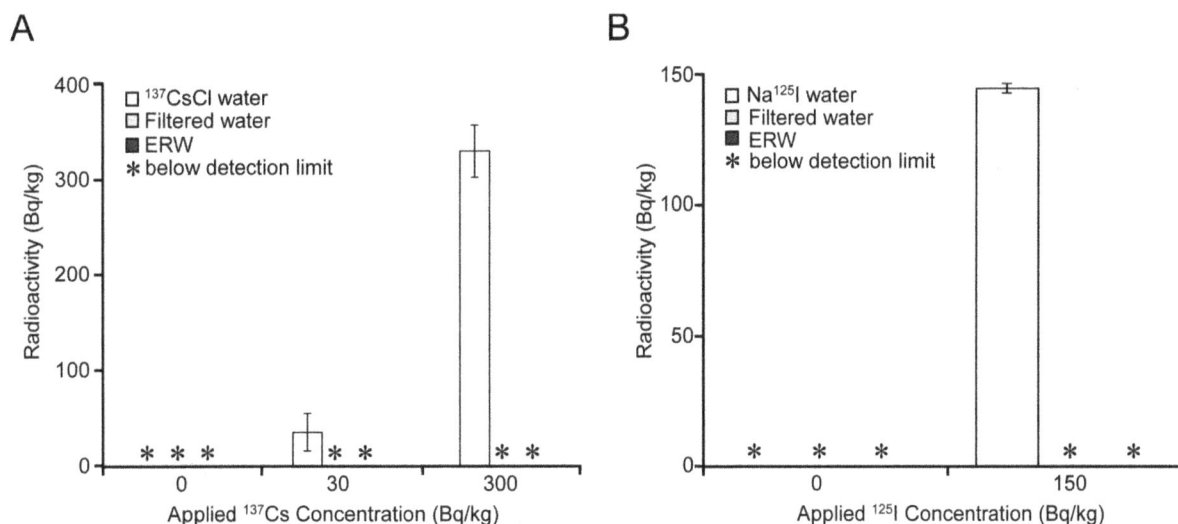

**Figure 5. Effects of electrolysis on filtered radioactive sample waters.** $^{137}$CsCl solutions of 30 and 300 Bq/kg were passed through the test apparatus and filtered waters were collected for measurement. Then, filtered water was passed through the electrolysis unit at the highest electrolysis level of 4 and ERW was collected for measurement. Collected waters were used to measure $^{137}$Cs counts by an AccuFLEX $\gamma$ ARC-7001 gamma counter (A and B). Using the same protocol, filtered water and ERW were collected for 150 Bq/kg of Na$^{125}$I solution. White bar: $^{137}$CsCl or Na$^{125}$I solutions, gray bar: Filtered $^{137}$CsCl or Na$^{125}$I solutions; black bar: ERWs of filtered $^{137}$CsCl or Na$^{125}$I solutions. Experiments were carried out in triplicate.

gen has been shown to exert radioprotective effects in both *in vitro* and *in vivo* studies [49–53]. Molecular hydrogen in ERW prepared from tap water suppressed neuroinflammation in mice [48], and extended the life span of *C. elegans* [54]. Additionally, molecular hydrogen was demonstrated to act as a neuroprotective agent and ROS scavenger [67]. Moreover, ERW produced from an electrolysis unit incorporating Pt-electrodes has been shown to contain 0.1–0.25 ppb Pt nanoparticles [39,54,68]. Pt nanoparticles exhibit protective effects that are attributed to their suppressing ROS production caused by UV-light-induced epidermal inflammation [69]. Synthetic Pt nanoparticles have been shown to scavenge ROS in cultured HeLa cells [70], to induce expression of antioxidant enzyme genes in rat skeletal muscle L6 cells [71], and to act as an SOD/catalase mimetic agent in human lymphoma cells [72]. Model ERW prepared from NaCl, KCl or NaOH solutions has been shown to exert beneficial effects such as anti-diabetic, anti-cancer, and life-span extension of nematodes because of its ROS scavenging ability in numerous *in vitro* and *in vivo* studies [73–78]. Therefore, molecular hydrogen and Pt nanoparticles dissolved in ERW could synergistically contribute to protect gastrointestinal damage caused by ingested radioactive foodstuffs. Furthermore, to maximize protective efficacy against radiation-induced gastrointestinal damage, the consumption of various supplemental foods such as naringin [42], probiotics

[57,66], Kefir [79], melatonin [80] and curcumin [81] are reported to be beneficial.

In conclusion, we demonstrated that radio-cesium and -iodine are efficiently removed by an apparatus containing a micro-carbon CM cartridge filter, prior to ingestion. We also suggest that the ERW produced by the test apparatus will provide maximum protection against accidentally and/or unconsciously ingested radionuclides because it contains dissolved hydrogen and Pt nanoparticles. Therefore, the test apparatus is considered to be a potential alternative tool to minimize radiation hazards caused by contaminated foodstuffs.

## Acknowledgments

The authors thank Nihon Trim Co. Ltd. for providing the Trim Ion NEO apparatus and an adjustable flow rate pump. The authors are also grateful to Ms. Yuri Fujimoto and Chika Kubota for their technical assistance.

## Author Contributions

Conceived and designed the experiments: TH NN KT SS. Performed the experiments: TH NN SS. Analyzed the data: TH NN KT SS. Contributed reagents/materials/analysis tools: TH NN SS. Wrote the paper: TH NN SS.

## References

1. Hamada N, Ogino H (2012) Food safety regulations: what we learned from the Fukushima nuclear accident. J Environ Radioact 111: 83–99.
2. Hamada N, Ogino H, Fujimichi Y (2012) Safety regulations of food and water implemented in the first year following the Fukushima nuclear accident. J Radiat Res 53: 641–671.
3. Tagami K, Uchida S (2011) Can we remove iodine-131 from tap water in Japan by boiling? – Experimental testing in response to the Fukushima Daiichi Nuclear Power Plant accident. Chemosphere 84: 1282–1284.
4. Amano H, Akiyama M, Chunlei B, Kawamura T, Kishimoto T, et al (2012) Radiation measurements in the Chiba Metropolitan Area and radiological aspects of fallout from the Fukushima Dai-ichi Nuclear Power Plants accident. J Environ Radioact 111: 42–52.

5. Koizumi A, Niisoe T, Harada KH, Fujii Y, Adachi A, et al (2013) $^{137}$Cs trapped by biomass within 20 km of the Fukushima Daiichi Nuclear Power Plant. Environ Sci Technol 47: 9612–9618.
6. Thakur P, Ballard S, Nelson R (2013) An overview of Fukushima radionuclides measured in the northern hemisphere. Sci Total Environ 458–460: 577–613.
7. Murakami M, Ohte N, Suzuki T, Ishii N, Igarashi Y, et al (2014) Biological proliferation of cesium-137 through the detrital food chain in a forest ecosystem in Japan. Sci Rep 4: 3599.
8. Nakanishi T, Matsunaga T, Koarashi J, Atarashi-Andoh M (2014) $^{137}$Cs vertical migration in a deciduous forest soil following the Fukushima Dai-ichi Nuclear Power Plant accident. J Environ Radioact 128: 9–14.
9. Momoshima N, Sugihara S, Ichikawa R, Yokoyama H (2012) Atmospheric radionuclides transported to Fukuoka, Japan remote from the Fukushima Dai-

ichi nuclear power complex following the nuclear accident. J Environ Radioact 111: 28–32.

10. Parajuli D, Tanaka H, Hakuta Y, Minami K, Fukuda S, et al (2013) Dealing with the Aftermath of Fukushima Daiichi Nuclear Accident: Decontamination of Radioactive Cesium Enriched Ash. Environ Sci Technol 47: 3800–3806.

11. Liu X, Chen G-R, Lee D-J, Kawamoto T, Tanaka H, et al (2014) Adsorption removal of cesium from drinking waters: A mini review on use of biosorbents and other adsorbents. Bioresour Technol Available: http://dx.doi.org/10.1016/j.biortech. Last accessed 2014.01.012.

12. Rana D, Matsuura T, Kassim MA, Ismail AF (2013) Radioactive decontamination of water by membrane processes – A review. Desalination 321: 77–92.

13. Song K-C, Lee HK, Moon H, Lee KJ (1997) Simultaneous removal of the radiotoxic nuclides Cs$^{137}$ and I$^{129}$ from aqueous solution. Sep Purif Technol 12: 215–227.

14. El-Kamash AM (2008) Evaluation of zeolite A for the sorptive removal of Cs$^+$ and Sr$^{2+}$ ions from aqueous solutions using batch and fixed bed column operations. J Hazard Mater 151: 432–445.

15. Borai EH, Harjula R, Malinen L, Paajanen A (2009) Efficient removal of cesium from low-level radioactive liquid waste using natural and impregnated zeolite minerals J Hazard Mater 172: 416–422.

16. Shakir K, Sohsah M, Soliman M (2007) Removal of cesium from aqueous solutions and radioactive waste simulants by coprecipitate flotation. Sep Purif Technol 54: 373–381.

17. Osmanlioglu AE (2007) Natural diatomite process for removal of radioactivity from liquid waste. Appl Radiat Isot 65: 17–20.

18. Hu B, Fugetsu B, Yu H, Abe Y (2012) Prussian blue caged in spongiform adsorbents using diatomite and carbon nanotubes for elimination of cesium. J Hazard Mater 217–218: 85–91.

19. Vipin AK, Hu B, Fugetsu B (2013) Prussian blue caged in alginate/calcium beads as adsorbents for removal of cesium ions from contaminated water. J Hazard Mater 258–259: 93–101.

20. Thammawong C, Opaprakasit P, Tangboriboonrat P, Sreearunothai P (2013) Prussian blue-coated magnetic nanoparticles for removal of cesium from contaminated environment. J Nanopart Res 15: 1689.

21. Dahiya S, Tripathi RM, Hegde AG (2008) Biosorption of heavy metals and radionuclide from aqueous solutions by pre-treated arca shell biomass. J Hazard Mater 150: 376–386.

22. Pangeni B, Paudyal H, Inoue K, Ohto K, Kawakita H, et al (2014) Preparation of natural cation exchanger from persimmon waste and its application for the removal of cesium from water. Chem Eng J 242: 109–116.

23. Ding D, Lei Z, Yang Y, Feng C, Zhang Z (2014) Selective removal of cesium from aqueous solutions with nickel (II)hexacyanoferrate (III) functionalized agricultural residue–walnut shell J Hazard Mater 270: 187–195.

24. Awual MR, Suzuki S, Taguchi T, Shiwaku H, Okamoto Y, et al (2014) Radioactive cesium removal from nuclear wastewater by novel inorganic and conjugate adsorbents. Chem Eng J 242: 127–135.

25. Valsala TP, Roy SC, Shah JG, Gabriela J, Raj K, et al. (2009) Removal of radioactive caesium from low level radioactive waste (LLW) streams using cobalt ferrocyanide impregnated organic anion exchanger. J Hazard Mater 166: 1148–1153.

26. Sengupta P, Dudwadkar NL, Vishwanadh B, Pulhani V, Rao Rekha, et al (2014) Uptake of hazardous radionuclides within layered chalcogenide for environmental protection. J Hazard Mater 266: 94–101.

27. Higaki S, Hirota M (2012) Decontamination Efficiencies of Pot-Type Water Purifiers for $^{131}$I, $^{134}$Cs and $^{137}$Cs in Rainwater Contaminated during Fukushima Daiichi Nuclear Disaster. PLoS ONE 7(5): e37184.

28. Goossens R, Delville A, Genot J, Halleux R, Masschelein WJ (1989) Removal of the typical isotopes of the Chernobyl fall-out by conventional water treatment. War Res 23: 693–697.

29. Kosaka K, Asami M, Kobashigawa N, Ohkubo K, Terada H, et al (2012) Removal of radioactive iodine and cesium in water purification processes after an explosion at a nuclear power plant due to the Great East Japan Earthquake. Water res 46: 4397–4404.

30. Shimura H, Itoh K, Sugiyama A, Ichijo S, Ichijo M, et al (2012) Absorption of Radionuclides from the Fukushima Nuclear Accident by a Novel Algal Strain. PLoS ONE 7(9): e44200.

31. Fukuda S, Iwamoto K, Atsumi M, Yokoyama A, Nakayama T, et al (2014) Global searches for microalgae and aquatic plants that can eliminate radioactive cesium, iodine and strontium from the radio-polluted aquatic environment: a bioremediation strategy. J Plant Res 127: 79–89.

32. Szabo J, Minamyer S (2014) Decontamination of radiological agents from drinking water infrastructure: A literature review and summary. Environ Int Available: http://dx.doi.org/10.1016/j.envint.2014.01.020.

33. Ministry of Health, Labour and Welfare, Japan. Information on the Great East Japan Earthquake-Water supply. Available: www.mhlw.go.jp/english/topics/2011eq/index. html. Accessed Jul. 18, 2013.

34. Ministry of Health, Labour and Welfare, Japan. Measurement results of radionuclides in foodstuffs (No. 522) Available: http://www.mhlw.go.jp/stf/houdou/2r9852000002oo2l-att/2r9852000002oo6v.pdf. Accessed Jul. 18, 2013.

35. Mizuno T, Kubo H (2013) Overview of active cesium contamination of fresh water fish in Fukushima and Eastern Japan. Sci Rep 3: 1742.

36. Nuclear Regulation Authority (NRA). Monitoring information of environmental radioactivity level: Readings of radioactivity level in drinking water by prefecture October–December, 2013, Accessed Mar. 13, 2014.

37. Ohta T, Mahara Y, Kubota T, Fukutani S, Fujiwara K, et al (2012) Prediction of groundwater contamination with $^{137}$Cs and $^{131}$I from the Fukushima nuclear accident in the Kanto district. J Environ Radioact 111: 38–41.

38. Sato I, Kudo H, Tsuda S (2011) Removal efficiency of water purifier and adsorbent for iodine, cesium, strontium, barium and zirconium in drinking water. J Toxicol Sci 36(6): 829–834.

39. Shirahata S, Hamasaki T, Teruya K (2012) Advanced research on the health benefit of reduced water. Trends Food Sci Technol 23: 124–131.

40. Ewing D, Jones SR (1987) Superoxide Removal and Radiation Protection in Bacteria. Arch Biochem Biophys 254(1): 53–62.

41. Ward JF (1988) DNA damage produced by ionizing radiation in mammalian cells: identities, mechanisms of formation, and reparability. Prog Nucleic Acid Res Mol Biol 35: 95–125.

42. Jagetia GC, Reddy TK (2005) Modulation of radiation-induced alteration in the antioxidant status of mice by naringin. Life Sci 77: 780–794.

43. Nunomura A, Honda K, Takeda A, Hirai K, Zhu X, et al (2006) Oxidative damage to RNA in neurodegenerative diseases. J Biomed Biotechnol 2006: Article ID 82323: 1–6.

44. Tanaka M, Chock PB, Stadtman ER (2007) Oxidized messenger RNA induces translation errors. Proc Natl Acad Sci USA 104(1): 66–71.

45. Radak Z, Zhao Z, Goto S, Koltai E (2011) Age-associated neurodegeneration and oxidative damage to lipids, proteins and DNA. Mol Aspects Med 32: 305–315.

46. Cerutti PA (1985) Prooxidant states and tumor promotion. Science 227: 375–381.

47. Gobbel GT, Bellinzona M, Vogt AR, Gupta N, Fike John R, et al (1998) Response of postmitotic neurons to X-irradiation: implications for the role of DNA damage in neuronal apoptosis. J Neurosci 18(1): 147–155.

48. Spulber S, Edoff K, Hong L, Morisawa S, Shirahata S, et al (2012) Molecular hydrogen reduces LPS-induced neuroinflammation and promotes recovery from sickness behaviour in mice. PLoS ONE 7(7): e42078.

49. Qian L, Cao F, Cui J, Wang Y, Huang Y, et al (2010) The potential cardioprotective effects of hydrogen in irradiated mice. J Radiat Res 51: 741–747.

50. Qian L, Cao F, Cui J, Huang Y, Zhou X, et al (2010) Radioprotective effect of hydrogen in cultured cells and mice. Free Radic Res 44(3): 275–282.

51. Terasaki Y, Ohsawa I, Terasaki M, Takahashi M, Kunugi S, et al (2011) Hydrogen therapy attenuates irradiation-induced lung damage by reducing oxidative stress. Am J Physiol Lung Cell Mol Physiol 301: L415–L426.

52. Chuai Y, Gao F, Li B, Zhao L, Qian L, et al (2012) Hydrogen-rich saline attenuates radiation-induced male germ cell loss in mice through reducing hydroxyl radicals. Biochem J 442: 49–56.

53. Ohno K, Ito M, Ichihara M, Ito M (2012) Molecular hydrogen as an emerging therapeutic medical gas for neurodegenerative and other diseases. Oxid Med Cell Longev 2012: Article ID 353152.

54. Yan H, Tian H, Kinjo T, Hamasaki T, Tomimatsu K, et al (2010) Extension of the lifespan of *Caenorhabditis elegans* by the use of electrolyzed reduced water. Biosci Biotechnol Biochem 74(10): 2011–2015.

55. Yoshinobu H, Arita S, Kawasaki S (2012) Molded activated charcoal and water purifier involving SAME. Patent application number: US20120132578.

56. Ministry of Health, Labour and Welfare, Japan. Available: http://www.mhlw.go.jp/shinsai_jo uhou/dl/leaflet_120329.pdf. Accessed Jul 18, 2013.

57. Christodouleas JP, Forrest RD, Ainsley CG, Tochner Z, Hahn SM, et al (2011) Short-term and long-term health risks of nuclear-power-plant accidents. N Engl J Med 364: 2334–41.

58. Ministry of Health, Labour and Welfare, Japan. Available: http://www.mhlw.go.jp/shingi/2002/10/s1007-5c.html. Accessed Jul 18, 2013.

59. Kametani K, Matsumura T, Naito M (1992) Separation of iodide and iodate by anion exchange resin and determination of their ions in surface water (In Japanese). Bunseki Kagaku 41: 337–341.

60. Watari K, Imai K, Ohmomo Y, Muramatsu Y, Nishimura Y, et al (1988) Simultaneous adsorption of Cs-137 and I-131 from water and milk on "metal ferrocyanide-anion exchange resin". J Nucl Sci Technol 25(5): 495–499.

61. Kamei D, Kuno T, Sato S, Nitta K, Akiba T (2012) Impact of the Fukushima Daiichi Nuclear Power Plant accident on hemodialysis facilities: An evaluation of radioactive contaminants in water used for hemodialysis. Ther Apher Dial1 6(1): 87–90.

62. Zhu Y-G, Smolders E (2000) Plant uptake of radiocaesium: a review of mechanisms, regulation and application. J Exp Bot 51(351): 1635–1645.

63. Kato H, Onda Y, Teramage M (2012) Depth distribution of $^{137}$Cs, $^{134}$Cs, and $^{131}$I in soil profile after Fukushima Dai-ichi Nuclear Power Plant accident. J Environ Radioact 111: 59–64.

64. Ministry of Health, Labour and Welfare (2013–595): Available: http://www.mhlw.go.jp/stf/houdou/2r9852000002wvi2.html.Accessed Jul. 18, 2013.

65. Ministry of Agriculture, Forestry and Fisheries: Available: http://www.maff.go.jp/j/kanbo/joho/saigai/s_chosa/hinmoku_kekka.html.Accessed Jul. 18, 2013.

66. Spyropoulos BG, Misiakos EP, Fotiadis C, Stoidis CN (2011) Antioxidant properties of probiotics and their protective effects in the pathogenesis of radiation-induced enteritis and colitis. Dig Dis Sci 56: 285–294.

67. Ohsawa I, Ishikawa M, Takahashi K, Watanabe M, Nishimaki K, et al (2007) Hydrogen acts as a therapeutic antioxidant by selectively reducing cytotoxic oxygen radicals. Nat Med 13(6): 688–694.

68. Yan H, Kinjo T, Tian H, Hamasaki T, Teruya K, et al (2011) Mechanism of the lifespan extension of *Caenorhabditis elegans* by electrolyzed reduced water Participation of Pt nanoparticles. Biosci Biotechnol Biochem 75(7): 1295–1299.

69. Yoshihisa Y, Honda A, Zhao Q-L, Makino T, Abe R, et al (2010) Protective effects of platinum nanoparticles against UV-light-induced epidermal inflammation. Exp Dermatol 19: 1000–1006.

70. Hamasaki T, Kashiwagi T, Imada T, Nakamichi N, Aramaki S, et al (2008) Kinetic analysis of superoxide anion radical-scavenging and hydroxyl radical-scavenging activities of platinum nanoparticles. Langmuir 24: 7354–7364.

71. Nakanishi H, Hamasaki T, Kinjo T, Yan Hanxu, Nakamichi N, et al (2013) Low concentration platinum nanoparticles effectively scavenge reactive oxygen species in rat skeletal L6 cells. Nano Biomed Eng 5(2): 76–85.

72. Yoshihisa Y, Zhao Q-L, Hassan MA, Wei Z-L, Furuichi M, et al (2011) SOD/catalase mimetic platinum nanoparticles inhibit heat-induced apoptosis in human lymphoma U937 and HH cells. Free Radic Res 45(3): 326–335.

73. Li Y, Hamasaki T, Nakamichi N, Kashiwagi T, Komatsu T, et al (2011) Suppressive effects of electrolyzed reduced water on alloxan-induced apoptosis and type 1 diabetes mellitus. Cytotechnology 63: 119–131.

74. Kim M-J, Kim HK (2006) Anti-diabetic effects of electrolyzed reduced water in streptozotocin-induced and genetic diabetic mice. Life Sci 79: 2288–2292.

75. Li Y, Nishimura T, Teruya K, Maki T, Komatsu T, et al (2002) Protective mechanism of reduced water against alloxan-induced pancreatic $\beta$-cell damage: Scavenging effect against reactive oxygen species. Cytotechnology 40: 139–149.

76. Ye J, Li Y, Hamasaki T, Nakamichi N, Komatsu T, et al (2008) Inhibitory effect of electrolyzed reduced water on tumor angiogenesis. Biol Pharm Bull 31(1): 19–26.

77. Yan H, Kashiwaki T, Hamasaki T, Kinjo T, Teruya K, et al (2011) The neuroprotective effects of electrolyzed reduced water and its model water containing molecular hydrogen and Pt nanoparticles. BMC Proc 5 (Suppl 8): 69–70.

78. Kinjo T, Ye J, Yan H, Hamasaki T, Nakanishi H, et al (2012) Suppressive effects of electrochemically reduced water on matrix metalloproteinase-2 activities and in vitro invasion of human fibrosarcoma HT1080 cells. Cytotechnology 64: 357–371.

79. Teruya K, Myojin-Maekawa Y, Shimamoto F, Watanabe H, Nakamichi N, et al (2013) Protective effects of the fermented milk kefir on X-ray irradiation-induced intestinal damage in B6C3F1 mice. Biol Pharm Bull 36(3): 352–359.

80. Vijayalaxmi, Reiter RJ, Tan D-X, Herman TS, Thomas CR (2004) Melatonin as a radioprotective agent: A review. Int J Radiat Oncol Biol Phys 59(3): 639–653.

81. Akpolat M, Kanter M, Uzal MC (2009) Protective effects of curcumin against gamma radiation-induced ileal mucosal damage. Arch Toxicol 83: 609–617.

# 15

# Variability of Carbon and Water Fluxes Following Climate Extremes over a Tropical Forest in Southwestern Amazonia

**Marcelo Zeri**[1]*, **Leonardo D. A. Sá**[2], **Antônio O. Manzi**[3], **Alessandro C. Araújo**[4], **Renata G. Aguiar**[5], **Celso von Randow**[1], **Gilvan Sampaio**[1], **Fernando L. Cardoso**[6], **Carlos A. Nobre**[7]

**1** Centro de Ciência do Sistema Terrestre, Instituto Nacional de Pesquisas Espaciais, Cachoeira Paulista, SP, Brazil, **2** Centro Regional da Amazônia, Instituto Nacional de Pesquisas Espaciais, Belém, PA, Brazil, **3** Instituto Nacional de Pesquisas da Amazônia (INPA), Manaus, Amazonas, Brazil, **4** Embrapa Amazônia Oriental, Belém, Pará, Brazil, **5** Universidade Federal de Rondônia, Porto Velho, Rondônia, Brazil, **6** Universidade Federal de Rondônia, Ji-Paraná, Rondônia, Brazil, **7** Secretaria de Políticas e Programas de Pesquisa e Desenvolvimento, Ministério da Ciência, Tecnologia e Inovação, Brasília, DF, Brazil

## Abstract

The carbon and water cycles for a southwestern Amazonian forest site were investigated using the longest time series of fluxes of $CO_2$ and water vapor ever reported for this site. The period from 2004 to 2010 included two severe droughts (2005 and 2010) and a flooding year (2009). The effects of such climate extremes were detected in annual sums of fluxes as well as in other components of the carbon and water cycles, such as gross primary production and water use efficiency. Gap-filling and flux-partitioning were applied in order to fill gaps due to missing data, and errors analysis made it possible to infer the uncertainty on the carbon balance. Overall, the site was found to have a net carbon uptake of $\approx$5 t C ha$^{-1}$ year$^{-1}$, but the effects of the drought of 2005 were still noticed in 2006, when the climate disturbance caused the site to become a net source of carbon to the atmosphere. Different regions of the Amazon forest might respond differently to climate extremes due to differences in dry season length, annual precipitation, species compositions, albedo and soil type. Longer time series of fluxes measured over several locations are required to better characterize the effects of climate anomalies on the carbon and water balances for the whole Amazon region. Such valuable datasets can also be used to calibrate biogeochemical models and infer on future scenarios of the Amazon forest carbon balance under the influence of climate change.

**Editor:** Dafeng Hui, Tennessee State University, United States of America

**Funding:** M. Zeri is grateful to São Paulo Research Foundation (FAPESP) — grant 2011/04101-0 — for the support during the preparation of this manuscript. Leonardo Sá is particularly grateful to CNPq — Conselho Nacional de Pesquisa e Desenvolvimento Tecnológico — for his research grant (process 303.728/2010-8). The funders had no role in study design, data collection and analysis, decision to publish, or preparation of the manuscript.

**Competing Interests:** The authors have declared that no competing interests exist.

* E-mail: marcelo.zeri@inpe.br

## Introduction

The intra-annual variability of carbon and water fluxes over forest and pasture sites in the Amazon region have been reported in many studies in the last several decades. The area covered by the world's largest tropical forest includes sites with evergreen species, semi-deciduous and transitions to *Cerrado*, among other classifications [1]. Sites with distinct vegetation types or topographies – and subjected to different sums of rainfall – are also different regarding the annual trends of fluxes of carbon, evapotranspiration and sensible heat flux. Southern sites (between latitudes $10°$ and $20°$ S) tend to have longer dry seasons while northern locations (between the Equator and $10°$ S) receive more rainfall due to the proximity to the Intertropical Convergence Zone (ITCZ), a migrating band of clouds and precipitation over the Equator, and the proximity of the Atlantic Ocean where the incoming air provides the moisture that forms precipitation over the Amazon [2]. Different rainfall patterns and annual cycles of air temperature, vapor pressure deficit and incoming solar radiation contribute to different trends in the fluxes of carbon, water and heat between the surface and the atmosphere [3,4].

Previous works on water and heat fluxes for a group of forest and savanna sites across the Amazon region revealed that evapotranspiration increases during the dry season in sites with higher annual precipitation and shorter dry seasons [5]. This apparent contradiction seems to be explained by the higher availability of incoming radiation and the hypothesized ability of trees for reaching water deep into the soil [1,6–9]. On the other hand, savanna and pasture sites were reported to have decreased evapotranspiration during the dry period [5,10]. For the carbon fluxes, while an increase in net ecosystem exchange (NEE) during the dry season is reported in some studies [11,12], others report a decrease of carbon assimilation during the same period [10].

The effects of droughts in 2005 and 2010 on the carbon balance of Amazonian forests have been extensively reported in recent years. A green-up effect following the drought of 2005 [13] was hypothesized to be related to the ability of trees to extract water using deep roots [14]. However, this effect was not observed in another study, which concluded that only 11–12% of Amazonian forests subjected to the drought exhibited greening during the dry season of 2005 [15]. The drought of 2010 was associated with low precipitation in 40% of the vegetated area and low water levels in several rivers of the Amazon basin, such as Rio Negro, near Manaus [16,17]. As a consequence, a decline of 7% in net primary production (NPP) between July and September of 2010 were reported in a remote sensing study which found that 0.5 Pg C were not sequestered in that year due to the dry conditions [18]. Climate change and droughts in the last decade were related to a decline in global NPP, specifically related to increases in air temperature over the Amazon, which increased autotrophic respiration [19]. Finally, deforestation was also found to play a role in drought events, caused by disturbances in evapotranspiration which affect other regions via regional circulation patterns [20].

In this study, fluxes of carbon dioxide and evapotranspiration were investigated using a dataset composed of seven years of measurements in a forest site within the Jaru Biological Reserve, in Brazil. The time series of fluxes reported here are the longest ever reported for a tropical forest in the southwestern region of the Amazon, enabling the investigation of the impacts on fluxes of extreme climatic events that affected the region, such as the droughts of 2005 and 2010 and the rainy year of 2009 [21–25]. The intra-annual variability of carbon flux and evapotranspiration was described using monthly averages and compared to common drivers of the carbon and water cycles, such as air temperature, vapor pressure deficit and incoming solar radiation (direct and diffuse).

The tower is located $\approx 14$ km from the site described in other studies [5,10,26], making it possible to test the spatial homogeneity of the forest-atmosphere exchanges of carbon and water by the similarity of some results, such as mean daily cycles of fluxes. The partitioning of net ecosystem productivity (NEP) in gross primary production (GPP) and ecosystem respiration ($R_e$) allowed the investigation of the intra-annual trends of those components of the carbon cycle as well as different metrics of water use efficiency [27–29].

## Site and Data

Measurements were carried out at the Jaru Biological Reserve, near the city of Ji-Paraná, Rondônia, Brazil (Figure 1). Authorization for field studies in this area was provided by IBAMA (National Institute of Environment and Renewable Resources). The tower was mounted in 2004 at the location marked with the red star in Figure 1 (10° 11' 21.2712" S, 61° 52' 15.1674" W, at 145 m above sea level), which is approximately 14 km south-southeast of the old location (orange star, zoomed in detail) up until November of 2002. The old tower, which was disassembled in 2002, was used in several experiments and studies [10,26,30–39] of the LBA (Large-Scale Biosphere-Atmosphere Experiment in Amazonia) project [30,34,40]. The forest was previously characterized as an open tropical rain forest with leaf area index ranging from 4 to 6 $m^2\,m^{-2}$ [30,41]. Trees are 35 m high, on average, but some reached up to 45 m. Soil depth at the old site ranged from 1 – 2 m and its texture was classified as sandy loam [30].

The new tower was equipped with an eddy covariance (EC) system and micrometeorological sensors. The EC system was installed at 63.4 m and included a 3D sonic anemometer (model Solent 1012R2, Gill Instruments, UK) and an open-path infra-red gas analyzer (IRGA) model LICOR 7500 (LICOR Inc., Nebraska, USA), both operating at 10.4 Hz. A wind vane placed at 62 m (model W200P, Vector Instruments, UK) was used for measurements of wind direction while a barometer (model PTB100A, Vaisala, Helsinki, Finland) was used for recording air pressure at 40 m. Wind speed was measured at several heights above and below the canopy using cup anemometers (model A100R, Vector

Instruments, UK) placed at 30, 41, 50.5 and 62.4 m. A vertical profile of thermohygrometers (model Temp107, Campbell, Logan, USA), for measurements of air temperature and relative humidity, was set up at heights 1.6, 11.2, 21.2, 33 and 49 m, while a model HMP45D (Vaisala) was placed at 61.5 m. Incoming and reflected shortwave solar radiation were measured with a pyranometer model CM21, from Kipp&Zonen (The Netherlands), while the incoming and emitted longwave components were measured with a pirgeometer model CG1 (Kipp&Zonen) mounted over arms extending from the tower. Both radiation sensors were installed at 57.5 m. The photosynthetically active component of solar radiation (PAR) was measured with a sensor model SKE 510 (Skye Instruments, UK), also mounted at 57.5 m. Lastly, a vertical profile of intakes for measurements of $CO_2$ concentration was setup in 2008 at the following heights: 62, 50, 34, 22, 12 and 2 m.

## Methodology

Fluxes were calculated using the eddy covariance technique [42–45], which is implemented in the software Alteddy (Jan Elbers, Alterra Group, Wageningen University, The Netherlands). The software was set up to apply the planar fit rotation [46,47] to the coordinate system in order to make the vertical velocity zero for different sectors of wind direction. The effects of humidity on the temperature measured by the sonic anemometer were corrected [48] while the influence of air density on the measurements from the infra-red gas analyzer were adjusted using the WPL correction [49–51]. In addition, known algorithms were applied to compensate for: a) losses in the high frequency end of the spectra due to spatial separation between sonic anemometer and IRGA [52–54]; b) the effects of heating of lenses in the open-path IRGA [55]. Finally, quality control of time series was based on the level of stationarity, i.e., the variability of statistical moments over time [56–58]. Stationarity was calculated [59] and summarized in flags from 1 to 9, which are proportional to the level of non-stationarity. For example, fluxes with flags ranging from 1 to 3 may have up to 50% of variability in their statistical moments during the period of 30 min used for averages. Data with flags 1–5 were accepted based on the good energy balance closure of this range, which was previously reported for the old Jaru forest site [26].

Recent developments in the dynamics of air flow past a sonic anemometer's body led to new findings about the errors of vertical velocity measurements. As a result, different configurations of an anemometer's transducers (orthogonal or non-orthogonal) may have an impact on fluxes [60,61]. Those errors can be on the order of $\approx 10\%$ for sensible heat flux and can propagate to other fluxes through direct measurements or corrections. Due to their recent nature and uncertain impact on fluxes, such corrections were not included in the calculations. However, we expect that the errors calculated in this work (random and gap-filling errors, described next) will account for most of the uncertainty on annual sums of evapotranspiration and carbon fluxes.

The balance of carbon in an ecosystem – the net ecosystem production, NEP – can be calculated as follows:

$$NEP = \overline{w'c'} + \int_0^{zi} \frac{\partial \overline{c}}{\partial t} dz \qquad (1)$$

where $w'$ and $c'$ are the departures from the mean of vertical velocity and concentration of $CO_2$, respectively, $zi$ is the measurement height and $z$ is the vertical coordinate. Positive

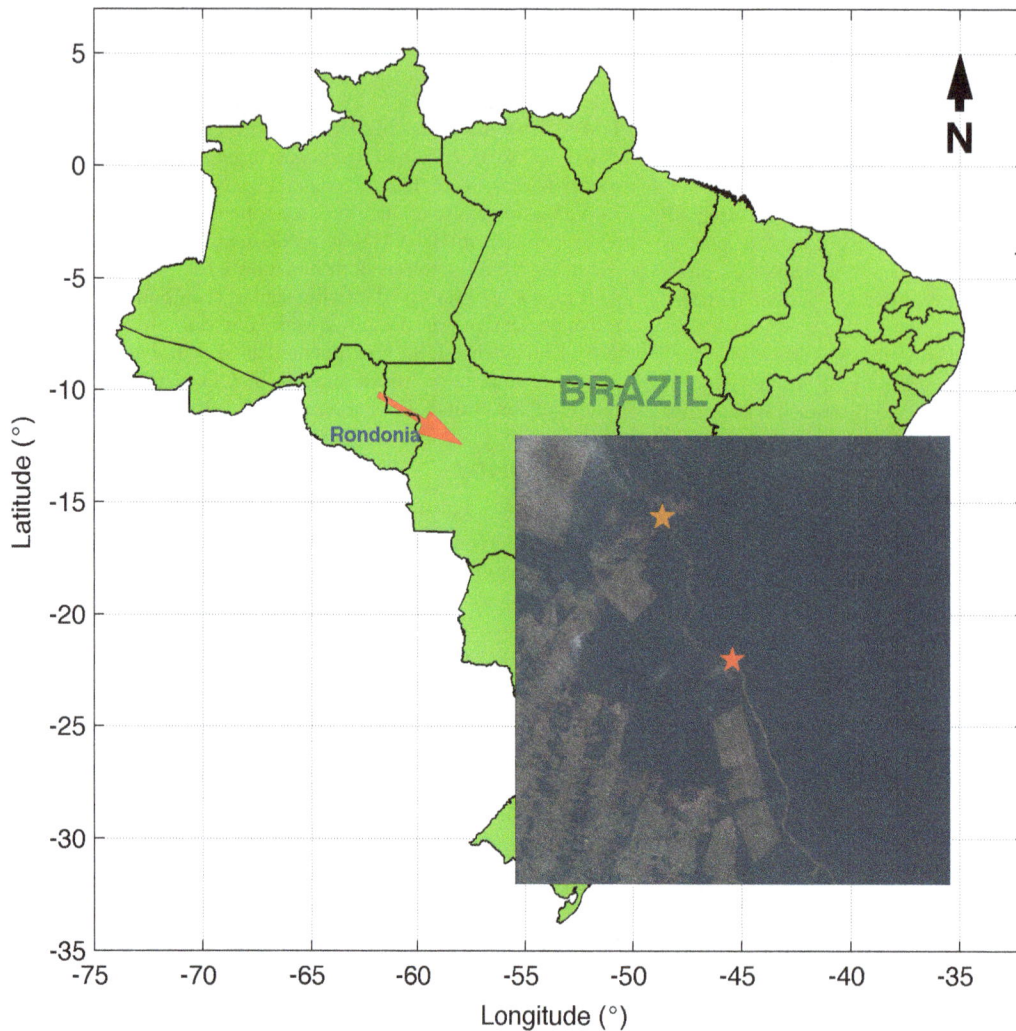

**Figure 1. Location of the tower marked with the red star in the detail.** The old tower was located at approximately 14 km northwest of the current location (orange star). Satellite picture recorded by Landsat 7 on October 1st 2002.

values of NEP denote accumulation of carbon by the ecosystem, according to the biological convention. The first term in Equation 1 is the eddy covariance flux, which accounts for the exchanges of carbon by the fast turbulent motions. The second term accounts for the storage of carbon below the measurement point at the top of the tower during conditions of low turbulent motions. The storage of carbon is usually calculated from a vertical profile of $CO_2$ concentrations above and below the canopy. The changes in concentration from one half-hour to the next are integrated vertically and contribute to a large fraction of NEP around sunrise due to the stratification of cold – and $CO_2$-enriched – air below the canopy during calm nighttime conditions.

The vertical profiles of $CO_2$ concentration were not available during the first four years of data, from 2004 to 2007. For this reason, an artificial time series of storage was calculated based on the average values of 2008 to 2010. First, the mean daily cycle of storage was calculated for each month from 2008 to 2010. Next, the daily cycles were grouped by month and averaged over the years. The resulting twelve diurnal cycles were then replicated to fill the respective month, creating a series of 30 minutes averages from January 1st to December 31st. The artificial series represented well the real data when comparing the annual impact on the

carbon balance: while the average annual sum of storage from 2008 to 2010 was an uptake of $\approx 1.7$ t C ha$^{-1}$ year$^{-1}$, the annual sum resulting from the artificial storage was of $\approx 1.9$ t C ha$^{-1}$ year$^{-1}$. The contribution of the artificial storage is likely to be small to the carbon balance of the first years since its magnitude is close to the average uncertainty derived from the error analysis and gap-filling, which was estimated to be $\approx \pm 1.7$ t C ha$^{-1}$ year$^{-1}$.

The validity of the eddy covariance method relies on the sufficient intensity of wind speed and turbulence in the surface layer, so that vertical exchanges can be averaged over several vortices passing by the tower [62]. The level of turbulence can be inferred by the value of $u_*$, the friction velocity, which is calculated as $u_* = \left[ \overline{(w'u')}^2 \right]^{1/4}$, where $\overline{w'v'}$ is the longitudinal vertical flux of momentum. The transversal component of the vertical flux $\overline{w'v'}$ is ignored after rotating the coordinate system to follow the average wind direction [47]. Nighttime conditions usually have light winds and low levels of turbulence, resulting in underestimated fluxes and high values $CO_2$ storage below the measuring height (Figure 2). The curves in Figure 2 are used to estimate the $u_*$-threshold, used to filter out nighttime or daytime fluxes which are later replaced by modeled values in the gap-filling analysis [63,64].

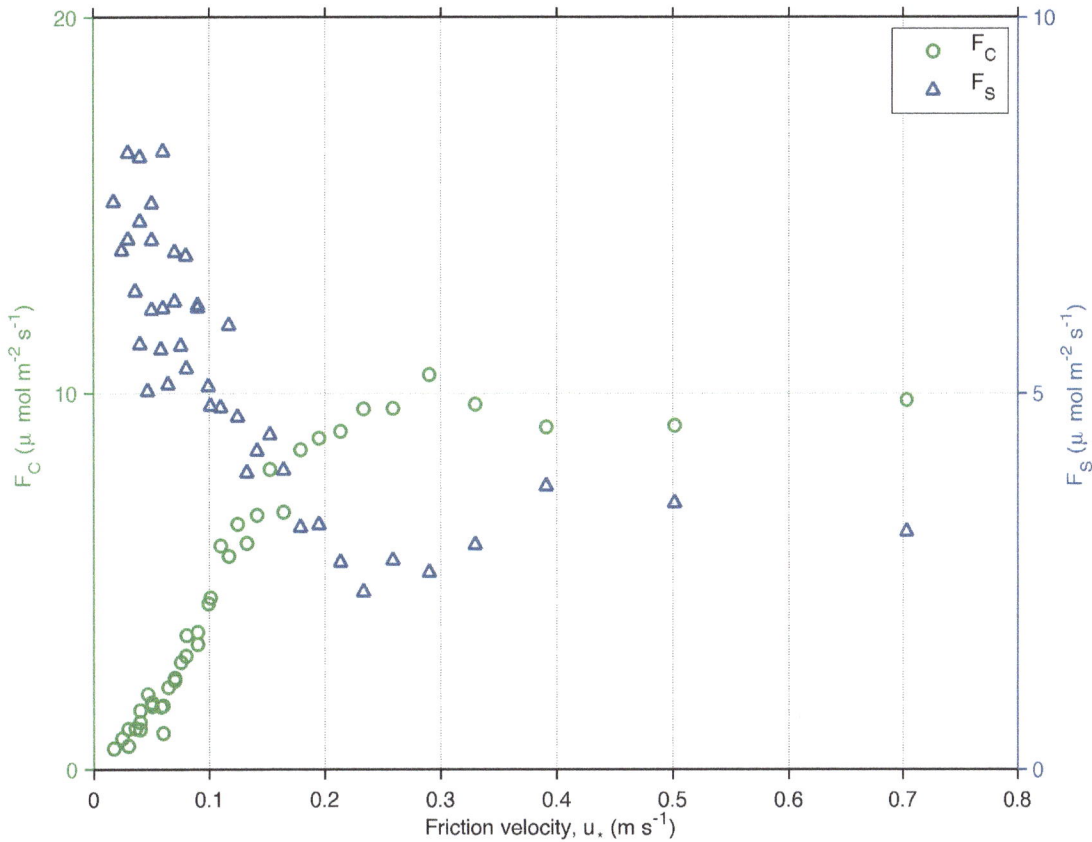

**Figure 2. CO₂-flux (F_C) and storage (F_S) plotted versus classes of friction velocity.** Atmospheric convention used to denote nocturnal respiration as positive. Only data from 2009 was used for the averages since this was the year with longer data availability of vertical profiles of $CO_2$.

However, the choice of the threshold can be subjective and may change the carbon balance depending on the fraction of data replaced by models [26,65]. Here, we chose the threshold as 0.1 m s⁻¹, a value that separates the top 60% of the storage values. A test of sensitivity to this choice was made and the results are presented in the section Results and discussion.

In recent years, two additional terms were proposed to equation 1 to account for horizontal and vertical advection, which are contributions to the flux caused by slopes in the terrain around the tower [66–71]. An inclined terrain can lead to horizontal and vertical transports of air that are usually small in flat areas and hence not considered in traditional eddy covariance applications. The effects of advection in Amazonian sites were already investigated over sites with complex topography [72,73]. An

indication of the importance of advection to a site is the curve of $CO_2$ storage versus friction velocity, as shown in Figure 2. If the storage is high for low values of u∗, then cold air is not being flushed out of the ecosystem by horizontal transports and thus advection is not occurring. Since this is the case in Figure 2, it is clear that the effects of advection can be ignored for this site.

Finally, the diffuse component of PAR – caused by scattering of light by clouds, aerosols, smoke or other particles [37,74,75] – was calculated using total PAR and the clearness index, an indicator of cloudiness used in solar radiation research. This index is defined as the ratio of global solar radiation at the surface over the radiation received from the Sun at the top of the atmosphere [76]. In addition, the calculations make use of characteristics of the site

**Table 1.** Accumulated rainfall and annual sums of total water use (TWU), gross primary production (GPP) and net ecosystem production (NEP) for each seasonal year (integration from September to August).

| Year | Rainfall (mm) | TWU (mm) | GPP (t C ha-1 yr-1) | NEP (t C ha-1 yr-1) |
|------|---------------|----------|---------------------|---------------------|
| 2004/2005 | 1552.8 | 1095.6±39.8 | 22.1±0.6 | 1.7±0.7 |
| 2005/2006 | 1683.8 | 1000.4±119.1 | 20.0±1.9 | −0.7±1.9 |
| 2006/2007 | 2114.4 | 1224.7±101.5 | 22.0±1.1 | 3.0±1.0 |
| 2007/2008 | 1975 | 1231.9±88.7 | 22.0±1.8 | 4.3±2.6 |
| 2008/2009 | 1964.8 | 1378.6±61.7 | 22.7±0.8 | 6.3±1.3 |
| 2009/2010 | 1861.4 | 1321.7±41.3 | 22.7±1.6 | 4.8±2.5 |

**Table 2.** Similar to Table 1, but using the calendar year, which uses data integrated from January to December.

| Year | Rainfall (mm) | TWU (mm) | GPP (t C ha$^{-1}$ yr$^{-1}$) | NEP (t C ha$^{-1}$ yr$^{-1}$) |
|------|---------------|----------|-------------------------------|-------------------------------|
| 2004 | 2181.8 | 968.8±20.0 | 29.9±1.2 | 1.9±1.6 |
| 2005 | 1315.2 | 1172.8±9.1 | 33.5±0.6 | 2.2±1.0 |
| 2006 | 2075.1 | 1153.4±48.9 | 29.7±1.9 | −1.2±2.1 |
| 2007 | 1942 | 804.3±22.4 | 35.7±1.1 | 4.8±1.1 |
| 2008 | 1782.2 | 1195.0±24.7 | 36.0±1.7 | 6.0±2.2 |
| 2009 | 2258.4 | 1498.7±13.4 | 35.4±0.9 | 10.4±1.2 |
| 2010 | 1551.2 | 961.7±27.2 | 38.7±1.8 | 7.4±2.5 |

such as solar elevation angle, latitude and declination of the Sun [76].

### Gap-filling and flux partitioning

Gaps in the time series of fluxes ($CO_2$, water vapor) and meteorological variables (air temperature, relative humidity, etc.)

need to be filled if annual sums of those variables are calculated. Short gaps – up to 1 hour – are filled by linear interpolation. Longer gaps are filled either by using the average daily cycle for each month, for meteorological variables, or by using other algorithms such as look-up tables, for fluxes. Initially, gaps in air temperature and other meteorological variables are filled since they are required in the gap-filling of fluxes. Next, it is applied an algorithm [26,64,77] to fill the fluxes of $CO_2$ and evapotranspiration. The algorithm is based on a "look-up table" approach, which tries to fill each gap with an average of good records taken on similar environmental conditions of net radiation, air temperature and vapor pressure deficit. The search starts within ±5 days from the gap and extends up to ±100 days, until it finds at least 5 records to average and fill the gap. In years with fewer long gaps, such as 2005 or 2009, the fraction of gaps due to the low quality flag or low u* was ≈40% before the filling process and 10% afterwards. For years with longer gaps due to instrument malfunction or maintenance, the fraction of missing data could reach up to 70% but the filling algorithm was still able to leave 10–20% of gaps. Those remaining gaps were filled with monthly mean daily cycles, for evapotranspiration, or with modeled fluxes for night and day, for $CO_2$ fluxes.

Daytime gaps were filled by light response curves of NEP versus PAR for two classes of air temperature: below and above 25°C.

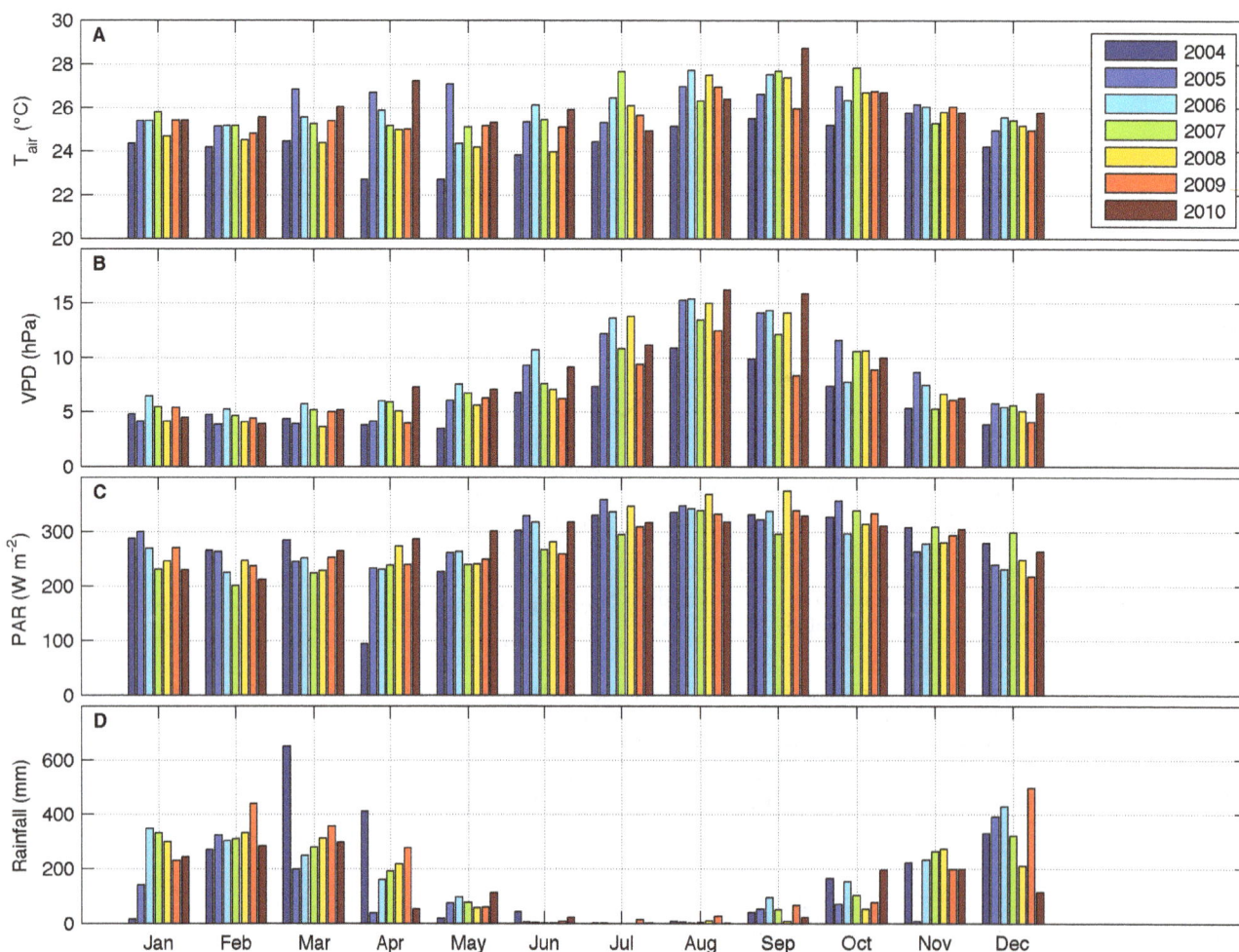

**Figure 3. Monthly averages (median) of meteorological drivers.** A: air temperature; B: vapor pressure deficit; C: maximum value of incoming PAR at noon; D: accumulated rainfall.

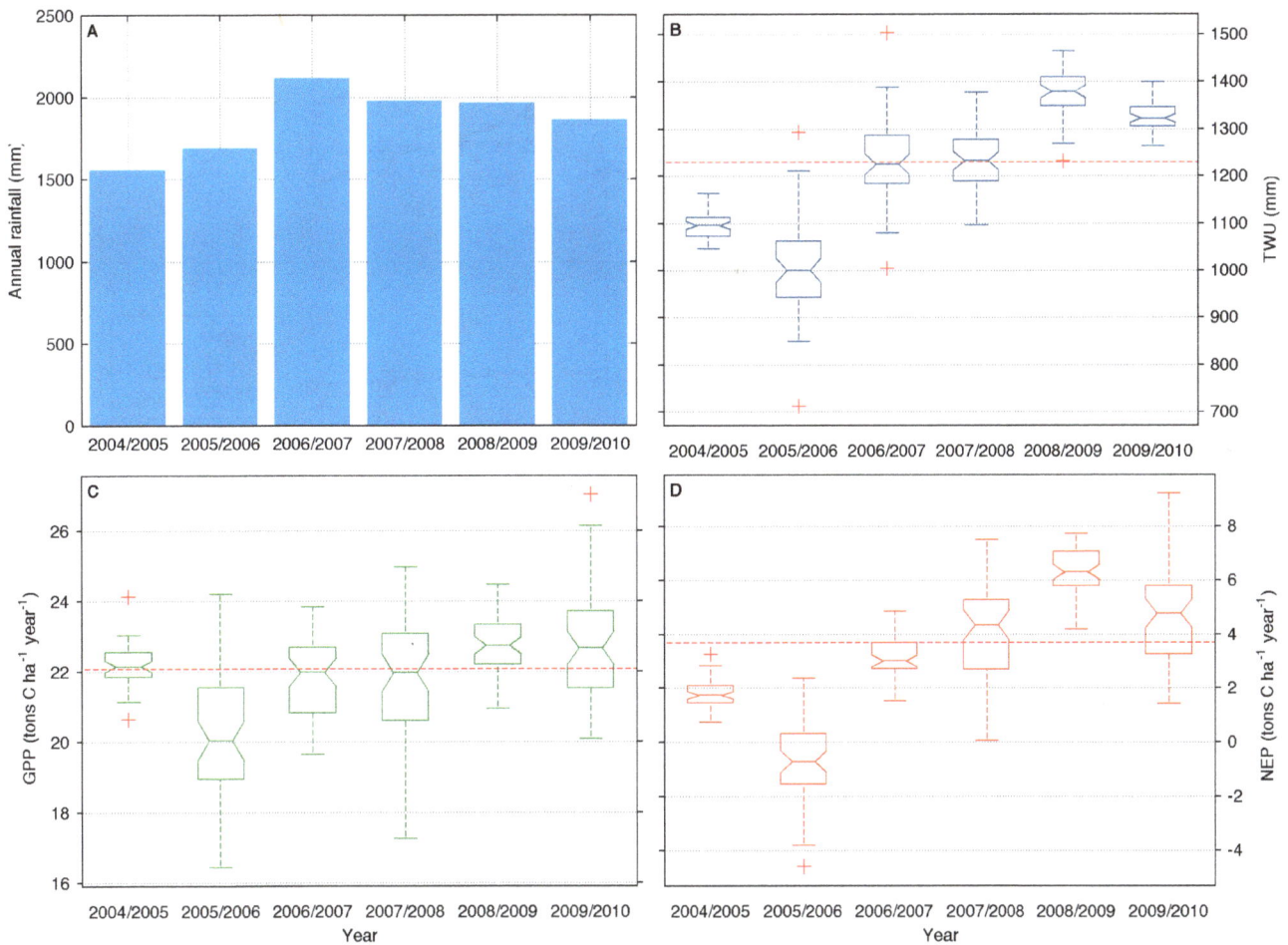

**Figure 4. Interannual variability of annual sums for precipitation (A), evapotranspiration, or total water used (B), gross primary production (C), and net ecosystem production (D).** Horizontal lines in panels B, C and D denote overall median. Labels in x-axis indicate seasonal year starting in September and ending in August (e.g. from Sep-2004 to Aug-2005).

Nighttime fluxes, or ecosystem respiration $R_e$, was modeled using an Arrhenius function [78]:

$$Re = R_{ref}\exp\left[E_0\left(\frac{1}{T_{ref} - T_0} - \frac{1}{T - T_0}\right)\right] \qquad (2)$$

where $R_{ref}$ is the respiration at the reference temperature $T_{ref}$, which was set to 293.15 K, $E_0$ is the activation energy and $T_0$ is a constant. The constant $T_0$ was set to 227.13 K, as previously reported [78]. The independent variable $T$ is referred to air temperature since no measurements of soil temperature were available at the site. The constants $E_0$ and $R_{ref}$ were calculated by using a non-linear least-squares regression method [64]. The ecosystem respiration calculated in Equation 2 in combination with the gap-filled NEP made it possible to calculate gross primary production (GPP) following the relation NEP = GPP − $R_e$. GPP was used in the analysis of intra-annual variability of fluxes and meteorological drivers, as well as in the calculation of monthly water use efficiency. For the analysis of annual sums (Table 1), the fluxes were integrated from September to August so that one cycle included full wet and dry seasons. This approach will be referred as the seasonal year to distinguish the periods in the discussions

that use the regular calendar year. Values computed using the regular calendar year are shown in Table 2.

Two metrics of water use efficiency were calculated using NEP, GPP and the amount of water used by the ecosystem in one year, i.e., total water use (TWU). TWU was calculated as the cumulative sum of evapotranspiration, which is also referred as the latent heat flux in atmospheric sciences. The first metric of water use efficiency was GWUE = GPP/TWU and the second was EWUE = NEP/TWU [27–29,79]. While GWUE measures the water use efficiency of the vegetation exclusively, EWUE measures the efficiency of the whole ecosystem, taking into account inputs and outputs of carbon.

The uncertainty in the annual fluxes was estimated by calculating the random ($\sigma_{rand}$) and gap-filling ($\sigma_{gap}$) errors. The random error was calculated from the variability of fluxes measured in successive days and under similar conditions [80]. Environmental variables such as net radiation, air temperature and vapor pressure deficit were used to select fluxes subjected to similar physical drivers. The random error $\sigma$ was estimated as the standard deviation of all differences, normalized by $\sqrt{2}$ [81]. Then, it was then averaged in several classes and a linear relationship with the magnitude of the $CO_2$ flux was found. Next, random noise from a normal distribution with mean 0 and standard deviation $\sigma$ was created and added to each class of flux,

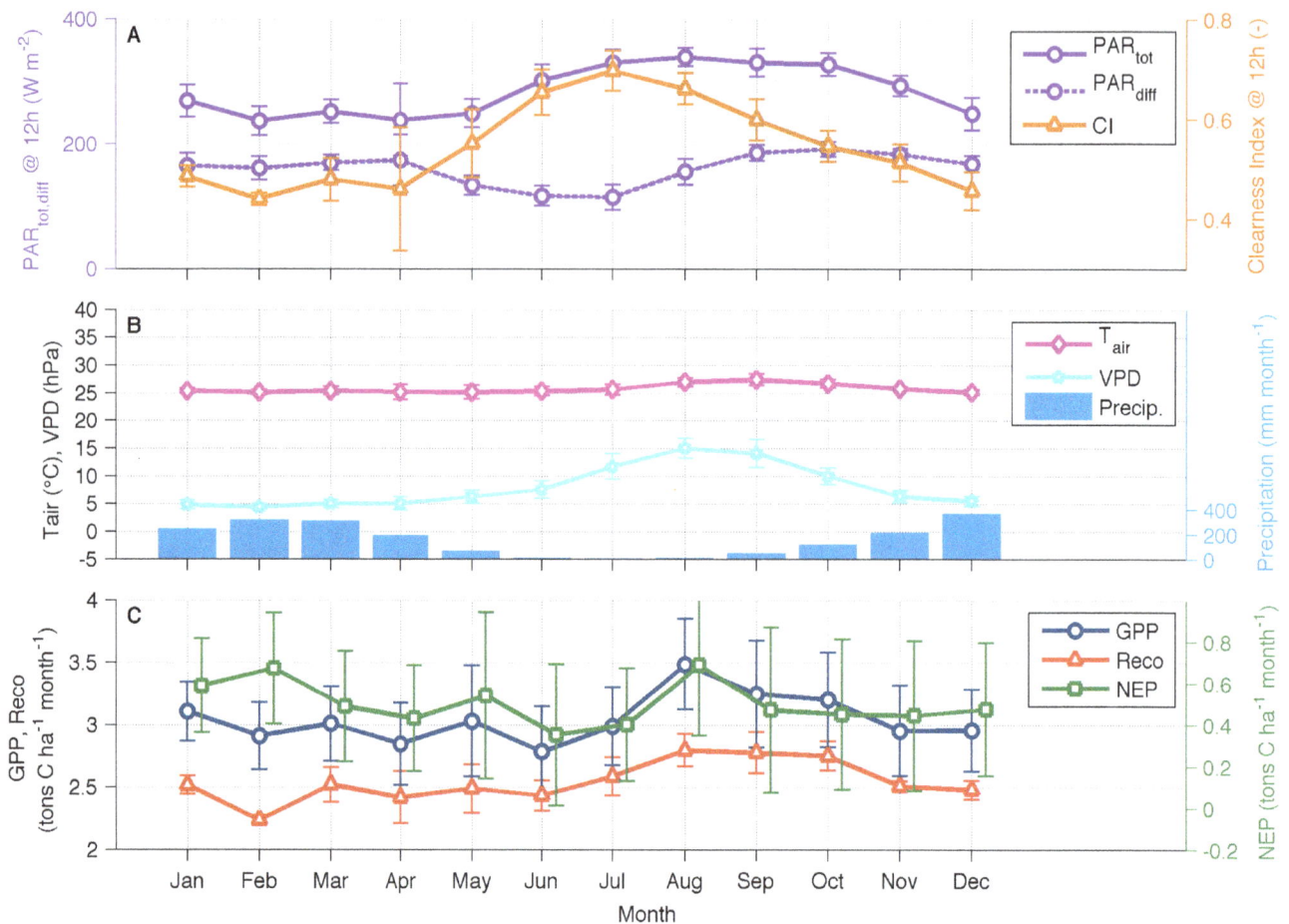

**Figure 5. Median intra-annual cycles of PAR (total and diffuse) at noon and clearness index (A); air temperature, vapor pressure deficit and precipitation (B); and gross primary production, net ecosystem production and ecosystem respiration (D).**

in order to have a synthetic flux with errors. This artificial flux was then gap-filled and its annual sum stored. Those steps were repeated 50 times, generating 50 versions of the original noisy flux, and the uncertainty due to the random error ($\sigma_{rand}$) was calculated as the standard deviation of all cumulative fluxes. The gap-filling error was calculated by inserting new gaps in the filled flux in the same proportion of the missing data found for day and night after filtering for high quality and turbulent conditions. First, the new random gaps were filled and the annual sum of the flux of $CO_2$ was calculated. Then, after 50 iterations, the gap-filling error $\sigma_{gap}$ was calculated as the standard deviation of all 50 annual sums.

The final error for the $CO_2$ flux was calculated as $\sqrt{\sigma_{rand}^2 + \sigma_{gap}^2}$.

## Results and Discussion

The most important meteorological drivers of fluxes (air temperature, vapor pressure deficit, radiation and precipitation) for the period of 2004 to 2010 are shown in Figure 3. On average, air temperature (Figure 3A) was $\approx 25°C$, from January to March and November to December, and $\approx 27°C$, from July to September. Vapor pressure deficit (Figure 3B) was highest, i.e. drier conditions, from June to October, the same period of the year when incoming PAR was maximal (Figure 3C). The high values of VPD and PAR are in synchronicity with the dry season at this site, as can be noticed by reduced precipitation in the period from May to October (Figure 3D). The remainder of the year is

characterized by abundant rainfall, contributing to lower values of air temperature, VPD and PAR, the latter caused by overcast skies.

Abnormal values for some variables are evident in some months when compared to the average of all seven years of data. The period of March to April of 2004 registered two of the three highest values of monthly accumulated precipitation in the dataset. It caused a minimum in PAR for April of 2004 most likely due to cloudy skies. For the following year, 2005, most of the wet season months (Jan, Mar, Apr, Sep, Oct, Nov) had the lowest values of rainfall, in accordance to the extensive drought that affected the Amazon region in this year [23]. Air temperature in March and May of 2005 was the highest among all years. On the other hand, the months of February, March, April and December of 2009 presented high values of accumulated precipitation.

The cumulative sums of rainfall, TWU, NEP and GPP (Figure 4, Table 1) help to explain the impacts of dry and wet years in the water and carbon cycles. The seasonal year was used in this figure, i.e., integration from September of 2004 to August of 2005. The median value of annual rainfall was 1913 mm, while the minimum was 1552.8 (2004/2005) and the maximum was 2114.4 (2006/2007). The minimum of rainfall was followed by another dry period in 2005/2006, which caused sharp drops in median values of TWU, GPP and NEP. As a result of this long drought, this forest was a net source of carbon for the period of 2005/2006, when NEP was $-0.7 \pm 1.9$ t C ha$^{-1}$ year$^{-1}$. The years that

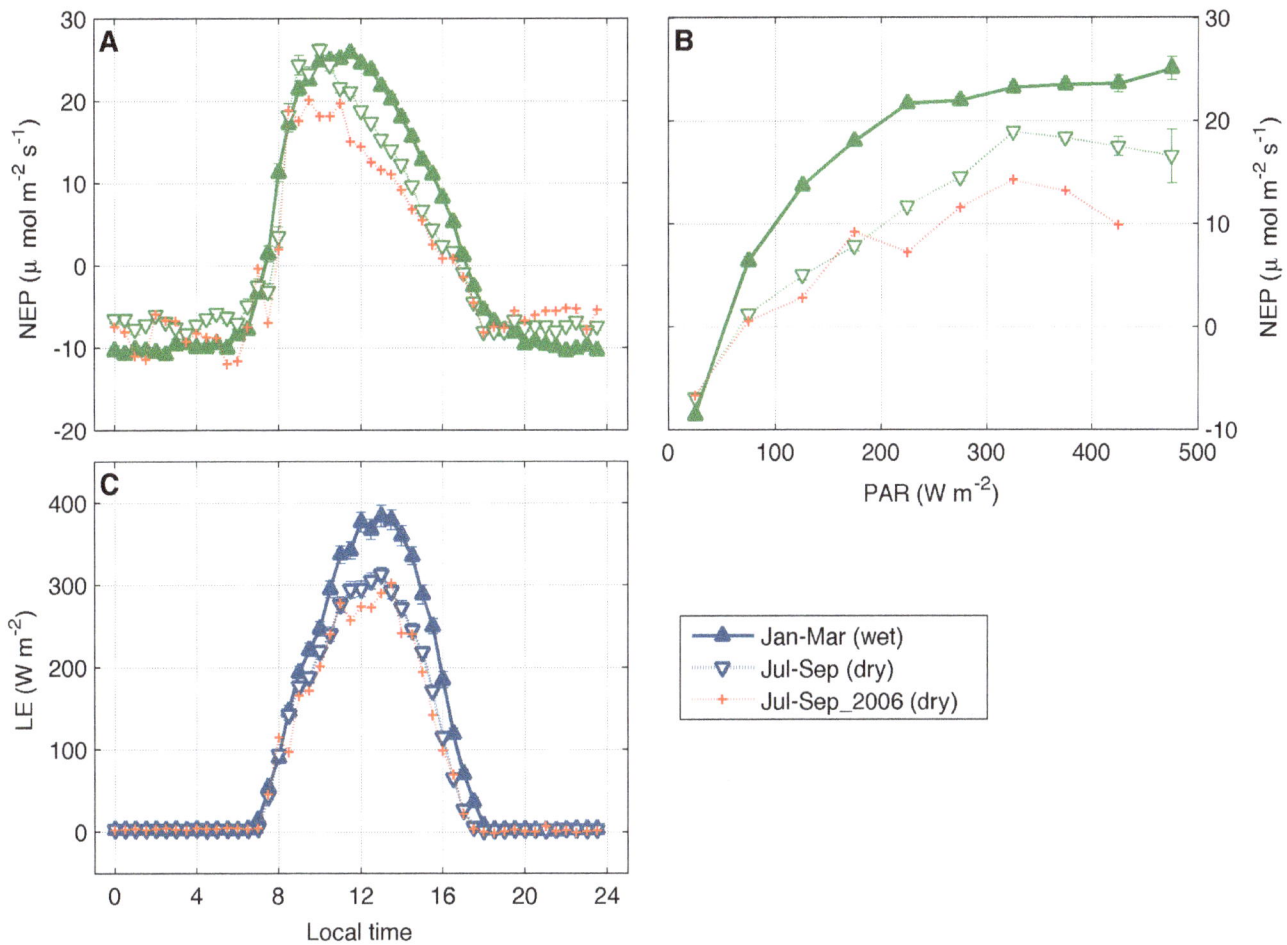

**Figure 6. Median daily cycles of net ecosystem production (A), NEP versus incoming PAR (B), and average daily cycles of latent heat flux (C).** Measured half-hourly values were averaged for hours of the day (A, C) and for classes of PAR.

followed this drought had above average precipitation, TWU, GPP and NEP. The peaks in TWU and NEP in 2008/2009 were likely influenced by the high availability of soil water and stand regeneration following three seasonal cycles with high precipitation [82]. The next cycle was characterized by a reduction in the annual rainfall caused by the drought of 2010 in the Amazon [21], which caused drops in TWU and NEP. However, those drops were smaller than the reduction in NEP and TWU immediately after the drought in 2004/2005. It is likely that a further reduction in NEP has occurred in 2011 (data not available), similar to the reduction in 2006.

The annual balances presented in Figure 4 are sensitive to the filtering used to remove nighttime conditions with low levels of turbulence. In general such filtering is based on a threshold of the friction velocity u*, below which values are replaced by modeled data in the gap-filling analysis. The results in Figure 4 are based on a u* threshold of 0.1 m s$^{-1}$. To test the sensitivity of this choice, the threshold was changed to 0.15 m s$^{-1}$ and the resulting carbon balance was calculated. The new threshold changed the annual carbon balance within 41% and 63% of the uncertainty generated by the error and gap-filling analysis, for 2004 and 2005, respectively. Hence, it is unlikely that a higher threshold would change the carbon balance beyond the uncertainty already determined.

The average intra-annual variability of some meteorological drivers helped to explain the variability of the carbon and water cycles (Figure 5). Total PAR, represented by the maximum at noon in Figure 5A, is highest in August, near the end of the dry season. The month of August is also the time of the year when the trend of clear-sky conditions reverses, as evident by the decrease in the clearness index and consequent increase of diffuse radiation. Despite the high values of $T_{air}$ and VPD in August, which forces leaves to close stomata and reduce photosynthesis, GPP and NEP present a temporary maximum at this time of the year. The increased ecosystem production and net uptake of carbon is most likely due to the combination of several factors, such as: the maximum in total PAR; the increase in diffuse PAR – which is known to be highly effective for NEP since light is able to reach leaves not directly exposed to the Sun [74,75,83]; new leaves being produced at the end of the dry season; and the availability of water via deep roots. It has been reported before that trees can reach water deep into the soil [7,14,84], maintaining high levels of productivity even during the dry season provided that the soil is recharged by rainfall each year during the wet season. Nonetheless, the peak in NEP and GPP was not sustained in September, on average, most likely due to the reduction in the availability of soil water at the end of the dry season.

Despite the temporary increase in NEP and GPP at the end of the dry season, this ecosystem has lower net uptake of carbon

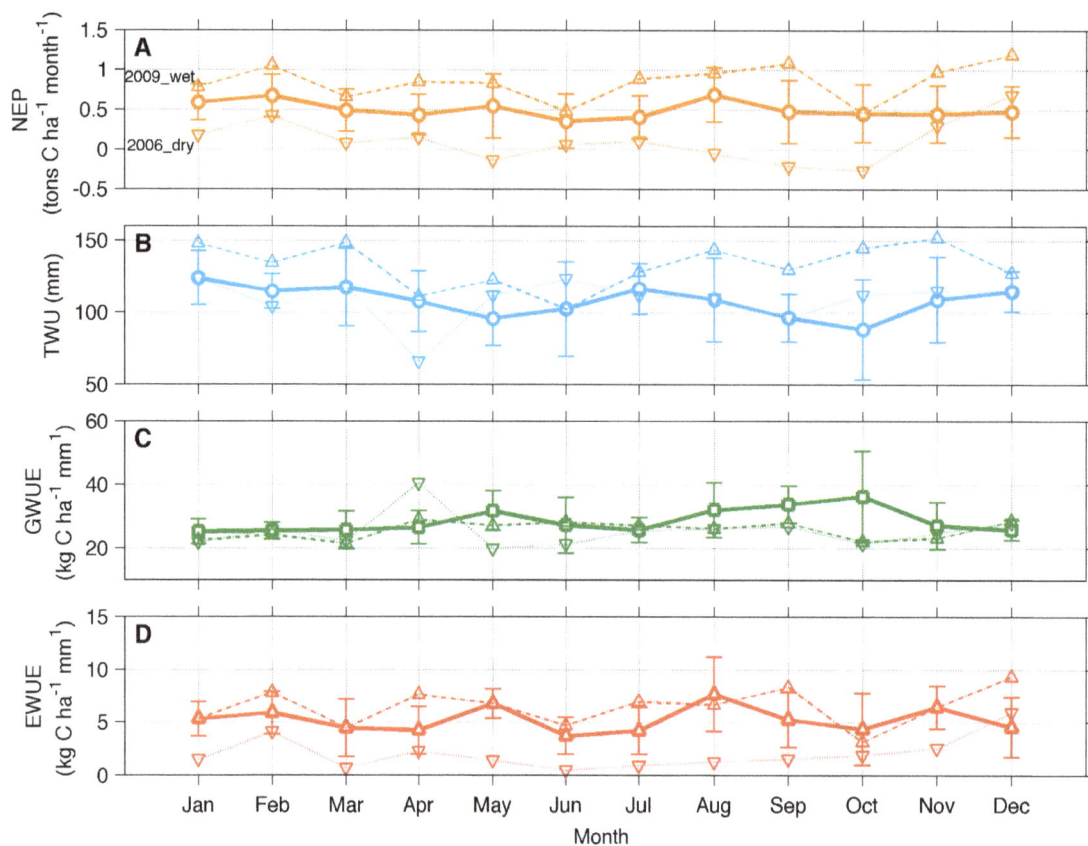

**Figure 7. Annual cycles of NEP (A), TWU (B), GWUE (C), and EWUE (D), for average conditions and for years under anomalous climate conditions (dry 2006 and wet 2009).** Vertical bars denote the interquartile range.

during this period, as can be seen in Figure 6, after averaging the daily cycle of NEP and evapotranspiration over a wet (Jan-Mar) and a dry period (Jul-Sep). Those periods were the same used in a previous accounting of the carbon balance for the old tower at this forest [10]. Daytime NEP is lowest from late morning to late afternoon, indicating lower net uptake of carbon (Figure 6A). The evolution of NEP versus PAR (Figure 6B) confirms the reduction in NEP during the dry season due to a combination of higher temperature, higher VPD and drier soils. In fact, the lower daytime evapotranspiration during the dry season confirms the lower vertical transport of water [85]. Nocturnal values of NEP during the dry season are less negative, indicating lower $CO_2$ emissions through soil respiration. This is caused by the drier soil, imposing limitations to microbial activity, which is responsible for $CO_2$ emissions from soils. Such lower emissions during nighttime are not enough to offset the lower uptake during daytime, causing the cumulative average NEP during the dry season (1.5 g C m$^{-2}$ day$^{-1}$) to be 21% lower than the cumulative for the wet season (1.9 g C m$^{-2}$ day$^{-1}$).

The dry season of 2006 presented a higher reduction on carbon and water fluxes compared to the usual reduction observed during a typical dry season (Figure 7). The impact of the drought of 2005 was strongest at this ecosystem during the dry season of 2006. Water limitations from September to November of 2005 likely reduced the recharge of soils to the next year. This impact is evident in the average daily cycle of NEP, its relation to PAR, and the daily cycle of latent heat flux (LE) in Figure 6. To further explore the effects of this climate extreme, the monthly variability of NEP and TWU for 2006 was compared with the average cycles

and with the wet year of 2009 (Figure 7A, B). In addition, the annual cycles of two water use efficiency ratios were investigated for both extreme years.

Monthly values of NEP were below average for most of 2006, a yearlong effect on fluxes likely caused by the water limitation experienced by this site in 2005 and during the first months of 2006 (February, March and April). This drought reduced net uptake of carbon most likely due to tree mortality and/or decomposition of fallen branches or trees. TWU was below or close to the average for the most part of the first semester of the year. The water use efficiencies (panels 7C,D) were also affected by the dry and wet years. For the dry year of 2006, most of the values of GWUE were below the average due to the decrease of GPP in that year (Figure 4C). For 2009, GPP did not strongly respond to the higher water availability and, when combined with the higher evapotranspiration (Figure 4B), resulted in a GWUE which was also below the average for most of the year. On the other hand, EWUE benefited from the wet year of 2009 with higher values above the monthly means, whereas the drought of 2006 contributed to lower values of EWUE. In conclusion, the drought that affected this forest in 2005 and 2006 strongly influenced the variability of carbon and water fluxes since it altered efficiencies of the ecosystem when using water and accumulating carbon.

## Conclusions

Previous works about the carbon balance of Amazon forest sites revealed that the annual uptake could vary from 1 to $\approx$6 t C ha$^{-1}$ year$^{-1}$ [10,86], while climate disturbances, such as droughts, may cause net release of carbon even during the wet season of the year [11]. The determination of the carbon balance of different forest ecosystems in the Amazon requires the long-term studies of fluxes in order to capture the typical conditions as well as transient influences such as droughts or floods. The time series of annual NEP analyzed in this work is the longest ever published for this ecosystem, making it possible to notice the influence of such extreme climate events.

According to results from the first year of measurements at this site [87] – 2004 – evapotranspiration during the dry season (from July to September) decreased by 20%, which is similar to the decrease of $\approx$ 22% found in this study when comparing the maximum values at midday for wet and dry seasons (Figure 6). In addition, the integration of carbon fluxes in that study resulted in an annual uptake of carbon of $\approx$5 t C ha$^{-1}$ year$^{-1}$, a value close to the range reported here (4.7±1.7 t C ha$^{-1}$ year$^{-1}$, Table 1).

The carbon flux at this site has a peculiar annual cycle with an increase in net uptake of carbon at the end of the dry season. This lack of synchronicity with the monthly-accumulated rainfall disrupts the positive correlation of the carbon cycle with precipitation. Instead, the effects of incoming solar radiation, new leaves with increased assimilation rates, and probable water use through deep roots [14,84], enable this ecosystem to increase its net carbon uptake at the end of the dry season. Long-term measurements of soil moisture and soil respiration would surely add valuable information to the carbon and water balances of this ecosystem.

The response of this ecosystem to the climate anomalies of the last decade is in agreement with the results found in many studies of the larger-scale impacts of the droughts of 2005 and 2010 [17–20]. The annual carbon balance of this forest (NEP) ranged from a net source, in 2006, to an increasing sink afterwards. A decline in NEP was observed in 2009/2010, probably caused by the drought in 2010. In spite of the typical increase in NEP at the end of the dry season, such mechanism was not enough to revert the long-term rainfall deficit experienced in 2005. In fact, the dry conditions affected the following seasonal year (2005/2006), contributing to turn the ecosystem into a carbon source. Therefore, it is unlikely that a green-up effect could have occurred over this forest [15,17].

Different parts of the Amazon have distinct annual cycles of meteorological drivers, longer or shorter dry seasons, and consequently different responses to changes in the region's atmospheric conditions [5,6,88,89]. Based on the variability in the results for NEP, the annual carbon balance is close to 5 t C ha$^{-1}$ year$^{-1}$, with significant changes from year to year depending on climatological conditions. Continuous monitoring of the biogeochemical cycles at this site as well as other ecosystems in the Amazon would help to identify the typical carbon balance, when the ecosystem is not under the influence of climate extremes. Moreover, long-term series of fluxes and meteorological drivers are crucial for the calibration of biogeochemical models that simulate the exchanges of energy and scalars between the biosphere and the atmosphere.

## Acknowledgments

The authors are grateful to Daniele S. Nogueira for improving the readability of the text.

## Author Contributions

Conceived and designed the experiments: LDAS AOM ACA CAN. Performed the experiments: MZ RGA CvR FLC. Analyzed the data: MZ LDAS RGA. Contributed reagents/materials/analysis tools: AOM ACA RGA FLC GS. Wrote the paper: MZ LDAS AOM.

## References

1. Betts AK, Dias MAFS (2010) Progress in Understanding Land-Surface-Atmosphere Coupling from LBA Research. J Adv Model Earth Sy 2. DOI: 10.3894/James.2010.2.6.

2. Costa MH, Foley JA (1999) Trends in the hydrologic cycle of the Amazon Basin. J Geophys Res Atmos 104: 14189–14198. DOI: 10.1029/1998JD200126.

3. Betts AK, Fisch G, von Randow C, Silva Dias MAF, Cohen JCP, et al. (2009) The Amazonian boundary layer and mesoscale circulations. In: Keller M, editor. Amazonia and Global Change. Washington, D.C.: AGU.

4. Betts AK (2007) Coupling of water vapor convergence, clouds, precipitation, and land-surface processes. J Geophys Res Atmos 112. DOI: 10.1029/2006jd008191.

5. da Rocha HR, Manzi AO, Cabral OM, Miller SD, Goulden ML, et al. (2009) Patterns of water and heat flux across a biome gradient from tropical forest to savanna in Brazil. J Geophys Res Biogeosci 114: G00B12. DOI: 10.1029/2007jg000640.

6. da Rocha HR, Goulden ML, Miller SD, Menton MC, Pinto LDVO, et al. (2004) Seasonality of Water and Heat Fluxes over a Tropical Forest in Eastern Amazonia. Ecol Appl 14: 22–32. DOI: 10.1890/02-6001.

7. Bruno RD, da Rocha HR, de Freitas HC, Goulden ML, Miller SD (2006) Soil moisture dynamics in an eastern Amazonian tropical forest. Hydrological Processes 20: 2477–2489. DOI: 10.1002/hyp.6211.

8. Costa MH, Biajoli MC, Sanches L, Malhado ACM, Hutyra LR, et al. (2010) Atmospheric versus vegetation controls of Amazonian tropical rain forest evapotranspiration: Are the wet and seasonally dry rain forests any different? J Geophys Res Biogeosci 115: G04021. DOI: 10.1029/2009jg001179.

9. Hutyra LR, Munger JW, Saleska SR, Gottlieb E, Daube BC, et al. (2007) Seasonal controls on the exchange of carbon and water in an Amazonian rain forest. J Geophys Res Biogeosci 112. DOI: 10.1029/2006JG000365.

10. von Randow C, Manzi AO, Kruijt B, de Oliveira PJ, Zanchi FB, et al. (2004) Comparative measurements and seasonal variations in energy and carbon exchange over forest and pasture in South West Amazonia. Theor Appl Clim 78: 5–26.

11. Saleska SR, Miller SD, Matross DM, Goulden ML, Wofsy SC, et al. (2003) Carbon in Amazon forests: unexpected seasonal fluxes and disturbance-induced losses. Science 302: 1554–1557. DOI: 10.1126/science.1091165.

12. Goulden ML, Miller SD, da Rocha HR, Menton MC, de Freitas HC, et al. (2004) Diel and seasonal patterns of tropical forest CO2 exchange. Ecol Appl 14: S42–S54.

13. Saleska SR, Didan K, Huete AR, da Rocha HR (2007) Amazon Forests Green-Up During 2005 Drought. Science 318: 612–612.

14. Nepstad DC, de Carvalho CR, Davidson EA, Jipp PH, Lefebvre PA, et al. (1994) The role of deep roots in the hydrological and carbon cycles of Amazonian forests and pastures. Nature 372: 666–669.

15. Samanta A, Ganguly S, Hashimoto H, Devadiga S, Vermote E, et al. (2010) Amazon forests did not green-up during the 2005 drought. Geophys Res Lett 37.

16. Lewis SL, Brando PM, Phillips OL, van der Heijden GMF, Nepstad D (2011) The 2010 Amazon Drought. Science 331: 554–554.

17. Xu L, Samanta A, Costa MH, Ganguly S, Nemani RR, et al. (2011) Widespread decline in greenness of Amazonian vegetation due to the 2010 drought. Geophys Res Lett 38: L07402. DOI: 10.1029/2011GL046824.

18. Potter C, Klooster S, Hiatt C, Genovese V, Castilla-Rubio JC (2011) Changes in the carbon cycle of Amazon ecosystems during the 2010 drought. Environmental Research Letters 6: 034024.

19. Zhao M, Running SW (2010) Drought-Induced Reduction in Global Terrestrial Net Primary Production from 2000 Through 2009. Science 329: 940–943.

20. Bagley JE, Desai AR, Harding KJ, Snyder PK, Foley JA (2013) Drought and Deforestation: Has land cover change influenced recent precipitation extremes in the Amazon? J Clim. DOI: 10.1175/JCLI-D-12-00369.1.

21. Marengo JA, Tomasella J, Alves LM, Soares WR, Rodriguez DA (2011) The drought of 2010 in the context of historical droughts in the Amazon region. Geophys Res Lett 38. DOI: 10.1029/2011gl047436.

22. Zeng N, Yoon JH, Marengo JA, Subramaniam A, Nobre CA, et al. (2008) Causes and impacts of the 2005 Amazon drought. Environmental Research Letters 3. DOI: 10.1088/1748-9326/3/1/014002.

23. Marengo JA, Nobre CA, Tomasella J, Oyama MD, Oliveira GS, et al. (2008) The drought of Amazonia in 2005. J Clim 21: 495–516. DOI: 10.1175/2007JCLI1600.1.

24. Marengo JA, Nobre CA, Tomasella J, Cardoso MF, Oyama MD (2008) Hydro-climate and ecological behaviour of the drought of Amazonia in 2005. Philos Trans R Soc Lond B Biol Sci 363: 1773–1778. DOI: 10.1098/rstb.2007.0015.

25. Marengo J, Tomasella J, Soares WR, Alves LM, Nobre CA (2012) Extreme climatic events in the Amazon basin. Theor Appl Clim 107: 73–85.

26. Zeri M, Sá LDA (2010) The impact of data gaps and quality control filtering on the balances of energy and carbon for a Southwest Amazon forest. Agric For Meteorol 150: 1543–1552. DOI: 10.1016/j.agrformet.2010.08.004.

27. Suyker AE, Verma SB (2009) Evapotranspiration of irrigated and rainfed maize-soybean cropping systems. Agric For Meteorol 149: 443–452. DOI: 10.1016/J.Agrformet.2008.09.010.

28. Zeri M, Hussain MZ, Anderson-Teixeira KJ, DeLucia EH, Bernacchi CJ (2013) Water use efficiency of perennial and annual bioenergy crops in central Illinois. J Geophys Res Biogeosci 118: 1–9. DOI: 10.1002/jgrg.20052.

29. Keenan TF, Hollinger DY, Bohrer G, Dragoni D, Munger JW, et al. (2013) Increase in forest water-use efficiency as atmospheric carbon dioxide concentrations rise. Nature. DOI: 10.1038/nature12291.

30. Andreae MO, Artaxo P, Brandao C, Carswell FE, Ciccioli P, et al. (2002) Biogeochemical cycling of carbon, water, energy, trace gases, and aerosols in Amazonia: The LBA-EUSTACH experiments. J Geophys Res Atmos 107: LBA 33-31–LBA 33-25. DOI: 10.1029/2001jd000524.

31. Bolzan MJA, Vieira PC (2006) Wavelet analysis of the wind velocity and temperature variability in the Amazon forest. Braz J Phys 36: 1217–1222. DOI: 10.1590/S0103-97332006000700018.

32. Bolzan MJA, Ramos FM, Sa LDA, Neto CR, Rosa RR (2002) Analysis of fine-scale canopy turbulence within and above an Amazon forest using Tsallis' generalized thermostatistics. J Geophys Res Atmos 107. DOI: 10.1029/2001JD000378.

33. Campanharo AS, Ramos FM, Macau EE, Rosa RR, Bolzan MJ, et al. (2008) Searching chaos and coherent structures in the atmospheric turbulence above the Amazon forest. Philos Transact A Math Phys Eng Sci 366: 579–589. DOI: 10.1098/rsta.2007.2118.

34. Andreae MO, Rosenfeld D, Artaxo P, Costa AA, Frank GP, et al. (2004) Smoking Rain Clouds over the Amazon. Science 303: 1337–1342.

35. von Randow C, Sa LDA, Gannabathula PSSD, Manzi AO, Arlino PRA, et al. (2002) Scale variability of atmospheric surface layer fluxes of energy and carbon over a tropical rain forest in southwest Amazonia - 1. Diurnal conditions. J Geophys Res Atmos 107: 1–12.

36. Kruijt B, Elbers JA, von Randow C, Araujo AC, Oliveira PJ, et al. (2004) The robustness of eddy correlation fluxes for Amazon rain forest conditions. Ecol Appl 14: S113.

37. Yamasoe MA, von Randow C, Manzi AO, Schafer JS, Eck TF, et al. (2006) Effect of smoke and clouds on the transmissivity of photosynthetically active radiation inside the canopy. Atmos Chem Phys 6: 1645–1656. DOI: 10.5194/acp-6-1645-2006.

38. Fisch G, Tota J, Machado LAT, Dias MAFS, Lyra RFD, et al. (2004) The convective boundary layer over pasture and forest in Amazonia. Theor Appl Clim 78: 47–59. DOI: 10.1007/S00704-004-0043-X.

39. Zeri M, Sá LDA (2011) Horizontal and Vertical Turbulent Fluxes Forced by a Gravity Wave Event in the Nocturnal Atmospheric Surface Layer Over the Amazon Forest. Boundary-Layer Meteorol 138: 413–431. DOI: 10.1007/s10546-010-9563-3.

40. Dias MAFS, Rutledge S, Kabat P, Dias PLS, Nobre C, et al. (2002) Cloud and rain processes in a biosphere-atmosphere interaction context in the Amazon Region. J Geophys Res Atmos 107. DOI: 10.1029/2001JD000335.

41. Kruijt B, Malhi Y, Lloyd J, Norbre AD, Miranda AC, et al. (2000) Turbulence statistics above and within two Amazon rain forest canopies. Boundary-Layer Meteorol 94: 297–331. DOI: 10.1023/A:1002401829007.

42. Montgomery RB (1948) Vertical Eddy Flux Of Heat In The Atmosphere. J Meteorol 5: 265–274. DOI: 10.1175/1520-0469(1948)005<0265:vefohi>2.0.co;2.

43. Swinbank WC (1951) The Measurement Of Vertical Transfer Of Heat And Water Vapor By Eddies In The Lower Atmosphere. J Meteorol 8: 135–145. DOI: 10.1175/1520-0469(1951)008<0135:tmovto>2.0.co;2.

44. Goulden ML, Munger JW, Fan SM, Daube BC, Wofsy SC (1996) Measurements of carbon sequestration by long-term eddy covariance: Methods and a critical evaluation of accuracy. Global Change Biol 2: 169–182. DOI: 10.1111/J.1365-2486.1996.Tb00070.X.

45. Baldocchi DD, Hicks BB, Meyers TP (1988) Measuring Biosphere-Atmosphere Exchanges of Biologically Related Gases with Micrometeorological Methods. Ecology 69: 1331–1340. DOI: 10.2307/1941631.

46. Wilczak JM, Oncley SP, Stage SA (2001) Sonic anemometer tilt correction algorithms. Boundary-Layer Meteorol 99: 127–150.

47. Kaimal JC, Finnigan JJ (1994) Atmospheric boundary layer flows: their structure and measurement. New York: Oxford University Press. 289 p.

48. Schotanus P, Nieuwstadt FTM, Debruin HAR (1983) Temperature-Measurement with a Sonic Anemometer and Its Application to Heat and Moisture Fluxes. Boundary-Layer Meteorol 26: 81–93. DOI: 10.1007/Bf00164332.

49. Webb EK, Pearman GI, Leuning R (1980) Correction of Flux Measurements for Density Effects Due to Heat and Water-Vapor Transfer. Q J Roy Meteorol Soc 106: 85–100.

50. Leuning R (2006) The correct form of the Webb, Pearman and Leuning equation for eddy fluxes of trace gases in steady and non-steady state, horizontally homogeneous flows. Boundary-Layer Meteorol 123: 263–267. DOI: 10.1007/s10546-006-9138-5.

51. Gu L, Massman WJ, Leuning R, Pallardy SG, Meyers T, et al. (2012) The fundamental equation of eddy covariance and its application in flux measurements. Agric For Meteorol 152: 135–148. DOI: 10.1016/j.agrformet.2011.09.014.

52. Philip JR (1963) The Damping of a Fluctuating Concentration by Continuous Sampling through a Tube. Aust J Phys 16.

53. Moore CJ (1986) Frequency response corrections for eddy correlation systems. Boundary-Layer Meteorol 37: 17–35.

54. Leuning R, King KM (1992) Comparison of eddy-covariance measurements of CO2 fluxes by open- and closed-path CO2 analysers. Boundary-Layer Meteorol 59: 297–311. DOI: 10.1007/bf00119818.

55. Burba GG, McDermitt DK, Grelle A, Anderson DJ, Xu L (2008) Addressing the influence of instrument surface heat exchange on the measurements of CO2 flux from open-path gas analyzers. Global Change Biol 14: 1854–1876. DOI: 10.1111/j.1365-2486.2008.01606.x.

56. Wilks DS (1995) Statistical Methods In The Atmospheric Sciences: An Introduction. San Diego: Academic Press. 467 p.

57. Foken T, Wichura B (1996) Tools for quality assessment of surface-based flux measurements. Agric For Meteorol 78: 83–105. DOI: 10.1016/0168-1923(95)02248-1.

58. Vickers D, Mahrt L (1997) Quality control and flux sampling problems for tower and aircraft data. J Atmos Oceanic Tech 14: 512–526.

59. Foken T, Goeckede M, Mauder M, Mahrt L, Amiro BD, et al. (2004) Post-field data quality control. In: Lee X, Massman WJ, Law B, editors. Handbook of micrometeorology: a guide for surface flux measurement and analysis. Dordrecht, The Netherlands: Kluwer Academic Publishers. pp. 181–208.

60. Kochendorfer J, Meyers T, Frank J, Massman W, Heuer M (2012) How Well Can We Measure the Vertical Wind Speed? Implications for Fluxes of Energy and Mass. Boundary-Layer Meteorol 145: 383–398. DOI: 10.1007/s10546-012-9738-1.

61. Frank JM, Massman WJ, Ewers BE (2013) Underestimates of sensible heat flux due to vertical velocity measurement errors in non-orthogonal sonic anemometers. Agric For Meteorol 171–172: 72–81. DOI: 10.1016/j.agrformet.2012.11.005.

62. Stull RB (1988) An Introduction to Boundary Layer Meteorology. Dordrecht, Boston, London: Kluwer Academic Publishers. 666 p.

63. Falge E, Baldocchi D, Olson R, Anthoni P, Aubinet M, et al. (2001) Gap filling strategies for defensible annual sums of net ecosystem exchange. Agric For Meteorol 107: 43–69.

64. Reichstein M, Falge E, Baldocchi D, Papale D, Aubinet M, et al. (2005) On the separation of net ecosystem exchange into assimilation and ecosystem respiration: review and improved algorithm. Global Change Biol 11: 1424–1439.

65. Anthoni PM, Freibauer A, Kolle O, Schulze ED (2004) Winter wheat carbon exchange in Thuringia, Germany. Agric For Meteorol 121: 55–67. DOI: 10.1016/S0168-1923(03)00162-X.

66. Finnigan J (1999) A comment on the paper by Lee (1998): "On micrometeorological observations of surface-air exchange over tall vegetation". Agric For Meteorol 97: 55–64. DOI: 10.1016/S0168-1923(99)00049-0.

67. Lee XH (1998) On micrometeorological observations of surface-air exchange over tall vegetation. Agric For Meteorol 91: 39–49. DOI: 10.1016/S0168-1923(98)00071-9.

68. Lee XH, Hu XZ (2002) Forest-air fluxes of carbon, water and energy over non-flat terrain. Boundary-Layer Meteorol 103: 277–301. DOI: 10.1023/A:1014508928693.

69. Aubinet M, Heinesch B, Yernaux M (2003) Horizontal and vertical CO2 advection in a sloping forest. Boundary-Layer Meteorol 108: 397–417. DOI: 10.1023/A:1024168428135.

70. Feigenwinter C, Bernhofer C, Vogt R (2004) The influence of advection on the short term CO2-budget in and above a forest canopy. Boundary-Layer Meteorol 113: 201–224. DOI: 10.1023/B:Boun.0000039372.86053.Ff.

71. Feigenwinter C, Bernhofer C, Eichelmann U, Heinesch B, Hertel M, et al. (2008) Comparison of horizontal and vertical advective CO2 fluxes at three forest sites. Agric For Meteorol 148: 12–24. DOI: 10.1016/J.Agrformet.2007.08.013.

72. de Araújo AC, Dolman AJ, Waterloo MJ, Gash JHC, Kruijt B, et al. (2010) The spatial variability of CO2 storage and the interpretation of eddy covariance fluxes in central Amazonia. Agric For Meteorol 150: 226–237. DOI: 10.1016/j.agrformet.2009.11.005.

73. Tóta J, Fitzjarrald DR, Staebler RM, Sakai RK, Moraes OMM, et al. (2008) Amazon rain forest subcanopy flow and the carbon budget: Santarém LBA-ECO site. J Geophys Res Biogeosci 113. DOI: 10.1029/2007jg000597.

74. Doughty CE, Flanner MG, Goulden ML (2010) Effect of smoke on subcanopy shaded light, canopy temperature, and carbon dioxide uptake in an Amazon rainforest. Global Biogeochem Cycles 24. DOI: 10.1029/2009gb003670.

75. Bai YF, Wang J, Zhang BC, Zhang ZH, Liang J (2012) Comparing the impact of cloudiness on carbon dioxide exchange in a grassland and a maize cropland in northwestern China. Ecol Res 27: 615–623. DOI: 10.1007/S11284-012-0930-Z.

76. Gu L, Fuentes JD, Shugart HH, Staebler RM, Black TA (1999) Responses of net ecosystem exchanges of carbon dioxide to changes in cloudiness: Results from two North American deciduous forests. J Geophys Res Atmos 104: 31421–31434. DOI: 10.1029/1999jd901068.

77. Zeri M, Anderson-Teixeira K, Hickman G, Masters M, DeLucia E, et al. (2011) Carbon exchange by establishing biofuel crops in Central Illinois. Agric Ecosyst Environ 144: 319–329. DOI: 10.1016/j.agee.2011.09.006.

78. Lloyd J, Taylor JA (1994) On the Temperature-Dependence of Soil Respiration. Funct Ecol 8: 315–323.

79. VanLoocke A, Twine TE, Zeri M, Bernacchi CJ (2012) A regional comparison of water use efficiency for miscanthus, switchgrass and maize. Agric For Meteorol 164: 82–95. DOI: 10.1016/j.agrformet.2012.05.016.

80. Richardson AD, Hollinger DY (2007) A method to estimate the additional uncertainty in gap-filled NEE resulting from long gaps in the CO2 flux record. Agric For Meteorol 147: 199–208.

81. Richardson AD, Hollinger DY, Burba GG, Davis KJ, Flanagan LB, et al. (2006) A multi-site analysis of random error in tower-based measurements of carbon and energy fluxes. Agric For Meteorol 136: 1–18.

82. Malhi Y, Pegoraro E, Nobre AD, Pereira MGP, Grace J, et al. (2002) Energy and water dynamics of a central Amazonian rain forest. J Geophys Res Atmos 107. DOI: 10.1029/2001JD000623.

83. Butt N, New M, Malhi Y, da Costa ACL, Oliveira P, et al. (2010) Diffuse radiation and cloud fraction relationships in two contrasting Amazonian rainforest sites. Agric For Meteorol 150: 361–368. DOI: 10.1016/J.Agrformet. 2009.12.004.

84. Wright IR, Nobre CA, Tomasella J, da Rocha HR, Roberts JM, et al. (1996) Towards a GCM surface parameterization of Amazonia. In: Gash JHC, editor. Amazonian Deforestation and Climate. New York: John Wiley. pp. 473–504.

85. Betts AK (2004) Understanding hydrometeorology using global models. Bull Am Meteorol Soc 85: 1673–1688. DOI: 10.1175/Bams-85-11-1673.

86. Grace J, Lloyd J, McIntyre J, Miranda AC, Meir P, et al. (1995) Carbon Dioxide Uptake by an Undisturbed Tropical Rain Forest in Southwest Amazonia, 1992 to 1993. Science 270: 778–780.

87. Aguiar RG (2005) Fluxos de massa e energia em uma floresta tropical no sudoeste da Amazônia [Masters Thesis]. Cuiabá, MT: Universidade Federal de Mato Grosso. 78 p.

88. Baker IT, Prihodko L, Denning AS, Goulden M, Miller S, et al. (2008) Seasonal drought stress in the Amazon: Reconciling models and observations. J Geophys Res Biogeosci 113. DOI: 10.1029/2007jg000644.

89. Fisher JB, Malhi Y, Bonal D, Da Rocha HR, De Araujo AC, et al. (2009) The land-atmosphere water flux in the tropics. Global Change Biol 15: 2694–2714. DOI: 10.1111/J.1365-2486.2008.01813.X.

# 16

# Exploring Novel Bands and Key Index for Evaluating Leaf Equivalent Water Thickness in Wheat Using Hyperspectra Influenced by Nitrogen

**Xia Yao, Wenqing Jia, Haiyang Si, Ziqing Guo, Yongchao Tian, Xiaojun Liu, Weixing Cao, Yan Zhu***

National Engineering and Technology Center for Information Agriculture, Jiangsu Key Laboratory for Information Agriculture, Nanjing Agricultural University, Nanjing, Jiangsu, P. R. China

## Abstract

Leaf equivalent water thickness (LEWT) is an important indicator of crop water status. Effectively monitoring the water status of wheat under different nitrogen treatments is important for effective water management in precision agriculture. Trends in the variation of LEWT in wheat plants during plant growth were analyzed based on field experiments in which wheat plants under various water and nitrogen treatments in two consecutive growing seasons. Two-band spectral indices [normalized difference spectral indices (NDSI), ratio spectral indices (RSI), different spectral indices (DSI)], and then three-band spectral indices were established based on the best two-band spectral index within the range of 350–2500 nm to reduce the noise caused by nitrogen and saturation. Then, optimal spectral indices were selected to construct models of LEWT monitoring in wheat. The results showed that the two-band spectral index $NDSI(R_{1204}, R_{1318})$ could be used for LEWT monitoring throughout the wheat growth season, but the model performed differently before and after anthesis. Therefore, further two-band spectral indices $NDSIb(R_{1445}, R_{487})$, $NDSIa(R_{1714}, R_{1395})$, and $NDSI(R_{1429}, R_{416})$, were constructed for the two developmental phases, with $NDSI(R_{1429}, R_{416})$ considered to be the best index. Finally, a three-band index $(R_{1429}-R_{416}-R_{1865})/(R_{1429}+R_{416}+R_{1865})$, which was superior for monitoring LEWT and reducing the noise caused by nitrogen, was formed on the best two-band spectral index $NDSI(R_{1429}, R_{416})$ by adding the 1,865 nm wavelenght as the third band. This produced more uniformity and stable performance compared with the two-band spectral indices in the LEWT model. The results are of technical significance for monitoring the water status of wheat under different nitrogen treatments in precision agriculture.

**Editor:** Liuling Yan, Oklahoma State University, United States of America

**Funding:** This work was supported by the National High-tech Research and Development Program of China (863 Program) (2013AA102404,2011AA100703), National Natural Science Foundation of China (31201130, 31201131), Special Program for Agriculture Science and Technology from Ministry of Agriculture in China (201303109), Science and Technology Support Plan of Jiangsu Province (BE2011351, BE2012302), Jiangsu agriculture science and technology innovation fund, SCX(12)3272, and Academic Program Development of Jiangsu Higher Education Institutions (PAPD). The funders had no role in study design, data collection and analysis, decision to publish, or preparation of the manuscript.

**Competing Interests:** The authors have declared that no competing interests exist.

\* E-mail: yanzhu@njau.edu.cn

## Introduction

Real-time, non-destructive monitoring of crop water content based on hyperspectra is an important area of research in precision irrigation in agriculture [1–7]. As a widely used measure of crop water status, leaf-equivalent water thickness (LEWT) and canopy-equivalent water thickness (CEWT) not only directly indicate crop water content, but also provide information for leaf area indices. Therefore,they can visually reflect crop water requirements and crop growth status. CEWT has been found to be linearly related to the vegetation water content (VWC), with an $R^2$ value of 0.87 for corn [8]. LEWT was shown to have a better linear correlation with reflectance than fuel moisture content (FMC) in the leaves of 10 plant species [9]. In cowpeas, beans, and sugar beet, the $R_{1300}/R_{1450}$ leaf water index (ratio of reflectance at 1,300 to 1,450 nm) displayed a characteristic logarithmic correlation with LEWT [10]. During the late period of wheat development (after anthesis), LEWT is more useful than FMC for assessing the water status of wheat. Several optimal water

indices for different stages of wheat development are available [11].

Many studies in recent decades have aimed to evaluate the LEWT using remote sensing. Detection of plant water stress through remote sensing has been proposed using indices based on the near-infrared (NIR, 700–1,300 nm) and the middle-infrared (MIR, 1,300–2,500 nm) wavelengths. Hunt et al. found the moisture stress index (MSI) linearly correlated with the $log_{10}$LEWT of *Quercus agrifolia* (sclerophyllous leaves) and *Liquidambar styraciflua* (hardwood deciduous tree leaves), in which the regression equations were different [12]. Ceccato et al. reported that shortwave infrared (SWIR, 1,400–2,500 nm) was sensitive to LEWT, but could not be used alone to determine LEWT because two other leaf parameters (internal structure and dry matter) also influence leaf reflectance in the SWIR [13]. A combination of SWIR and NIR was necessary to determine LEWT. Gao proposed a new vegetation index, the normalized difference water index (NDWI), for the remote sensing of CEWT from space [14]. This index was constructed based on two narrow bands centered

**Table 1.** Basic information about different field experiments.

| Exp. Number | Year and Variety | Treatment (water treatments (W): %; nitrogen rates (N): kg/hm²)) | Spectrum and sample data | Soil (yellow brown soil) | Environmental conditions during the growing seasons |
|---|---|---|---|---|---|
| Exp. 1 (Calibration data set) | 2010–2011 Yangmai 18 | **4 W:** W1 (9.5–10.5), W2 (15.5–16.5), W3 (21.5–22.5), W4 (29.5–30.5). **2 N:** N1 (150), N2 (300) | Jointing (3.25), Booting (4.1), Heading (4.12), Anthesis (4.20), Filling (5.6) | Organic matter: 15.5 g kg⁻¹, Total N: 1.1 g kg⁻¹, Available P: 50.8 mg kg⁻¹, Available K: 89.6 mg kg⁻¹. | Mean temperature: 10.84°C Maximum temperature:15.75Minimum temperature:5.92Mean Diurnal temperature: 9.83°C Sunshine hours: 1302.5 hPrecipitation:155.14 mm |
| Exp. 2 (Validation data set) | 2011–2012 Yangmai 18 | **3 W:** W1 (13.5–14.5), W2 (21.5–22.5), W3 (29.5–30.5). **3 N:** N1 (90), N2 (180), N3 (270) | Jointing (3.24), Booting (4.3), Heading (4.11), Anthesis (4.16), Filling (4.22) | Organic matter: 14.8 g kg⁻¹, Total N: 1.1 g kg⁻¹, Available P: 50.4 mg kg⁻¹, Available K: 88.9 mg kg⁻¹. | Mean temperature: 10.98°C Maximum temperature:14.92Minimum temperature:7.04Mean Diurnal temperature: 7.88°C Sunshine hours: 999.003 hPrecipitation:329.51 mm |

near 860 nm and 1,240 nm and was used successfully to detect CEWT and LEWT in cotton and trees [15–18]. Based on the difference in reflectance between 945 nm and 975 nm and using Beer's law, Liu et al. calculated the radiation-equivalent water thickness of leaves (RLEWT) [19]. The authors demonstrated that RLEWT was significantly correlated with LEWT. Zarco et al. estimated LEWT from canopy-level reflectance with the simple ratio water index (SRWI), which had a strong correlation with LEWT [20]. Other researchers have proposed the normalized difference infrared index (NDII) [8,21] and the water index (WI) [22] to estimate LEWT and CEWT.

In addition to these two-band spectral indices, three-band spectral indices have been proposed to evaluate other growth parameters of plants. Schneider et al. [23] and Stow et al. [24] found the visible atmospherically resistant index (VARI) to be minimally sensitive to atmospheric effects and strongly related to live fuel moisture (LFM). Li et al. [25] and Jie et al. [26] found VARI-700 to be significantly correlated with yield at the whole development stages in cotton. Wang et al. constructed three-band vegetation indices, $(R_{\lambda 1} - R_{\lambda 2} + 2 \times R_{\lambda 3})/(R_{\lambda 1} - R_{\lambda 2} - 2 \times R_{\lambda 3})$ and $(R_{\lambda 1} - R_{\lambda 2} - R_{\lambda 3})/(R_{\lambda 1} + R_{\lambda 2} + R_{\lambda 3})$, to reduce the saturation observed in two-band vegetation indices [27]. They demonstrated that the models for leaf nitrogen content (LNC) using $(R_{924} - R_{703} + 2 \times R_{423})/(R_{924} + R_{703} - 2 \times R_{423})$ were stable and accurate and more effective than other published vegetation indices.

Since the value of $R_{445}$ is constant until the total chlorophyll content drops below 0.04 mol m⁻², Sims et al. added $R_{445}$ to $ND_{705}$ and $SR_{705}$ and developed new three-band spectral indices. The modified $ND_{705}$ and $SR_{705}$ ($mND_{705}$, $mSR_{705}$) were used to predict leaf pigment content, and both were insensitive to species and leaf structure variation [28]. Tian et al. developed a blue nitrogen index ($R_{434}/(R_{496}+R_{401})$) and estimated the canopy leaf nitrogen concentration of rice [29].

Water and nitrogen are the main limiting factors in plant growth [30–32], and they interact in a complex manner. Previous studies have shown that water and nitrogen are indirectly related (within plants) through chlorophyll and cellulose [33]. It may be effective to construct spectral indices that include a waveband sensitive to chlorophyll or cellulose, which might indirectly eliminate the impact of nitrogen on LEWT monitoring with good performance.

In this study, two experiments were conducted on winter wheat with different water and nitrogen treatments in two consecutive growing seasons. The objectives of the study were: (1) to determine hyperspectral bands that were sensitive to LEWT but insensitive to

nitrogen in wheat, (2) to develop new spectral indices for monitoring LEWT in wheat, and (3) to quantify the relationships between LEWT and the new spectral indices to reliably estimate LEWT. These results may provide a technical approach to effectively monitoring of plant water status while minimizing the noise from nitrogen in precision wheat management.

## Materials and Methods

### 2.1 Design of field experiments

Two experiments were conducted at the Pailou Experiment Station at Nanjing Agricultural University, China (118°15′E, 32°1′N). These experiments involved different water (W) and nitrogen (N) treatments. The wheat cultivar Yangmai 18 was grown in two consecutive seasons from November 2010 to June 2011 (a low rainfall season) and from November 2011 to June 2012 (a high rainfall season). Seeds were sown on 1 November 2010 and 3 November 2011 at a density of 180 plants per m² with a plot size of 10 m² (2.5 m×4 m). Each experiment employed a randomized complete block design with three replications. The data from Experiment 1 were used to derive the monitoring models, while the data from Experiment 2 were used to evaluate the models. More details about the W and N treatments, sampling procedures, and environmental conditions are given in Table 1.

There were 24 plots in 2010–2011and 27 in 2011–2012. Each plot was constructed from cement with identical length (3 m), width (3 m) and depth (1 m). A transparent plastic sheet was placed over the plots to a height of 3 m above the ground to prevent natural rainfall from flowing into the plot and disturbing the experimental water treatments. From greening until harvest, the runoff from the plastic covers was discharged into a cement drainage channel outside of the plots. From early jointing, the volumetric soil water content was measured with a TRIME-PICO TDR (TRIME-PICO, IMKO, Germany) portable soil moisture speed measuring device in each plot using the five-point method at 4:00 p.m. local time every day. If necessary, water was added immediately to the soil to maintain the designed water content. The date for obtaining spectrum and plant samples was jointing, booting, anthesis, filling.

### 2.2. Data measurements

**2.2.1. Measurement of leaf hyperspectral reflectance.** Leaf spectral measurements were taken using an accessory of the ASD Field Spec Pro spectrometer (Analytical Spectral Devices, Boulder, CO, USA). The accessory comprises a handheld leaf folder spectral detector with its own light source, which is designed to reduce the effect of time of day, weather, atmospheric vapor pressure, or soil background on the readings. The ASD spectrometer is operated in the 350–2,500 nm spectral region, with a sampling interval of 1.4 nm and spectral resolution of 3 nm between 350 and 1,050 nm, and a sampling interval of 2 nm and spectral resolution of 10 nm between 1,050 and 2,500 nm.

Ten wheat plants were randomly sampled from each plot. The top four leaves on each plant were identified (numbered L1 to L4 starting from the top), and their spectral reflectance was measured. The spectral reflectance averaged over 10 plants at each leaf position was taken as the reflectance for that leaf position. Before the measurement of leaf reflectance in each plot, a standard whiteboard (Labsphere, North Sutton, NH, USA) was used to calibrate the spectral reflectance of the leaves.

**2.2.2. Determination of leaf equivalent water thickness.** After the leaves were measured for spectral reflectance, they were stored in pre-weighed valve bags in a jar of liquid nitrogen. The leaves were transferred to the laboratory for the determination of leaf area using a portable leaf area meter (LAI-3000, Licor, NE, USA) and their fresh and dry weights were obtained. LEWT was calculated as follows:

$$LEWT = (WF\text{-}WD)/(DW * A) * 10000 (\mu m),$$

where $W_F$ is the leaf fresh weight (g), $W_D$ is the leaf dry weight (g), $D_W$ is the water density value (g/cm$^3$), and A is the leaf area (cm$^2$).

**2.2.3 Determination of leaf nitrogen content.** For each sample, the leaf dry weight was determined by oven-drying the leaves at 80°C to a constant weight. The LNC was determined on a dry weight basis (g 100 g$^{-1}$) using the micro-Kjeldahl method.

## 2.3. Data analysis

The LEWT values and the corresponding leaf spectral reflectance for each sampling date were analyzed with MATLAB 8.2 and Excel 2011 software. The two-band spectral indices considered included the normalized difference spectral index (NDSI, equation (1)), the ratio spectral index (RSI, equation (2)), and the difference spectral index (DSI, equation (3)). To determine the wavelengths in the two-band spectral indices, all combinations of wavelengths within the range of 350–2,500 nm at 1 nm intervals were evaluated according to the criteria presented below:

$$NDSI = (R_{\lambda1}\text{-}R_{\lambda2})/(R_{\lambda1}+R_{\lambda2}) \qquad (1)$$

$$RSI = R_{\lambda1}/R_{\lambda2} \qquad (2)$$

$$DSI = R_{\lambda1}\text{-}R_{\lambda2}, \qquad (3)$$

where $R_{\lambda i}$ is spectral reflectance at the wavelength of $\lambda i$.

For the three-band spectral indices, the best wavelengths were combined with all possible third wavelengths within the range of 350–2,500 nm at 1 nm intervals. The principle of retaining the best wavebands of the three-band indices has been applied in previous studies [26–27]. Three classical three-band spectral indices based on the NDSI were considered and constructed (equations (4) to (6)).

$$(R_{\lambda1}\text{-}R_{\lambda2})/(R_{\lambda1}+R_{\lambda2}\text{-}R_{\lambda3}) \qquad (4)$$

$$(R_{\lambda1}\text{-}R_{\lambda2}\text{-}R_{\lambda3})/(R_{\lambda1}+R_{\lambda2}+R_{\lambda3}) \qquad (5)$$

$$(R_{\lambda1}\text{-}R_{\lambda2}+2R_{\lambda3})/(R_{\lambda1}+R_{\lambda2}\text{-}2R_{\lambda3}) \qquad (6)$$

The criteria for evaluating the best wavelengths were the coefficient of determination ($R^2$), standard error (SE), and relative root mean square error (RRMSE) [25].

The RRMSE was calculated as follows:

$$RRMSE = \sqrt{\frac{1}{n} \times \sum_{i=1}^{n} (P_i - O_i)^2} \times \frac{100}{\bar{O}_i} \qquad (7)$$

where $P_i$ and $O_i$ are the predicted and observed values, respectively, $\bar{O}_i$ is the observed mean value, and $n$ is the number of samples.

## Results

### 3.1 Variation in LEWT

Fig. 1A shows how LEWT varied under the same nitrogen level and different water treatments in Exp. 1. There was a greater difference at anthesis and filling than at booting, jointing, or heading due to water treatment, with LEWT values in the order of W4 > W3 > W2 > W1, which is normal for irrigation scenarios. During the late stages of plant development, evaporation increased with the air temperature, which resulted in greater differences in LEWT among water treatments. Drought stress (W1, W2) was found to accelerate plant development, and LEWT value was higher than in W3 (normal amounts of water), while the W4 treatment was found to enhance the acceleration of plant development.

Fig. 1B displays LEWT for the W3 treatment for leaf L1 at various nitrogen levels. LEWT was greater at N2 than at N1, which was also the case for the W1, W2, and W4 treatments. Due to the synergistic effects of water and nitrogen on the growth processes of wheat, high-level nitrogen treatment promoted water uptake, resulting in a greater LEWT compared to low nitrogen levels [32–33]. Similar results were obtained for L2, L3, and L4 in Exp. 2.

### 3.2 Variations in leaf hyperspectral reflectance

Differences in LEWT and LNC values significantly affected the leaf hyperspectral reflectance of wheat. Fig. 2A shows leaf reflectance at different LEWT levels for N2 during the booting stage in Exp. 1. Leaf reflectance in the visible range was not significantly affected by LEWT. However, in the NIR range, especially at the central wavelengths of 900 nm, 1,200 nm, 1,400 nm, 1,450 nm, and 1,930 nm, leaf reflectance significantly decreased as LEWT increased. This result suggests that nitrogen levels influenced the variations in leaf hyperspectral reflectance. Thus the reflectance spectrum used for monitoring crop water may contain some information due to the presence of nitrogen, which should be eliminated or reduced.

Fig. 2B shows the reflectance at various LNC levels for the W4 treatment. N levels and W treatments markedly influenced the characteristics of spectral reflectance, with different spectral responses in the various waveband regions. With increasing N levels, the reflectance decreased in the visible bands and increased in the near infrared, with obvious differences among the four N

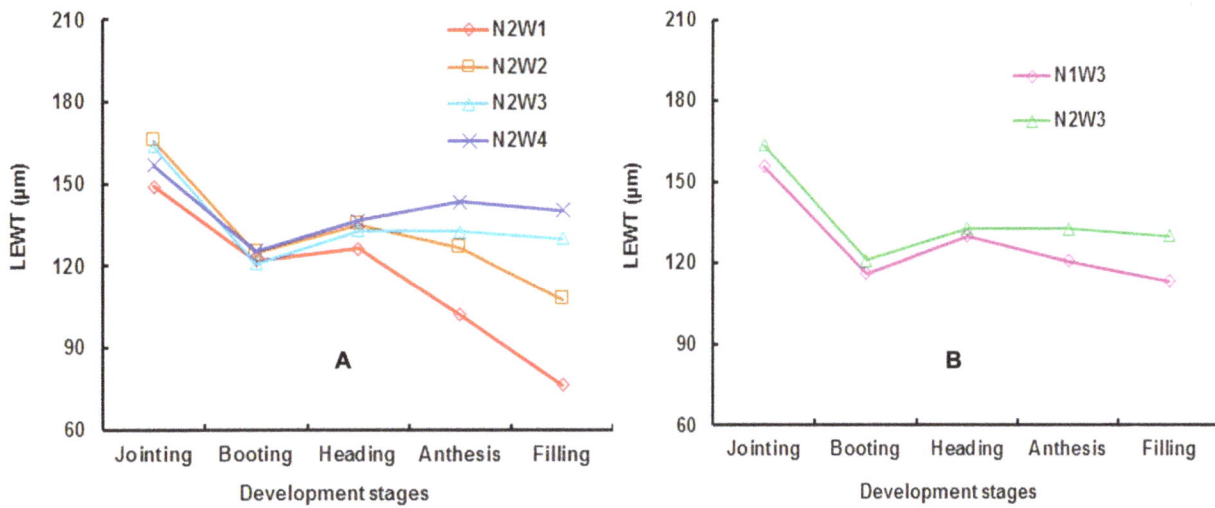

**Figure 1. Changes of LEWT under different water treatments at N2 (A) and varied nitrogen treatments at W3 (B) in Exp. 1.**

levels. This result implies that these spectral regions are relatively sensitive to the growth status of wheat at different N levels. In the visible range, differences in reflectance appeared mainly at wavelengths of 450 nm and 700 nm, the central wavelengths of pigment absorption. The reflectance increased as LNC increased in the near infrared, particularly at 970 nm and 1,200 nm, the central wavelengths of water absorption.

## 3.3 Correlation of LEWT and LNC with the original and first derivative hyperspectra in wheat leaves

The correlation of LEWT and LNC with the original spectrum and its first derivative spectrum was determined based on the observed data from Exp.1. Fig. 3 shows that LEWT and the original spectral reflectance were positively correlated in the visible region and negatively correlated in the NIR, with peaks of the absolute value of the correlation at 970 nm, 1,200 nm, 1,395 nm, 1,450 nm, 1,850 nm, and 2,200 nm. The strongest correlation was at 1,395 nm ($R_{1395} = -0.7576$). This result indicates that LEWT markedly influenced the reflectance of wheat leaves, with

different spectral responses in the various waveband regions. In general, the correlation between LNC and the original spectral reflectance generally had the opposite sign to the correlation of LEWT.

Fig. 3A shows that the largest 10% of $|R|_{LEWT}$ and $|R|_{LNC}$ overlapped at 1,365–1,424 nm and at 1,839–1,869 nm in the NIR region. The largest 20% of $|R_{LEWT}|$ and $|R_{LNC}|$ overlapped in two regions of the visible bands, 526–566 nm and 701–723 nm (Fig. 3A). These results indicate that the wavelengths sensitive to both LEWT and LNC were mainly in the visible and NIR range. However, the largest 20% of $|R_{LEWT}|$ did not overlap anywhere, with the smallest values of $|R_{LNC}|$, suggesting that no wavelength was sensitive to LEWT and insensitive to LNC in the same region.

The first derivatives of the original spectrum and LEWT were also strongly correlated in the NIR range, especially at the central wavelengths of 1,890 nm, 1,300 nm, and 1,400 nm (Fig. 3B). The largest 10% of $|R_{LEWT}|$ overlapped with $|R_{LNC}|$ at 770 nm (Fig. 3B). In the original spectrum, there was no overlap between the largest 20% of $|R_{LEWT}|$ and the smallest values of $|R_{LNC}|$.

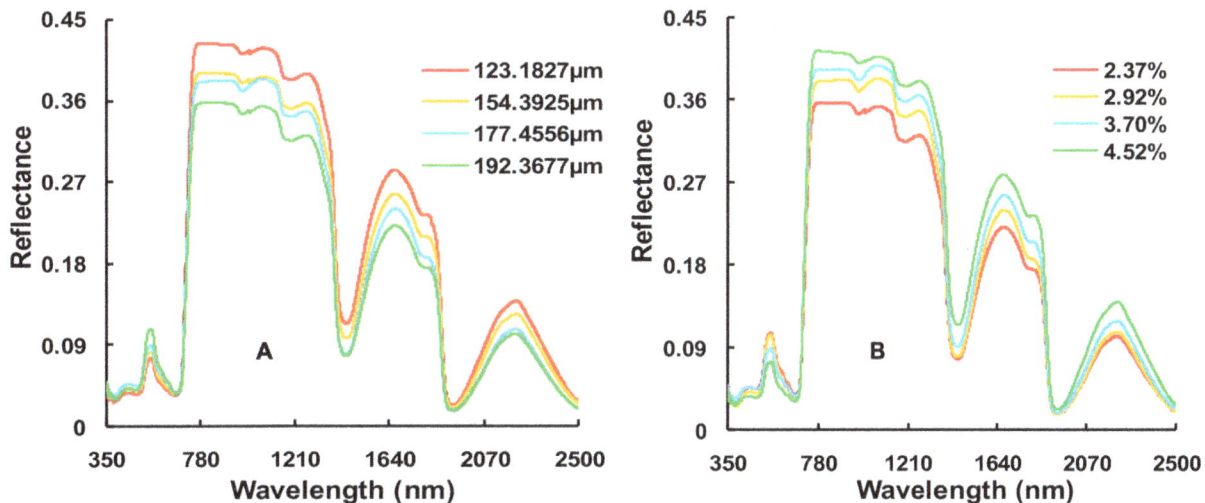

**Figure 2. Reflectance of leaves under different LEWT at N2 (A) and LNC at W4 (B) in Exp. 1.**

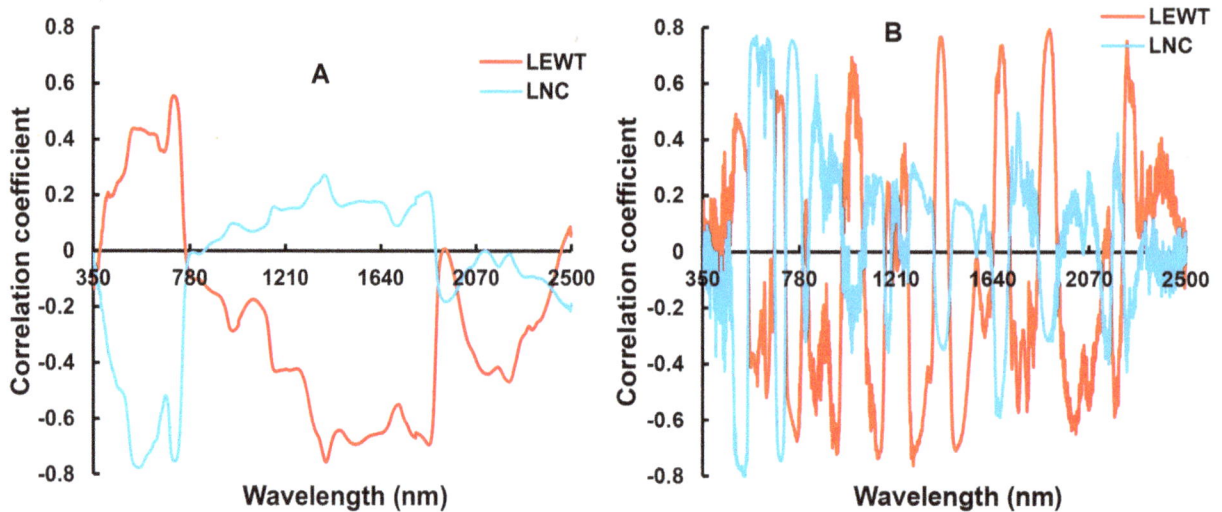

**Figure 3. The correlation of LEWT and LNC to original spectral reflectance (A) and its first derivative (B) in Exp. 1.**

No wavelength in the first derivative was sensitive to LEWT and insensitive to LNC simutaneously.

## 3.4 The relationships of spectral indices to LEWT

**3.4.1. Relating two-band spectral indices to LEWT during the entire growth period.** Contour maps of $R^2$ values for the relationship of LEWT during the full growth period with the two wavelengths in the two-band indices were constructed (not shown). The results for NDSI and RSI were similar, while $R^2$ values were lower for DSI. The top 10% of $R^2$ values for NDSI, using the calibration data, were mainly in the ranges of $\lambda 1 = 1,000–1,320$ nm and $\lambda 2 = 1,300–1,350$ nm (Fig. 4A). Using the validation data, the top 10% of $R^2$ values were in the ranges of $\lambda 1 = 1,205–1,315$ nm and $\lambda 2 = 1,287–1,337$ nm (Fig. 4B). Finally, NDSI ($R_{1204}$, $R_{1318}$) was selected based on the overlapping region. This index had an $R^2$ value of 0.708 and an SE of 15.5053 μm (Fig. 5A). Fig. 5B shows the relationship between the predicted and observed LEWT. There are clear differences in the fit to the data before and after anthesis, which may be due to metabolism

(especially the nitrogen content) and changes in position of the plants.

Further statistical analysis was conducted to determine whether the two linear relationships could be grouped. Thereby, a linear parallel curve analysis with grouped data was used to determine whether the critical curves for LEWT differed before and after anthesis [34]. The results showed that the F value for the LEWT before and after anthesis was significantly different from that of the residual. The F value for NDSI($R_{1204}$, $R_{1318}$)×LEWT before anthesis and NDSI($R_{1204}$, $R_{1318}$)×LEWT after anthesis also differed significantly from that of the residual (Table 2).

**3.4.2. Relating NDSI ($R_{\xi 1}$, $R_{\xi 2}$) to LEWT before and after anthesis.** Based on the above results, the developmental stages were considered separately in further studies of the relationship between LEWT and the NDSI ($R_{\xi 1}$, $R_{\xi 2}$). The optimal wavelengths for the period before anthesis were 1,445 and 487, giving the spectral index NDSIb($R_{1445}$, $R_{487}$) (NDSI before anthesis). This index had an $R^2$ value of 0.8137 and an SE of 10.6503 μm. The optimal wavelengths for the period after anthesis

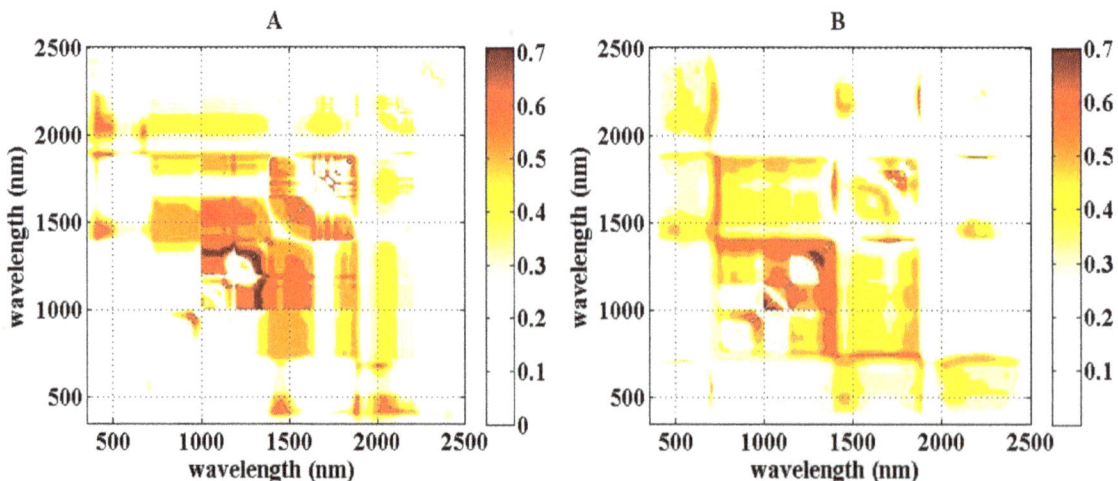

**Figure 4. Contour map of coefficients of determination ($R^2$) for linear relationship between NDSI and LEWT (A: calibration dataset, B: validation dataset).**

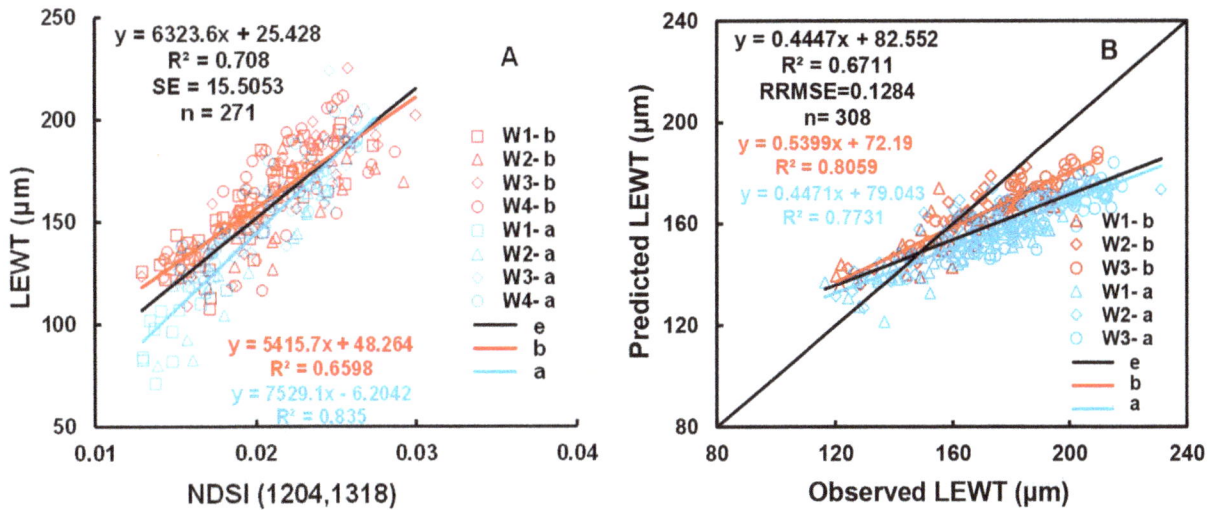

**Figure 5. Quantitative relationships between NDSI ($R_{1204}$, $R_{1318}$) and LEWT (A); the 1:1 relationship between the predicted and observed LEWT based on NDSI ($R_{1204}$, $R_{1318}$) (B).**

were 1,714 and 1,395, giving the spectral index NDSIa($R_{1714}$, $R_{1395}$) (NDSI after anthesis). For this index, the $R^2$ value was 0.8622 and the SE was 12.2442 μm (Fig. 6). The independent dataset from Exp. 2 was used to test the models for LEWT before and after anthesis (Fig. 7). The results showed that NDSIb and NDSIa predicted LEWT more accurately than did the spectral index NDSI ($R_{1204}$, $R_{1318}$) for the entire growth period and the spectral index NDSI($R_{1429}$, $R_{416}$) of the common sensitive areas. In the optimal spectral indices NDSIb($R_{1445}$, $R_{487}$) and NDSIa($R_{1714}$, $R_{1395}$), 1,445 nm and 1,395 nm were the most sensitive wavelengths to LEWT, as explained above. The bands at 487 nm and 1,714 nm were sensitive to chlorophyll and lignin.

The indices NDSIb($R_{1445}$, $R_{487}$) and NDSIa($R_{1714}$, $R_{1395}$) contain four wavelengths, which will increase the cost of manufacturing-self. This motivated us to search for common two-band wavelengths before and after anthesis. The central waveband ranges of NDSI before and after anthesis differed significantly, and the top 10% of the optimal values of $R^2$ had no common area (Figures were omitted). However, the top 15% of $R^2$ values before (Fig. 8A) and after (Fig. 8B) anthesis had two common areas, with the first $\lambda 1$ in the range of 405–420 nm and $\lambda_2$ in the range of 1,418–1,477 nm and the second $\lambda 1$ in the range of 408–443 nm and $\lambda_2$ in the range of 1,876–1,886 nm (Fig. 8C). Based on the $R^2$, SE, and RRMSE of the calibration and validation data, the spectral index NDSI($R_{1429}$, $R_{416}$) was selected for the periods both before and after anthesis. The calibration of LEWT both before and after anthesis had an $R^2$ value of 0.6776

and an SE of 16.2914, and the prediction using the validation data had an $R^2$ value of 0.3967 and an RRMSE of 0.1243 (Fig. 9, Table 3).

**3.4.3. Relating three-band spectral indices based on NDSI($R_{1429}$, $R_{416}$) to LEWT.** The third band ($\lambda 3$) can help to reduce both the effect of nitrogen and the effect of saturation. The best three-band index, using the best wavelengths for NDSI, was ($R_{1429}-R_{416}-R_{1865}$)/($R_{1429}+R_{416}+R_{1865}$). The third band (at 1,865 nm) located in the SWIR is close to the sensitive band of cellulose (1,880 nm) (Table 3). However, the performance of the model for the three-band index for NDSIb($R_{1445}$, $R_{487}$) and NDSIa($R_{1714}$, $R_{1395}$). The cost of developing the equipment in the future will decrease as its use on a large scale becomes more feasible due to the need for fewer wavelengths. Moreover, the three-band index model displayed a marked improvement compared with the model based on NDSI($R_{1429}$, $R_{416}$) (Fig. 9, 10).

**3.4.4. Comparison of the performance of the new and previous models.** Five popular existing spectral indices that are used to predict crop water content were compared with the spectral indices developed in the current study. As shown in Table 3, of the five indices, WI, MSI, and NDII yielded the best predictions of LEWT, and WI performed especially well. However, the two spectral indices developed in this study, NDSIb and NDSIa, performed considerably better than all of the existing indices. This result may be due to the fact that the existing spectral indices are based on the relative water content of leaves and whole plants of various crops, whereas the spectral indices developed in the current study were constructed based on the LEWT in wheat.

**Table 2.** Simple grouping linear analysis of the model between NDSI and LEWT in wheat before and after anthesis.

| Developmental stage | | | NDSI ($R_{1204}$,$R_{1318}$)×developmental stage | | | Residual | | |
|---|---|---|---|---|---|---|---|---|
| df | MS | F | df | MS | F | df | MS | F |
| 1 | 7478.7 | 37.34* | 1 | 4202.6 | 20.98* | 267 | 200.3 | 6.76 |

**Notes:**
**df:** degrees of freedom.
**MS:** mean square.
**\*:** $P<0.05$.

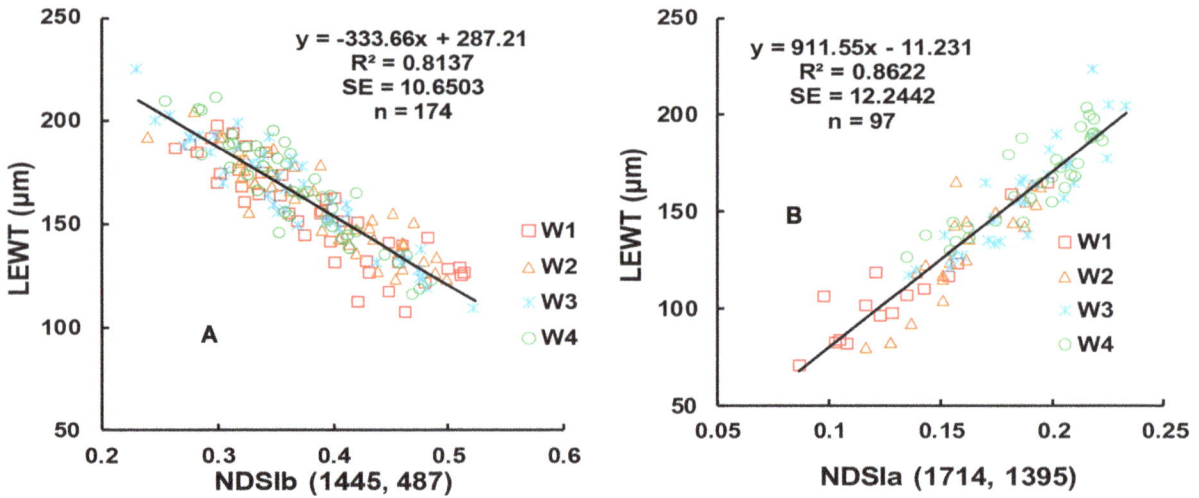

**Figure 6.** Contour map of the linear relationship with top 15% $R^2$ between NDSI and LEWT before (A) and after (B) anthesis, and the common area between A and B (C).

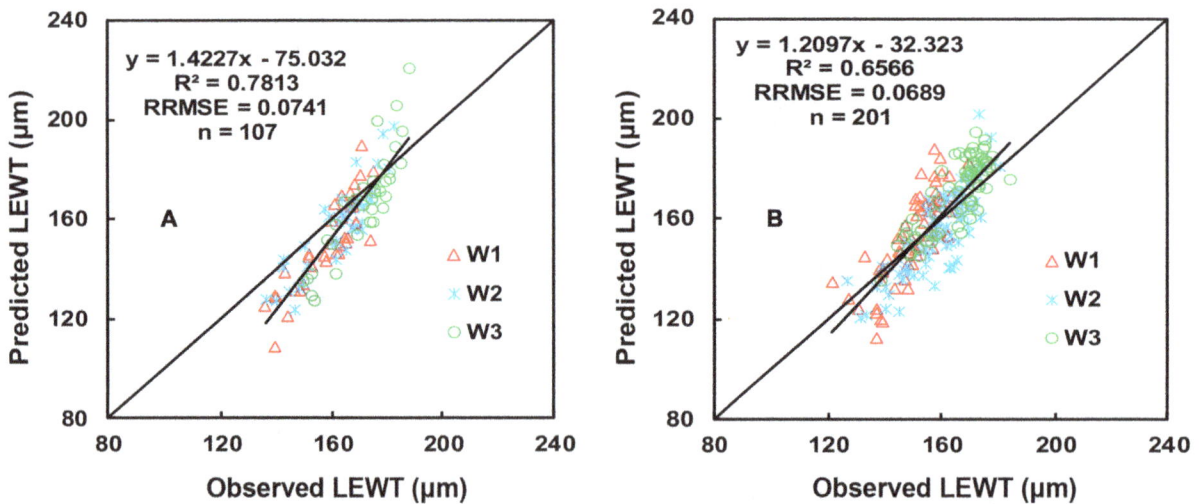

**Figure 7.** Quantitative relationships between $NDSIe$ ($R_{1445}$, $R_{487}$) and LEWT before anthesis (A); quantitative relationships between $NDSIt$ ($R_{1714}$, $R_{1395}$) to LEWT after anthesis (B).

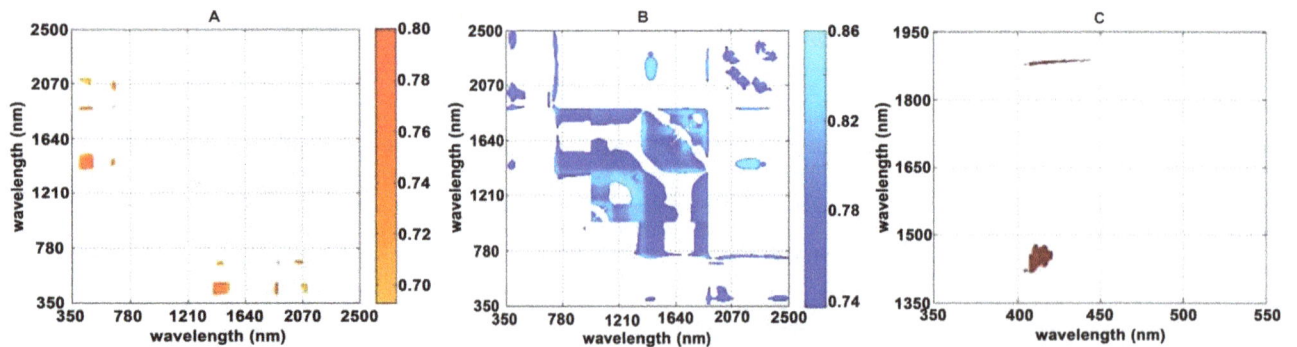

**Figure 8.** The 1:1 relationship between the predicted and observed LEWT based on $NDSIb$ ($R_{1445}$, $R_{487}$) for wheat before anthesis (A); the 1:1 relationship between the predicted and observed LEWT based on $NDSIa$ ($R_{1714}$, $R_{1395}$) for wheat after anthesis (B).

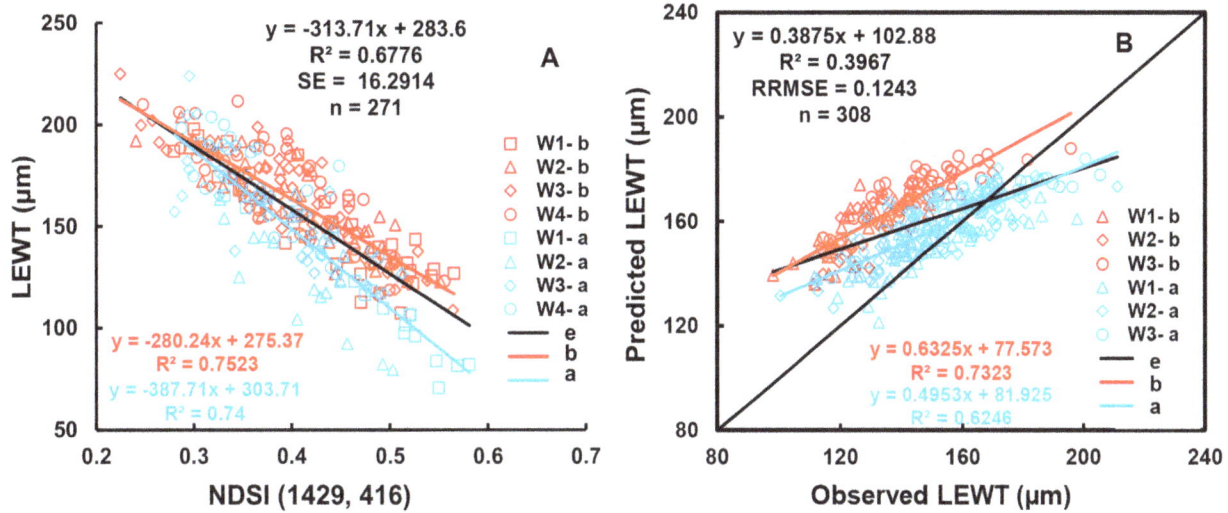

**Figure 9. Quantitative relationships between $(R_{1429}-R_{416})/(R_{1429}+R_{416})$ and LEWT (A); the 1:1 relationship between the predicted and observed LEWT on $(R_{1429}-R_{416})/(R_{1429}+R_{416})$ in wheat (B).**

## Discussion

### 4.1. Response of leaf spectral reflectance to water and nitrogen treatments in wheat

Previous studies of the response of leaf spectra to different water conditions and nitrogen levels have shown that the sensitivity of leaf spectral reflectance to water content peaks at the central wavelengths of 970 nm, 1,200 nm, 1,400 nm, 1,450 nm, 1,730 nm, 1,930 nm, 2,100 nm, and 2,500 nm [35]. Within the range of 1,300–2,500 nm, spectral reflectance is affected by water in the plants, which directly absorbs radiation. Within the range of 400–1,300 nm, reflectance is influenced by changes in the internal structure of the leaf, which is a result of their different water content ratios [7,36]. The influence of nitrogen on spectral reflectance occurs mainly in the range of the visible and NIR wavebands; and the wavebands most sensitive to nitrogen are in the range of 530–560 nm [37–38]. In the present study, leaf hyperspectral reflectance at different LEWT levels in the N2 treatment did not change significantly in the visible band range. However, reflectance significantly decreased as LEWT increased within the NIR range, especially at 900 nm, 1,200 nm, 1,400 nm, 1,450 nm, 1,730 nm, and 1,930 nm. Leaf hyperspectral reflectance increased with increasing LNC within the NIR range, and it decreased considerably at the central wavelengths of water absorption. Therefore, to monitor plant water content, some sensitive wavelength and spectral indices that are strongly correlated with LEWT, but less well correlated with LNC, should be extracted with consideration of the fact that the spectrum reflectance is affected by both water and nitrogen.

### 4.2. Two-band spectral indices sensitive to LEWT in wheat

The systematic analysis of two-band spectral indices sensitive to LEWT was conducted for the entire growth. NDSI is more closely related to LEWT than RSI or DSI. NDSI($R_{1204}$, $R_{1318}$) was the optimal spectral index for predicting LEWT in wheat during the whole growth period. As earlier analyses have indicated, differences in LNC before and after anthesis might have an impact on the accuracy of the LEWT model throughout plant growth. Therefore, separate indices were proposed for the periods before and after anthesis, namely, NDSIb($R_{1445}$, $R_{487}$) and

NDSIa($R_{1714}$, $R_{1395}$), respectively. The wavelengths 1,445 nm and 1,395 nm were the most sensitive to LEWT, while the wavelengths 487 nm and 1,714 nm were sensitive to chlorophyll and lignin, respectively. This indicated that LEWT models before and after anthesis were established separately, while the impact of nitrogen was eliminated by the use of wavebands that were sensitive to chlorophyll or lignin, as reported previous [33].

### 4.3. Three-band spectral indices sensitive to LEWT in wheat

Three-band indices contain less noise due to chlorophyll or lignin than two-band spectral indices [23–24,28]. They also display less saturation than two-band spectral indices [27]. Therefore, to reduce the noise caused by nitrogen before and after anthesis in the LEWT model, the three-band spectral index $(R_{1429}-R_{416}-R_{1865})/(R_{1429}+R_{416}+R_{1865})$ was developed based on the best overall two-band spectral index NDSI($R_{1429}$, $R_{416}$). Although the model based on the three-band index was a poorer predictor than those based on NDSI ($R_{1204}$, $R_{1318}$) or on NDSIb($R_{1445}$, $R_{487}$) and NDSIa($R_{1714}$, $R_{1395}$), the calibration and validation on the resulting scatter plot was centralized, and the LEWT model was more uniform and stable with a lower RRMSE. The results also showed that the three-band spectral indices $(R_{1429}-R_{416}-R_{1865})/(R_{1429}+R_{416}+R_{1865})$ clearly reduced the noise due to nitrogen in the LEWT model throughout the whole period of wheat growth, with an $R^2$ value of 0.7255and RRMSE of 0.0689 for LEWT validation, and an $R^2$ value of 0.0849 and an RRMSE of 39.7950 for LNC validation.

## Conclusions

Previous studies have shown that the transportation of amino acids from leaves to grains after anthesis leads to physiological and biochemical changes in the organizational structures of leaves, which affect LEWT monitoring based on leaf hyperspectral reflectance. In this study, we demonstrated that different water and nitrogen treatments affected the variation in LEWT and the leaf hyperspectral reflectance of wheat within the 350–2,500 nm range. Furthermore, the top 10% of the maximum $|R_{LEWT}|$ and $|R_{LNC}|$ values were found to share common wavelength ranges.

**Table 3.** Relationships of spectral indices to LEWT (n = 271) and testing performances (n = 308) in wheat.

| Spectral index | | Growth stage | Calibration (n = 271) | | Validation to LEWT (n = 308) | | Validation to LNC (n = 308) | |
|---|---|---|---|---|---|---|---|---|
| | | | $R^2$ | SE | $R^2$ | RRMSE | $R^2$ | RRMSE |
| 2-band | NDSI ($R_{1204}$, $R_{1318}$) | e | 0.7080 | 15.5053 | 0.6711 | 0.1284 | 0.1451 | 42.6089 |
| | NDSIb ($R_{1445}$, $R_{487}$) | b | 0.8137 | 10.6503 | 0.7813 | 0.0741 | 0.2221 | 36.0804 |
| | NDSIa ($R_{1395}$, $R_{1714}$) | a | 0.8622 | 12.2442 | 0.6566 | 0.0689 | 0.2272 | 40.0445 |
| | NDSI ($R_{1429}$, $R_{416}$) | e | 0.6776 | 16.2914 | 0.3967 | 0.1243 | 0.2075 | 35.9918 |
| 3-band | **($R_{1429}$−$R_{416}$−$R_{1865}$)/($R_{1429}$+$R_{416}$+$R_{1865}$)** | **e** | **0.6905** | **15.9613** | **0.7255** | **0.0689** | **0.0849** | **39.7950** |
| | ($R_{1429}$−$R_{416}$)/($R_{1429}$+$R_{416}$−$R_{1883}$) | e | 0.7093 | 15.4695 | 0.3860 | 0.0994 | 0.1632 | 36.9978 |
| | ($R_{1429}$−$R_{416}$+2×$R_{1435}$)/($R_{1429}$+$R_{416}$−2×$R_{1435}$) | e | 0.6260 | 17.5466 | 0.3375 | 0.1117 | 0.1858 | 36.0329 |
| Previous | WI ($R_{900}$, $R_{970}$) [22] | e | 0.5260 | 19.7528 | 0.6480 | 0.2241 | 0.1518 | 47.2203 |
| | SRWI ($R_{858}$, $R_{1240}$) [20] | e | 0.1625 | 26.2569 | 0.4932 | 0.0614 | 0.1315 | 39.7522 |
| | MSI ($R_{1600}$, $R_{820}$) [12,13] | e | 0.4982 | 20.3248 | 0.5206 | 0.2915 | 0.0599 | 50.0599 |
| | NDWI ($R_{850}$, $R_{1240}$) [14] | e | 0.1604 | 26.2907 | 0.4954 | 0.0613 | 0.1295 | 39.7553 |
| | NDII ($R_{850}$, $R_{1650}$) [8,21] | e | 0.2388 | 25.0332 | 0.5104 | 0.2953 | 0.0562 | 50.1623 |

**Notes:**
b represents before anthesis,
a represents after anthesis,
e represents the entire growth period.

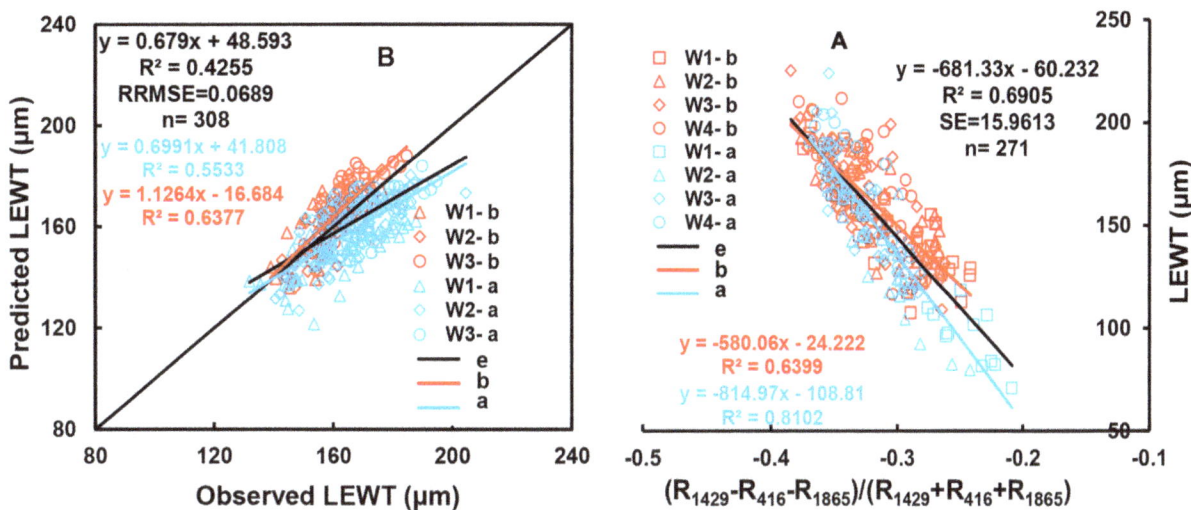

**Figure 10. Quantitative relationships between** $(R_{1429}-R_{416}-R_{1865})/(R_{1429}+R_{416}+R_{1865})$ **between LEWT in wheat (A); the 1:1 relationship between the predicted and observed LEWT on** $(R_{1429}-R_{416}-R_{1865})/(R_{1429}+R_{416}+R_{1865})$ **(B).**

Based on this study and previous reports, when monitoring LEWT, the noise from the LNC should be considered. The model based on a three-band index $(R_{1429}-R_{416}-R_{1865})/(R_{1429}+R_{416}+R_{1865})$ described in this paper performed well for monitoring LEWT, with a higher predictability and stability for water content and lower noise levels due to nitrogen under various water and nitrogen treatments.

## References

1. Tian YC, Cao WX, Jiang D, Zhu Y, Xue LH (2005) Relationship between canopy reflectance and plant water content in rice under different soil water and nitrogen conditions. Journal of Plant Ecology 29: 318–323.
2. Shen Y, Niu Z, Wang W, Xu YM (2005) Establishment of leaf water content models based on derivative spectrum variables. Geography and Geo-Information Science 21: 16–19.
3. Liu XJ, Tian YC, Yao X, Cao WX, Zhu Y (2012) Monitoring leaf water content based on hyperspectra in rice. Scientia Agricultura Sinica 45: 435–442.
4. Zhang JH, Guo WJ, Yao FM (2007) The study on vegetation water content estimating model based on remote sensing technique. Journal of Basic Science and Engineering 15: 45–53.
5. Champagne CM, Staenz K, Bannari A, McNairn H, Deguise JC (2003) Validation of a hyperspectral curve-fitting model for the estimation of plant water content of agricultural canopies. Remote Sensing of Environment 87: 148–160.
6. Wang JH, Zhao CJ, Guo XW, Tian QJ (2001) Study on the water status of the wheat leaves diagnosed by the spectral reflectance. Scientia Agricultura Sinica 34: 104–107.
7. Carter GA (1991) Primary and secondary effects of water content on the spectral reflectance of leaves. American Journal of Botany 78: 916–924.
8. Yilmaz MT, Hunt Jr ER, Jackson TJ (2008) Remote sensing of vegetation water content from equivalent water thickness using satellite imagery. Remote Sensing of Environment 112: 2514–2522.
9. Dong JJ, Niu Z, Shen Y, Yuan JG (2006) Comparison of the methods of obtaining leaf water content by using reflectance data. Acta Agricultural Universitatis Jiangxiensis 28: 587–591.
10. Seelig HD, Hoehn A, Stodieck L, Klaus D, Adams III W, et al. (2008) Relations of remote sensing leaf water indices to leaf water thickness in cowpea, bean, and sugarbeet plants. Remote Sensing of Environment 112: 445–455.
11. Wang P, Wu JJ, Nie JL, Kong FM, Ding HY, et al. (2010) A comparatively study of the capabilities of different vegetation water indices in monitoring water status of wheat. Remote Sensing for Land & Resources 3: 97–100.
12. Hunt ER, Rock BN (1989) Detection of changes in leaf water content using near-and middle-infrared reflectances. Remote Sensing of Environment 30: 43–54.
13. Ceccato P, Flasse S, Tarantola S, Jacquemoud S, Grégoire JM (2001) Detecting vegetation leaf water content using reflectance in the optical domain. Remote Sensing of Environment 77: 22–33.
14. Gao BC (1996) NDWI—normalized difference water index for remote sensing of vegetation liquid water from space. Remote Sensing of Environment 58: 257–266.
15. Zhang L, Zhou ZG, Zhang GW, Meng YL, Chen BL, et al. (2012) Monitoring the leaf water content and specific leaf weight of cotton (Gossypiumhirsutum L.) in saline soil using leaf spectral reflectance. European Journal of Agronomy 41: 103–117.
16. Maki M, IshiahraM, Tamura M (2004) Estimation of leaf water status to monitor the risk of forest fires by using remotely sensed data. Remote Sensing of Environment 90: 441–450.
17. Jackson TJ, Chen D, Cosh M, Li F, Anderson M, et al. (2004) Vegetation water content mapping using Landsat data derived normalized difference water index for corn and soybeans. Remote Sensing of Environment 92: 475–482.
18. Verbesselt J, Somers B, Lhermitte S, Jonckheere I, Van Aardt J, et al. (2007) Monitoring herbaceous fuel moisture content with SPOT vegetation time-series for fire risk prediction in savanna ecosystems. Remote Sensing of Environment 108: 357–368.
19. Liu LY, Wang JH, Zhang YJ, Huang WJ (2007) Detection of leaf EWT by calculating REWT from reflectance spectra. Journal of Remote Sensing 11: 289–295.
20. Zarco-Tejada PJ, Rueda C, Ustin S (2003) Water content estimation in vegetation with MODIS reflectance data and model inversion methods. Remote Sensing of Environment 85: 109–124.
21. Hardisky MA, Klemas V, Smart RM (1983) The influences of soil salinity, growth form, and leaf moisture on the spectral reflectance of Spartinaalterniflora canopies. Photogrammetric Engineering & Remote Sensing 49: 77–83.
22. Penuelas J, Filella I, Biel C, Serrano L, Save R (1993) The reflectance at the 950–970 nm region as an indicator of plant water status. International Journal of Remote Sensing 14: 1887–1905.
23. Schneider P, Roberts D, Kyriakidis P (2008) A VARI-based relative greenness from MODIS data for computing the Fire Potential Index. Remote Sensing of Environment 112: 1151–1167.
24. Stow D, Niphadkar M, Kaiser J (2005) MODIS-derived visible atmospherically resistant index for monitoring chaparral moisture content. International Journal of Remote Sensing 26: 3867–3873.
25. Bai L, Wang J, Jiang GY, Yang P, Sun SJ (2008) Study on hyperspectral remote sensing date of cotton in estimating yield of arid region in china. Scientia Agricultura Sinica 41: 2499–2505.
26. Hou XJ, Jiang GY, Bai L, Wang JC, Ling HB, et al. (2008) Relationship between cotton yield components and their hyperspectral remote sensing characteristics. Remote Sensing Information: 96: 10–16.
27. Wang W, Yao X, Yao XF, Tian YC, Liu XJ, et al. (2012) Estimating leaf nitrogen concentration with three-band vegetation indices in rice and wheat. Field Crops Research 129: 90–98.

## Author Contributions

Conceived and designed the experiments: XY WQJ HYS ZQG YCT XJL WXC YZ. Performed the experiments: XY WQJ HYS ZQG YCT XJL WXC YZ. Analyzed the data: XY WQJ HYS ZQG YCT XJL WXC YZ. Contributed reagents/materials/analysis tools: XY WQJ HYS ZQG YCT XJL WXC YZ. Wrote the paper: XY WQJ HYS ZQG YCT XJL WXC YZ.

28. Sims DA, Gamon JA (2002) Relationships between leaf pigment content and spectral reflectance across a wide range of species, leaf structures and developmental stages. Remote Sensing of Environment 81: 337–354.

29. Tian YC, Yang J, Yao X, Zhu Y, Cao WX (2010) A newly developed blue nitrogen index for estimating canopy leaf nitrogen concentration of rice. Chinese Journal of Applied Ecology 21: 966–972.

30. Zhao QZ, Gao TM, Yin CY, Ning HF, Lu Q (2006) Effects of moisture on content of major nutrients in soil and upland rice plants. Agricultural Research in the Arid Areas 24: 61–65.

31. Verasan V, Phillips RE (1978) Effects of soil water stress on growth and nutrient accumulation in corn. Agronomy Journal 70: 613–618.

32. Tilling AK, O'Leary GJ, Ferwerda JG, Jones SD, Fitzgerald GJ, et al. (2007) Remote sensing of nitrogen and water stress in wheat. Field Crops Research 104: 77–85.

33. Pu RL, Gong P (2000) Hyperspectral remote sensing and its applications: Higher Education Press. pp: 194–202.

34. Ziadi N, Brassard M, Bélanger G, Claessens A, Tremblay N, et al. (2008) Chlorophyll measurements and nitrogen nutrition index for the evaluation of corn nitrogen status. Agronomy Journal 100: 1264–1273

35. Cheng T, Rivard B, Sánchez-Azofeifa A (2010) Spectroscopic determination of leaf water content using continuous wavelet analysis. Remote Sensing of Environment 115(2): 659–670

36. Chen HB, Li JH (2010) Advances on crop water content diagnosis based on spectral reflectance. Water Saving Irrigation 8: 69–72.

37. Yao X, Zhu Y, Tian YC, Feng W, Cao WX (2010) Exploring hyperspectral bands and estimation indices for leaf nitrogen accumulation in wheat. International Journal of Applied Earth Observation and Geoinformation 12: 89–100.

38. Zhu Y, Zhou D, Yao X, Tian Y, Cao W (2007) Quantitative relationships of leaf nitrogen status to canopy spectral reflectance in rice. Australian Journal of Agricultural Research 58: 1077–1085.

# 17

# Physiological and Proteomic Analyses of *Saccharum* spp. Grown under Salt Stress

**Aline Melro Murad[1,2], Hugo Bruno Correa Molinari[3], Beatriz Simas Magalhães[1], Augusto Cesar Franco[4], Frederico Scherr Caldeira Takahashi[5], Nelson Gomes de Oliveira-Júnior[2], Octávio Luiz Franco[1,2]*, Betania Ferraz Quirino[1,3]***

1 Genome Sciences and Biotechnology Program, Universidade Católica de Brasília, Brasília, Distrito Federal, Brazil, 2 Centro de Análises Proteômicas e Bioquímicas, Universidade Católica de Brasília, Brasília, Distrito Federal, Brazil, 3 Embrapa-Agroenergy, Brasília, Distrito Federal, Brazil, 4 Department of Botany, Universidade de Brasília, Brasília, Distrito Federal, Brazil, 5 Department of Ecology, Universidade de Brasília, Brasília, Brazil

## Abstract

Sugarcane (*Saccharum* spp.) is the world most productive sugar producing crop, making an understanding of its stress physiology key to increasing both sugar and ethanol production. To understand the behavior and salt tolerance mechanisms of sugarcane, two cultivars commonly used in Brazilian agriculture, RB867515 and RB855536, were submitted to salt stress for 48 days. Physiological parameters including net photosynthesis, water potential, dry root and shoot mass and malondialdehyde (MDA) content of leaves were determined. Control plants of the two cultivars showed similar values for most traits apart from higher root dry mass in RB867515. Both cultivars behaved similarly during salt stress, except for MDA levels for which there was a delay in the response for cultivar RB867515. Analysis of leaf macro- and micronutrients concentrations was performed and the concentration of $Mn^{2+}$ increased on day 48 for both cultivars. In parallel, to observe the effects of salt stress on protein levels in leaves of the RB867515 cultivar, two-dimensional gel electrophoresis followed by MS analysis was performed. Four proteins were differentially expressed between control and salt-treated plants. Fructose 1,6-bisphosphate aldolase was down-regulated, a germin-like protein and glyceraldehyde 3-phosphate dehydrogenase showed increased expression levels under salt stress, and heat-shock protein 70 was expressed only in salt-treated plants. These proteins are involved in energy metabolism and defense-related responses and we suggest that they may be involved in protection mechanisms against salt stress in sugarcane.

**Editor:** Tianzhen Zhang, Nanjing Agricultural University, China

**Funding:** Authors are also grateful to Conselho Nacional de PEsquisa e Desenvolvimento and Fundação de Amparo a Pesquisa do Distrito Federal for financial support. The funders had no role in study design, data collection and analysis, decision to publish, or preparation of the manuscript.

**Competing Interests:** The authors have declared that no competing interests exist.

* E-mail: ocfranco@gmail.com (OLF); betania.quirino@embrapa.br (BFQ)

## Introduction

Sugarcane (*Saccharum* spp.) is a semi-perennial monocot that can be propagated vegetatively by culms [1,2]. Its cultivation occurs in more than 80 tropical and subtropical countries [3,4]. Sugar and bioethanol are the main products obtained from sugarcane and Brazil is one of the largest sugarcane producers of the world [5,6].

Crop irrigation is essential in arid and semi-arid regions. However, when inappropriately applied, it may result in environmental degradation [7]. Soil salinization has been reported to be one of the causes of soil degradation, menacing productive lands under irrigated agriculture. According to FAO, it is estimated that 34 million hectares (i.e., 11% of the irrigated area) are affected by some level of salinization [8]. The cost of soil salinization to agriculture is estimated to be approximately US$ 12 billion a year. However, this value is expected to increase [9].

High concentrations of salt reduce osmotic potential in soil solution and promote drought stress in plants, which explains the fact that drought and salt stress cause similar symptoms in plants. Salinity imposes diffusive and metabolic limitations to photosynthesis, affects cell growth by restricting water uptake and cell turgor, resulting in increasing accumulation of $Na^+$ and $Cl^-$ ions

inside the cell [10–12]. Accumulation of $Na^+$ and $Cl^-$ ions severely inhibits many photosynthetic enzymes among others and triggers the production of reactive oxygen species (ROS) [13], which can cause plant damage and, in severe cases, death [14]. In an attempt to overcome the toxic effects caused by salinity, plants use various defense mechanisms such as the production of compatible osmolytes (i.e., aminoacids, sugars, and alcohols). These osmolytes balance the osmotic pressure within the cell [15–17], thus maintaining root water uptake, plant water balance and photosynthetic activity. They also play a role in membrane and protein protection and scavenging of reactive oxygen species. There is also increased production of certain proteins in response to salt stress, such as superoxide dismutase [10,18] that eliminates ROS excess, and heat-shock proteins [19] that are responsible for maintaining the correct folding of proteins.

According to the sugarcane cultivar census in Brazil held by the Centro de Tecnologia Canavieira (CTC) [20], the RB (Brazilian Republic) cultivars represent approximately 50% of sugarcane planted in Brazil. Cultivars RB855536 and RB867515 are respectively the second and seventh in farmers' preference, due to traits such as high productivity, erect culms and resistance to diseases [20,21]. Both cultivars are derived from interspecific hybridizations between *Saccharum officinarum* and *S. spontaneum*. Farmers consider cultivar RB867515 more drought-stress tolerant when compared to cultivar RB855536, although the scant experimental evidence is inconclusive [21]. In fact, water deficit is one of the major factors limiting sugarcane productivity [22]. Given the similarity between drought and salinity responses, we hypothesized that RB867515 would be salt tolerant when compared to RB855536. We assessed the salinity tolerance of the two cultivars by measuring photosynthesis, water potential, macro- and micronutrients and lipid peroxidation of leaves and biomass allocation in response to a long-term period of salt stress (48 days). Additionally, a proteomic approach was used to identify salt stress-induced proteins in cultivar RB867515 that may have biotechnological potential.

## Results

### Photosynthesis and leaf water potential

In both cultivars, RB855536 and RB867515, photosynthetic rates of control and salt-treated plants significantly decreased after 48 days of salt stress (Figure 1A and 1B). However, there were no statistically significant differences between the two cultivars, indicating that the varieties RB855536 and RB867515 behaved similarly with respect to net photosynthesis during salt stress.

Leaf water potential of RB855536 and RB867515 plants subjected to salinity became more negative from day 15 until the end of the experiment (Figure 1C and 1D). At day 48, the water potential of control plants remained at values similar to those of previous timepoints, while salt-treated plants showed a

sharp decrease in leaf water potential compared to that of day 15. However, there were no differences in leaf water potential between salt-stressed plants of the two cultivars.

### Biomass allocation and malondialdehyde (MDA) content

Salt treated RB855536 and RB867515 plants showed a reduction in shoot dry mass in comparison to control plants (Figure 2A and 2B). Similar results were obtained for roots (Figure 2C and 2D). Comparing the dry mass of controls between the two cultivars, no significant difference was observed for shoots. However, RB867515 control plants showed significantly more root dry mass than RB855536. In relation to malondialdehyde content, cultivar RB855536 plants subjected to salt stress showed a statistically significant increase in lipid peroxidation (MDA) levels from day 10 to 48, with a slight decrease of MDA levels for this last day (Figure 2E). For cultivar RB867515 (Figure 2F), up to day 10, both control and salt-treated plants, showed low values of MDA. However, levels of MDA showed a statistically significant increase in salt-treated plants at day 15 and a decrease at the 48, when MDA levels were similar in control and salt-treated plants. Therefore, MDA levels increased in leaves of salt-stressed plants of both cultivars; however, there was a delay in response for cultivar RB867515 in comparison to cultivar RB855536.

### Macro- and micronutrient leaf concentrations

No significant change in leaf concentrations was observed for any of the macro and micronutrients tested (results not shown), except for manganese. On day 48, there was a statistically significant reduction in $Mn^{2+}$ concentration values in control and

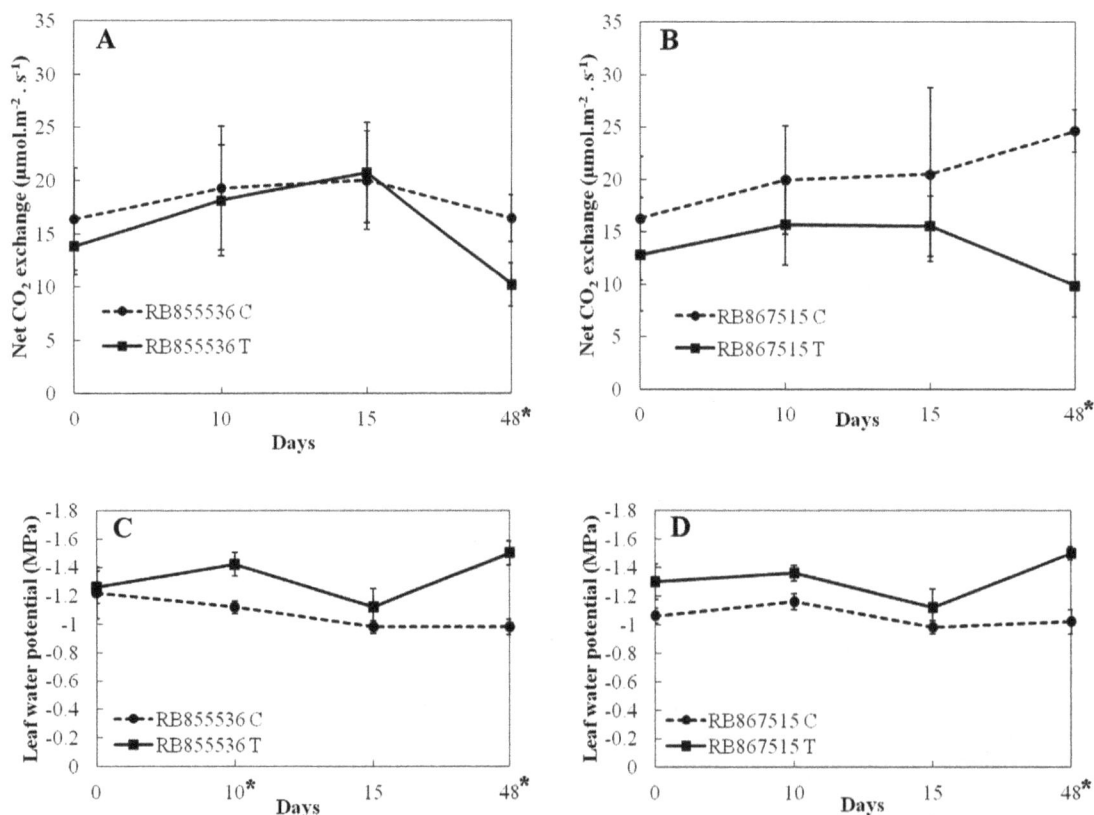

**Figure 1. Net $CO_2$-exchange ($\mu mol.m^{-2}s^{-1}$) for (A) cultivar RB855536 and (B) cultivar RB867515 over time.** Leaf water potential (MPa) in sugarcane leaves during 48 days of salt treatment for (C) cultivar RB855536; and (D) cultivar RB867515. Values are presented as mean ± SD (n = 6). "●" are control plants and "■" are salt-treated plants. *Significant at p≤0.05.

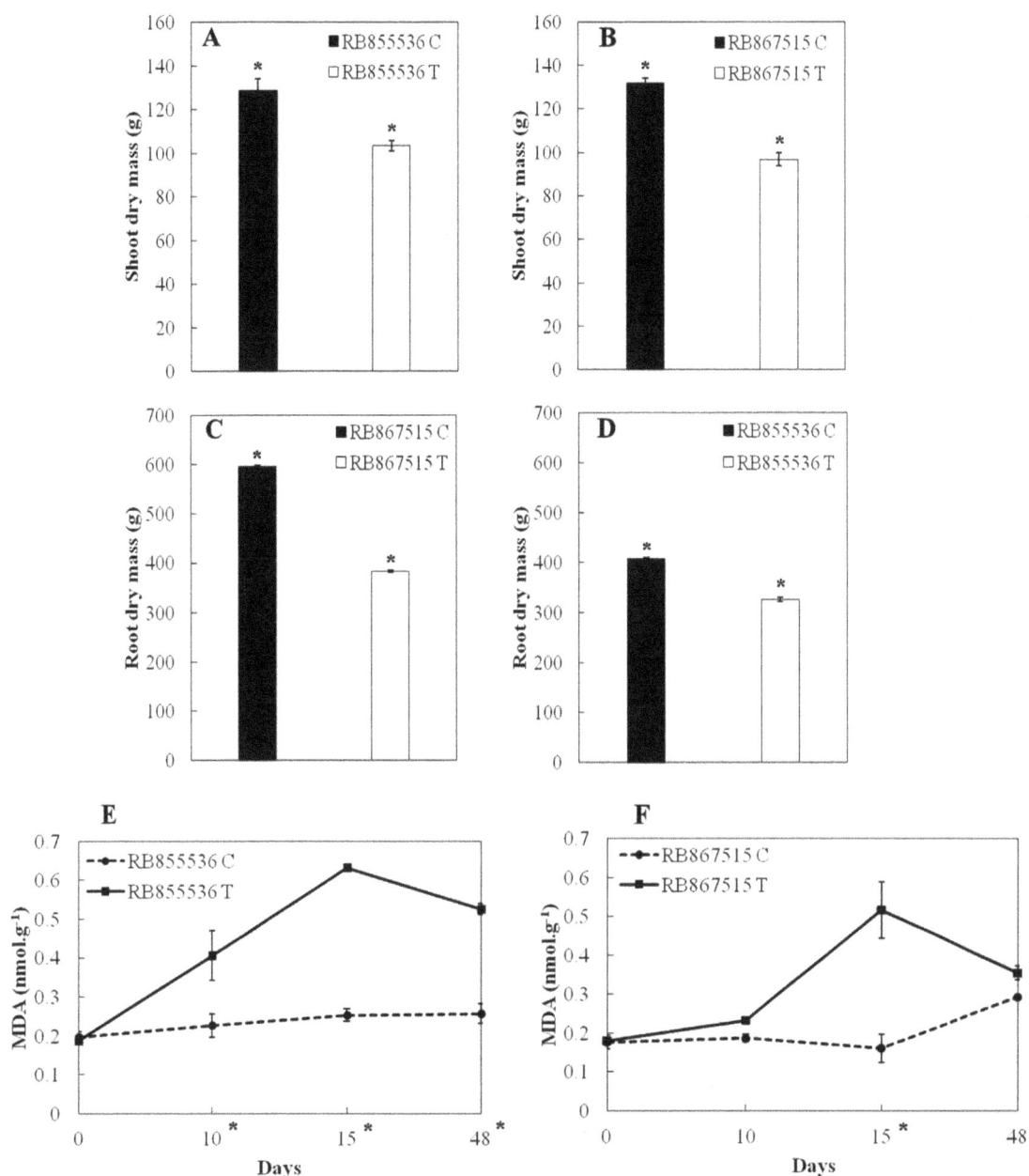

**Figure 2. Shoot dry mass for cultivars(A) RB855536 and (B) RB867515; and root dry mass for cultivars (C) RB855536 and (D) RB867515 after being subjected to 48 days of salt stress (100 mM NaCl).** Lipid peroxidation levels (MDA) in sugarcane leaves during 48 days of salt stress (100 mM NaCl) for (E) cultivar RB855536 and (F) RB867515. Values are presented as mean ± SD (n = 6 plants). "●" are control plants and "■" are salt-treated plants. *Significant at p≤0.05.

salt-treated plants for both cultivars (Figure S1A and S1B in File S1).

## 2-DE analysis of proteins in the sugarcane cultivar RB867515

To identify proteins that are expressed during salt stress in cultivar RB867515, the protein expression profiles of leaves of plants watered with distilled water and with 100 mM of NaCl solution for 48 days were compared using bidimensional protein electrophoresis (Figure 3). Although at days 15 and 48, cultivar RB867515 showed significant changes in some physiological parameters, day 48 was chosen for protein expression analysis

due to the greater differences in physiological parameters between water and salt-treated plants, such as a decline in net photosynthesis and leaf water potential (Figure 1). Proteins for both salt-treated and control plants were found mostly in the 4 to 7 pI range. After the second dimension was run, replicates of gels were compared for reproducibility. The gels with highest $r^2$ were used to make the reference gels for control ($r^2 = 0.85$) and salt-treated plants ($r^2 = 0.84$). Comparison of control and salt-treated plants reference gels allowed the identification of proteins that showed at least a 1.5-fold differential expression between gels. Twelve proteins were selected from gels of salt-treated plants and eight proteins were selected from gels of water-treated plants. From a

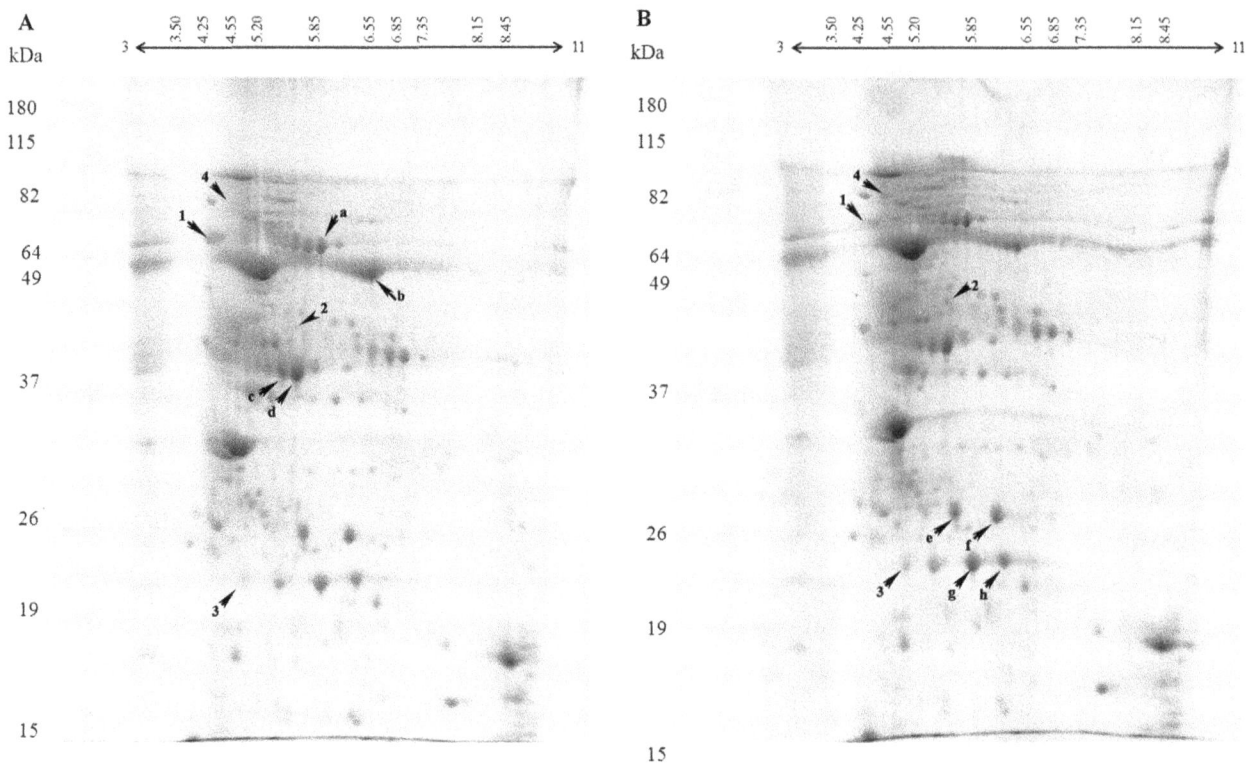

**Figure 3. Two-dimensional gel electrophoresis patterns of proteins extracted from sugarcane leaves of the RB867515 cultivar watered (A) with distilled water and (B) after being subjected to 100 mM NaCl for 48 days.** The strips used were 13 cm long, with a non-linear pH gradient of 3-11, stained with Coomassie G-250. The proteins indicated by numbers (1-4) correspond to those showing at least 1.5-fold difference in expression levels between the two different treatments; proteins indicated with letters (a-h) represent proteins with no difference in expression profile.

total of twenty selected proteins, twelve were identified (i.e., four showed difference in protein expression and eight were used as control proteins) (Figures S2 to S10 in File S1). Low concentration of proteins in spots precluded identification of the remaining proteins. As shown in Table 1, the four differentially expressed proteins successfully identified were: (1) fructose 1,6-bisphosphate aldolase that was down-regulated in salt-treated plants, (2) germin-like protein that was up-regulated in salt-treated plants and (3) glyceraldehyde 3-phosphate dehydrogenase that was up-regulated in salt-treated plants, and (4) a heat-shock 70 protein that was found only in salt-treated plants (Figure 4A and 4B). Eight additional proteins that showed no change in expression levels were chosen as controls. These were identified as another isoform of fructose 1,6-bisphosphate aldolase, RUBISCO large subunit, ATP synthase CF1 α subunit, 23 kDa polypeptide of PS II oxygen evolving complex and another isoform of germin-like protein.

## Discussion

### Physiological and biochemical analysis

Many physiological functions in plants are affected by soil salinity and the effects of prolonged stress were observed in sugarcane leaves. Both sugarcane cultivars showed a decrease in their photosynthetic rates during the experiment. In spite of being considered drought tolerant, and the fact that there are similarities between drought and salt-stress responses, RB867515 did not behave as a salt tolerant cultivar in controlled experiments. However, the decrease in photosynthetic rates may also have been a response caused by the decrease of water potential for both

cultivars. Stepien and Kłobus [23] working with several concentrations of NaCl in cucumber (*Cucumis sativus* L.) observed a decrease in net photosynthesis due to increasing water deficit. According to Suzuki and Esteves [24], salinity can affect net photosynthesis by changing mesophyll cells' structure and by reducing water availability, thus decreasing the water potential. The presence of salts in the soil solution leads to decreased osmotic potential of the solution, inducing a shortage of water in plants which accounts for the resemblance between drought and salt-stress responses [11].

Decrease in growth of both shoots and roots are a well-known effect of increased salinity. In our experiments, salinity reduced shoots' and roots' mass, affecting both cultivars similarly. Interestingly, however, greater root mass was observed in cultivar RB867515 water-treated control plants in comparison to cultivar RB855536. Since cultivar RB867515 is considered by farmers to be more drought tolerant than cultivar RB855536, the fact that RB867515 had a more developed root system in control plants could help to explain this observation. A deeper root system may lead to drought tolerance as the plant has access to water in deeper layers of soil. In theoretical studies on the potential yield of sugarcane in São Paulo, van der Berg and collaborators [25] observed that the higher the root volume per layer, higher is also the potential yield of the crop for sugarcane plants of first and second cuts. Moreover, the authors also showed that the yield tends to increase with increasing volume of roots. Morris and Tai [26] tested 12 varieties of sugarcane in different water regimes and observed the effect on the development of roots and leaves. The amount of roots in the upper layers was higher in comparison to

Physiological and Proteomic Analyses of Saccharum spp. Grown under Salt Stress

165

**Table 1.** Cultivar RB867515 sugarcane leaf proteins differentially expressed between water-treated and salt-treated plants.

| Spot # | Accession # Gene Index/NCBI | Specie | Protein ID | Sequence coverage (%) | Mowse Score | Theoretical pI/ MM (kDa) | Experimental pI/ MM (kDa) | Peptide sequences |
|---|---|---|---|---|---|---|---|---|
| 1 | TC113485/ACG36798 | Saccharum officinarum | Fructose 1,6-bisphosphate aldolase | 37.6 | 67.0 | 3.50/72.24 | 5.95/42.014 | IVDILVEQGIVPGIK |
| 2 | CA117405/ABF93789.1 | S. officinarum | Glyceraldehyde 3-P-dehydrogenase | 22.9 | 43.0 | 10.94/54.99 | 5.33/59.51 | TGITADDVNAAFR |
| 3 | TC121006/ABG46233 | S. officinarum | Germin-like protein | 20.8 | 49.0 | 5.00/22.36 | 4.60/21.815 | VFLFDDAQVKPGVDGLGLSAAR |
| 4 | -/CAC16169 | Zea mays | HSP70 | 19.6 | 44.3 | 4.10/75.78 | 6.23/23.415 | SSVHDVVLVGGSTR |
| a | -/AAN27981 | Hordeus stenostachys | RUBISCO large subunit | 19.9 | 31.6 | 7.00/67.56 | 6.04/52.766 | DTDILAAFR |
| b | TC112694/YP_054628.1 | S. officinarum | ATP synthase CF1 α subunit | 53.4 | 43.9 | 5.89/71.08 | 6.32/55.294 | IAQIPVSEAYLGR |
| c | -/ACM78035 | Triticum aestivum | Fructose 1,6-bisphosphate aldolase | 26.0 | 27.4 | 5.07/53.58 | 5.48/39.238 | AAQDALLLR |
| d | TC113485/ACM78035 | S. officinarum | Fructose 1,6-bisphfosphate aldolase | 27.8 | 113.0 | 5.38/40.00 | 5.39/40.245 | EAAYYQQGAR |
| e | -/ABY52939.1 | Oryza sativa | Polypeptide 23 kDa of PS II | 16.2 | 21.0 | 5.49/26.00 | 8.28/28.315 | QYYSWVLTR |
| f | CA184061/ABY52939.1 | S. officinarum | Polypeptide 23 kDa of PS II | 16.2 | 75.0 | 6.50/26.00 | 8.66/26.939 | TNTDYLAYSGDGFK |
| g | TC121006/ABG46233 | S. officinarum | Germin-like protein | 15.0 | 21.0 | 5.88/22.56 | 5.96/21.815 | PGVDGLGLSAAR |
| h | TC121006/ABG46233 | S. officinarum | Germin-like protein | 16.3 | 66.0 | 6.63/23.20 | 6.49/21.815 | AAVTPAFVGEFPGVDGLGISAAR |

Proteins were trypsin digested and identified by MALDI-ToF/ToF MS. "1–4" are differentially expressed proteins and "a–h" are proteins with no difference in protein expression.

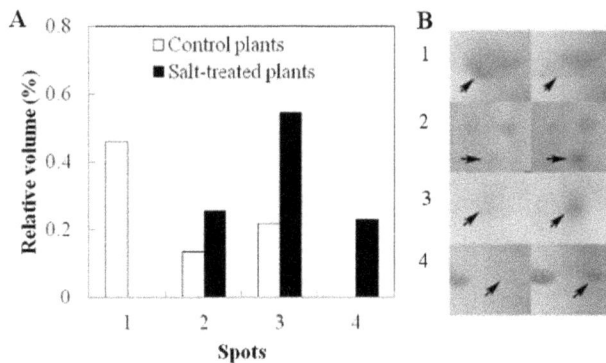

**Figure 4. Relative volume of protein spots corresponding to differentially expressed proteins.** (A) Quantification of protein expression and (B) Image of protein spots on gels. The left panel shows protein expression on water-treated control plants and right panel shows protein expression on salt-treated plants at day 48. Proteins were identified by mass spectrometry as being (1) fructose 1,6-bisphosphatealdolase; (2) glyceraldehyde-3-phostate dehydrogenase; (3) germin-like protein (4) HSP70. Bars show the mean values of replicate spots.

the lower layers. However, the diameter of the roots was increased in the lower layers. This result is in agreement with those found by Laclau and Laclau [27], where the greatest amount of roots with smaller diameter was found in the upper layers of soil in irrigated culture and the largest amount of roots with greater diameter in the deeper layers in rainfed crops. The maximum depth of sugarcane roots, however, is not a consensus. Smith *et al.* [28] report that root water uptake activity is restricted to a depth of 1.5–2.0 m, but Evans [29] observed this activity at a depth of 6.0 m for sugarcane roots. The size and distribution of the root system of plants is deeply affected by the availability of water in soil, which causes differences in the ability of crops to exploit resources in the lower soil layers [28]. Tolerance of sugarcane to water deficit in places where water is present in deep soil layers may imply an increase in root mass, length and diameter of the root [30]. It is important to note, though, that unlike drought, salt is expected to stress plants continually, from the time of emergence. Although drought tolerance was not tested here and the roots of both cultivars were similarly susceptible to salt, the results obtained suggest that farmer's observations that cultivar RB867515 is more tolerant to drought than RB855536 may be due to its greater root mass.

Reactive oxygen species (ROS) oxidize membranes with the increase of abiotic stresses like salt and drought. Active oxygen species cause deterioration of lipid membranes in plant cells and the levels of peroxidation are measured in terms of MDA content [31]. The RB867515 cultivar resisted to salt-stress conditions until day 15, in contrast to the RB855536 cultivar which showed an increase in MDA levels starting on day 10. However, on day 48, a decrease in MDA levels was observed for both cultivars. This decrease could be due to the presence of detoxification enzymes acting under the ROS [32]. Moradi and Ismail [33] observed that, under different levels of salt stress, rice tolerant to salinity responded to this stress producing lipid peroxidation, however with no statistical differences between tolerant and control plants. Our results were similar to those reported by Shao *et al.* [34]. Working with ten wheat genotypes and several water deficit levels, they were able to separate them according to the production of anti-oxidant enzymes and the production of MDA in each level of stress (mild, moderate and severe). The varieties which showed a

greater production of MDA had a lower production of anti-oxidant enzymes. Moreover, in the genotypes that showed an increased production of enzymes, the production of MDA was lower. Although we did not directly test enzyme activity, these results highlight the importance of antioxidant enzyme activity in plants to adverse actions of salt stress, indicating the presence of different pathways to adapt to water stress.

## Proteomic analysis

The study of global patterns of protein expression via various proteomics techniques has gained a lot of attention in recent years. Assessment of mRNA expression has an important caveat which is that multiple layers of regulation of gene expression can lead to situations where mRNA expression levels are not mirrored by protein expression levels [35]. Given that the protein is the active biomolecule in the cell, studying the proteome becomes crucial.

Few proteomic studies have been performed using sugarcane. In studies of sugarcane under the related abiotic stress of drought, Jangpromma *et al.* [36] described an increased expression of an 18 kDa protein. In other report by Jangpromma *et al.* [37], the 18 kDa protein, named p18, was similar to heat-shock proteins or dehydrins and they hypothesized that it may have an important function in protecting the plant against drought, once this protein may help to protect specific cell structures by binding water molecules. Also, p18 may be a stress-inducible heat shock protein, protecting cells from stress injury and helping the folding of new proteins. According to MS/MS, the p18 may be a hydrophilic protein. Hydrophilic proteins are usually charged, which allows them to interact with water or other hydrophilic/polar molecules or to act as molecular chaperones, preventing damaged protein aggregation. Zhou and collaborators [38] verified a change in the expression pattern of proteins in sugarcane leaves submitted to osmotic stress induced by PEG, and reported an increase of two proteins (i.e., 22 kDa protein and RuBisCO small subunit) and the decrease of the other two (i.e., isoflavone reductase-like protein and delta chain of ATP synthase). RuBisCO (ribulose-1,5-bisphosphate carboxylase/oxygenase) catalyses the reaction of D-ribulose 1,5 -bisphosphate and atmospheric $CO_2$ to form one molecule of 3-phosphoglycerate and one of phosphoglycolate, being an essential enzyme of the Calvin cycle. The presence of salt in the soil interferes with root water uptake changing the plant water status, increasing leaf water potential and reducing stomatal conductance, therefore reducing photosynthesis. Found in large amounts in the plant leaves, RuBisCO is crucial to provide adequate photosynthetic rates, especially during the salt stress [39]. Ngamhui *et al.* [40] in their work with drought stress and two-dimensional electrophoresis of sugarcane leaf proteins used 13 cm strips ranging from pI 4-7 and identified more than 300 proteins with differences in their expression; and successfully sequenced 19, among them proteins related to photosynthesis, ROS detoxification and defense proteins. The fact they used strips with pI ranging from 4 to 7 may be one of the reasons why these authors identified several proteins that responded to stress. The proteins identified in the present work about salt stress were also in this pI range. The strips used in this study were 13 cm, pI ranging 3-11. Strip selection may have made it difficult to identify proteins that respond to salt stress as they would be compressed in the acidic side of the strip.

A recent study, complementary to this one, addresses changes in sugarcane roots subjected to salt stress [41]. In this study, plants of the same sugarcane varieties used in the present study were cultivated for 45 days and then treated with nutrient solution containing 200 mM NaCl. Samples were harvested for analysis at 2 h and 72 h after treatment. This protocol is in contrast to the one used in the present work where 4 month old plants were treated with water or 100 mM NaCl solution for 48 days.

In this work we have identified four proteins in cultivar RB867515 leaves that respond to salt stress: Fructose 1,6-bisphosphate aldolase (Figure S2 in File S1) was down-regulated, a glyceraldehyde 3-phosphate dehydrogenase and a germin-like protein (Figure S3 and S4 in File S1, respectively) showed increased expression levels under salt stress, and a heat-shock protein 70 (Figure S5 in File S1) was expressed only in salt-treated plants. Our proteome analysis was reproducible, however, the identified changes in protein expression pattern should be confirmed in the future by an alternative technique such as Western blot. The proteins identified are involved in energy metabolism and defense-related responses and their possible participation in helping plants tolerate salt stress is discussed below.

## Proteins involved in energy metabolism

The adaptation of plants to stress is associated with changes in the expressed complement of proteins. It follows that proteomic studies can contribute significantly to the understanding of the relationship between protein abundance and plant acclimation to a stressful environment. Current data indicate more than 2170 identified proteins that respond to stress from 34 plant species, of which 940 or so were identified in leaves of different plants, including the Poaceae family [42,43], which includes sugarcane. Understanding how the plant responds to stress at a proteomic level, together with data from physiology and biochemistry, can provide directions for how to obtain cultivars with resistance to abiotic stresses such as salinity [44] in sugarcane breeding programs.

Differences in physiological parameters between water-treated and salt-treated plants, such as decrease in photosynthetic rate and water potential, prompted the proteomic analysis of cultivar RB867515 at day 48. The proteins identified for the cultivar RB867515 were involved in energy metabolism processes and are known from studies with other plant species to be early responders of abiotic stresses such as salinity [42,45]. Fructose-1,6-bisphosphate aldolase (EC 4.1.2.13) is a key enzyme of the energy metabolism, which catalyses the cleavage of β-fructose-1,6-phosphate into D-glyceraldehyde-3-phosphate and dihydroxyacetone phosphate in glycolysis, in addition to the reverse reaction during gluconeogenesis. In this work, it was found that in RB867515 salt-treated plants there was a decrease in the expression of fructose-1,6-bisphosphate aldolase (Figure 4, spot number 1). Sobhanian et al. [46] observed similar results when the halophyte grass Aeluropus lagopoides was submitted to increased salt levels, some photosynthesis-related proteins, such as fructose 1,6-bisphosphate aldolase, showed decreased expression levels. Different results were obtained by Salekdeh and colleagues [47], who observed that this enzyme increased its expression by 40% in salt-stress rice leaves. Abbasi and Komatsu [48] also observed in the rice leaf sheath an increased expression of FBP aldolase, under different stresses such as cold, salinity and drought, indicating that the plant responded to stressful stimuli by overexpressing this enzyme. There are different reports of a decrease in fructose 1,6-bisphosphate aldolase [49,50]. According to Chaves et al. [51], the expression of some proteins from the Calvin cycle and photorespiration (e. g. fructose bisphosphatealdolase) are differently affected by abiotic stresses (e.g., salt and drought) [52–54]. These differences may be due to the plant's ability to react differently to imposed stress conditions and the need for growth and compartmentalization of metabolites resulting from the photosynthetic process. Therefore, the response of increased or decreased

fructose 1,6-bisphosphate aldolase expression may be due to genetic factors of the plant. It is to be noted though that FBP aldolase was sequenced from different spots on gels (Figure S6 in File S1), indicating that perhaps the presence of isoforms may explain some of the different results in protein expression reported in the literature [45]. Also, FBP aldolase may be an important function in ion vacuole compartmentalization. Barkla et al. [55] demonstrate that this enzyme can interact directly with and active the ATPase-depended $H^+$ presents in vacuolar membrane, stimulating its ATP binding and hydrolysis activity, important step for salt import into the vacuole, helping the plant cell eliminate the excess of ions $Na^+$ and $Cl^-$ of the cytoplasm.

Another important enzyme in the energy metabolism is glyceraldehyde-3-phosphate dehydrogenase (EC 1.2.1.12), which was up-regulated in salt-treated plants (Figure 4, spot number 3). This enzyme belongs to the family of dehydrogenases and catalyzes the oxidation of glyceraldehyde 3-phosphate to 1,3-biphosphoglycerate in the glycolytic pathway, in a reaction that produces ATP. This enzyme is also present in the nucleus where it has important roles in gene transcription, DNA replication, DNA repair and RNA export [56]. According to Yang et al. [57] overexpression of the glycolysis pathway enzyme is of paramount importance for the increase of soluble sugars accumulation, as well as for providing more energy needed for the plant under stress, and therefore, is an indicator of stress tolerance. The increase in glyceraldehyde-3-phosphatedehydrogenase and the reduction of fructose 1,6-bisphosphate aldolase has been reported for cucumber by Du et al. [58], in which these proteins may have altered the activity of the glycolysis pathway, hence the accumulation of soluble sugars would be lower. Sobhanian et al. [46] also found a decrease of fructose bisphosphate aldolase in the halophyte grass Aeluropus lagopoides, but they also observed an increase in glyceraldehyde-3-phosphate dehydrogenase in response to salinity. The increased expression of this protein may reflect the pattern of carbon flux in response to a reduction in photosynthesis and high demand for the osmotic regulation in the leaves caused by salinity.

## Germin-like protein

Initially described in wheat seeds, germin-like proteins (GLP) have several functions [59,60] such as receptors and detoxification enzymes [54,61,62]. According to Woo et al. [63], germin is an apoplastic, glycosylated enzyme with resistance to heat, degradation by proteases and hydrogen peroxide. This resistance may be due to its similarity to desiccation tolerant proteins present in seeds. Germin-like proteins may function as reactive oxygen species (ROS)-scavengers and have a common structure with "true-germin" protein family members, as β-jellyrolls monomers united in a trimer of dimers (homohexamers), with a single manganese ion per monomer. This structure is also similar to other plant ROS-removing enzymes such as Mn-dependent superoxide dismutase (Mn-SOD) [64]. The catalytic processes of these enzymes depend only on the presence of a manganese ion bound between the monomers, with no involvement of other cofactors or specific changes in amino acid residues [63]. Ngamhui and colleagues identified two enzymes involved in ROS detoxification, among them a CuZn-SOD in sugarcane leaves of Thai drought-tolerant cultivars [40].

A GLP with increased protein expression in salt-treated plants was identified in cultivar RB867515 (Figure 4, spot number 2). According to Zimmermman et al. [65], germin-like proteins belong to a multigene family – e.g. groups of genes from the same organism encoding proteins with similar sequences either in its full length or limited to some specific domain. This would explain the presence of different germin-like proteins with the same molecular mass although with a different pI. The identification of a protein that uses manganese ions during catalysis is interesting because among all macro and micronutrients studied, manganese was the only micronutrient that showed a statistically significant increase in salt-treated plants at 48 days (Figure S1 in File S1). Increased expression of enzymes such as GLP that are responsible for ROS-detoxification together with high levels of manganese could be one of the mechanisms that enables cultivar RB867515 to withstand the adverse conditions of salt stress.

## Defense-related proteins

Molecular chaperones are key components to cellular homeostasis under normal and adverse conditions of growth. They are responsible for protein folding, translocation and degradation processes in normal cell functioning. Most molecular chaperones are stress proteins, many of them identified as heat-shock proteins (HSPs) [66,67]. Protein spot 4 (Figure 4) present only in salt-stressed plants has been identified as a HSP 70. Aghaei et al. [68] investigated the behavior of two contrasting varieties of potatoes (i.e., salt-sensitive and salt-tolerant) in response to 90 mM of salt and found that the overexpression of HSPs occurred only in salt-stress tolerant potato plants. They concluded that HSPs could be considered part of the mechanism that confers salt tolerance in potatoes. In grape, Grimplet et al. [69] demonstrated HSP60 expression under drought stress. Studies performed previously by Tiroli and Ramos [70] identified the production of HSP70 in grapes during harvest, which may be considered a stress. After harvest and during ripening, plants undergo a period of dehydration, which can be considered as the main stress factor, initiating the production of proteins responsible for cell turgor and protection against oxidative stress [71] and defense-related proteins, such as heat shock proteins [72]. In sugarcane, there have been reports of the expression of small HSPs (sHSP). Tiroli and Ramos [70] identified a class I sHSP in sugarcane that responded to high temperature stress using ESTs from the sugarcane database. Similar results were obtained by Tiroli-Cepeda and Ramos [19], when they observed that high temperatures induced protein aggregation. Sugarcane plants exposed to high temperatures induced the expression of sHSP class I proteins, which led to increased activity of chaperones in the cell to help previously existing proteins return to function and newly synthesized proteins achieve correct folding. Rodrigues et al. [73], despite not having used a proteomic approach, observed an increase in three types of HSP (17.2, 70 and 101) in drought stress tolerant Brazilian sugarcane cultivars. Recently, Ngamhui et al. [40] using drought-tolerant Thai sugarcane cultivars described a class IHSP of 16.9 kDa that was up-regulated in sugarcane leaves under drought stress for five days. The expression of HSPs occurred mainly in tolerant plants, a similar result found in this work with RB867515 salt-treated plants, demonstrating that these proteins may participate in the protection of sugarcane against salt stress.

Experiments in which chrysanthemum HSP70 gene was over-expressed in *Arabidopsis thaliana* showed that the increasing in HSP70 expression led to a remarkable tolerance to heat, drought and salinity [74]. Salinity also increases the peroxidation of membranes, and an increase in MDA concentration. The presence of oxidized lipids led to the increase in peroxidase activity. Song and co-workers [74] also noted that membrane damage caused by the action of ROS was lower in plants overexpressing HSP70 compared to the wild type, indicating that the presence of these proteins may be crucial to minimize the damage caused by salinity in plants.

In conclusion, the increase in the glycolytic pathway proteins, such as glyceraldehyde 3-P dehydrogenase, could help the carbon flux through the Calvin cycle leading to an increase in sucrose production [75] and contribute to plant stress tolerance. HSP70 identified in the RB867515 variety, together with GLP (Figure S11 in File S1), may alleviate the damage caused by oxidation, especially in chloroplasts, which might partially contribute to reducing the damage caused by stress, since the decrease in MDA concentrations was observed for day 48 in sugarcane leaves. HSP70 and GLP may protect sugarcane plants against protein unfolding and membrane peroxidation, contributing to the tolerance of sugarcane to moderate salt stress.

## Materials and Methods

### Plant material

Sugarcane cultivars RB855536 and RB867515 were acquired from RIDESA. Culms of both cultivars were grown in vermiculite for a month and then transplanted to 25 cm in diameter pots with drainage holes containing a mix of soil/manure/sand (4:2:1, w/w/w). At 4-months, six plants of each cultivar were treated with 100 mM of NaCl solution and other six plants of each cultivar were watered with distilled water (control group) up to field capacity every day in the morning during a period of 48 days. All twenty four plants were randomly arranged.

### Measurements of net photosynthesis and water potential

Net photosynthesis measurements were performed in the morning (8–10 am) after watering. Photosynthesis of the middle third of fully expanded +1 leaves (the first leaf, from top to bottom, with visible sheath, see Figure S12 in File S1) of each repetition was measured using a portable photosynthesis system LICOR-6400 (Li-COR, Lincon, NE, EUA) at 0, 10, 15 and 48 days. After net photosynthesis was measured, +1 leaves were introduced into a Scholander pressure chamber [76]. The applied pressure was increased by increments of 0.2 MPa using nitrogen gas until the xylem sap became visible in the leaf lamina surface. This pressure was considered as the xylem water potential. This analysis was performed at 0, 10, 15 and 48 days of salt stress.

### Determination of dry mass and malondialdehyde content (MDA)

At day 48, shoots and roots of both control and salt-treated cultivars were separated and dried in an oven at 80 °C with forced air circulation and weighted using a digital scale until the mass values became constant. The MDA content of sugarcane leaves was determined according to Hodges et al. [77]. Briefly, one hundred milligrams of ground leaves were homogenized with 6.5 mL of 80% ethanol (v/v) and centrifuged for 10 minutes at 16,100 $g$. A total of 1 mL of this extract was transferred to a new tube and 1 mL of 0.65% thiobarbituric acid (TBA) (w/v) in 20% trichloroacetic acid (TCA) (w/v) were added followed by incubation at 95 °C for 25 min. Samples were then transferred to ice for 10 minutes and centrifuged for 10 minutes at 16,100 g. The supernatant was transferred to a new tube and absorbance was read at 532 nm and 600 nm. MDA equivalents in nmol.g $FM^{-1}$ were obtained using the following equation: MDA (nmol.g $FM^{-1}$) = $[(A_{532}-A_{600})/155000] \times 10^6$.

### Quantification of leaf nutrient concentrations

For the analysis of macro and micro nutrients, approximately 1 mg of leaf tissue (leaf +1) of each replicate (six plants from controls and six from salt-treated plants) of both cultivars previously pulverized in liquid $N_2$ were placed in an oven at 65

°C for 72 hours. Leaf concentrations of P, K, Mg, S, Al, B, Cu, Fe, Mn and Zn were determined in the Analytical Chemistry Laboratory of Embrapa Cerrados (CPAC) by the technique of optical emission spectrometry with inductively coupled argon plasma in a Thermo Jarrell Ash spectrometer model IRIS/AP. Leaf nitrogen concentrations were determined by the colorimetric method described by Kjeldahl [78].

## Protein extraction from leaf material and quantification

At day 48, fully expanded +1 and +2 leaves of six plants (Figure S2 in File S1) from each treatment (control and salt-treated plant) of the RB867515 cultivar were harvested and ground to a fine powder with liquid nitrogen using a mortar and pestle. Total protein extraction was performed according to Wang et al. [79]. To remove the photosynthetic pigments, 10 g of powdered leaves were homogenized with 10 mL of 100% chilled acetone, followed by centrifugation at 23,500 g for 5 min at 4 °C. The supernatant was discarded and the procedure was repeated three times. The pellet was then homogenized with 10% trichloroacetic acid (TCA) in chilled acetone (w/v), followed by centrifugation at 23,500 g for 5 min at 4 °C. The supernatant was discarded and this step was repeated three times. The pellet was homogenized in 10% TCA in cold distilled water (w/v), centrifuged (23,500 g for 5 min at 4 °C), and the supernatant discarded, this step was repeated three times. The pellet was resuspended with 80% chilled acetone (v/v), centrifuged (23,500 g for 5 min at 4 °C), and the supernatant again discarded. This step was repeated three times. The final pellet was then homogenized with 10 mL of buffered phenol (Sigma-Aldrich) and 10 mL of solubilization buffer containing 30% sucrose (w/v), 2% SDS (w/v), 0.1 M Tris-HCl buffer (pH 8.0) and 5% β-mercaptoethanol (v/v). This solution was vortex-mixed and centrifuged at 23,500 g, for 5 min at 4 °C. The upper phase (phenolic) was collected, and proteins were precipitated by adding 3 times the volume of a solution containing 0.1 M ammonium acetate in 100% of cold methanol (w/v) overnight at −80°C. Samples were then centrifuged at 23,500 g for 5 min at 4 °C. The pellet obtained was washed twice with 0.1 M ammonium acetate in 100% of cold methanol and twice with chilled 100% acetone. After complete evaporation of acetone at room temperature, the pellet was resuspended in 2% SDS (w/v), 5% glycerol (v/v), 50 mM Tris-HCl buffer (pH 6.8). Protein quantification was performed using the methodology of Lowry et al. [80], and the RC DC protein assay kit (BioRad).

## 2-DE and comparative proteome analysis

Seven hundred micrograms of sugarcane leaf proteins (in triplicate for each treatment) were precipitated on ice for 1 h using a 10% TCA (final concentration) solution. After precipitation, proteins were centrifuged at 16,100 g for 20 min at 4 °C. The supernatant was discarded and the pellet washed three times with cold 100% acetone (each wash was followed by centrifugation at 16,100 g for 20 minutes at 4 °C). The pellet was solubilized in a hydration solution (8 M urea, 2% CHAPS, 0.5% IPG buffer with a trace of bromophenol blue and 65 mM DTT) and applied onto a IPG (immobilized pH gel) 13 cm non-linear strip (pH 3-11) (GE Healthcare) by incubation for 16 h at room temperature. The strips were then submitted to isoelectric focusing (IEF) using an Ettan IPGphor 3 (GE Healthcare) apparatus until it accumulated 53,250 v.h$^{-1}$. After IEF, strips were equilibrated in solutions of DTT (50 mM Tris-HCl buffer (pH 8.8), 6 M urea, 30% glycerol, 2% SDS, 1% dithiothreitol-DTT, a trace of bromophenol blue) and iodoacetamide (50 mM Tris-HCl buffer (pH 8.8), 6 M urea, 30% glycerol, 2% SDS, 1% iodoacetamide, a trace of bromophenol blue) for 15 min each, placed on top of 12% polyacrylamide

gels according to Laemmli [81] and sealed with agarose solution (25 mM Tris-HCl buffer (pH 8.3), 192 mM glycine, 0.1% SDS, 0.5% agarose and a trace of bromophenol blue). Gels were run until the bromophenol blue reached the end of the gel using the following parameters for electrophoresis: (1) 30 min at 600 V, 90 mA, 100 W and (2) 8 h at 700 V, 240 mA, 100 W. After electrophoresis, gels were fixed in distaining solution (50% distilled water, 40% methanol, 10% acetic acid) for at least 1 h and stained overnight according to Neuhoff et al. and Kang et al. [82,83], with Coomassie Blue G-250 (BioRad) (0.1% Coomassie Blue G-250, 2% phosphoric acid, 10% ammonium sulfate and 20% methanol).

## Image acquisition and data analysis

Images of gels were obtained using a scanner HP Scanjet 8290 on photography mode. Analysis of the images was performed using BioNumerics software version 5.10 (Applied Maths, Belgium). Before analysis, all images were converted to gray scale, using the following parameters of the program: TIFF format with OD of 8 bits, 500 kbits, 47% of spot contrast, 75% of spot separation, 25 pixels of spot size and 3 pixels for minimum spot size. For the normalization of gels, one reference gel was created to generate a standard gel, determined by the molecular mass markers (Y), isoelectric point (x) and intensity (z) of spots. After normalization, spots of each gel were connected to the reference gel previously created. Other values remained below the default values of the software. For comparison, the gels had their equivalent spots connected and identified numerically and then the values of volume generated by the program for each spot were used for calculations of correlation. All possible comparisons with the values of volume for each treatment were performed. The gel with the highest correlation coefficient with the other two repetitions was considered representative and was chosen for comparison with the corresponding representative gel from the other treatment (see Figure S13 in File S1). Spot were selected for MALDI-TOF/TOF MS identification using the criteria of 2-fold increased or decreased expression levels.

## In-gel digestion and desalinization of digested proteins

The gel spots selected for identification were excised from the three replicate gels and pooled into 1.5 mL tubes. Protein digestion was performed using the methodology described by Shevchenko et al. [84,85], with modifications. Gel slices were distained overnight and washed with 50% ethanol (v/v) three times, with a 15 min interval between washes. After discarding the ethanol solution, 300 μL of 100% acetonitrile (ACN) was added until gels exhibited a white color. Next, the ACN was removed and 50 μL of 100 mM ammonium bicarbonate and 10 mM DTT were added and tubes were incubated in a water bath at 56 °C for 30 min. After this, the liquid was removed, 50 μL of 100 mM ammonium bicarbonate and 55 mM iodoacetamide were added and tubes were left at room temperature for 90 min. The solution was discarded and the gel slices were washed twice with 100 μL 100 mM ammonium bicarbonate with a 10 min interval between washes. After this, 100% ACN was added until gel slices turned white. Acetonitrile was removed, and tubes were kept at room temperature until the remaining acetonitrile evaporated. Next, tubes were put on ice and 45 μL of 50 mM ammonium bicarbonate containing 5 mM CaCl$_2$ and 5 μL of trypsin gold (Promega) were added to each tube for 45 minutes and then incubated at 37 °C for 24 h. The liquid was then transferred to a new tube and dried in a cold speed vacuum. After in gel digestion, proteins were desalted with PerfectPure C-18 columns coupled tips (Eppendorf), according to manufacturer's instructions. Desalted proteins were dried in a speed vacuum and solubilized in 10 µL of ultrapure water.

## Identification of proteins through MALDI-TOF/TOF MS, NCBI and Gene index database

Proteins previously digested and desalted were prepared for MALDI-ToF/ToF mass spectrometry analysis using an Ultraflex III instrument (BrukerDaltonics, Billerica, MA). Three microliters of an α-cyan 4-hydroxicynnamic acid saturated solution (1% [w/v] α-cyano-4-hydroxycinnamic acid, 3% [vol/vol] trifluoroacetic acid, and 50% [v/v] acetonitrile) were added to 1 µL of the resuspended sample and applied onto a MALDI target plate in triplicate. Samples were dried at room temperature and the mass spectrometer was operated in reflective mode to obtain the mass spectral profile of peptide fragments generated by trypsin digestion. MS/MS spectra for selected peptides from each protein (around 60 peptides in total) were acquired in LIFT mode. Protein identification proceeded by peptide mass fingerprinting (PMF) and peptide *de novo* sequencing. The peptides masses obtained per protein digestion were compared to the non-redundant plant NCBI database with MASCOT software (MASCOT version 2.2, *Matrix Science*, London) assuming carboxyamidomethylation of cystein and methyonine oxidation as modifications. In parallel, the sequences obtained from the MS/MS spectra were compared to the non-redundant plant NCBI database and Gene Index database (http://compbio.dfci.harvard.edu/tgi/), using organism, max score and max identity as criteria of protein selection.

## Statistical analysis

The photosynthetic rate, water potential and manganese concentration data were analyzed by linear mixed models using individuals as random factors to allow the analysis of repeated measurements over time. The fixed variables of these models were measurement day, treatments, and cultivar type. The measurement day was handled as a categorical variable because of the relatively low number of levels and the lack of a clear linear relationship with the response variables. Instead of using a full factorial model, only the meaningful interactions for this experiment were evaluated. These were day:treatment (the effect of treatment could be different along days), cultivar:treatment (cultivars could respond in different manner to treatments), and cultivar:day (cultivars could have different time dynamics). Since a constant difference among treatments along the entire experiment, including the initial day, was not expected, the main effect treatment was omitted from the models. All models were checked by visual inspection of the residual plots. For some models heteroscedasticity related to day was observed, and under these circumstances variance functions were included in the model. The differences in aerial and root dry mass at the end of the experiment among treated and control experiments were evaluated by t tests. The MDA values were analyzed only by direct observation of descriptive statistics since there were not enough leaves available for biological replicates of treatments and cultivars. All statistical procedures were carried out using the software R version 2.15.2 [86] and the mixed models analysis used also the package nlme [87]. A significance level of 0.05 was used in all tests.

## Supporting Information

**File S1** Contains the following files: **Figure S1**. Manganese concentration in sugarcane leaves (mg.kg-1) of cultivar RB855563 (A) and cultivar RB867515 (B) at various timepoints. "•" are control plants and "■" are salt-treated plants; Values are presented as mean ± SD (n = 6). * Significant at p≤0.05. **Figure S2**. MALDI-ToF/ToF spectrum sequence of fructose 1,6-bisphosphate aldolase (1) of cultivar RB867515 sugarcane leaves treated with 100 mM NaCl for 48 days. **Figure S3**. MALDI-ToF/ToF spectrum sequence of glyceraldehyde 3-P-dehydrogenase (2) of cultivar RB867515 sugarcane leaves treated with 100 mM NaCl for 48 days. **Figure S4**. MALDI-ToF/ToF spectrum sequence of germin-like protein (3) of cultivar RB867515 sugarcane leaves treated with 100 mM NaCl for 48 days. **Figure S5**. MALDI-ToF/ToF spectrum sequence of heat shock protein 70 (HSP 70) (4) of cultivar RB867515 sugarcane leaves treated with 100 mM NaCl for 48 days. **Figure S6**. MALDI-ToF/ToF spectrum sequence of fructose 1,6-bisphosphate aldolase of cultivar RB867515 sugarcane leaves treated with 100 mM NaCl for 48 days. **Figure S7**. MALDI-ToF/ToF spectrum sequence of RUBISCO of cultivar RB867515 sugarcane leaves treated with 100 mM NaCl for 48 days. **Figure S8**. MALDI-ToF/ToF spectrum sequence of ATP synthase subunit α of cultivar RB867515 sugarcane leaves treated with 100 mM NaCl for 48 days. **Figure S9**. MALDI-ToF/ToF spectrum sequence of 23 kDa photosystem II of cultivar RB867515 sugarcane leaves treated with 100 mM NaCl for 48 days. **Figure S10**. MALDI-ToF/ToF spectrum sequence of 23 kDa photosystem II of cultivar RB867515 sugarcane leaves treated with 100 mM NaCl for 48 days. **Figure S11**. Schematic diagram of identified proteins in sugarcane leaves proteome in response to salinity stress. Proteins in stars: up-regulated under saline conditions (100 mM NaCl). Proteins in crosses: expressed only in salt-treated plants under saline conditions (100 mM NaCl). Proteins underlined: down-regulated under saline conditions (100 mM NaCl). Arrows: putative influences on metabolic processes. **Figure S12**. Sugarcane leaves numbering system proposed by Kuijper (1915), with modifications. Leaves +1+2+3 are fully expanded and photosynthetically active. **Figure S13**. Experimental design for comparison and selection of proteins differentially expressed between replicates of control and salt-treated plant gels of sugarcane cultivar RB867515.

## Acknowledgments

We thank Embrapa Genetic Resources and Biotechnology for allowing access to their greenhouse facility and Mass Spectrometry Laboratory, Embrapa Cerrados for macro and micronutrient analyses and Universidade Católica de Brasília (UCB) for allowing access to their equipment.

## Author Contributions

Conceived and designed the experiments: AMM HBCM OLF BFQ. Performed the experiments: AMM BSM NGO OLF BFQ. Analyzed the data: AMM HBCM BSM ACF FSCT NGO OLF BFQ. Contributed reagents/materials/analysis tools: OLF BFQ. Wrote the paper: AMM HBCM BSM ACF FSCT NGO OLF BFQ.

## References

1. Borba MMZ, Bazzo AM (2009) Estudo Econômico Do Ciclo Produtivo Da Cana-De-Açúcar Para Reforma De Canavial, Em Área De Fornecedor Do Estado De São Paulo. Available: http://www.sober.org.br/palestra/13/1169.pdf. 12 May 2014.

2. Saciloto RFZ (2003) Inserção do gene PR5K em cana-de-açúcar visando induzir resistência ao fungo da ferrugem Puccinia melanocephala. Fisiologia e Bioquímica de Plantas. Piracicaba-SP: Escola Superior de Agricultura Luiz de Queiroz. pp. 74.

3. Miranda JR (2008) História da cana-de-açúcar. Campinas-SP. 168 p.

4. Molinari HBC, Marur CJ, Daros E, Campos MKFd, Carvalho JFRPd, et al. (2007) Evaluation of the stress-inducible production of proline in transgenic

sugarcane (*Saccharum* spp.): osmotic adjustment, chlorophyll fluorescence and oxidative stress. Physiol Plantarum 130: 218–229.

5. Cheavegatti-Gianotto A, Abreu HMCd, Arruda P, Filho JCB, Burnquist WL, et al. (2011) Sugarcane (*Saccharum* X *officinarum*): A Reference Study for the Regulation of Genetically Modified Cultivars in Brazil. Trop Plant Biol 4: 62–89.

6. FAO (2009) Food and Agriculture Organization of the United Nations-FAOSTAT. Available: http://faostat.fao.org/site/567/default.aspx#ancor. 05 March 2013.

7. Amorim JRAd (2009) Salinidade em áreas irrigadas: origem do problema, consequências e possíveis soluções. Embrapa Tabuleiros Costeiros Available: http://www.infoteca.cnptia.embrapa.br/handle/doc/661398. 12 May 2014.

8. FAO (2011) The state of the world's land and water resources for food and agriculture (SOLAW) - Managing systems at risk. London: The Food and Agriculture Organization of the United Nations and Earthscan. 47 p. Available: http://www.fao.org/docrep/015/i1688e/i1688e00.pdf.

9. Ghassemi F, Jakeman AJ, Nix HA (1995) Salinisation of Land and Water Resources: human causes, extent, management and case studies. Canberra, Australia: University of New South Wales Press Ltd. 526 p.

10. Moons A, Gielen J, Vandekerckhove J, VanDerStraeten D, Gheysen G, et al. (1997) An abscisic-acid- and salt-stress-responsive rice cDNA from a novel plant gene family. Planta: 443–454.

11. Verslues PE, Agarwal M, Katiyar-Agarwal S, Zhu JH, Zhu JK (2006) Methods and concepts in quantifying resistance to drought, salt and freezing, abiotic stresses that affect plant water status. Plant J 45: 523–539.

12. Solari LI, Johnson S, DeJong TM (2006) Relationship of water status to vegetative growth and leaf gas exchange of peach (*Prunus persica*) trees on different rootstocks. Tree Physiol 26: 1333–1341.

13. Takahashi S, Murata N (2008) How do environmental stresses accelerate photoinhibition? Trends Plant Sci. 13: 178–182.

14. Tunçturk M, Tunçturk R, Yasar F (2008) Changes in micronutrients, dry weight and plant growth of soybean (*Glycine max* L. Merrill) cultivars under salt stress. Afr J Biotechnol 7: 1650–1654.

15. Sahi C, Singh A, Blumwald E, Grover A (2006) Beyond osmolytes and transporters: novel plant salt-stress tolerance-related genes from transcriptional profiling data. Physiol Plantarum 127: 1–9.

16. Sairam RK, Tyagi A (2004) Physiology and molecular biology of salinity stress tolerance in plants. Curr Sci 86: 407–421.

17. Grover A, Mittal D, Negi M, Lavania D (2013) Generating high temperature tolerant transgenic plants: Achievements and challenges. Plant Sci 205–206: 38–47.

18. Molinari HBC, Marur CJ, Bespalhok-Filho JC, Kobayashi AK, Pileggi M, et al. (2004) Osmotic adjustment in transgenic citrus rootstock Carrizo citrange (*Citrus sinensis* Obs. X *Poncirus trifoliata* L. Raf.) overproducing proline. Plant Sci 167: 1375–1381.

19. Tiroli-Cepeda AO, Ramos CHI (2010) Heat causes oligomeric disassembly and increases the chaperone activity of small heat shock proteins from sugarcane. Plant Physiol Biochem 48: 108–116.

20. Hoffmann HP, Santos EGD, Bassinello AI, Vieira MAS (2008) Variedades RB de Cana-de-açúcar. Araras-SP: 136 p. Available: http://pmgca.dbv.cca.ufscar.br/dow/VariedadesRB_2008.pdf.

21. Gava GJdC, Silva MdA, Silva RCd, Jeronimo EM, Cruz JCS, et al. (2011) Produtividade de três cultivares de cana-de-açúcar sob manejos de sequeiro e irrigado por gotejamento. Rev Bras Eng Agríc Ambient 15: 250–255.

22. Silva MdA, Silva JAGd, Enciso J, Sharma V, Jifon J (2008) Yield components as indicators of drought tolerance of sugarcane. Sci Agric 65: 620–627.

23. Stepien P, Klobus G (2006) Water relations and photosynthesis in *Cucumis sativus* L. leaves under salt stress. Biol Plantarum 50: 610–616.

24. Esteves BdS, Suzuki MS (2008) Efeito da salinidade sobre as plantas. Oecologia Brasiliensis 12: 662–679.

25. van den Berg M, Burrough PA, Driessen PM (2000) Uncertainties in the appraisal of water availability and consequences for simulated sugarcane yield potentials in São Paulo State, Brazil. Agric Ecosyst Environ 81: 43–55.

26. Morris DR, Tai PYP (2004) Water Table Effects on Sugarcane Root and Shoot Development. J Am Soc Sugar Cane Technol 24: 41–59.

27. Laclau PB, Laclau J-P (2009) Growth of the whole root system for a plant crop of sugarcane under rainfed and irrigated environments in Brazil. Field Crop Res 114: 351–360.

28. Smith DM, Inman-Bamber NG, Thorburn PJ (2005) Growth and function of the sugarcane root system. Field Crop Res 92: 169–183.

29. Evans H (1936) The root-system of the sugar-cane: II. Some typical root-systems. Empire J Exp Agr 4: 208–221.

30. Chopart J-L, Mézo LL, Brossier J-L (2009) Spatial 2D Distribution and Depth of Sugarcane Root System in a Deep Soil. International Symposium "Root Research and Applications" RootRAP. Boku – Vienna, Austria. pp. 1–4.

31. Gunes A, Inal A, Alpaslan M, Eraslan F, Bagci EG, et al. (2007) Salicylic acid induced changes on some physiological parameters symptomatic for oxidative stress and mineral nutrition in maize (*Zea mays* L.) grown under salinity. J Plant Physiol 164: 728–736.

32. Hakeem KR, Khan F, Chandna R, Siddiqui TO, Iqbal M (2012) Genotypic variability among soybean genotypes under NaCl stress and proteome analysis of salt-tolerant genotype. Appl Biochem Biotech 168: 2309–2329.

33. Moradi F, Ismail AM (2007) Responses of photosynthesis, chlorophyll fluorescence and ROS-Scavenging systems to salt stress during seedling and reproductive stages in rice. Ann Bot 99: 1161–1173.

34. Shao H-B, Chu L-Y, Wu G, Zhang J-H, Lu Z-H, et al. (2007) Changes of some anti-oxidative physiological indices under soil water deficits among 10 wheat (*Triticum aestivum* L.) genotypes at tillering stage,. Colloids Surf., B 54: 143–149.

35. Quirino BF, Candido ES, Campos PF, Franco OL, Kruger RH (2010) Proteomic approaches to study plant-pathogen interactions. Phytochemistry 71: 351–362.

36. Jangpromma N, Kitthaisong S, Daduang S, Jaisil P, Thammasirirak S (2007) 18 kDa protein accumulation in sugarcane leaves under drought stress conditions. KMITL (STJ) 7: 44–54.

37. Jangpromma N, Kitthaisong S, Lomthaisong K, Daduang S, Jaisil P, et al. (2010) A Proteomics Analysis of Drought Stress-Responsive Proteins as Biomarker for Drought-Tolerant Sugarcane Cultivars. Am J Biochem Biotechnol 6: 89–102.

38. Zhou G, Yang L-T, Li Y-R, Zou C-L, Huang L-P, et al. (2011) Proteomic Analysis of Osmotic Stress-Responsive Proteins in Sugarcane Leaves. Plant Mol Biol Rep, 10.1007/s11105-011-0343-0: 1–11.

39. Parry MA, Andralojc PJ, Scales JC, Salvucci ME, Carmo-Silva AE, et al. (2013) Rubisco activity and regulation as targets for crop improvement. J Exp Bot. 64: 717–730.

40. Ngamhui N-o, Akkasaeng C, Zhu YJ, Tantisuwichwong N, Roytrakul S, et al. (2012) Differentially expressed proteins in sugarcane leaves in response to water deficit stress. Plant Omics 5: 365–371.

41. Pacheco CM, Pestana-Calsa MC, Gozzo FC, Mansur Custodio Nogueira RJ, Menossi M, et al. (2013) Differentially delayed root proteome responses to salt stress in sugar cane varieties. J Proteome Res 12: 5681–5695.

42. Song Y, Zhang C, Ge W, Zhang Y, Burlingame AL, et al. (2011) Identification of NaCl stress-responsive apoplastic proteins in rice shoot stems by 2D-DIGE. J Proteomics 74: 1045–1067.

43. Kosová K, Vítámvás P, Prášil IT, Renaut J (2011) Plant proteome changes under abiotic stress - Contribution of proteomics studies to understanding plant stress response. J Proteomics 74: 1301–1322.

44. Sobhanian H, Aghaei K, Komatsu S (2011) Changes in the plant proteome resulting from salt stress: Toward the creation of salt-tolerant crops? J Proteomics 74: 1323–1337.

45. Kosová K, Prášil IT, Vítámvás P (2013) Protein Contribution to Plant Salinity Response and Tolerance Acquisition. Int J Mol Sci. 14: 6757–6789.

46. Sobhanian H, Motamed N, Jazii FR, Nakamura T, Komatsu S (2010) Salt Stress Induced Differential Proteome and Metabolome Response in the Shoots of *Aeluropus lagopoides* (Poaceae), a Halophyte C4 Plant. J Proteome Res 9: 2882–2897.

47. Salekdeh GH, Siopongco J, Wade LJ, Ghareyazie B, Bennett J (2002) A proteomic approach to analyzing drought- and salt-responsiveness in rice. Field Crop Res 75: 199–219.

48. Abbasi FM, Komatsu S (2004) A proteomic approach to analyze salt-responsive proteins in rice leaf sheath. Proteomics 4: 2072–2081.

49. Ghaffari A, Gharechahi J, Nakhoda B, Salekdeh GH (2014) Physiology and proteome responses of two contrasting rice mutants and their wild type parent under salt stress conditions at the vegetative stage. J Plant Physiol 171: 31–44.

50. Lv DW, Subburaj S, Cao M, Yan X, Li X, et al. (2014) Proteome and phosphoproteome characterization reveals new response and defense mechanisms of *Brachypodium distachyon* leaves under salt stress. Mol. Cell Proteomics 13: 632–652.

51. Chaves MM, Flexas J, Pinheiro C (2009) Photosynthesis under drought and salt stress: regulation mechanisms from whole plant to cell. Ann Bot 103: 551–560.

52. Harmer SL, Hogenesch JB, Straume M, Chang HS, Han B, et al. (2000) Orchestrated transcription of key pathways in *Arabidopsis* by the circadian clock. Science 290: 2110–2113.

53. Bernier F, Berna A (2001) Germins and germin-like proteins: Plant do-all proteins. But what do they do exactly? Plant Physiol Biochem 39: 545–554.

54. Davidson RM, Reeves PA, Manosalva PM, Leach JE (2009) Germins: A diverse protein family important for crop improvement. Plant Sci 17: 499–510.

55. Barkla BJ, Vera-Estrella R, Hernandez-Coronado M, Pantoja O (2009) Quantitative proteomics of the tonoplast reveals a role for glycolytic enzymes in salt tolerance. Plant Cell. 21: 4044–4058.

56. Hara M, Cascio M, Sawa A (2006) GAPDH as a sensor of NO stress. Biochim Biophys acta 1762: 502–509.

57. Yang F, Xiao X, Zhang S, Korpelainen H, Li C (2009) Salt stress responses in *Populus cathayana* Rehder. Plant Sci: 669–677.

58. Du C-X, Fan H-F, Guo S-R, Tezuka A, Li J (2010) Proteomic analysis of cucumber seedling roots subjected to salt stress. Phytochemistry 71: 1450–1459.

59. Dumas B, Sailland A, Cheviet JP, Freyssinet G, Pallett K (1993) Identification of barley oxalate oxidase as a germin-like protein. C R Acad Sci 316: 793–798.

60. Thomson EW, Lane BG (1980) Relation of protein synthesis in imbibing wheat embryos to the cell-free translational capacities of bulk mRNA from dry and imbibing embryos. J Biol Chem 255: 5964–5970.

61. Ohmiya A, Tanaka Y, Kadowak K, Hayashi T (1998) Cloning of genes encoding auxin-binding proteins (ABP19/20) from peach: significant peptide sequence similarity with germin-like proteins,. Plant Cell Physiol. 39: 492–499.

62. Ono M, Sage-Ono K, Inoue M, Kamada H, Harada H (1996) Transient increase in the level of mRNA for a germin-like protein in leaves of the short-day

plant *Pharbitis nil* during the photoperiodic induction of flowering. Plant Cell Physiol. 37: 855–886.

63. Woo E-J, Dunwell JM, Goodenough PW, Marvier AC, Pickersgill RW (2000) Germin is a manganese containing homohexamer with oxalate oxidase and superoxide dismutase activities. Nat Struct Mol Biol 7: 1036–1040.

64. Lane BG, Dunwell JM, Rag JA, Schmitt MR, Cuming AC (1993) Germin, a Protein Marker of Early Plant Development, Is an Oxalate Oxidase. J Biol Chem 268: 12239–12242.

65. Zimmermann G, Bäumlein H, Mock H-P, Himmelbach A, Schweizer P (2006) The Multigene Family Encoding Germin-Like Proteins of Barley. Regulation and Function in Basal Host Resistance. Plant Physiol 142: 181–192.

66. Lindquist S (1986) The heat-shock response. Annu Rev Biochem 55: 1151–1191.

67. Lindquist S, Craig EA (1988) The heat-shock proteins. Annu Rev Genet 22: 631–677.

68. Aghaei K, Ehsanpour AA, Komatsu S (2008) Proteomic analysis of potato under salt stress. J Proteome Res 7: 4858–4868.

69. Grimplet J, Wheatley M, Jouira H, Deluc L, Cramer G, et al. (2009) Proteomic and selected metabolite analysis of grape berry tissues under well-watered and water-deficit stress conditions. Proteomics 9: 2503–2528.

70. Tiroli AO, Ramos CHI (2007) Biochemical and biophysical characterization of small heat shock proteins from sugarcane: Involvement of a specific region located at the N-terminus with substrate specificity. Int J Biochem Cell 39: 818–831.

71. Zamboni A, Minoia L, Ferrarini A, Tornielli GB, Zago E, et al. (2008) Molecular analysis of post-harvest withering in grape by AFLP transcriptional profiling. J Exp Bot. 59: 4145–4159.

72. Deytieux C, Geny L, Lapaillerie D, Claverol S, Bonneu M, et al. (2007) Proteome analysis of grape skins during ripening. J Exp Bot. 58: 1851–1862.

73. Rodrigues FA, Laia MLd, Zingaretti SM (2009) Analysis of gene expression profiles under water stress in tolerant and sensitive sugarcane plants. Plant Sci 176: 286–302.

74. Song A, Zhu X, Chen F, Gao H, Jiang J, et al. (2014) A chrysanthemum heat shock protein confers tolerance to abiotic stress. Int J Mol Sci. 15: 5063–5078.

75. Tada Y, Kashimura T (2009) Proteomic analysis of salt-responsive proteins in the mangrove plant, *Bruguiera gymnorhiza*. Plant Cell Physiol. 50: 439–446.

76. Scholander PF, Hammel HT, Hemmingsen EA, Bradstreet ED (1964) Hydrostatic pressure and osmotic potential in leaves of mangroves and some other plants. Proc Natl Acad Sci U. S. A 52: 119–125.

77. Hodges DM, DeLong JM, Forney CF, Prange RK (1999) Improving the thiobarbituric acid-reactive-substances assay for estimating lipid peroxidation in plant tissues containing anthocyanin and other interfering compounds. Planta 207: 604–611.

78. Nogueira ARDA (2005) Manual de Laboratório: Solo, Água Nutrição Vegetal, Nutrição Animal e Alimentos. In: E P. Sudeste , editor editors. Manual de Laboratório: Solo, Água Nutrição Vegetal, Nutrição Animal e Alimentos. São Carlos-SP.

79. Wang W, Scali M, Vignani R, Spadafora A, Sensi E, et al. (2003) Protein extraction for two-dimensional electrophoresis from olive leaf, a plant tissue containing high levels of interfering compounds. Electrophoresis 24: 2369–2375.

80. Lowry O, Rosebrough N, Farr A, Randall RJ (1951) Protein measurement with the folin phenol reagent. J Biol Chem 193: 265–275.

81. Laemmli UK (1970) Cleavage of structural proteins during the assembly of the head of bacteriophage T4. Nature 15: 680–685.

82. Neuhoff V, Arold N, Taube D, Ehrhardt W (1988) Improved staining of proteins in polyacrylamide gels including isoelectric focusing gels with clear background at nanogram sensitivity using Coomassie Brilliant Blue G-250 and R-250. Electrophoresis 9: 255–262.

83. Kang D, Gho YS, Suh M, Kang C (2002) Highly Sensitive and Fast Protein Detection with Coomassie Brilliant Blue in Sodium Dodecyl Sulfate-Polyacrylamide Gel Electrophoresis. Bull Korean Chem Soc 23: 1511–1512.

84. Shevchenko A, Wilm M, Vorm O, Mann M (1996) Mass spectrometric sequencing of proteins silver-stained polyacrylamide gels. Anal Chem 68: 850–858.

85. Shevchenko A, Tomas H, Breve JH, Olsen JV, Mann M (2007) In-gel digestion for mass spectrometric characterization of proteins and proteomes. Nat Protoc 1: 2856–2560.

86. Team RC (2012) R: A language and environment for statistical computing. R Foundation for Statistical Computing. Vienna, Austria.

87. Pinheiro J, Bates D, DebRoy S, Sarkar D (2012) nlme: Linear and Nonlinear Mixed Effects Models.

# 18

# Water Consumption Characteristics and Water Use Efficiency of Winter Wheat under Long-Term Nitrogen Fertilization Regimes in Northwest China

**Yangquanwei Zhong, Zhouping Shangguan***

State Key Laboratory of Soil Erosion and Dryland Farming on the Loess Plateau, Northwest A & F University, Yangling, Shaanxi, P.R. China

## Abstract

Water shortage and nitrogen (N) deficiency are the key factors limiting agricultural production in arid and semi-arid regions, and increasing agricultural productivity under rain-fed conditions often requires N management strategies. A field experiment on winter wheat (*Triticum aestivum* L.) was begun in 2004 to investigate effects of long-term N fertilization in the traditional pattern used for wheat in China. Using data collected over three consecutive years, commencing five years after the experiment began, the effects of N fertilization on wheat yield, evapotranspiration (ET) and water use efficiency (WUE, i.e. the ratio of grain yield to total ET in the crop growing season) were examined. In 2010, 2011 and 2012, N increased the yield of wheat cultivar Zhengmai No. 9023 by up to 61.1, 117.9 and 34.7%, respectively, and correspondingly in cultivar Changhan No. 58 by 58.4, 100.8 and 51.7%. N-applied treatments increased water consumption in different layers of 0–200 cm of soil and thus ET was significantly higher in N-applied than in non-N treatments. WUE was in the range of 1.0–2.09 kg/m$^3$ for 2010, 2011 and 2012. N fertilization significantly increased WUE in 2010 and 2011, but not in 2012. The results indicated the following: (1) in this dryland farming system, increased N fertilization could raise wheat yield, and the drought-tolerant Changhan No. 58 showed a yield advantage in drought environments with high N fertilizer rates; (2) N application affected water consumption in different soil layers, and promoted wheat absorbing deeper soil water and so increased utilization of soil water; and (3) comprehensive consideration of yield and WUE of wheat indicated that the N rate of 270 kg/ha for Changhan No. 58 was better to avoid the risk of reduced production reduction due to lack of precipitation; however, under conditions of better soil moisture, the N rate of 180 kg/ha was more economic.

**Editor:** Raffaella Balestrini, Institute for Plant Protection (IPP), CNR, Italy

**Funding:** The study was sponsored by the National Natural Science Foundation of China (41390463, 61273329) and the Important Direction Project of Innovation of CAS (KZCX2-YW-JC408). The funders had no role in study design, data collection and analysis, decision to publish, or preparation of the manuscript.

**Competing Interests:** The authors have declared that no competing interests exist.

* E-mail: shangguan@ms.iswc.ac.cn

## Introduction

Northwest China is a vast semi-arid area with average annual precipitation in the range of 300–600 mm and more than 90% of the land is cropland [1]. This means that water is the primary factor limiting crop yields. In addition, world food demand is expected to double during 2005–2050 [2], thus it is important to increase food production with lower water use [3], particularly in water shortage regions. Currently, water stress and nutrient deficits are the main factors limiting primary production in arid and semi-arid environments [4–8]. Therefore, many rain-fed farming experts have focused on how to increase crop water use efficiency (WUE, i.e. the ratio of grain yield to total ET in the crop growing season) by irrigation and fertilization.

In the 1990 s, many studies on effects of limited irrigation on crop yields and WUE showed that by reducing irrigation volume, crop yield could be generally maintained and product quality improved [9–13],and appropriate irrigation management can increase crop yield and WUE [14–16]. There are several sources of soil water in irrigated or high water-table areas, however, precipitation is the only source of soil water for crop growth in many rain-fed farming systems of arid and semi-arid regions.

Therefore, new methods need to be devised to improve WUE in this non-irrigated farming system.

N fertilization is a common practice to increase grain production, but its performance depends on soil water status [17–19]. The importance of increasing crop yield and improving soil quality through fertilization has been confirmed. The increasing use of N fertilizer could significantly increase maize production [8,20], and already affects a large proportion of the world's food production [21,22]. Fan et al. [1] reported that inorganic N and phosphorus (P) fertilization increased grain yields by 50–60% in China, and reports from Europe showed that N fertilizers can increase crop yield significantly [23]. N fertilization is well known to improve soil fertility [24,25]; however, using excessive N fertilizer can decrease the N utilization rate, which not only causes a huge waste of resources and economic losses, but can also adversely impact the environment [26–28]. Balancing the N rate, WUE and yield is an important problem in dryland farming systems. Better understanding of interactions among precipitation, fertilization and crops production is essential for efficient utilizations of water resources and N fertilizers, and sustainable food productions in rain-fed cropping systems experiencing climate change [1]. Long-term fertilization experiments are

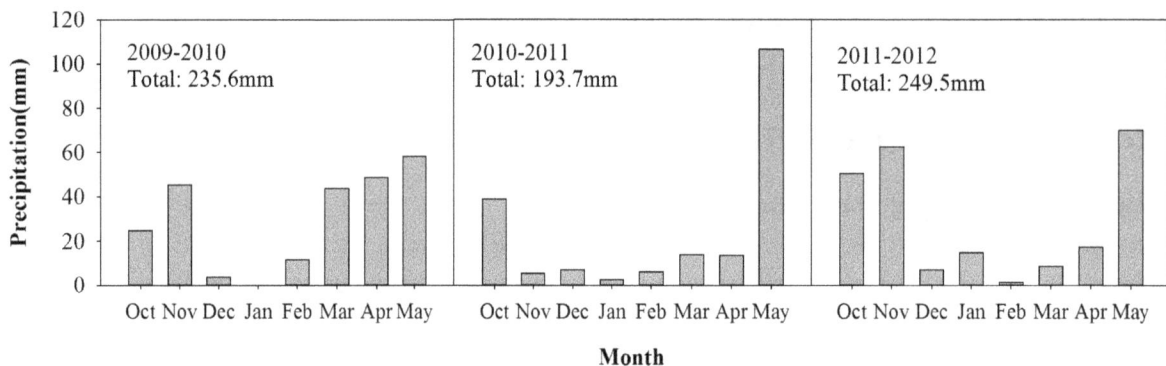

**Figure 1. Monthly and total precipitation during the 2009–2012 wheat growing seasons.** Monthly precipitation of wheat growing seasons in 2009–2010, 2010–2011, 2011–2012 in Shaanxi, Yangling.

valuable to follow crop yield, soil fertility, WUE and risk management over time [29,30]. Various long-term experiments have examined how to increase yield and WUE of wheat, using irrigation, organic or inorganic fertilizer, soil tillage and crop management [31–33]. However, few experiments have been done on evapotranspiration (ET) and WUE under circumstances with only N fertilizer and without irrigation in northwest China. With China's urbanization, increasing numbers of farmers have abandoned farms to urban construction and this has led to a loss of labor. Thus, most farmland in northwest China region still uses traditional cropping practices that all fertilizer applied once prior to planting [1], lack of careful management of irrigation and other tasks. Kang et al. [14] reported that difference in yield and WUE are also related to regional variability in environment and crop varieties, so information specific to a region is needed for developing and refining the agricultural performance in this region. In these circumstances, it is very important to determine the advantages and disadvantages of long-term N fertilization on yield of different varieties.

This study examined two different water-sensitive cultivars of winter wheat (*Triticum aestivum* L.) to investigate effects of N fertilizers on crop yield, ET and WUE, using the most common management of farmers in northwest China. The objectives were (1) to investigate impacts of traditional long-term N fertilization on yields of two different water-sensitive wheat cultivars; (2) to examine the effect of N fertilizers on total ET, and soil water consumption from different soil layers of the two cultivars; and (3) to establish relationships among crop yield, WUE and ET and determine optimum N fertilizer rates in northwest China. This study may compensate for some of the lack of long-term influence only N fertilizer on crop production, and the results should provide guidelines to farmers in the region on choosing appropriate cultivars and obtaining high yields with appropriate N application.

## Materials and Methods

### Experiment site and climatic conditions

The study commenced in October 2004 in an experiment field of the Institute of Soil and Water Conservation of the Northwest A & F University, Yangling, Shaanxi (34°17′56″N, 108°04′7″E). Located on the southern boundary of the Loess Plateau, the experiment site has a temperate and semi-humid climate with a mean annual temperature of 13°C and a mean annual precipitation of 632 mm, of which about 60% occurs during July–September.

### Experiment design

The study adopted a randomized block design with three replications. Two winter wheat (*Triticum. aestivum* L.) cultivars were used: Zhengmai No. 9023 (ZM) is water sensitive and poorly drought-tolerant and Changhan No. 58 (CH) is drought-tolerant and suitable for drought prone environments. The thousand-kernel weights of ZM and CH were 43.58 and 43.61 g, respectively. N treatments were applied at five rates: 0, 90, 180, 270 and 360 kg/ha (N0, N90, N180, N270 and N360, respectively). Plot size was 2 m×3 m with 20 rows (15-cm spaces) of wheat sown at 90 seeds/row. Wheat was sown in early October and harvested in early June the following year. The seeding rate was 130 kg/ha. Immediately before sowing, the fertilizer was evenly spread on the soil surface and then incorporated into the upper 15 cm soil by chiseling. N was applied as urea and P (75 kg $P_2O_5$/ha) as super phosphate. No potassium fertilizer was applied, and the site was ploughed to bury weeds before sowing.

### Measurements

In all treatments, the volumetric soil water content was measured every 10 cm for 0–100 cm of soil and every 20 cm for 100–300 cm with a neutron moisture meter (CNC100, Super Energy, Nuclear Technology Ltd., Beijing, China). The 3-m-long neutron gauge access tube was buried vertically in the center of each plot at the beginning of the study. Soil water was measured during the first week of every wheat growing month except January and February. If any precipitation occurred just before or during the measurement period, then measurements were postponed for several days until the soil moisture attained a normal degree. The yields and the thousand-kernel weights of wheat in all plots were measured at harvest time in early June. Since this paper aimed to test the cumulative effects of N fertilization, we chose data from three wheat growing years with different precipitation characteristics: 2009–2010, 2010–2011 and 2011–2012.

### Calculation and statistics

ET of winter wheat was calculated using the following equation [33]:

$$ET = \Delta S + P + I - R - D$$

**Table 1.** Nitrogen effects on wheat yield and thousand-kernel weights of two cultivars in three years.

| Varieties | Treatments | 2010 Yield (kg/ha) | | 2010 Thousand-kernel (g) | 2011 Yield (kg/ha) | | 2011 Thousand-kernel (g) | 2012 Yield (kg/ha) | | 2012 Thousand-kernel (g) |
|---|---|---|---|---|---|---|---|---|---|---|
| ZM | N0 | 4716 | c | 49.5 a | 2974 | c | 47.0 a | 5906 | bc | 49.1 a |
| | N90 | 6355 | ab | 42.5 b | 5499 | ab | 42.2 ab | 7272 | a | 42.3 bcd |
| | N180 | 7597 | a | 42.3 b | 6355 | ab | 42.1 ab | 7953 | a | 41.6 cd |
| | N270 | 7527 | a | 42.5 b | 6482 | ab | 40.5 b | 7923 | a | 41.1 d |
| | N360 | 7519 | a | 43.9 b | 6390 | ab | 41.2 ab | 7512 | a | 40.4 d |
| CH | N0 | 4162 | c | 46.1 ab | 3391 | c | 42.2 ab | 5407 | c | 46.5 ab |
| | N90 | 5862 | ab | 42.3 b | 4886 | b | 41.1 ab | 6926 | ab | 45.5 abc |
| | N180 | 6594 | ab | 42.3 b | 5879 | ab | 41.9 ab | 7777 | a | 40.4 d |
| | N270 | 6334 | ab | 37.9 c | 6748 | a | 40.5 b | 8199 | a | 40.1 d |
| | N360 | 6126 | b | 37.1 c | 6808 | a | 40.3 b | 8018 | a | 38.6 d |

Values are means of three replicates for each treatment. Different letters indicate statistical significance at P<0.05 within the same column.

Where $\Delta S$ is soil water storage change, P is precipitation, I is irrigation rate, R is surface runoff and D is deep water percolation (all in mm).

No irrigation was used and so $I = 0$. Precipitation in the three growing seasons is shown in Fig. 1, and measured surface runoff was negligible during these years. Deep percolation was calculated as the difference between soil moisture content and field moisture capacity when the soil water content at this depth was more than the field water holding capacity. In the study site, deep water percolation did not occur.

WUE was defined as follows:

$$WUE = Y/ET$$

Where Y is grain yield (kg/ha).

All data concerned were analyzed by SPSS 16.0 Statistical software. ANOVA was adopted to determine whether treatments were significantly different at P<0.05. Duncan's multiple range test was used to differentiate treatment means at P<0.05.

## Results and Discussion

### Wheat yield

The yields and thousand-kernel weights of the two wheat cultivars at the different N rates in the different years are presented in Table 1. Grain yield was in the range of 2.94–7.92 t/ha for ZM and 3.39–8.19 t/ha for CH. Grain yields of the two cultivars both differed significantly between N0 and the other N rates, and increased as N rates increased, but did not differ significantly among the N-applied treatments. The lack of significant differences may due to yield in three replications being affected by other factors in field experiment. Usually the significance of increase yield is hard to attain in agricultural research, Morell et al. [36] reported that grain yield mostly 1000 kg/ha higher than control, but still have no statistical difference. The yields of wheat slightly decreased at N360, except for yield of CH in 2011. In 2010, 2011 and 2012, the wheat yields of ZM increased by up to 61.1, 118.0 and 34.7%, respectively, in the N-applied treatments compared to treatment without N fertilization; and corresponding yields of CH increased by up to 58.4, 100.8 and 51.7%. The highest yields of ZM and CH were both for treatments of N180, N270 and N270 in 2010, 2011 and 2012, respectively. Thus, N application significantly increased yields of wheat, but an excessive N rate had no positive effect on grain yield. Previous research has shown similar results with N fertilizer application significantly increasing maize and wheat yield compared to unfertilized treatments [14,15]. Bassoa et al. [23] examined the long-term wheat response to N in rain-fed Mediterranean environments, and showed that yield response was stronger for 120 than 60 and 90 kg N/ha. Many other studies have demonstrated a parabolic relationship between N and grain yield, i.e. when N rate surpassed a certain threshold, the grain yield greatly decreased. In China, there have been many experiments on different wheat cultivars and fertilizer regimes that have shown the maximum N rate is 150–225 kg/ha. At excessive N rates, the leaf protein and chlorophyll contents decrease, and then photosynthesis also decreases [34]. Tinsina et al. [35] also showed that wheat yield was higher at 120 than 180 kg/ha in Bangladesh. Morell et al. [36] showed no additional wheat yield responses to N fertilizers at N rates >100 N kg/ha. All these studies demonstrated that N application could increase wheat yield, but excessive N had no yield benefit.

At the same N rates, the yields of the two cultivars did not differ significantly. In 2010, the rainfall, which was evenly distributed

**Table 2.** Water consumption from soil ($\Delta S$) or precipitation and its ratio to total evapotranspiration (ET) and WUE.

| Year | Varieties | Treatments | Evapotranspiration (ET)(mm) | | Soil water consumption ($\triangle S$) | | Precipitation | | WUE (kg/m³) | |
|------|-----------|------------|------------------------------|---|-------------------|----------------|------------------|----------------|-------------|---|
|      |           |            |                              |   | Amount | Ratio | Amount | Ratio |  |  |
|      |           |            |                              |   | (mm) | (%) | (mm) | (%) |  |  |
| 2009–2010 | ZM | N0 | 325.72 | e | 90.12 | 27. 7 | 235.6 | 72.3 | 1.45 | de |
|  |  | N90 | 348.58 | de | 112.98 | 32.4 |  | 67.6 | 1.83 | abcd |
|  |  | N180 | 361.57 | bcd | 125.97 | 34.8 |  | 65.2 | 2.09 | a |
|  |  | N270 | 385.25 | abc | 149.65 | 38.9 |  | 61.2 | 1.96 | ab |
|  |  | N360 | 385.59 | abc | 149.99 | 38.9 |  | 61.1 | 1.97 | ab |
|  | CH | N0 | 321.39 | e | 85.79 | 26.7 |  | 73.3 | 1.29 | e |
|  |  | N90 | 352.69 | cde | 117.09 | 33.2 |  | 66.8 | 1.66 | bcde |
|  |  | N180 | 373.65 | bcd | 138.05 | 36.9 |  | 63.1 | 1.76 | abc |
|  |  | N270 | 411.98 | a | 176.38 | 42.8 |  | 57.2 | 1.48 | cde |
|  |  | N360 | 395.49 | ab | 159.89 | 40.4 |  | 59.6 | 1.52 | cde |
| 2010–2011 | ZM | N0 | 298.40 | d | 104.70 | 35.1 | 193.7 | 64.9 | 1.00 | d |
|  |  | N90 | 323.28 | bc | 129.58 | 40.1 |  | 59.9 | 1.70 | ab |
|  |  | N180 | 344.98 | ab | 151.28 | 43.8 |  | 56.2 | 1.84 | ab |
|  |  | N270 | 345.60 | ab | 151.90 | 44.0 |  | 56.1 | 1.88 | a |
|  |  | N360 | 337.41 | ab | 143.71 | 42.6 |  | 57.4 | 1.89 | a |
|  | CH | N0 | 308.48 | cd | 114.78 | 37.2 |  | 62.8 | 1.10 | cd |
|  |  | N90 | 344.01 | ab | 150.31 | 43.7 |  | 56.3 | 1.42 | bc |
|  |  | N180 | 351.92 | a | 158.22 | 45.0 |  | 55.0 | 1.67 | ab |
|  |  | N270 | 351.75 | a | 158.05 | 44.9 |  | 55.0 | 1.92 | a |
|  |  | N360 | 344.63 | ab | 150.93 | 43.8 |  | 56.2 | 1.84 | ab |
| 2011–2012 | ZM | N0 | 373.92 | c | 124.42 | 33.3 | 249.5 | 66.7 | 1.58 | a |
|  |  | N90 | 412.28 | bc | 162.78 | 39.5 |  | 60.5 | 1.77 | a |
|  |  | N180 | 456.45 | a | 206.95 | 45.3 |  | 54.7 | 1.75 | a |
|  |  | N270 | 442.06 | ab | 192.56 | 43.6 |  | 56.4 | 1.80 | a |
|  |  | N360 | 440.44 | ab | 190.94 | 43.3 |  | 56.6 | 1.71 | a |
|  | CH | N0 | 381.35 | c | 131.85 | 34.6 |  | 65.4 | 1.43 | a |
|  |  | N90 | 464.53 | a | 215.03 | 46.3 |  | 53.7 | 1.50 | a |
|  |  | N180 | 469.35 | a | 219.85 | 46.8 |  | 53.5 | 1.66 | a |
|  |  | N270 | 461.05 | a | 211.55 | 45.9 |  | 54.1 | 1.79 | a |
|  |  | N360 | 441.13 | ab | 191.63 | 43.4 |  | 56.6 | 1.82 | a |

Values are means of three replicates for each treatment. Different letters indicate statistical significance at P<0.05 within the same column. $\Delta S$ has the same significance as total ET.

through the growing season, provided a more favorable environment for ZM, the water-sensitive cultivar, and so its yields were higher than those of CH for all treatments. In 2011, precipitation was the lowest in the whole growing season of all years, despite 106 mm of rainfall in May when wheat filled its seeds. However, such high rainfall at seed filling was unfavorable for wheat yield. Sheng and Wang [37] found that high soil-water contents at the seed-filling stage of wheat can result in lower thousand-kernel weights and grain yields. So drought during the growing season and too much rainfall from the seed-filling to the ripening stages led to the lower yield of wheat in 2011 compared to 2010 and 2012. Before the 2011–2012 growing season, the summer of 2011 received a lots of rain, 672.7 mm during June–September, larger

than 421.6 mm in 2009 and 436.7 mm in 2010 summer. This caused total water consumption in 2011–2012 to be higher than previously and so the yields of wheat were the highest among the three years, although precipitation did not differ greatly from the growing season of 2009–2010. Shangguan et al. [38] reported that the fallow efficiencies, expressed as the ratio of soil water accumulation to precipitation received during the period of fallow, were important for yield in the next growing season. The importance of soil-water storage during the fallow period for increasing grain yields of post-fallow crops are supported by many studies on dryland including the Southern Great Plains in the USA [39–41] and the Loess Plateau [38]. In 2011 and 2012, the yield of ZM was higher than that of CH at N0, N90 and N180, and

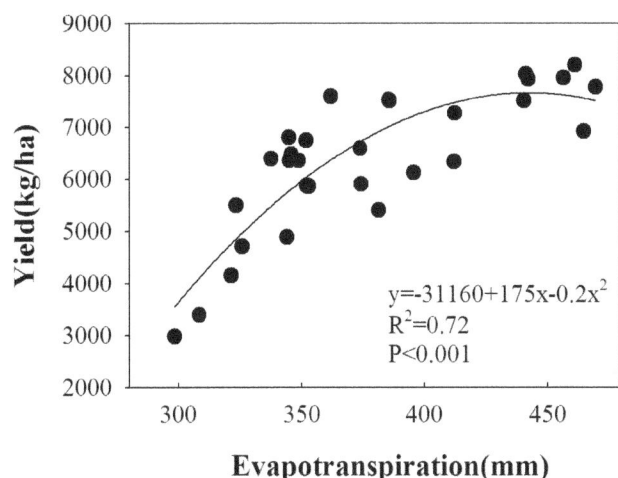

Figure 2. Relationship between wheat evapotranspiration (ET) (mm) and grain yield (Y) (kg/ha) for winter wheat in northwest China. The relationship between ET and yield is shown by the equation.

drought-tolerant CH showed higher yield only at the higher N rate in dry years. This is because CH was developed in recent years and prefers high fertilizer levels – consequently its cultivation has greatly expanded in northwest areas. CH is sensitive to N, and high rates of N result in higher yields; however, in contrast ZM is a poorly drought-tolerant cultivar but can produce higher yield at lower N rates. These cultivar characteristics have been demonstrated by many physiological indices in our previous studies [42]. Overall, the water consumption characteristics differed between the wheat varieties, leading to the different production performance. N180 resulted in higher yields of ZM in 2010 when rainfall was evenly distributed over the growing season. However, when rainfall was unevenly distributed or there was a lack of rainfall in the growing season, appropriate increases in the amount of N for CH could result in higher wheat yield to avoid the risk of reduced production.

The thousand-kernel weights of wheat decreased with increased N rates, consistent with many other research results [34]. There were two reasons for this: first, N can increase numbers of wheat tillers and panicles as well as flag leaf photosynthetic rates, but large and thick leaves would shade one another, affecting starch assimilation and transportation to kernels and resulting in lower thousand-kernel weights; secondly, N application could delay the flowering stage of wheat, thereby shortening the grain-filling stage and leading to lower thousand-kernel weights.

## Water consumption characteristics

**Total ET.** In dryland farming, ET is supplied partly from precipitation in the growing season and partly from soil-water storage before planting. However, the relative contribution between precipitation and crop-consumed soil water to ET differs significantly among crops.

The total ET, ΔS or rainfall and its ratio to total ET at the different N rates are shown in Table 2. Total ET behaved differently between years and cultivars. Total ET was in the range of 298.40–442.46 mm for ZM and 361.57–469.35 mm for CH. Total ET was significantly higher in the N-applied than non-N treatments, except for treatment N90. Total ET were highest in 2011–2012 of CH and ZM. In the N-applied treatments, the ET of ZM increased by up to 18.4, 15.8 and 22.1% in 2009–2010,

2010–2011 and 2011–2012, respectively, and correspondingly for CH by 28.0, 14.1 and 23.1%. The ET slightly decreased for N360, indicating that ET could not increase further if too much N was applied. Zhou et al. [15] showed that N fertilizer application decreased water storage in 0–200 cm of soil and particularly so after wheat harvesting. Hunsaker et al. [43] showed that wheat ET was significantly higher than in low N treatments. One explanation for N increasing ET of wheat is that N promotes wheat to grow more and produce longer roots, enabling more soil water to be absorbed; another explanation is that N fertilization increases the leaf area index and transpiration rates of wheat [44]. However, too much N makes soil environments stressful by increasing N concentration in soil solution, thus preventing roots from absorbing water. The total ET in this study was considerably lower than that reported for the southern high plains of the USA [45,46] and the North China Plain [47], but was close to that for the Loess Plateau [48]. These differences are likely due to different climatic conditions, like temperature and precipitation and also attributed to different field management.

In all the experiment years, CH had higher ET than ZM but not significantly at the same N rates – probably due to different characteristics of the varieties. As a drought-tolerant cultivar with long roots, CH is capable of absorbing deep soil water, thus presenting higher ET than ZM. A deep-growing root system will favor taking up deep soil water under water-limited conditions. Research on dryland crops has shown that deep soil-water utilization is probably limited by root density [11,48,49]. The rainfall was less during 2010–2011 than 2009–2010 and 2011–2012 growing seasons so that the ratio of soil water consumption was higher in the former than the other two years (Table 2). As the N application rates increased, the ratios also increased, indicating that N application helped plants utilize deeper soil water.

The relationship between grain yields and seasonal ET was best described by a quadratic function obtained by regression analysis $(Y = -31160 + 175x - 0.2x^2$; Fig. 2). Grain yield did not increase when ET exceeded a certain critical value, e.g. about 430 mm in the present study. Grain yield required a minimum ET of 244 mm for winter wheat (Fig. 2). This minimum ET value is higher than the 84 mm for wheat in the North China Plain [47] and 156 mm in the Mediterranean region [12], as well as higher than the 206 mm of dryland and irrigated wheat reported by Musick et al. [41] in US southern plains. These differences are likely due to such different climates and crop management. This result may indicate that the crop yield in this area will more relies on precipitation and soil water storage.

**Water consumptions in the different soil layers.** Water consumption in the different layers of the soil profile in 40-cm increments is plotted with depth in Fig 3. N applications had a significant effect on ΔS, as well as on ET (Table 2), since N application increased water consumption in the different layers above 200 cm, except during the 2011–2012 growing season for both cultivars. In 2011–2012, at 200 cm soil depth, the N treatments still had higher soil water consumption than N0 treatment, likely due to the large amount of rainfall in this year. A similar result was found by Zhou et al. [15], with N fertilizer application decreasing water storage at soil depths of <200 cm after wheat harvesting. N application increased water consumption in the different soil layers; however, in the same soil layers, water consumption did not differ significantly among the different N rates except for some layers of CH.

Soil water consumption clearly changed with the different N rates (Fig. 3). The trends of water consumption were similar in all treatments as soil layers became deeper. N application increased water consumption in all soil layers. Generally, water consumption

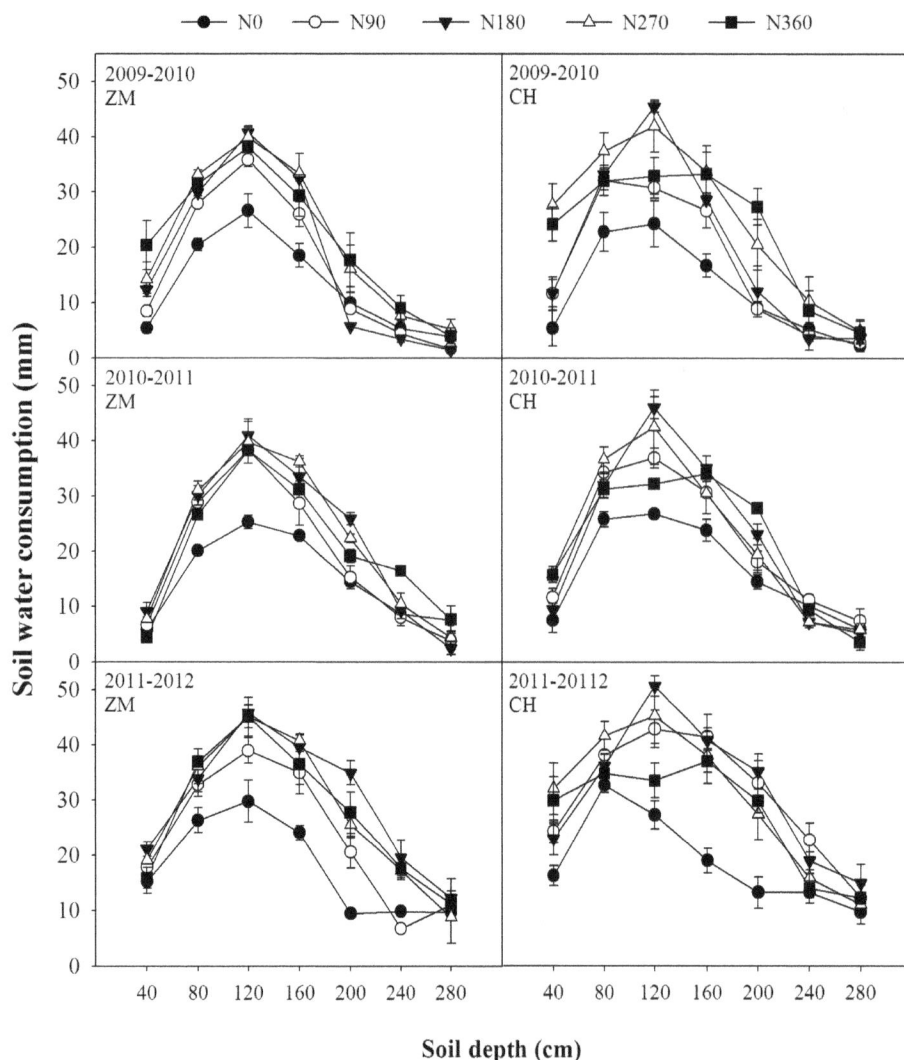

**Figure 3. Wheat evapotranspiration (ET) (mm) of two cultivars in different soil layers and different nitrogen (N) treatments in three years.** Water ET trends as soil depth increased with influence of N fertilizer for two cultivars in three years. Standard error bars are also shown.

of CH was higher than that of ZM, which was true of the total soil water consumption in all layers (Table 2). Water was mainly consumed in layers of 40–160 cm deep, with the highest water consumption for 100–140 cm. Water was stably absorbed for soil layers <120 cm in the N0 treatment, and <160 cm in the N-applied treatments, shown by ΔS of N treatments at 160 cm being higher than for the non-N treatment at 120 cm. This was probably because N application could promote roots to grow longer and stronger, and which were able to absorb deeper soil water. In addition, the N-applied treatments had greater effects on CH than ZM (Fig. 3), showing that CH was sensitive to low N.

## WUE

WUE was in the range of 1–2.09 kg/m$^3$ for ZM and 1.1–1.92 kg/m$^3$ in the three years (Table 2). As the N rates increased the WUE increased, and at N360 the WUE increased slightly but not significantly. Zhou et al. [15] reported that grain yields and WUE did not significantly differ between N rates of 120 and 240 kg/ha. The above indicated that excessive N application had no favorable effect on WUE. In 2009–2010 and 2010–2011, the N-applied treatments significantly improved WUE. However, in

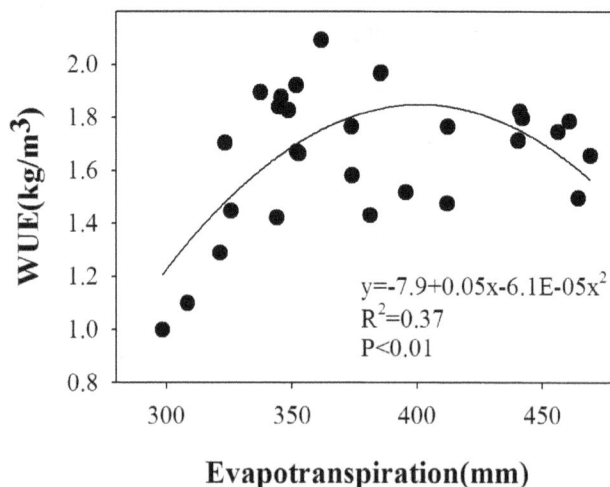

**Figure 4. Relationship between evapotranspiration (ET) (mm) and WUE (Y) for winter wheat in northwest China.** The relation between ET and WUE is shown by the equation.

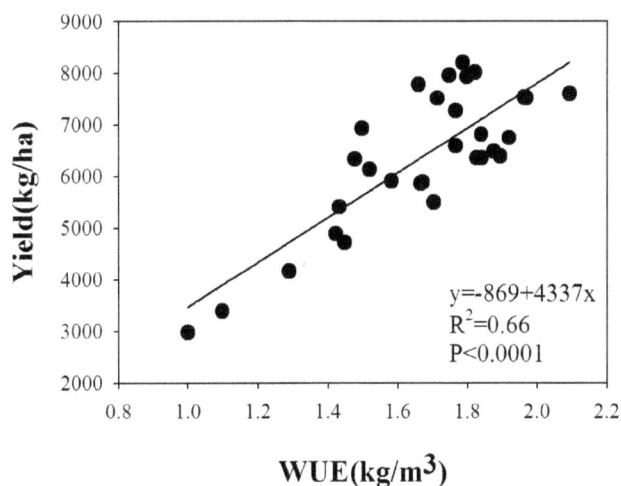

**Figure 5. Relationship between WUE and grain yield (Y) for winter wheat in northwest China.** The relation between WUE and yield could be deduced from the liner equation.

2011–2012, WUE did not significantly differ between the N-applied and non-N treatments. Compared to 2009–2010 and 2010–2011, the WUE of the non-N treatment was higher in 2011–2012. This may be due to the higher soil water content before sowing and the higher rainfall in the growing season in 2012 that increased the grain yield in the non-N treatment. Consequently higher WUE in non-N treatment reduced the difference between non-N and N-applied treatments.

The WUEs obtained in the present study were higher than those of irrigated winter wheat (0.40–0.88 kg/m³) [45,46] and of irrigated wheat in the US southern plains (0.82 kg/m³) [41], as well as higher than these of irrigated wheat in the Loess Plateau (0.73–0.93 kg/m³) [14], demonstrating that in dryland farming systems N fertilizer can be a useful way to increase WUE. However, the results of the present study were similar to those of winter wheat in the North China Plain (0.84–1.39 kg/m³) [12] and of N-fertilized winter wheat in Yangling, Shaanxi (0.8–1.5 kg/m³) [15]. These differences are caused by different climate or water, fertilizer and crop management. N fertilizer application significantly increases the yield and WUE of both wheat and

maize, indicating that N fertilizer application is an effective way to increase grain yield in the study region. Deng et al. [50] reviewed four published research reports and found that N fertilizers increased WUE of wheat and potato by an average of 20% in north central and northwest China.

Regression analysis produced a quadratic relationship between ET and WUE (Fig. 4) and correlations were calculated between WUE and wheat yields (Fig. 5). The yields increased linearly with WUE: WUE reached a maximum value at the ET of 401 mm and then decreased (Fig. 4). However, the maximum WUE did not correspond to the maximum grain yield in the study – the higher WUE means that the crop can gain high yield using less water. This is an important method to obtain a balance between higher yields and lower water supplies of wheat in arid and semi-arid regions by increasing its WUE. In the present study, although there were significant differences between N treatments, both cultivars had relatively higher WUEs at the N rate of 270 kg/ha in a dry year.

## Conclusions

N fertilization affected the grain yields, thousand-kernel weights, ET and WUE of the two different water-sensitive wheat cultivars, ZM and CH. The most common pattern of farming in northwest China was used in the present study, with long-term different rates of N fertilization and no irrigation during the wheat growing season. We concluded that (1) in this dryland farming system, increased N fertilization resulted in higher wheat yields in a situation of low precipitation; the drought-tolerant CH showed a yield advantage in a drought environment with high N fertilizer rates; (2) N application affected water consumption in the different soil layers, and promoted absorption and utilization of water from deeper soil layers; and (3) comprehensive consideration of yield and WUE of wheat indicated that the N rate of 270 kg/ha for CH was better to avoid the risk of reduced production due to lack of precipitation; however, under conditions of better soil moisture, the N rate of 180 kg/ha was more economic.

## Author Contributions

Conceived and designed the experiments: YZ ZS. Performed the experiments: YZ. Analyzed the data: YZ. Contributed reagents/materials/analysis tools: YZ. Wrote the paper: YZ ZS.

## References

1. Fan TL, Stewart BA, Wang YG, Luo JJ, Zhou GY (2005) Long-term fertilization effects on grain yield, water-use efficiency and soil fertility in the dry land of Loess Plateau in China. Agr Ecosyst Environ 106: 313–329.

2. Borlaug NE (2009) Foreword. Food Sec. 1, 1–11

3. Perry C, Steduto P, Allen RG, Burt CM (2009) Increasing productivity in irrigated agriculture: agronomic constraints and hydrological realities. Agr Water Manage 96: 1517–1524

4. Li SX, Wang ZH, Malhi SS, Li SQ, Gao YJ, et al. (2009) Nutrient and water management effects on crop production, and nutrient and water use efficiency in dry land areas of China. Adv Agron 102: 223–265.

5. Hooper DU, Johnson L (1999) Nitrogen limitation in dryland ecosystems: responses to geographical and temporal variation in precipitation. Biogeochemistry 46: 247–293.

6. Rockström J, De Rouw A (1997) Water, nutrients and slope position in on-farm pearl millet cultivation in the Sahel. Plant Soil 195: 311–327.

7. Austin AT (2011) Has water limited our imagination for arid land biogeochemistry? Trends Ecol Evol 26: 229–235.

8. Zand-Parsa S, Sepaskhah A, Ronaghi A (2006) Development and evaluation of integrated water and nitrogen model for maize. Agr Water Manage 81: 227–256.

9. Li YS (1982) Evaluation of field soil moisture condition and the ways to improve crop water use efficiency in Weibei region. Journal of Agronomy Shananxi 2: 1–8. (in Chinese)

10. Shan L (1983) Plant water use efficiency and dryland farming production in North West of China. Newslett Plant Physiology 5: 7–10. (in Chinese)

11. Hamblin AP, Tennant D (1987) Root length density and water uptake in cereals and grain legumes: how well are they correlated. Crop Pasture Sci 38(3): 513–527.

12. Zhang H, Oweis T (1999) Water-yield relations and optimal irrigation scheduling of wheat in the Mediterranean region. Agr Water Manage 38: 195–211.

13. Zhang J, Sui X, Li J, Zhou D (1998) An improved water use efficiency for winter wheat grown under reduced irrigation. Field Crops Res 59: 91–98.

14. Kang SZ, Zhang L, Liang YL, Hu XT, Cai HJ, et al. (2002) Effects of limited irrigation on yield and water use efficiency of winter wheat in the loess plateau of China. Agr Water Manage 55: 203–216.

15. Zhou JB, Wang CY, Zhang H, Dong F, Zheng XF, et al. (2011) Effect of water saving management practices and nitrogen fertilizer rate on crop yield and water use efficiency in a winter wheat–summer maize cropping system. Field Crops Res 122. 157–163.

16. Guo YQ, Wang LM, He XH, Zhang Y, Chen SY, et al. (2008) Water use efficiency and evapotranspiration of winter wheat and its response to irrigation regime in the north China plain. Agr Forest Meteorol 148: 1848–1859.

17. Halvorson AD, Nielsen DC, Reule CA (2004) Nitrogen management nitrogen fertilization and rotation effects on no-till dry land wheat production. Agron J 96: 1196–1201.

18. Turner NC (2004) Agronomic options for improving rain fall-use efficiency of crops in dryland farming systems. J Exp Bot 55: 2413–2415.

19. Turner NC, Asseng S (2005) Productivity, sustainability, and rainfall-use efficiency in Australian rainfed Mediterranean agricultural systems. Aust J Agric Res 56, 1123–1136 56: 1123–1136.

20. Kirda C, Topcu S, Kaman H, Ulger AC, Yazici A, et al. (2005) Grain yield response and N-fertilizer recovery of maize under deficit irrigation. Field Crops Res 93: 132–141.

21. Pimentel D, Hurd L, Bellotti A, Forster M, Oka I, et al. (1973) Food production and the energy crisis. Science 182: 443–449.

22. Erisman JW, Sutton MA, Galloway J, Klimont Z, Winiwarter W (2008) How a century of ammonia synthesis changed the world. Nat Geosci 1: 636–639.

23. Bassoa B, Cammarano D, Troccoli A, Chen DL, Joe T (2010) Long-term wheat response to nitrogen in a rainfed Mediterranean environment: Field data and simulation analysis. Eur J Agron 33: 132–138.

24. Hai L, Li XG, Li FM, Suo DR, Guggenberger G (2010) Long-term fertilization and manuring effects on physically-separated soil organic matter pools under a wheat-wheat-maize cropping system in an arid region of China. Soil Biol Biochem 42: 253–259.

25. Malhi S, Nyborg M, Goddard T, Puurveen D (2011) Long-term tillage, straw management and N fertilization effects on quantity and quality of organic C and N in a Black Chernozem soil. Nutr Cycl Agroecosys 90(2): 227–241.

26. Godfray HCJ, Beddington JR, Crute IR, Haddad L, Lawrence D, et al. (2010) Food security: the challenge of feeding 9 billion people. Science 327: 812–818.

27. Schindler D, Hecky R (2009) Eutrophication: more nitrogen data needed. Science 324: 721–722.

28. Hvistendahl M (2010) China's push to add by subtracting fertilizer. Science 327: 801–801.

29. Dawe D, Dobermann A, Moya P, Abdulrachman S, Bijay S, et al. (2000) How widespread are yield declines in long-term rice experiments in Asia. Field Crops Res 66: 175–193.

30. Regmi AP, Ladha JK, Pathak H, Pasuquin E, Bueno C, et al. (2002) Yield and soil fertility trends in a 20-year rice–rice–wheat experiment in Nepal. Soil Sci Soc Am J 66: 857–867.

31. Huifang H, Jiayin S, Dandan Z, Xuanqi L (2012) Effect of irrigation frequency during the growing season of winter wheat on the water use efficiency of summer maize in a double cropping system. Maydica 56(2).

32. Shen JY, Zhao DD, Han HF, Zhou XB, Li QQ (2012) Effects of straw mulching on water consumption characteristics and yield of different types of summer maize plants. Plant Soil Environ 58(4): 161–166.

33. Zhou XB, Chen YH, Ouyang Z (2011) Effects of row spacing on soil water and water consumption of winter wheat under irrigated and rain-fed conditions. Plant Soil Environ 57(3): 115–121.

34. Shangguan Z, Shao M, Dyckmans J (2000) Effects of nitrogen nutrition and water deficit on net photosynthetic rate and chlorophyll fluorescence in winter wheat. Aust J Plant Physiol 156(1): 46–51.

35. Tinsina J, Singh U, Badaruddin M, Meisiner C, Amin MR (2001) Cultivar, nitrogen, and water effects on productivity and nitrogen-use efficiency and balance for rice-wheat sequences of Bangladesh. Field Corps Res 72: 143–161.

36. Morell FJ, Lampurlane J, Alvaro FJ, Martne C (2011) Yield and water use efficiency of barley in a semiarid Mediterranean agro-ecosystem: Long-term effects of tillage and N fertilization. Soil Till Res 117: 76–84.

37. Sheng HD, Wang PH (1985) The relationship between the weight of 1000-seeds and soil water content in winter wheat season. Acta University of Agricultural Boreali occidentalis 13: 73–79. (in Chinese)

38. Shangguan ZP, Shao MA, Lei TW, Fan TL (2002) Runoff water management technologies for dryland agriculture on the Loess Plateau of China. Int J Sust Dev World 9: 341–350.

39. Johnson WC (1964) Some observations on the contribution of an inch of seeding-time soil moisture to wheat yields in the Great Plains. Agron J 56: 29–35.

40. Unger PW (1972) Dryland winter wheat and grain sorghum cropping systems, northern High Plains of Texas. Texas Agricultural Experiment Station 11–26.

41. Musick JT, Jones OR, Stemart BA, Dusek DA (1994) Water-yield relationships for irrigated and dry land wheat in the US southern plains. Agron J 86: 980–986.

42. Zhang XC, Shangguan ZP (2007) Effects of application nitrogen on photosynthesis and growth of different drought resistance winter wheat cultivars. Chinese Journal of Eco-Agriculture. 15(6). (in Chinese)

43. Hunsaker DJ, Kimball BA, Pinter Jr P, Wall G, LaMorte RL, et al. (2000) $CO_2$ enrichment and soil nitrogen effects on wheat evapotranspiration and water use efficiency. Agr Forest Meteorol 104: 85–105

44. Rahman MA, Chikushi J, Saifizzaman M, Lauren JG (2005) Rice straw mulching and nitrogen response of no-till wheat following rice in Bangladesh. Field Crops Res. 91: 71–81.

45. Howell TA, Steiner JL, Schneider AD, Evett SR (1995) Evapotranspiration of irrigated winter wheat-southern high plains. Trans ASAE 38: 745–759

46. Schneider AD, Howell TA (1997) Methods, amount, and timing of sprinkler irrigation for winter wheat. Trans ASAE 40: 137–142

47. Zhang H, Wang X, You M, Liu C (1999) Water-yield relations and water use efficiency of winter wheat in the north China plain. Irrigation Sci 19: 37–45

48. Jupp AP, Newman EI (1987) Morphological and anatomical effects of severe drought on the roots of Lo1ium-perene L. New Phytol 105: 393–402

49. McIntyre BD, Riha SJ, Flower DJ (1995) Water uptake by pearl millet in a semiarid environment. Field Crop Res 43: 67–76

50. Deng XP, Shan L, Zhang HP, Turner NC (2006) Improving agricultural water use efficiency in arid and semiarid areas of China. Agr Water Manage 80: 23–40.

# 19

# Variability of Root Traits in Spring Wheat Germplasm

**Sruthi Narayanan[1], Amita Mohan[2], Kulvinder S. Gill[2], P. V. Vara Prasad[1]***

**1** Department of Agronomy, Kansas State University, Manhattan, Kansas, United States of America, **2** Department of Crop and Soil Sciences, Washington State University, Pullman, Washington, United States of America

## Abstract

Root traits influence the amount of water and nutrient absorption, and are important for maintaining crop yield under drought conditions. The objectives of this research were to characterize variability of root traits among spring wheat genotypes and determine whether root traits are related to shoot traits (plant height, tiller number per plant, shoot dry weight, and coleoptile length), regions of origin, and market classes. Plants were grown in 150-cm columns for 61 days in a greenhouse under optimal growth conditions. Rooting depth, root dry weight, root: shoot ratio, and shoot traits were determined for 297 genotypes of the germplasm, Cultivated Wheat Collection (CWC). The remaining root traits such as total root length and surface area were measured for a subset of 30 genotypes selected based on rooting depth. Significant genetic variability was observed for root traits among spring wheat genotypes in CWC germplasm or its subset. Genotypes Sonora and Currawa were ranked high, and genotype Vandal was ranked low for most root traits. A positive relationship ($R^2 \geq 0.35$) was found between root and shoot dry weights within the CWC germplasm and between total root surface area and tiller number; total root surface area and shoot dry weight; and total root length and coleoptile length within the subset. No correlations were found between plant height and most root traits within the CWC germplasm or its subset. Region of origin had significant impact on rooting depth in the CWC germplasm. Wheat genotypes collected from Australia, Mediterranean, and west Asia had greater rooting depth than those from south Asia, Latin America, Mexico, and Canada. Soft wheat had greater rooting depth than hard wheat in the CWC germplasm. The genetic variability identified in this research for root traits can be exploited to improve drought tolerance and/or resource capture in wheat.

**Editor:** Manoj Prasad, National Institute of Plant Genome Research, India

**Funding:** This research was funded by Kansas Wheat Alliance, Coordinated Agricultural Project Grant no. 2011-68002-30029 (Triticeae- CAP) from the USDA National Institute of Food and Agriculture, and USAID Feed the Future Innovation Lab on Climate-Resilient Wheat. The funders had no role in study design, data collection and analysis, decision to publish, or preparation of the manuscript.

**Competing Interests:** The authors have declared that no competing interests exist.

* Email: vara@ksu.edu

## Introduction

Wheat (*Triticum* spp.) is one of the most important food crops in the world in terms of the area harvested, production, and productivity [1]. Wheat is grown in a wide variety of environments from tropical to temperate. Although wheat has a wide range of climatic adaptability, its productivity is limited by several abiotic stresses. Among those stresses, drought is the most widespread limitation to wheat productivity under dry-land conditions. Consequently, developing drought-tolerant wheat genotypes has been the focus of many wheat improvement programs. Root traits are critical for soil exploration and water and nutrient uptake, and are important for crop improvement under drought conditions [2], [3].

The effectiveness of a deep root system in maintaining yield under drought conditions has been confirmed by simulation studies across several years and environments in the USA [4]. A deep root system helps the plant to avoid drought stress by extracting water stored in deep soil layers (reviewed by [5] and [6]). Total root length was associated with drought tolerance in wheat because it affects the distribution of roots in the soil and influences the amount of water uptake [7]. Increased root diameter was associated with drought tolerance in rice (*Oryza sativa* L.) because thicker roots have large xylem vessels with increased axial conductance and are more efficient in penetrating deep soil layers to extract water [8], [9]. Root length density (RLD) increases the prolificacy of the root system, and was the most important trait for increased phosphorus uptake in wheat [10]. Root length density in the active root zone (0–30 cm soil depth) was correlated with water and nutrient uptake and yield under water-sufficient and water-limited conditions in chick pea (*Cicer arietinum* L.) [11], [12], [13]. Root dry weight and root: shoot ratio were positively correlated with drought tolerance in rice [14]. Fine root production in response to soil drying contributed to drought tolerance in turf grass (*Festuca arundinacea* Schreb.) [15]. Fine roots increase water and nutrient absorption because they increase root surface area per unit mass [16]. Fine roots constitute the major component of the root systems and are the most active part of the root system in extracting water and nutrients [17], [18], [19].

Despite the importance of root traits in drought tolerance, little work has been done to include drought-adaptive root traits in breeding for drought-tolerant wheat varieties. Most wheat improvement programs have concentrated on above-ground components, particularly for decreasing plant height and increasing harvest index. Crop breeding programs have largely ignored root traits, mainly because of the difficulties associated with root recovery and evaluating root traits in situ. In addition, large phenotypic plasticity of root traits in response to changes in soil conditions, and lack of high-throughput and cost-effective

**Table 1.** Analyses of variance results on effects of year (Y), genotype (G), and Y×G interaction and range for various root and shoot traits.

| Traits[†] | df (G)[‡] | P values | | | Range |
|---|---|---|---|---|---|
| | | Y | G | Y*G | |
| Rooting depth (cm) | 296 | <.0001 | <.0001 | 0.1003 | 77–202 |
| Plant height (cm) | 296 | <.0001 | <.0001 | <.0001 | 11–60 |
| Shoot dry weight (g) | 296 | <.0001 | 0.0098 | 0.2849 | 0.17–6.2 |
| Root dry weight (g) | 296 | <.0001 | <.0001 | 0.2263 | 0.23–7.6 |
| Root: shoot ratio | 296 | <.0001 | 0.0341 | 0.0567 | 0.18–4.1 |
| Tiller number per plant | 296 | 0.2104 | <.0001 | 0.9986 | 1–14 |
| Total root length (cm) | 29 | <.0001 | 0.0412 | 0.0834 | 1692–9094 |
| Total root surface area (cm²) | 29 | <.0001 | 0.0034 | 0.0545 | 184–1435 |
| Root volume (cm³) | 29 | <.0001 | 0.0021 | 0.0064 | 1.6–18 |
| Average root diameter (cm) | 29 | <.0001 | <.0001 | 0.0641 | 0.35–1.4 |
| **Root traits in 0–30 cm soil depth** | | | | | |
| Length (cm) | 29 | <.0001 | 0.0089 | 0.0501 | 1166–2484 |
| Surface area (cm²) | 29 | <.0001 | 0.0027 | 0.1907 | 144–447 |
| Root length density (cm cm⁻³) | 29 | <.0001 | 0.0082 | 0.0501 | 0.857–1.83 |
| **Traits of fine roots with diameter <0.25 mm** | | | | | |
| Length (cm) | 29 | <.0001 | 0.0463 | 0.1454 | 1005–4540 |
| Surface area (cm²) | 29 | <.0001 | 0.0455 | 0.1196 | 41–195 |
| Volume (cm³) | 29 | 0.0020 | 0.0275 | 0.0852 | 0.16–0.76 |

[†]Rooting depth, plant height, shoot dry weight, root dry weight, root: shoot ratio, and tiller number were estimated for all the 297 genotypes of the Cultivated Wheat Collection. Other traits were estimated only within the subset of 30 genotypes.
[‡]Degrees of freedom for genotype.

screening techniques make root studies highly challenging [2], [20], [21]. As a result, limited information is available on genetic variability of root traits in wheat. Exploring genetic variability of root traits could assist wheat improvement programs in developing varieties with desired root traits for drought tolerance or target environments. An understanding of the relationship of root traits to the shoot traits that contribute to grain yield is also essential to achieve improvements in productivity.

The region of origin of crop plants has implications in plant breeding as they act as potential centers to locate useful genes. Region of origin may provide useful sites for germplasm exploration to identify traits that improve productivity [22]. The adaptation profiles of domesticated plants well reflect their region of origin [23]. The agro-climatic conditions of specific regions might influence the evolution of adaptive root traits in crop plants. However, the influence of region of origin on root traits is not investigated in wheat.

Based on kernel hardness and color, wheat genotypes can be classified into different market classes. Suitability of each market class to a location depends largely on rainfall, temperature, and soil conditions. Recent findings suggested that market classes of wheat differed for coleoptile length and effect of coleoptile length

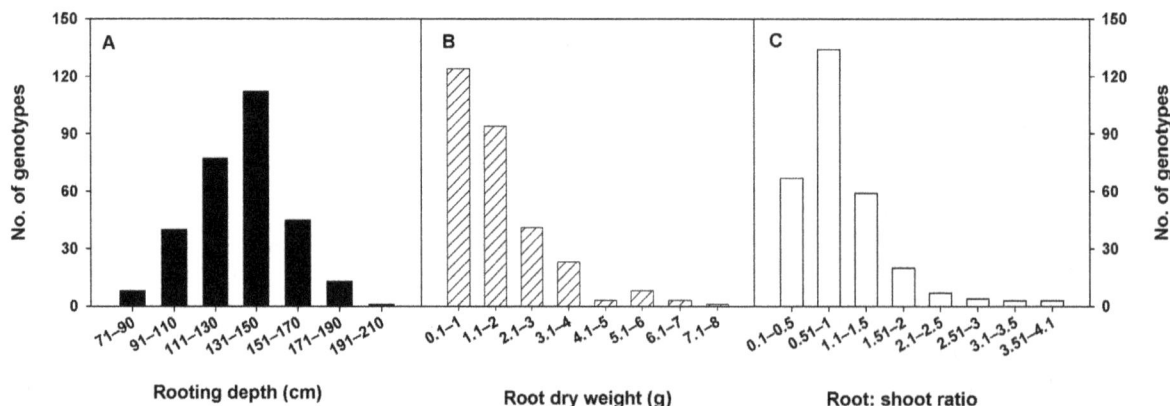

**Figure 1. Distribution of rooting depth, root dry weight, and root: shoot ratio among 297 spring wheat genotypes of the Cultivated Wheat Collection.**

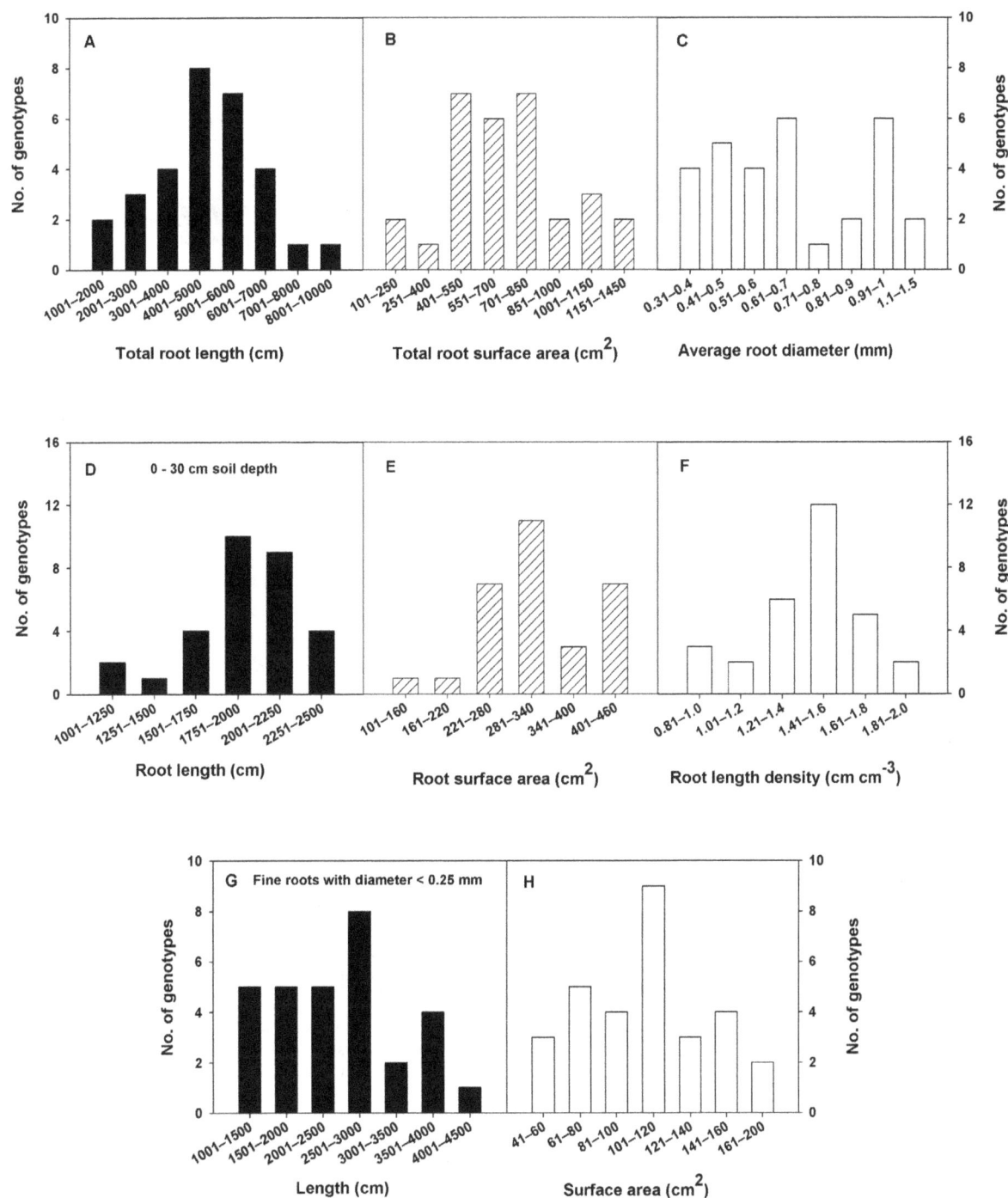

**Figure 2. Distribution of major root traits within the subset of 30 spring wheat genotypes.** The represented traits were measured for the 15 deepest and 15 smallest roots. Root length density is the ratio of root length in 0–30 cm depth of root system to the volume of the 30-cm section of the PVC column.

on seedling emergence [24]. This suggests that differences may exist among market classes for traits contributing to productivity. However, no studies have investigated the differences in root traits among different market classes of wheat.

The objectives of this research were to (i) characterize variability of root traits among spring wheat genotypes, (ii) determine whether root traits are related to plant height, shoot dry weight, tiller number per plant, and coleoptile length, and (iii) determine

whether the regions of origin and market classes of genotypes have any influence on root traits.

## Materials and Methods

### Germplasm

The germplasm used in this study was Cultivated Wheat Collection (CWC) [24] consisting of 297 spring wheat genotypes (Table S1) released since 1901. The germplasm was developed

**Table 2.** Spring wheat genotypes that were ranked high and low for rooting depth, root dry weight, and root: shoot ratio in the Cultivated Wheat Collection of 297 genotypes.

| | Rooting depth (cm) | Root dry weight (g) | Root: shoot ratio |
|---|---|---|---|
| Highest 10[†] | Florence Aka Quality (202±26)[‡a] | YSCA-1 (7.6±5.9)[a] | Olaf (4.1±2.2)[a] |
| | Sonora (190±19)[a] | Florence Aka Quality (6.6±4.6)[a] | Whitebird (4.0±3.3)[a] |
| | Marfed (186±24)[a] | Wilbur (6.1±2.2)[a] | Ramona 50 (3.9±2.9)[ab] |
| | Idaho 61M3404 (183±43)[a] | Schlanstedt (6.1±4.2)[a] | Schlanstedt (3.4±2.4)[abc] |
| | Lemhi 66 (183±16)[a] | Challis (5.8±3.5)[a] | Redchaff (3.4±1.9)[abc] |
| | Union (183±8)[a] | Kenya Kwale (5.5±4.1)[a] | Kenya Kwale (3.1±1.8)[abc] |
| | Walladay (182±31)[a] | Kinney (5.5±3.7)[a] | Eden (3.0±1.6)[abc] |
| | Sakha 69 (180±25)[a] | Utac (5.4±3.1)[a] | Faislabad 83 (2.9±2.5)[abc] |
| | Currawa (179±17)[a] | Pacific Bluestem (5.32.7)[a] | WA 7175 (2.8±2.1)[abc] |
| | Lemhi (176±20)[a] | Pirsabak 85 (5.1±2.3)[a] | Wells (2.3±1.6)[bc] |
| Lowest 10 | Vandal (96±30)[b] | ND 66 (0.35±0.10)[b] | White Marquis (0.27±0.015)[d] |
| | MN 6616M (93±18)[b] | ND 22 (0.35±0.15)[b] | White Federation (0.26±0.09)[d] |
| | Era (89±17)[b] | Faislabad 83 (0.35±0.15)[b] | Era (0.26±0.17)[d] |
| | Cumhuriyet 75 (87±18)[b] | Wells (0.34±0.19)[b] | Scarlet (0.24±0.10)[d] |
| | Yecora Rojo (85±27)[b] | MN 6616M (0.33±0.15)[b] | Vanna (0.22±0.07)[d] |
| | Marquis (84±25)[b] | Era (0.33±0.15)[b] | Ceres (0.22±0.09)[d] |
| | Sonora 64 (84±18)[b] | Vandal (0.32±0.12)[b] | Peak (0.22±0.09)[d] |
| | ND 287 (81±16)[b] | ND 287 (0.25±0.03)[b] | Sonora 64 (0.19±0.09)[d] |
| | Baw898 (78±15)[b] | Sonora 64 (0.23±0.11)[b] | McKay (0.19±0.05)[d] |
| | Hope (77±6)[b] | Calidad (0.23±0.09)[b] | Wadual (0.18±0.11)[d] |
| LSD | 46 | 2.7 | 1.6 |

[†]Genotypes were ranked based on the numerical values of root traits.
[‡]Values in parentheses are means ± standard errors of the respective traits. Values followed by different letters are significantly different according to a LSD test at P< 0.05.

using the seed material obtained from the Germplasm Resources Information Network (GRIN), International Maize and Wheat Improvement Center (CIMMYT), and Washington State University Historical Collection. These genotypes represent cultivars from 27 different countries (Table S1); Egypt (2), Libya (1), Lebanon (1), Armenia (1), Turkey (3), Iraq (1), Jordan (1), USA (190), Canada (14), Mexico (31), India (6), Pakistan (8), Nepal (2), Bangladesh (1), Australia (9), Argentina (6), Chile (2), Brazil (1), Colombia (3), Guatemala (1), Paraguay (4), Uruguay (1), Russia (2), Kenya (1), South Africa (1), Japan (1), and Germany (1). Two genotypes were not confirmed of their origin. The genotypes from USA represent the most popular cultivars during each 5-year interval since 1950 from all major breeding programs in the country. Genotypes in CWC germplasm also represented four different market classes of wheat: soft white spring (SWS), soft red spring (SRS), hard white spring (HWS), and hard red spring (HRS) (Table S1).

## Experimental Details

This research was conducted in controlled environment facilities (greenhouse) at the Department of Agronomy, Kansas State University, Manhattan, KS. Two independent experiments (2011 and 2012) were conducted to evaluate the variability of root traits among spring wheat genotypes. The greenhouse was equipped with an automated sulfur vaporizer (Rosemania, Franklin, TN) that vaporized sulfur for 1 h between 23:00 and 24:00 h. Sulfur vaporization was done from the start of the experiment as a preventive measure against powdery mildew. Plants were grown in

PVC columns with inside diameter of 7.5 cm and height of 150 cm. The columns had plastic caps at the bottom with a central hole of 0.5 cm diameter for drainage. Rooting medium was Turface MVP (PROFILE Products LLC, Buffalo Grove, IL), which had a bulk density of $576.66 \pm 32$ kg m$^{-3}$. Turface is calcined, non-swelling illite and silica clay, and allows easy separation of roots. The rooting medium was fertilized with Osmocote (Scotts, Marysville, OH), a controlled-release fertilizer with 19:6:12 N:P$_2$O$_5$:K$_2$O, respectively, at 4 g per column before sowing. A systemic insecticide, Marathon 1% G (a.i.: Imidacloprid: 1–[(6–Chloro–3–pyridinyl)methyl]–N–nitro–2–imidazolidinimine; OHP, Inc., Mainland, PA) was applied at 1 g per column before sowing to control sucking pests. Twenty seeds of each genotype were weighed before sowing to estimate seed size (individual seed weight). Three seeds of a single genotype were sown at 4 cm depth in each column on 28 December 2011 and 6 December 2012. After emergence, plants were thinned to one plant per column, which was maintained until harvest. Plants were irrigated daily (0.9±0.1 L per day) through an automated drip irrigation system until harvest to avoid water stress. Emissions from drip-tubes were examined weekly for proper water delivery. Irrigation was provided three times per day at 06:00, 12:00, and 18:00 h. Plants were maintained under optimum temperature (24/ 14°C, daytime maximum/nighttime minimum) conditions from sowing to harvest at a photoperiod of 16 h. The fungicide, Bumper 41.8 EC (a.i.: Propiconazole: 1–[[2–(2,4 dichlorophenyl)– 4–propyl–1,3–dioxolan–2–yl]Methyl]–1H–1,2,4–triazole; 1.2 mL L$^{-1}$; Makhteshim Agan of North America, Inc., Raleigh, NC) was

**Table 3.** Ranking of spring wheat genotypes for major root traits within the subset of 30 genotypes.

| | Total root length (cm) | Total root surface area (cm²) | Average root diameter (mm) | Fine roots with diameter less than 0.25 mm Length (cm) | Fine roots with diameter less than 0.25 mm Surface area (cm²) | Root traits in 0–30 cm soil depth Length (cm) | Root traits in 0–30 cm soil depth Surface area (cm²) | Root length density (cm cm⁻³) |
|---|---|---|---|---|---|---|---|---|
| Highest 10† | Sonora (9094±1762)‡[a] | | Onas (1.41±0.42)[a] | Sonora (4540±1100)[a] | Sonora (195±34)[a] | Sel 90 (2484±151)[a] | Pilcraw (447±36)[a] | Sel 90 (1.83±0.02)[a] |
| | Currawa (7249±1446)[ab] | | CI014953 (1.05±0.21)[ab] | Currawa (3823±967)[ab] | Currawa (164±29)[ab] | WA 6101 (2471±35)[a] | Hybrid 123 (438±70)[ab] | WA 6101 (1.82±0.01)[a] |
| | Pilcraw (6785±921)[abc] | | Awned Onas (1.03±0.38)[ab] | WA 6101 (3694±1516)[abc] | WA 6101 (153±54)[abc] | II-58-60 (2356±218)[ab] | CI014953 (437±91)[ab] | II-58-60 (1.73±0.03)[ab] |
| | Hyper (6649±2415)[abc] | | Pilcraw (1.01±0.30)[bcd] | Marfed (3586±1091)[abcd] | Marfed (150±39)[abcd] | McVEY (2307±392)[abc] | Currawa (432±38)[abc] | McVEY (1.7±0.06)[abc] |
| | WA 6101 (6364±2180)[abcd] | | Sonora (0.97±0.28)[bcd] | Hyper (3583±1489)[abcd] | Hyper (149±55)[abcd] | Sonora (2221±119)[abcd] | WA 6101 (423±50)[abcd] | Sonora (1.63±0.01)[abcd] |
| | Marfed (6133±1799)[abcde] | | Hybrid 123 (0.97±0.27)[bcd] | Lemhi 66 (3303±1200)[abcde] | Lemhi 66 (141±44)[abcde] | Federation 67 (2215±286)[abcd] | Onas (419±46)[abcd] | Federation 67 (1.63±0.03)[abcd] |
| | Lemhi 66 (5970±2041)[abcde] | | Currawa (0.97±0.32)[bcd] | Pilcraw (3083±523)[abcdef] | Pilcraw (138±14)[abcde] | PITIC62 (2205±190)[abcd] | Sonora (415±45)[abcd] | PITIC62 (1.62±0.06)[abcd] |
| | Onas (5848±1586)[abcde] | Awned Onas (834±191)[bcdefg] | Union (0.93±0.26)[bcde] | Union (2995±409)[abcdefg] | Union (134±16)[abcde] | Pilcraw (2141±85)[abcde] | Hyper (376±44)[abcde] | Pilcraw (1.57±0.07)[abcde] |
| | Union (5844±707)[abcde] | WA 6101 (831±229)[edefg] | Kitt (0.86±0.22)[bcdef] | Andes-56 (2846±1081)[abcdefgh] | Awned Onas (121±15)[abcdef] | Currawa (2115±241)[abcde] | Awned Onas (344±47)[abcdef] | Currawa (1.55±0.02)[abcde] |
| | CI014953 (5824±1961)[abcde] | Lemhi 66 (775±260)[cdefg] | Sel 90 (0.81±0.23)[bcdefg] | Awned Onas (2753±466)[abcdefgh] | Onas (120±36)[abcdef] | Hyper (2082±157)[abcde] | Peak 72 (343±93)[abcdef] | Hyper (1.53±0.01)[abcde] |
| Intermediate 10 | Awned Onas (5531±817)[abcde] | Andes-56 (771±377)[cdefg] | Peak 72 (0.74±0.27)[bcdefgh] | CI014953 (2633±978)[abcdefgh] | Andes-56 (117±38)[abcdef] | Awned Onas (2066±155)[abcde] | Union (340±68)[abcdef] | Awned Onas (1.52±0.01)[abcde] |
| | Andes-56 (5262±2029)[bcde] | Marfed (762±218)[cdefgh] | Hyper (0.68±0.21)[bcdefgh] | Fielder (2605±931)[abcdefgh] | CI014953 (115±37)[abcdef] | Marfed (2055±209)[abcde] | Kitt (338±33)[abcdef] | Marfed (1.51±0.02)[abcde] |
| | Federation 67 (5065±3065)[bcdef] | Hybrid 123 (736±164)[cdefghi] | Marfed (0.67±0.15)[bcdefgh] | Federation 67 (2577±1583)[bcdefgh] | Fielder (110±36)[bcdef] | Marquis (2030±471)[abcde] | Federation 67 (335±54)[abcdef] | Marquis (1.49±0.05)[abcde] |
| | McVEY (4729±2527)[bcdef] | Conley (735±379)[cdefghi] | Langdon (0.66±0.15)[bcdefgh] | Onas (2550±684)[bcdefgh] | Federation 67 (109±63)[bcdef] | Kitt (1989±149)[abcdef] | Sel 90 (329±22)[abcdef] | Kitt (1.46±0.02)[abcdef] |
| | Copper (4672±1936)[bcdef] | Langdon (688±376)[cdefghi] | PITIC62 (0.64±0.15)[cdefgh] | McVEY (2536±1341)[bcdefgh] | McVEY (105±51)[bcdef] | Fielder (1976±458)[abcdef] | McVEY (322±65)[abcdef] | Fielder (1.45±0.04)[abcdef] |
| | Fielder (4539±1714)[bcdef] | Federation 67 (671±377)[cdefghi] | WA 6101 (0.64±0.13)[cdefgh] | Copper (2445±1001)[bcdefgh] | Sel 90 (103±33)[bcdef] | Peak 72 (1971±218)[abcdef] | Penjamo T 62 (308±64)[abcdef] | Peak 72 (1.45±0.02)[abcdef] |
| | Conley (4500±2285)[bcdef] | McVEY (636±359)[cdefghi] | Lemhi (0.61±0.21)[defgh] | Zak (2363±685)[bcdefgh] | Copper (102±37)[bcdef] | Zak (1969±210)[abcdef] | II-58-60 (306±44)[abcdef] | Zak (1.45±0.02)[abcdef] |
| | Zak (4485±1311)[bcdef] | Copper (620±297)[cdefghi] | II-58-60 (0.56±0.19)[efgh] | II-58-60 (2224±535)[bcdefgh] | Zak (101±23)[bcdef] | Hybrid 123 (1927±168)[abcdef] | PITIC62 (305±27)[abcdef] | Hybrid 123 (1.42±0.02)[abcdef] |
| | Saunders (4319±2719)[bcdef] | Saunders (617±446)[cdefghi] | Penjamo T 62 (0.54±0.11)[efgh] | Sel 90 (2196±583)[bcdefgh] | II-58-60 (97±21)[bcdef] | Union (1916±193)[abcdefg] | Lemhi 66 (294±54)[bcdef] | Union (1.41±0.02)[abcdefg] |

**Table 3.** Cont.

| | Total root length (cm) | Total root surface area (cm²) | Average root diameter (mm) | Fine roots with diameter less than 0.25 mm | | Root traits in 0–30 cm soil depth | | |
| --- | --- | --- | --- | --- | --- | --- | --- | --- |
| | | | | Length (cm) | Surface area (cm²) | Length (cm) | Surface area (cm²) | Root length density (cm cm⁻³) |
| | Sel 90 (4091±1056)[bcdef] | Zak (606±206)[cdefghi] | Zak (0.52±0.11)[fgh] | Saunders (2065±1109)[bcdefgh] | Saunders (88±13)[bcdef] | Penjamo T 62 (1906±201)[abcdefg] | Marfed (290±43)[cdef] | Penjamo T 62 (1.40±0.06)[abcdefg] |
| Lowest 10 | Hybrid 123 (4012±713)[bcdef] | Fielder (536±221)[defghi] | Andes-56 (0.52±0.09)[fgh] | Conley (1940±1063)[bcdefgh] | Conley (84±42)[bcdef] | Calidad (1848±499)[abcdef] | Andes-56 (288±77)[defg] | Calidad (1.36±0.13)[abcdefg] |
| | Il-58-60 (4000±889)[bcdef] | Sel 90 (530±135)[defghi] | Copper (0.48±0.09)[fgh] | Penjamo T 62 (1866±810)[cdefgh] | Hybrid 123 (83±13)[cdef] | Saunders (1811±478)[abcdefgh] | Fielder (279±92)[defg] | Saunders (1.33±0.05)[abcdefgh] |
| | Langdon (3468±1304)[cdef] | Peak 72 (490±222)[efghi] | Fielder (0.47±0.14)[fgh] | Hybrid 123 (1852±280)[cdefgh] | Penjamo T 62 (79±30)[cdef] | Andes-56 (1769±250)[bcdefgh] | Zak (262±28)[efg] | Andes-56 (1.30±0.04)[bcdefgh] |
| | Penjamo T 62 (3291±1265)[cdef] | Il-58-60 (476±92)[fghi] | Conley (0.47±0.06)[fgh] | PITIC62 (1783±312)[cdefgh] | PITIC62 (79±13)[cdef] | Lemhi 66 (1698±152)[bcdefgh] | Saunders (260±97)[efg] | Lemhi 66 (1.25±0.02)[bcdefgh] |
| | PITIC62 (3290±489)[cdef] | Kitt (464±81)[ghi] | Federation 67 (0.45±0.03)[gh] | Marquis (1734±839)[defgh] | Marquis (72±29)[ef] | Copper (1660±289)[cdefgh] | Langdon (259±108)[efg] | Copper (1.22±0.06)[cdefgh] |
| | Kitt (2935±500)[def] | Penjamo T 62 (445±147)[fghi] | McVEY (0.41±0.02)[h] | Langdon (1486±548)[efgh] | Kitt (66±14)[ef] | Onas (1619±390)[defgh] | Conley (248±94)[efg] | Onas (1.19±0.03)[defgh] |
| | Marquis (2909±1335)[def] | PITIC62 (428±60)[fghi] | Saunders (0.39±0.04)[h] | Kitt (1365±191)[fghe] | Langdon (63±20)[f] | CI014953 (1528±336)[efgh] | Copper (242±80)[efg] | CI014953 (1.12±0.05)[efgh] |
| | Peak 72 (2552±764)[ef] | Marquis (327±151)[ghi] | Vandal (0.36±0.02)[h] | Peak 72 (1175±222)[fgh] | Peak 72 (52±11)[f] | Vandal (1323±341)[fgh] | Marquis (242±68)[efg] | Vandal (0.97±0.05)[fgh] |
| | Calidad (1848±499)[f] | Calidad (210±66)[hi] | Marquis (0.36±0.01)[h] | Calidad (1083±287)[gh] | Calidad (45±9)[f] | Langdon (1246±320)[gh] | Calidad (210±66)[fg] | Langdon (0.92±0.03)[gh] |
| | Vandal (1692±650)[f] | Vandal (184±68)[i] | Calidad (0.35±0.02)[h] | Vandal (1005±413)[h] | Vandal (41±14)[f] | Conley (1166±277)[h] | Vandal (144±34)[g] | Conley (0.86±0.03)[h] |
| LSD | 3605 | 553 | 0.39 | 1950 | 80 | 673 | 144 | 0.47 |

†Genotypes were ranked based on the numerical values of root traits.
‡Values in parentheses are means of the respective traits. Values followed by different letters are significantly different according to a LSD test at $P<0.05$.

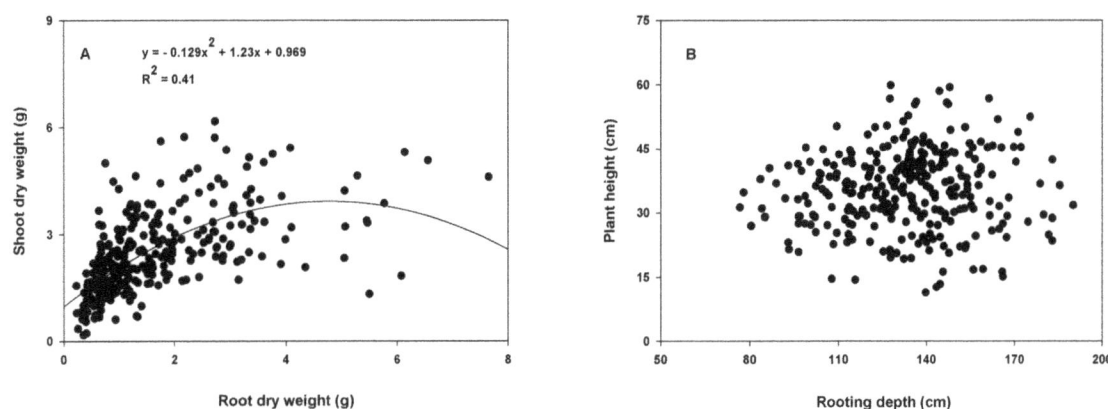

**Figure 3. Relationship between root and shoot traits of 297 spring wheat genotypes of the Cultivated Wheat Collection.** Slope of the regression line was not significant in Fig. B.

also applied at 20 d after sowing to prevent powdery mildew. The insecticide and fungicide treatments helped to maintain the plants without any pest or pathogen problems until harvest. Plants were harvested at 61 d after sowing when more than 50% of the population reached flowering stage.

## Data Collection

Shoot traits measured in this study were plant height, number of tillers per plant, shoot dry weight, and coleoptile length (see the measurement details below). These traits were measured on all 297 genotypes in the CWC germplasm. Height, tiller number, and growth stage of all plants were recorded 1 d before harvest. Plant height was determined as the distance between Turface level and the last leaf ligule. At harvest, the PVC columns were gently inverted at about 140° to let the contents of columns (Turface and plants with the entire root system) slip down to the ground. Roots were carefully separated from the Turface without any breakage in the root system. The shoot of each plant was separated by cutting at the base of the stem. After removing shoots, roots were laid on a flat surface and stretched to measure their length (from the base of the stem to the tip of the root system) as an estimate of rooting depth. Rooting depth was measured using the above procedure for all 297 genotypes in the CWC germplasm. The root system was then washed, placed between moist paper towels, sealed in Ziploc bags (S.C. Johnson & Sons, Inc. Racine, WI), transported to the laboratory, and stored at 4°C. Fifteen genotypes that were ranked the highest and 15 genotypes that were ranked the lowest for rooting depth were selected for further complete root analyses. This subset of 30 genotypes represented cultivars from Australia, Turkey, USA, Mexico, and Canada. The subset included genotypes representing the four market classes, SWS, SRS, HWS, and HRS. Root system of each of these 30 genotypes was stretched and sliced into 30-cm long portions. Each portion was submerged in a water bath (20 cm×15 cm×2 cm) to maximize separation of roots and to minimize their overlap, and scanned using an Epson photo scanner (Epson Perfection V700 with 6400 dpi resolution) (Epson, Long Beach, CA). Images of scanned roots of the 30 genotypes were analyzed using WinRHIZO Pro image analysis system (Regent Instruments, Inc., Quebec City, QC) to estimate total root length (sum of the lengths of all roots in the root system), total root surface area, root volume, average root diameter, length, surface area and RLD of roots in 0–30 cm soil depth, fine root (roots with diameter <0.25 mm) length, fine root surface area, and fine root volume [25], [26]. Root length density in each 30-cm depth of root system was calculated as the ratio of

root length to the volume of 30-cm section of the PVC column, and it represented RLD in each 30 cm of soil depth [22]. After scanning, root systems were packed in paper bags for drying. Roots and shoots of all 297 genotypes were dried to constant weight at 60°C for determining dry weight. Root: shoot ratio for each of the 297 genotypes was calculated as the ratio of root dry weight to shoot dry weight.

Coleoptile length was measured according to the procedure of [24]. Fifteen uniform-sized seeds of each of the 297 genotypes with no physical damage were placed in the middle of a moist germination paper (Heavy Germination paper #SD 7615L; Anchor Paper Co., Saint Paul, MN), about 1 cm apart with the germ end down. The germination paper was then folded vertically in half with the seed placed in the crease, and the folded half was again folded horizontally four times and placed in a plastic tray with holes at the base to drain excess water. The plastic trays were then placed inside a completely darkened box and kept in a growth chamber at a constant temperature of 22°C. After 10 d, the coleoptile length of 10 randomly-selected seedlings of each genotype was recorded to the nearest millimeter measuring from the base of the seed to the tip of the coleoptile.

## Statistical Analyses

The experimental design was a randomized complete block in 2011 (Experiment 1) and 2012 (Experiment 2) for the greenhouse studies. There were two blocks (replications) in both years. Analysis of variance was performed on genotypes using the GLM procedure in SAS (Version 9.2, SAS Institute) for root and shoot traits. The probability threshold level ($\alpha$) was 0.05. Genotype was treated as a fixed effect, and replication nested within year was treated as a random effect. Genotype, replication, and year were used as class variables. Separation of means was done using the LSD test (P<0.05). The CORR procedure in SAS was used to find out the correlation between different root and shoot traits. Pearson correlation coefficient was used as a measure of degree of correlation between root and shoot traits. The REG procedure in SAS was used to regress root traits against shoot traits.

## Results

### Genetic Variability of Root and Shoot Traits

Significant variability was observed for root and shoot traits among spring wheat genotypes in the CWC germplasm or its subset (Table 1). Because there was no significant interaction between genotype and year for most of the traits, data were pooled

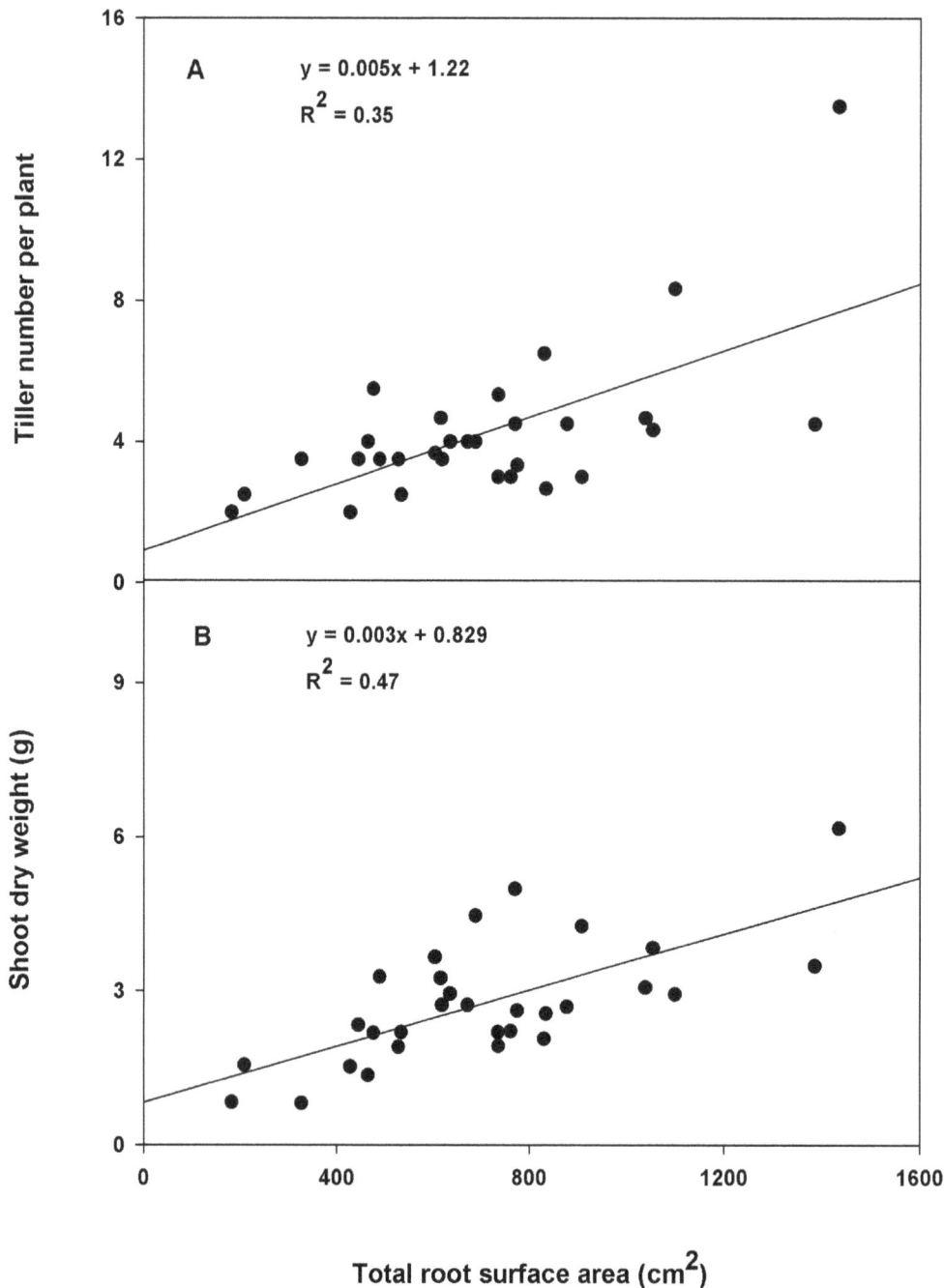

**Figure 4. Relationship of total root surface area with tiller number per plant and shoot dry weight within the subset of 30 spring wheat genotypes.**

across years. More than 100% variation was observed between minimum and maximum values of all root traits (Table 1). Range of major root traits was 77–202 cm, rooting depth and 0.23–7.6 g, root dry weight in the CWC germplasm and 1692–9094 cm, total root length and 184–1435 $cm^2$, total root surface area in the subset (Table 1). Extent of variability for different root traits among genotypes in the CWC germplasm and the subset is shown in Fig. 1 and 2, respectively. Ranking of genotypes based on the numerical values of different root traits in the CWC germplasm and the subset are given in Table 2 and 3, respectively. Genotypes Sonora and Currawa had increased rooting depth (Table 2), total root length, total root surface area, average root diameter, fine

root length, and fine root surface area (Table 3). Similarly, genotypes Vandal and Marquis had decreased rooting depth (Table 2), total root length, total root surface area, average root diameter, fine root length, and fine root surface area (Table 3). Genotypes Sonora and Currawa were also ranked high and genotype Vandal was also ranked low for total root length, total root surface area, and RLD in the 0–30 cm depth of soil (Table 3). Genotype Florence Aka Quality had increased rooting depth and root dry weight (Table 2). Genotypes Federation 67 and McVEY had decreased root diameter, but increased root length and RLD in 0–30 cm depth of soil (Table 3).

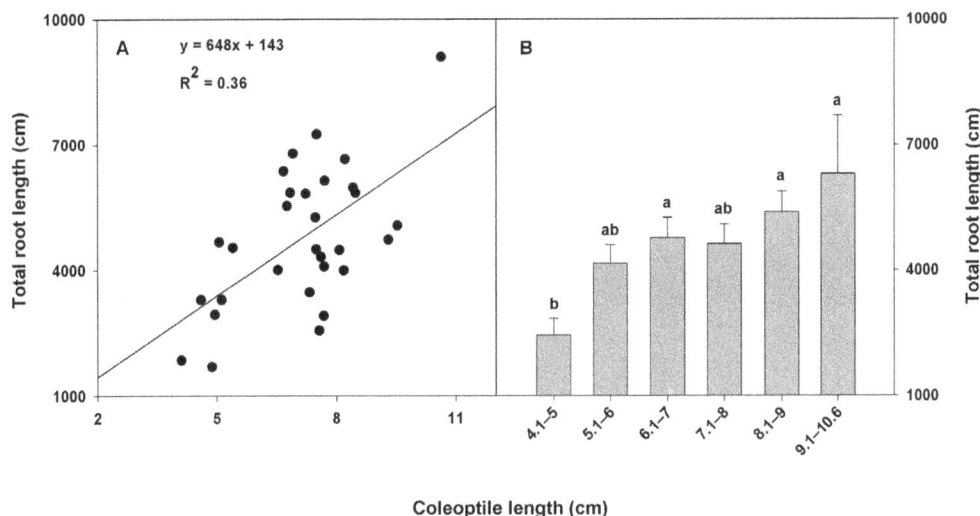

**Figure 5. Relationship between coleoptile length and total root length within the subset of 30 spring wheat genotypes.** Error bars (Fig. B) represent standard errors.

## Relationship between Root and Shoot Traits

A positive relationship (coefficient of determination [$R^2$]≥0.35) was found between root dry weight and shoot dry weight within the CWC germplasm (Fig. 3A) and between total root surface area and tiller number (Fig. 4A); total root surface area and shoot dry weight (Fig. 4B); and total root length and coleoptile length within the subset (Fig. 5A). A correlation coefficient of >0.50 was observed for the correlation of shoot dry weight with root dry weight and rooting depth within the CWC germplasm and shoot dry weight with total root surface area, and root volume; tiller number with total root surface area and root volume; and coleoptile length with total root length and total root surface area within the subset (Table 4, 5). Slope of the regression between plant height and rooting depth was not significant within the CWC germplasm (Fig. 3B). In addition, plant height did not show correlation with most root traits within the CWC germplasm (Table 4) or the subset (Table 5). Coleoptile length had a significant effect on total root length within the subset ($P<0.001$; Fig. 5B). Genotypes with longer coleoptiles (>8 cm) had significantly greater total root length than genotypes with shorter coleoptiles (≤5 cm; Fig. 5B).

## Region of Origin, Market Class, and Root Traits of Wheat Genotypes

The CWC germplasm included genotypes originating from 27 different countries. Country of origin had significant effect on rooting depth ($P<0.05$). When genotypes in the CWC germplasm were categorized into eight regional groups based on their country of origin, significant difference in the mean rooting depth was observed among the eight regions ($P<0.05$; Fig. 6A). The wheat genotypes collected from Australia, Mediterranean, and west Asia regions had greater rooting depth than those collected from south Asia, Latin America, Mexico, and Canada. West Asia (Fig. 6A) included Armenia (one genotype), Turkey (three genotypes), Iraq (one genotype), and Jordan (one genotype), which encompass the region where wheat originated. This shows that the six genotypes collected from the center of origin of wheat had deep root systems. When genotypes from USA were classified into 10 groups based on their state of origin, there was not much variation in the mean rooting depth among different groups (Fig. 6C). However,

genotypes from Oregon had greater rooting depth than genotypes from other states such as North Dakota, Colorado, Arizona, and Minnesota. When genotypes in the CWC germplasm were categorized into four different market classes, they differed in rooting depth ($P<0.0001$). Soft wheat had greater rooting depth than hard wheat (Fig. 7). Soft white spring wheat had the largest rooting depth among the market classes evaluated in this research.

## Discussion

Considerable genetic variability was observed for root traits in the CWC germplasm or its subset. The extent of genetic variability is indicated by the large range observed for root traits. Because roots followed a zigzag pattern of growth within the columns, in many cases rooting depth attained values that exceeded column height. The $P>0.05$ for genotype-by-year interaction (Table 1) implies that genotypes had similar responses in both years for root traits. Plants were at flag leaf, booting, spike emergence, or flowering stages at the time of harvest. However, data analysis showed that except on plant height and tiller number, growth stage had no effect ($P>0.05$) on any of the root and shoot traits measured on the 297 genotypes in the CWC germplasm (data not shown).

Genotypes Sonora and Currawa were ranked high and genotype Vandal was ranked low for most root traits in the CWC germplasm or its subset. The contrasting genotypes for root traits identified in this study (Table 2, 3) offer useful plant materials that can be included in wheat improvement programs. Genotypes Federation 67 and McVEY were ranked in the lowest one third of genotypes for average root diameter and in the top one third of genotypes for root length and RLD in the upper soil profile (0–30 cm; Table 3). Decreased root penetration due to decreased root diameter [9] in genotypes Federation 67 and McVEY might have resulted in increased spreading behavior, which was manifested in terms of increased root length and RLD in the upper soil profile. Small root diameter and xylem vessels can enhance grain yield in wheat under terminal drought stress conditions because these traits help to conserve sufficient soil water for grain filling stage [5], [27].

In the present research, total root surface area showed a positive correlation with tiller number and shoot dry weight (Fig. 4). Previous reports in other cereals have suggested that water and

**Table 4.** Correlations among various root and shoot traits of 297 spring wheat genotypes of the Cultivated Wheat Collection.

| Traits | Root dry weight | Shoot dry weight | Root: shoot ratio | Plant height | Tiller number | Coleoptile length | Seed size |
|---|---|---|---|---|---|---|---|
| Rooting depth | 0.43†****‡ | 0.51**** | −0.13* | NS§ | 0.28**** | 0.16** | NS |
| Root dry weight | | 0.57**** | 0.33**** | NS | 0.46**** | 0.14* | NS |
| Shoot dry weight | | | −0.18** | 0.36**** | 0.55**** | 0.13* | NS |
| Root: shoot ratio | | | | −0.35**** | NS | NS | NS |
| Plant height | | | | | NS | NS | NS |
| Tiller number | | | | | | 0.17** | NS |
| Coleoptile length | | | | | | | NS |

†Values in each cell represent Pearson correlation coefficient.
‡*, **, ***, and **** indicate significance at 0.05, 0.01, 0.001, and 0.0001 probability levels, respectively.
§Not significant at 0.05 probability level.

**Table 5.** Correlations among various root and shoot traits within the subset of 30 spring wheat genotypes.

| Traits | Total root surface area | Root volume | Average root diameter | Rooting depth | Root dry weight | Shoot dry weight | Root: shoot ratio | Plant height | Tiller number | Coleoptile length | Seed size |
|---|---|---|---|---|---|---|---|---|---|---|---|
| Total root length | 0.88†****‡ | 0.73**** | 0.51** | 0.79**** | 0.59**** | 0.47** | NS§ | NS | NS | 0.66*** | NS |
| Total root surface area | | 0.94**** | 0.70**** | 0.68**** | 0.87**** | 0.69**** | NS | NS | 0.61**** | 0.51** | NS |
| Root volume | | | 0.78**** | 0.60*** | 0.81**** | 0.69**** | NS | NS | 0.54** | 0.43* | NS |
| Average root diameter | | | | 0.55** | 0.57*** | NS | NS | NS | 0.38* | NS | NS |

†Values in each cell represent Pearson correlation coefficient.
‡*, **, ***, and **** indicate significance at 0.05, 0.01, 0.001, and 0.0001 probability levels, respectively.
§Not significant at 0.05 probability level.

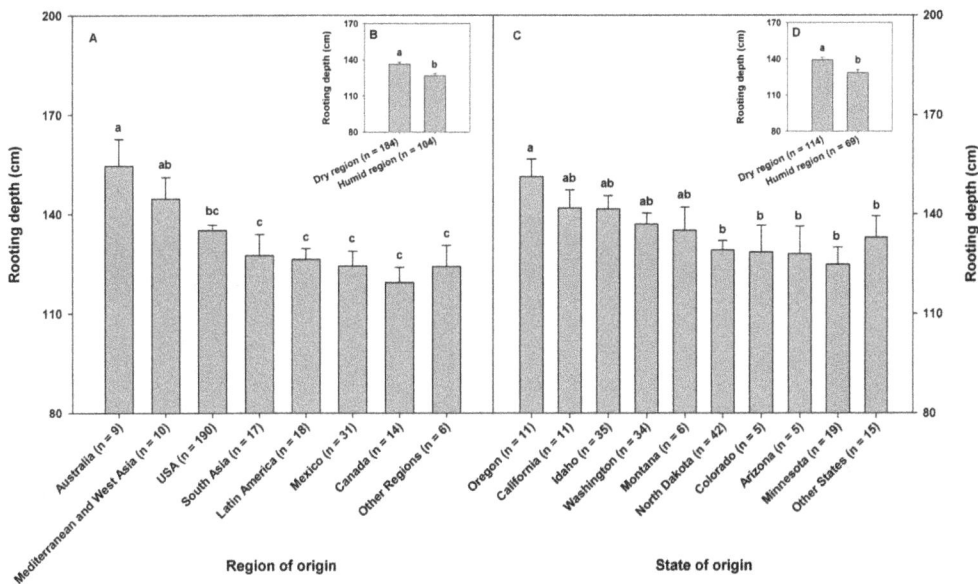

**Figure 6. Rooting depth of spring wheat genotypes originating from different wheat growing regions in the world (Fig. A), and different states in the USA (Fig. C).** Fig. A represents 295 genotypes of the Cultivated Wheat Collection (297 genotypes) because two genotypes were not confirmed of their origin. Fig. C represents 183 of the 190 genotypes originated from USA because seven genotypes were not confirmed of their state of origin. Mediterranean – Egypt, Libya, and Lebanon; West Asia – Armenia, Turkey, Iraq, and Jordan; South Asia – India, Pakistan, Nepal, and Bangladesh; Latin America – Argentina, Chile, Brazil, Colombia, Guatemala, Paraguay, and Uruguay; Other Regions – Russia, Japan, Germany, Kenya and South Africa. Other States – Indiana, Nebraska, Nevada, Oklahoma, South Dakota, Utah, Vermont, and Wisconsin. Dry region in Fig. B included Argentina, Armenia, Australia, Chile, Egypt, Germany, Iraq, Japan, Jordan, Kenya, Lebanon, Libya, Mexico, Pakistan, South Africa, Turkey, and USA states of Arizona, California, Colorado, Idaho, Montana, Nebraska, Nevada, Oregon, Utah, and Washington. Similarly, humid region in Fig. B included Bangladesh, Brazil, Canada, Colombia, Guatemala, India, Nepal, Paraguay, Russia, Uruguay, and USA states of Indiana, Minnesota, North Dakota, Oklahoma, South Dakota, Vermont, and Wisconsin. Two genotypes with unknown country of origin and seven genotypes with unknown states of origin in the USA were not included in Fig. B. Error bars represent standard errors.

nutrient uptake from the soil is proportional to contact area between root surface and soil [28], [29]. This indicates that resource uptake increases with root surface area. The increased resource uptake through increased root surface area might have helped the plant to produce more tillers. The increased tiller number, which leads to increased shoot biomass production, might be the reason for increased shoot dry weight.

The positive correlation between root dry weight and shoot dry weight ($R^2 = 0.41$; Fig. 3A) observed in this research is consistent with reports on other crops [25], [30]. The increased resource capture achieved through increased root mass might have contributed to the increased shoot dry weight. In turn, the surplus of photoassimilates as a result of increased shoot growth might be allocated to roots that increased root dry weight. However, the amount of resource uptake by different genotypes was not quantified in this study to evaluate its effects on root and shoot dry weights.

The absence of correlation between plant height and root traits (Table 4, 5; Fig. 3B) observed in this research is supported by previous reports in wheat [31], chickpea [30], or field pea (*Pisum sativum* L.) [25]. It is reported in field pea that plant height is not expected to have a correlation with total root length and weight because total root length is determined by number and length of lateral roots [25]. Reports suggest that root length and weight are predominately controlled by different sets of genes compared to that of shoot length [31]. Some studies have reported that decreased plant height genes had no impact on root diameter [32], and root dry weight [33]. Even though plant height was not correlated with root traits and tiller number, a negative correlation was found between plant height and root: shoot ratio (Table 4, 5).

This may be due to increased shoot biomass production and therefore, increased shoot dry weight by tall plants.

The positive relationship between coleoptile length and total root length ($R^2 = 0.43$; Fig. 5) has important practical implications. Selection for a deep and prolific root system on the basis of total root length is not easy because it is difficult to measure roots in situ. In addition, direct selection for total root length is a destructive process and prevents selection. Therefore, nondestructive selection criteria for improved root traits are important. Because total root length and coleoptile length show a positive linear relationship, selecting genotypes with increased coleoptile length might result in genotypes with increased root length. Selection for coleoptile length is easy, non-destructive, and involves high heritability ($h^2 > 0.70$) [34]. A long coleoptile enables sowing at greater soil depths where moisture is available [35], and improves seedling vigor and stand establishment [36].

When countries or USA states of origin of all 297 genotypes were broadly classified into dry or humid regions (Köppen-Geiger climate classification) [37], it had significant influence on the relationship between coleoptile length and rooting depth ($P < 0.05$ for the effect of 'coleoptile length-by-region' interaction on rooting depth). Coleoptile length and rooting depth had a positive linear relationship with $R^2 = 0.11$, in the dry region (Fig. 8). This implies that rooting depth increases with coleoptile length in the dry regions. Deep roots increase soil water extraction from deep soil layers where moisture is available [5] and longer coleoptiles improve stand establishment and vigor in deep-sown crop in the dry areas [36]. Therefore, both of these traits provide adaptational advantages to genotypes grown under soil moisture limited environments.

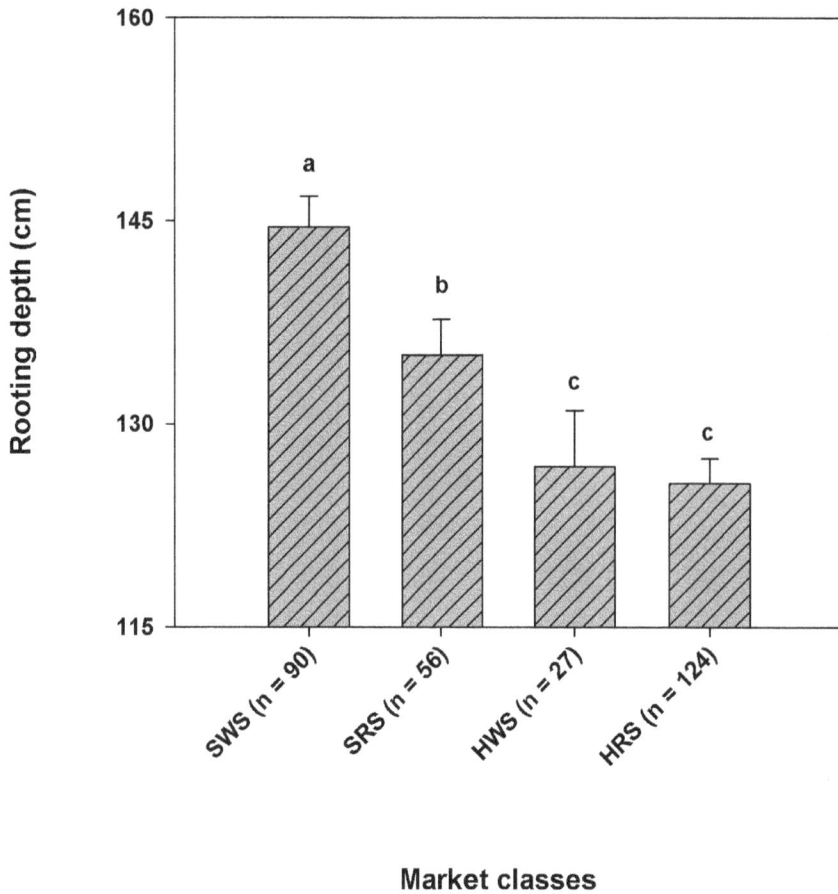

**Figure 7. Rooting depth of spring wheat genotypes belonging to different market classes within the Cultivated Wheat Collection.** Error bars represent standard errors. SWS – soft white spring; SRS – soft red spring; HWS – hard white spring; HRS – hard red spring.

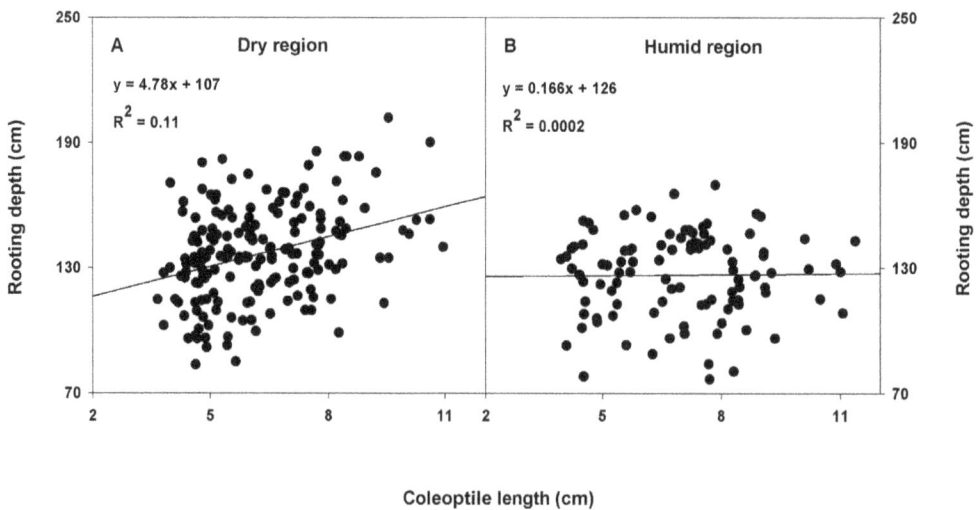

**Figure 8. Relationship between coleoptile length and rooting depth among spring wheat genotypes originated from dry (n = 184) or humid regions (n = 104).** Dry region included Argentina, Armenia, Australia, Chile, Egypt, Germany, Iraq, Japan, Jordan, Kenya, Lebanon, Libya, Mexico, Pakistan, South Africa, Turkey, and USA states of Arizona, California, Colorado, Idaho, Montana, Nebraska, Nevada, Oregon, Utah, and Washington. Similarly, humid region included Bangladesh, Brazil, Canada, Colombia, Guatemala, India, Nepal, Paraguay, Russia, Uruguay, and USA states of Indiana, Minnesota, North Dakota, Oklahoma, South Dakota, Vermont, and Wisconsin. Two genotypes with unknown country of origin and seven genotypes with unknown states of origin in the USA were not included in these figures.

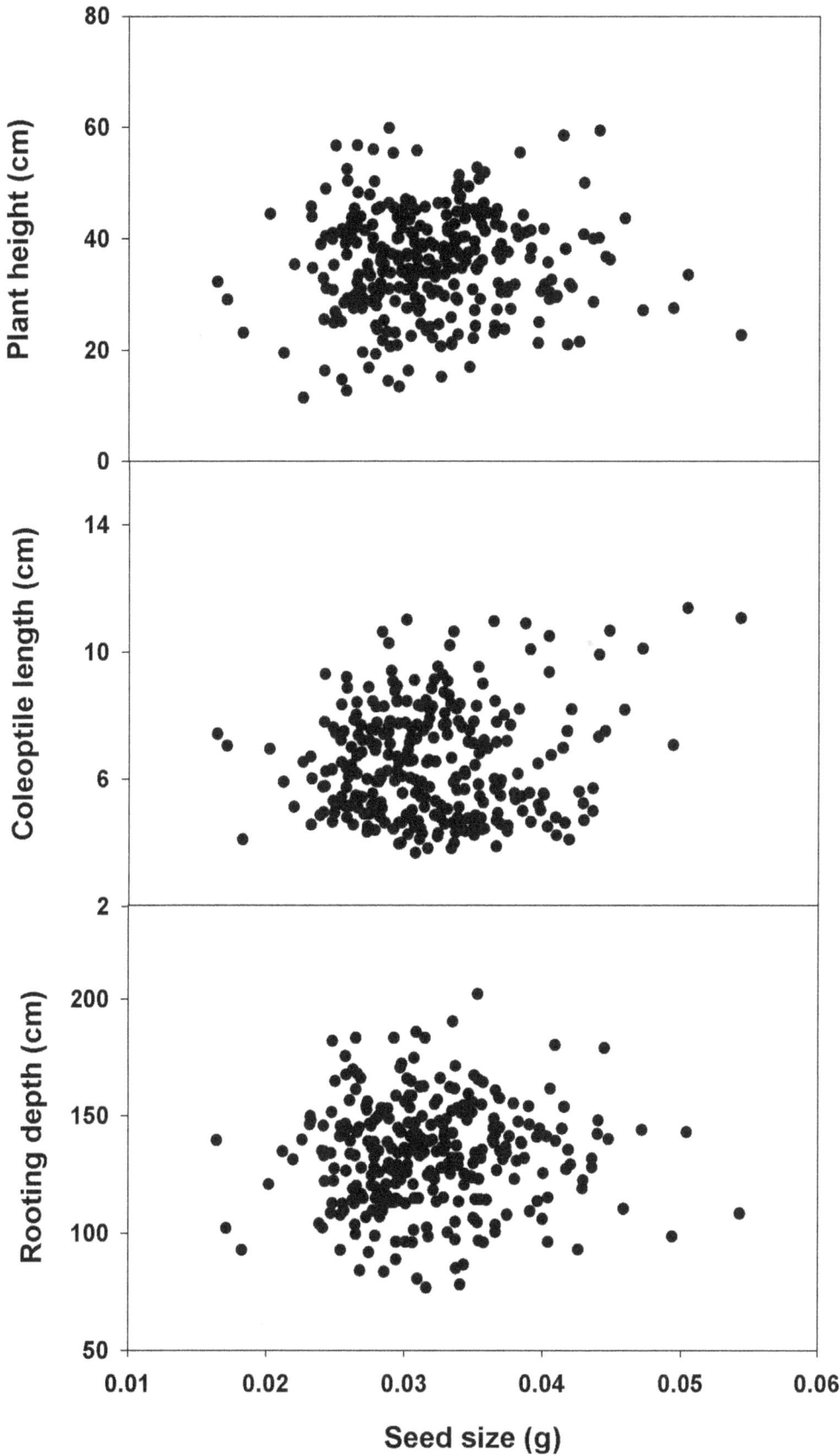

**Figure 9. Relationship of seed size with plant height, coleoptile length, and rooting depth of 297 spring wheat genotypes of the Cultivated Wheat Collection.**

None of the root traits evaluated in this research showed a correlation with seed size within the CWC germplasm or its subset (Table 4, 5). Slope was not significant when plant height, coleoptile length, and rooting depth were regressed against seed size within the CWC germplasm (Fig. 9). This shows that seeds with increased size may not always produce longer coleoptile, deeper roots, or taller plants. This result is in agreement with previous reports suggesting that seed size has no influence on coleoptile length and seedling emergence in wheat [24], [38], [39], [40]. However, contradictory reports also exist in literature that suggested a positive correlation between seed size and coleoptile length [41], [42], [43].

This research found that the geographic regions from which wheat genotypes originated had significant impacts on rooting depth (Fig. 6A). The wheat genotypes collected from Australia, Mediterranean, and west Asia regions possessed greater rooting depth compared with those collected from south Asia, Latin America, Mexico, and Canada. However, we acknowledge that different geographic regions were not represented by equal number of genotypes. Genotypes originated from dry regions had greater rooting depth than those from humid regions (Fig. 6B, D). Growth environments for wheat in Mediterranean, Australia, and west Asia regions are much drier than they are in the other regions such as south Asia, Latin America, or Canada [22], [44]. Maximum utilization of stored soil moisture is important for the dry environments in Australia, Mediterranean or west Asia regions [2]. Plant root systems in these regions are adapted to thrive on the available soil moisture and not deplete it before maturity [44]. Wheat depending on stored soil moisture needs a root system that reaches the deep soil profile [5], [45]. Thus, wheat genotypes that evolved in those drier areas might have adapted by increasing rooting depth to capture water from the deeper layers of soil. The wheat genotypes collected from Australia also had larger root diameter (mean ± SD, 1.1±0.31 mm) than those collected from other regions. Large diameter is an important trait of plant roots that helps them to penetrate deep soil layers, which is evident from the positive correlation between average root diameter and rooting depth in the present study (Table 5). Even though mean rooting depth did not show much variation among different states within the USA, a gradual decrease in rooting depth was noticed from west to east (Fig. 6C). This could be associated with general trends of increasing precipitation and decreasing temperature from west to east in the USA. The increased rooting depth may be an adpational trait of genotypes grown in comparatively drier areas of western USA to improve water absorption. We also observed differences in rooting depth among different market classes of

wheat (Fig. 7). The greater rooting depth of soft wheat compared with hard wheat may be an inherent characteristic of soft wheat genotypes.

In summary, significant genetic variability was observed for root traits in the CWC germplasm or its subset. Genotypes Sonora and Currawa were ranked high and genotype Vandal was ranked low for most root traits. A strong positive relationship ($R^2 \geq 0.35$) was found between (1) root dry weight and shoot dry weight within the CWC germplasm and between total root surface area and tiller number; total root surface area and shoot dry weight; and total root length and coleoptile length within the subset. There was no correlation between plant height and most root traits within the CWC germplasm or its subset. Region of origin of wheat genotypes had significant impact on rooting depth in the CWC germplasm. The wheat genotypes collected from Australia, Mediterranean, and west Asia regions had greater rooting depth than those collected from south Asia, Latin America, Mexico, and Canada. Rooting depth differed among market classes of wheat genotypes in the CWC germplasm. Soft wheat had greater rooting depth than hard wheat in the CWC germplasm. The genetic variability identified in this research for root traits can be exploited to improve drought tolerance and/or resource capture in wheat. Our future research will evaluate drought tolerance of the contrasting genotypes identified in this study for root traits under controlled environment and field conditions.

## Acknowledgments

We thank Austin Hughes, Prudhvi Tej Sri Adhibatla, Sheila Ngao, Prakarsh Tiwari, George Paul and Predeesh Chandran for help in data collection. This publication is Kansas Agriculture Experiment Station Contribution No. 14-030-J.

## Author Contributions

Conceived and designed the experiments: PVVP SN. Performed the experiments: SN. Analyzed the data: SN PVVP. Wrote the paper: SN PVVP. Contributed the germplasm: KSG AM. Contributed data on coleoptile length: AM KSG.

## References

1. FAO (2011) FAOSTAT. Available at http://faostat3.fao.org/faostat-gateway/go/to/download/Q/QC/E. FAO, Rome, Italy.

2. Manschadi AM, Hammer GL, Christopher JT, deVoil P (2008) Genotypic variation in seedling root architectural traits and implications for drought adaptation in wheat (Triticum aestivum L.). Plant Soil 303: 115–129.

3. Gupta PK, Balyan HS, Gahlaut V, Kulwal PL (2012) Phenotyping, genetic dissection, and breeding for drought and heat tolerance in common wheat: Status and prospects. In: Janick J, editor. Plant breeding reviews, volume 36. John Wiley & Sons, Inc., Hoboken, New Jersey, USA. 85–168.

4. Sinclair TR (1994) Limits to crop yield? In: Boote KJ, editor. Physiology and determination of crop yield. ASA, CSSA, and SSSA, Madison, Wisconsin, USA. 509–532.

5. Ludlow MM, Muchow RC (1990) A critical evaluation of traits for improving crop yields in water-limited environments. Adv Agron 43: 107–153.

6. Boyer JS (1996) Advances in drought tolerance in plants. Adv Agron 56: 187–218.

7. Manschadi AM, Christopher J, deVoil P, Hammer GL (2006) The role of root architectural traits in adaptation of wheat to water-limited environments. Funct Plant Biol 33: 823–837.

8. Fukai S, Cooper M (1995) Development of drought-resistant cultivars using physiomorphological traits in rice. Field Crops Res 40: 67–87.

9. Clark LJ, Price AH, Steele KA, Whalley RR (2008) Evidence from near-isogenic lines that root penetration increases with root diameter and bending stiffness in rice. Funct Plant Biol 35: 1163–1171.

10. Manske GGB, Ortiz-Monasterio JI, Van Ginkel M, Gonzalez RM, Rajaram S, et al. (2000) Traits associated with improved P-uptake efficiency in CIMMYT' semidwarf spring bread wheat grown on an acid andisol in Mexico. Plant Soil 221: 189–204.

11. Krishnamurthy L, Johansen C, Ito O (1996) Genetic variation in root system development and its implications for drought resistance in chickpea. In: Ito O, Johansen C, Adu-Gyamfi JJ, Katayama K, Kumar Rao JVDK, et al. editors. Roots and nitrogen in cropping systems of the semi-arid tropics, Culio Corporation, Tsukuba, Japan. 234–250.

12. Krishnamurthy L, Johansen C, Sethi SC (1999) Investigation of factors determining genotypic differences in seed yield of non-irrigated and irrigated chickpeas using a physiological model of yield determination. J Agron Crop Sci 183: 9–17.

13. Kashiwagi J, Krishnamurthy L, Crouch JH, Serraj R (2006) Variability of root length density and its contributions to seed yield in chickpea (*Cicer arietinum* L.) under terminal drought stress. Field Crops Res 95: 171–181.

14. Champoux MC, Wang G, Sarkarung S, Mackill DJ, O'Toole JC, et al. (1995) Locating genes associated with root morphology and drought avoidance in rice via linkage to molecular markers. Theor Appl Genet 90: 969–981.

15. Huang B, Fry JD (1998) Root anatomical, morphological, and physiological responses to drought stress for tall fescue cultivars. Crop Sci 38: 1017–1022.

16. Eissenstat DM (1992) Costs and benefits of constructing roots of small diameter. J Plant Nutr 15: 763–782.

17. Smucker AJM (1984) Carbon utilization and losses by plant root systems. In: Barber SA, Boulden DR, editors. Roots, nutrient and water influx, and plant growth. Special Publication no. 49. ASA, CSSA, and SSSA, Madison, Wisconsin, USA. 27–46.

18. Hodge A, Robinson D, Griffiths BS, Fitter AH (1999) Why plants bother: root proliferation results in increased nitrogen capture from an organic patch when two grasses compete. Plant Cell Environ 22: 811–820.

19. Pierret A, Moran CJ, Doussan C (2005) Conventional detection methodology is limiting our ability to understand the roles and functions of fine roots. New Phytol 166: 967–980.

20. Poorter H, Nagel O (2000) The role of biomass allocation in the growth response of plants to different levels of light, $CO_2$, nutrients and water: a quantitative review. Aust J Plant Physiol 27: 595–607.

21. Fitter AH (2002) Characteristics and functions of root systems. In: Waisel Y, Eshel A, Kafkafi U, editors. Plant roots: the hidden half. Marcel Dekker, New York, USA. 249–259.

22. Kashiwagi J, Krishnamurthy L, Upadhyaya HD, Kishana H, Chandra S, et al. (2005) Genetic variability of drought-avoidance root trait in the mini core germplasm collection of chickpea (*Cicer arietinum* L.). Euphytica 146: 213–222.

23. Jones H, Leigh FJ, Mackay I, Bower MA, Smith LMJ, et al. (2008) Population-based resequencing reveals that the flowering time adaptation of cultivated barley originated east of the Fertile Crescent. Mol Biol Evol 25: 2211–2219.

24. Mohan A, Schillinger WF, Gill KS (2013) Wheat seedling emergence from deep planting depths and its relationship with coleoptile length. PLoS ONE 8(9).

25. McPhee K (2005) Variation for seedling root architecture in the core collection of pea germplasm. Crop Sci 45: 1758–1763.

26. Singh V, van Oosterom EJ, Jordan DR, Hunt CH, Hammer GL (2011) Genetic variability and control of nodal root angle in sorghum. Crop Sci 51: 2011–2020.

27. Richards RA, Passioura JB (1989) A breeding program to reduce the diameter of the major xylem vessel in the seminal roots of wheat and its effect on grain yield in rain-fed environments. Aust J Agr Res 40: 943–950.

28. Caassen N, Barber SA (1976) Simulation model for nutrient uptake from soil by a growing plant system. Agron J 68: 961–964.

29. Yoshida S, Hasegawa S (1982) The rice root system: its development and function. In: Drought resistance in crops, with emphasis on rice. IRRI, Los Banos, Laguna, Philippines. 97–114.

30. Serraj R, Krishnamurthy L, Kashiwagi J, Kumar J, Chandra S, et al. (2004) Variation in root traits of chickpea (*Cicer arietinum* L.) grown under terminal drought. Field Crops Res 88: 115–127.

31. Sanguineti MC, Li S, Maccaferri M, Corneti S, Rotondo F, et al. (2007) Genetic dissection of seminal root architecture in elite durum wheat germplasm. Ann Appl Biol 151: 291–305.

32. Wojciechowski T, Gooding MJ, Ramsay L, Gregory PJ (2009) The effects of dwarfing genes on seedling root growth of wheat. J Exp Bot 60: 2565–2573.

33. Bush MG, Evans LT (1988) Growth and development in tall and dwarf isogenic lines of spring wheat. Field Crops Res 18: 243–270.

34. Rebetzke GJ, Ellis MH, Bonnett DG, Richards RA (2007) Molecular mapping of genes for coleoptile growth in bread wheat (*Triticum aestivum* L.). Theor Appl Genet 114: 1173–1183.

35. Schillinger WF, Donaldson E, Allan RE, Jones SS (1998) Winter wheat seedling emergence from deep sowing depths. Agron J 90: 582–586.

36. Rebetzke GJ, Richards RA, Fettell NA, Long M, Condon AG, et al. (2007) Genotypic increases in coleoptile length improves stand establishment, vigour and grain yield of deep-sown wheat. Field Crops Res 100: 10–23.

37. Peel MC, Finlayson BL, McMahon TA (2007) Updated world map of the Köppen-Geiger climate classification. Hydrol Earth Syst Sci 11: 1633–1644.

38. Kaufmann ML (1968) Coleoptile length and emergence in varieties of barley, oats, and wheat. Canad J Plant Sci 48: 357–361.

39. Mian AR, Nafziger ED (1992) Seed size effects on emergence, head number, and grain yield of winter wheat. Journal of Production Agriculture 5: 265–268.

40. Chastian T, Ward K, Wysocki D (1995) Stand establishment response of soft white winter wheat to seedbed residue and seed size. Crop sci 35: 213–218.

41. Cornish P, Hindmarsh S (1988) Seed size influences the coleoptile length of wheat. Aust J Exp Agr 28: 521–524.

42. Botwright TL, Rebetzke GJ, Condon AG, Richards RA (2001) Influence of variety, seed position and seed source on screening for coleoptile length in bread wheat (*Triticum aestivum* L.). Euphytica 119: 349–356.

43. Nik MM, Babaeian M, Tavassoli A (2011) Effect of seed size and genotype on germination characteristic and seed nutrient content of wheat. Sci Res Essays 6: 2019–2025.

44. Blum A (2005) Drought resistance, water-use efficiency, and yield potential–are they compatible, dissonant, or mutually exclusive? Aust J Agr Res 56: 1159–1168.

45. Manske GGB, Vlek PLG (2002) Root architecture-wheat as a model plant. In: Waisel Y, Eshel A, Kafkafi U, editors. Plant roots: the hidden half. Marcel Dekker, New York, USA. 249–259.

# 20

# Temporal Variations in the Abundance and Composition of Biofilm Communities Colonizing Drinking Water Distribution Pipes

**John J. Kelly[1]\***, **Nicole Minalt[1]**, **Alessandro Culotti[2]**, **Marsha Pryor[3]**, **Aaron Packman[2]**

1 Department of Biology, Loyola University Chicago, Chicago, Illinois, United States of America, 2 Department of Civil and Environmental Engineering, Northwestern University, Evanston, Illinois, United States of America, 3 Pinellas County Utilities Laboratory, Largo, Florida, United States of America

## Abstract

Pipes that transport drinking water through municipal drinking water distribution systems (DWDS) are challenging habitats for microorganisms. Distribution networks are dark, oligotrophic and contain disinfectants; yet microbes frequently form biofilms attached to interior surfaces of DWDS pipes. Relatively little is known about the species composition and ecology of these biofilms due to challenges associated with sample acquisition from actual DWDS. We report the analysis of biofilms from five pipe samples collected from the same region of a DWDS in Florida, USA, over an 18 month period between February 2011 and August 2012. The bacterial abundance and composition of biofilm communities within the pipes were analyzed by heterotrophic plate counts and tag pyrosequencing of 16S rRNA genes, respectively. Bacterial numbers varied significantly based on sampling date and were positively correlated with water temperature and the concentration of nitrate. However, there was no significant relationship between the concentration of disinfectant in the drinking water (monochloramine) and the abundance of bacteria within the biofilms. Pyrosequencing analysis identified a total of 677 operational taxonomic units (OTUs) (3% distance) within the biofilms but indicated that community diversity was low and varied between sampling dates. Biofilms were dominated by a few taxa, specifically *Methylomonas*, *Acinetobacter*, *Mycobacterium*, and Xanthomonadaceae, and the dominant taxa within the biofilms varied dramatically between sampling times. The drinking water characteristics most strongly correlated with bacterial community composition were concentrations of nitrate, ammonium, total chlorine and monochloramine, as well as alkalinity and hardness. Biofilms from the sampling date with the highest nitrate concentration were the most abundant and diverse and were dominated by *Acinetobacter*.

**Editor:** Ahmed Moustafa, American University in Cairo, Egypt

**Funding:** This work was supported by Water Research Foundation grant 4259 to Aaron Packman and John Kelly. The funders had no role in study design, data collection and analysis, decision to publish, or preparation of the manuscript.

**Competing Interests:** The authors have declared that no competing interests exist.

\* E-mail: jkelly7@luc.edu

## Introduction

The pipes that are used to transport drinking water through municipal drinking water distribution systems (DWDS) are challenging habitats for microorganisms. The transported water generally contains chemical disinfectants such as chlorine or chloramine, as well as very low concentrations of organic carbon and inorganic nutrients [1]. Despite these challenges, microbes frequently colonize the interior surfaces of DWDS pipes [1]. Indeed the pipe surfaces may represent the best possible microbial habitats within DWDS, as previous research has shown that surface attachment can enable bacteria to grow in oligotrophic habitats due to the accumulation of nutrients at the solid-liquid interface [2,3]. In addition, biofilm formation can provide bacteria with protection against chemical disinfectants [4–6].

Relatively little is known about the species composition and ecology of biofilms within DWDS. Obtaining samples from below-ground pipes is difficult and expensive [7], and as a result most of the work that has been done on drinking water biofilms has been based on model systems run in the laboratory [6,8,9]. While these studies have provided valuable insight into biofilm formation

within drinking water, model systems often differ in significant ways from actual DWDS, including duration of biofilm growth, temporal variability, water flow conditions, diversity of pipe materials and the presence or absence of disinfectants. Additionally, most of the work on microbes within DWDS has focused on classical pathogens such as *Vibrio cholerae* and *Salmonella typhi*, emerging pathogens such as *Campylobacter jejuni* and *Legionella pneumophila*, or indicator organisms for fecal contamination, such as coliform bacteria [1]. Many of these studies have used culture-based techniques [1], which are able to assess only a small fraction of natural microbial diversity [10]. In contrast, recent studies using molecular approaches have demonstrated the predominance of nonpathogenic bacterial species within drinking water biofilms [7,11,12].

Information regarding the composition and ecology of biofilms within DWDS is valuable for several reasons. First, it improves our general understanding of microbial life in oligotrophic habitats, including built environments. Secondly, biofilms in DWDS are a concern for public health as they have been shown to harbor and protect pathogens from disinfectants and increase pathogen persistence in DWDS [13]. Thirdly, there is evidence that the activities of nitrifying microorganisms in DWDS can decrease monochloramine concentrations [14], which could lead to increased microbial growth and possibly increased persistence and transport of pathogens [15]. Finally, the presence of biofilms in DWDS can promote pipe corrosion [16] and cause taste and odor problems in the water [3]. Understanding the microbial composition and development of DWDS biofilms can suggest strategies for management of these problems.

We report here the analysis of biofilms found within pipe samples collected five times over an 18 month period from a DWDS in Pinellas County, FL, USA. Sections of below-ground pipe were cut and transported to the lab, and biofilm communities within the pipes were analyzed by heterotrophic plate counts and tag pyrosequencing of 16S rRNA genes.

## Materials and Methods

Pipe samples were collected from the municipal drinking water distribution system in Pinellas County, FL, USA, which is operated by Pinellas County Utilities (PCU). The water supply for this system is a blend of groundwater and treated surface water, with desalinated seawater being used periodically as needed. Before entering the DWDS water is treated at the Tampa Bay Water Treatment Plant (TBW) by a four stage process: 1) clarification using the ACTIFLO system (Kruger Inc., Cary, NC), 2) ozone disinfection, 3) biologically active filtration, and 4) disinfection with chlorine. Disinfection within the PCU DWDS is based on maintenance of a chloramine residual, although the utility does switch to chlorine disinfection for a brief period in the summer each year to limit biofilm growth. The switch to chlorine occurred once during our sampling period, specifically between August 1 and September 11, 2011. The average flow through the system during our sampling period was 54.9 million gallons per month.

Sections of six-inch ductile-iron pipe from the main line in the Seminole, FL region within the Pinellas County DWDS were collected periodically over an 18 month period during planned replacement events. Specifically, pipe sections were collected on February 20, 2011, July 20, 2011, December 20, 2011, March 20, 2012, and August 8, 2012. All pipe samples were collected at approximately 9 am, before peak demand which occurs at approximately 10 am. The water main that we sampled was approximately 40–45 years old. The average water age for the main line during our sampling period was 3 to 4 days and the average velocity was 0.4 c.f.s. The sampling location is downstream of a large above ground storage (AGS) tank that was permanently shut down on June 19, 2012 due to concerns about nitrification occurring within the tank. A key outcome of the shut-down of this water tank was decreased water age at our sampling location for the August 2012 sampling date.

On each sampling date the road surface and soil above the water main were excavated using a backhoe. Soil was cleared from around the pipe by hand using a shovel and any soil adhering to the exterior of the pipe was removed using a brush or cloth. The exterior of the pipe was disinfected by pouring a 10% bleach solution over the pipe and simultaneously wiping the pipe exterior with a cloth saturated with 10% bleach solution. A section approximately 1 ft. long was cut from the 6-inch diameter pipe using a scoring-type pipe cutter. The pipe section was capped at one end using a flexible PVC cap with adjustable clamps. The capped pipe section was filled with dechlorinated water from the same water main, which was collected in a plastic carboy and dechlorinated on-site using sodium thiosulfite ($\sim 2$ g/g $Cl_2$). The pipe section was filled until overflowing, capped at the other end,

placed in a cooler with ice packs, and shipped overnight to Northwestern University, Evanston, IL. Water was also collected from the main for chemical analysis and transported to the Pinellas County Utilities lab in a cooler.

## Sample Processing

In the laboratory one cap was removed from the pipe section and the water was carefully poured out. The interior of the pipe section was rinsed gently with filter-sterilized tap water to remove unattached or settled solids. Biofilm samples were collected from three separate, evenly-spaced sections of the pipe interior. For each section, biofilm material was collected by scraping a 5.5 cm wide band around the entire interior circumference of the pipe with a sterile spatula, resulting in a sampling area of 263.3 $cm^2$. The collected biofilm material was transferred to a sterile 10 ml vial and 3 ml ultrapure water was added. The suspension was homogenized by vortexing and large particles such as corrosion byproducts were allowed to settle out of suspension. From this suspension 100 µl was used for the plate count assay (see method below) and 2 ml was used for molecular analysis of the biofilm communities (see methods below). The 2 ml for molecular analysis was transferred to a 2 ml microcentrifuge tube and centrifuged at 10,000×g for 10 minutes. The supernatant was then discarded and the remaining biofilm pellet was stored at -20°C.

## Water Chemistry

All water chemistry analyses were done by Pinellas County Utilities. Analyses were performed based on either EPA Methods or Standard Methods for the Examination of Water and Wastewater [17]. The specific methods used for each assay and results of the water chemistry assays are listed in Table 1. Total haloacetic acid (HAA) concentrations in the source water were determined periodically by EPA method 552.2.

## Plate Count Assay

R2A agar was purchased as a dried powder (Fisher Scientific, Pittsburgh, PA) and prepared according to the manufacturer's instructions. Biofilm suspensions were serially diluted from $10^2$ to $10^4$ in ultrapure water and 100 µl of all dilutions were spread on R2A agar plates. Plates were incubated at 37°C for 36 hours and counted. Counts were normalized based on the surface area of the pipe from which the biofilm had been collected.

## Molecular Analysis of Biofilm Communities

DNA was extracted from the frozen biofilm pellets using the Power Biofilm DNA Kit (MoBio Laboratories, Carlsbad, CA) according to the manufacturer's instructions and stored at $-20°C$. For tag pyrosequencing of bacterial 16S rRNA genes the extracted DNA was sent to Research and Testing Laboratory (Lubbock, TX). Polymerase chain reaction (PCR) amplification was performed using primers 530F and 1100R [18]. The 530F primer was chosen in order to obtain sequences for the V4 hypervariable region, which has been shown to provide species richness estimates comparable to those obtained with the nearly full-length 16S rRNA gene [19]. Sequencing reactions utilized a Roche 454 FLX instrument (Roche, Indianapolis, IN) with Titanium reagents. Sequences were processed using MOTHUR software [20]. Briefly, any sequences containing ambiguities or homopolymers longer than 8 bases were removed. Remaining sequences were individually trimmed to retain only high quality sequence reads and sequences were aligned based on comparison to the SILVA-compatible bacterial alignment database available within MOTHUR. Aligned sequences were trimmed to a uniform length

**Table 1.** Water chemistry.

| Analyte | Method | Sampling Date | | | | |
|---|---|---|---|---|---|---|
| | | February 2011 | July 2011 | December 2011 | March 2012 | August 2012 |
| Temperature (°C) | SM 2550 B | 20.6 | 29.1 | 20.2 | 23.8 | 28.5 |
| pH | SM 4500 H-B | 7.6 | 7.58 | 7.65 | 7.52 | 7.69 |
| Total Organic Carbon (mg L$^{-1}$) | SM 5310-C | 1.8 | 1.9 | 2.2 | 2 | 2.3 |
| Total Phosphorous as P (mg L$^{-1}$) | EPA 365.4 | 0.35 | 0.24 | 0.41 | 0.34 | 0.31 |
| Nitrate as N (mg L$^{-1}$) | EPA 300.0 | 0.08 | 0.11 | 0.09 | 0.08 | 0.43 |
| Nitrite as N (mg L$^{-1}$)[1] | EPA 300.0 | BD | BD | BD | BD | BD |
| Free Ammonia as N (mg L$^{-1}$) | SM 4500 NH3-F | 0.19 | 0.42 | 0.21 | 0.16 | 0.14 |
| Total chlorine (mg L$^{-1}$) | SM 4500 CL-G | 2.9 | 1.6 | 2.5 | 3 | 3 |
| Monochloramine (mg L$^{-1}$) | SM 4500 CL-G | 3.2 | 1.6 | 2.2 | 2.6 | 2.7 |
| Alkalinity as CaCO$_3$ (mg L$^{-1}$) | SM 2320 B | 170 | 190 | 170 | 180 | 150 |
| Calcium Hardness (mg L$^{-1}$) | SM 2340 B | 210 | 202 | 202 | 205 | 192 |
| Specific Conductance (umhos cm$^{-1}$) | SM 2510 B | 549 | 521 | 429 | 447 | 477 |
| Aluminum (mg L$^{-1}$)[2] | EPA 200.7-DW | BD | BD | BD | BD | BD |
| Calcium (mg L$^{-1}$) | EPA 200.7-DW | 84.3 | 80.7 | 81 | 81.9 | 65.1 |
| Iron (mg L$^{-1}$) | EPA 200.7-DW | 0.055 | 0.079 | 0.381 | 0.046 | 0.014 |
| Magnesium (mg L$^{-1}$) | EPA 200.7-DW | 7.35 | 6.51 | 6.39 | 6.53 | 7.14 |

BD = below detection limit.
[1]detection limit 0.02 mg L-1.
[2]detection limit 0.015 mg L-1.

of 127 bases and chimeric sequences were removed using UCHIME [21] run within MOTHUR. Sequences were grouped into phylotypes by comparison to the SILVA-compatible bacterial alignment database and algal chloroplast and mitochondrial sequences were removed from the data set. To avoid any biases associated with different numbers of sequences in each of the samples we randomly subsampled a total of 6,808 sequences from each sample, and used these subsampled sequences for all downstream analyses. Sequences were clustered into operational taxonomic units (OTUs) based on 97% sequence identity using the average neighbor algorithm. Rarefaction curves were produced using MOTHUR. The total OTU richness in each sample was calculated based on the Chao1 richness estimator [22]. The diversity of each sample was assessed based on the Shannon index [23] calculated using the Primer software package (Primer V.5, Primer-E Ltd., Plymouth, UK). The community composition of the individual samples was compared by using MOTHUR to calculate distances between sites based on the theta index [24]. The significance of differences in theta index scores between sites was assessed by analysis of molecular variance (AMOVA) run within MOTHUR. PC-ORD v. 6.08 (MjM Software, Gleneden Beach, Oregon, USA) was used to ordinate the theta index distance matrix via non-metric multidimensional scaling (nMDS) and to determine correlations between the water chemistry data and the axes in the nMDS ordination.

## Statistical Analyses

Plate count data and diversity scores were analyzed by one-way analysis of variance (ANOVA) based on sampling date and pairwise comparisons were made by Tukey's post hoc test. Correlations were assessed by determining Pearson product-moment correlation coefficients and Bonferroni-corrected probabilities. Correlation between the relative abundance of *Methylomo-*

*nas* sequences and the concentration of total HAA in the source water was based on average abundance data for each biofilm sampling date and the HAA concentration from the closest source water sampling date. All statistical analyses were run using Systat 13 (Systat Software, Inc., San Jose, CA) and p values less than 0.05 were considered to be significant.

## Data Sharing

All of the sequence data analyzed in this paper can be downloaded from the National Center for Biotechnology Information (NCBI) Sequence Read Archive (SRA) with accession number SRP038002.

## Results

### Water Source

The source water for the PCU DWDS is a blend of groundwater, surface water, and desalinated seawater. An approximately equal mix of groundwater and surface water was the most common during the study period, although there were some significant variations in the relative proportions of source waters (Fig. 1). For the four weeks preceding our February 2011 sampling date, the source water averaged 38% Groundwater, 48% surface water and 14% seawater. Between June 20, 2011 and March 20, 2012, a period which included three of our sampling dates, the source water averaged 50% Groundwater, 50% surface water and 0% seawater. Prior to our August 2012 sampling date there were some dramatic shifts in source waters. Between April 14, 2012 and June 9, 2012 the water was predominantly groundwater, with an average of 73% of the water coming from groundwater over that period. Finally, between July 6, 2012 and August 8, 2012 the water was predominantly surface water, with

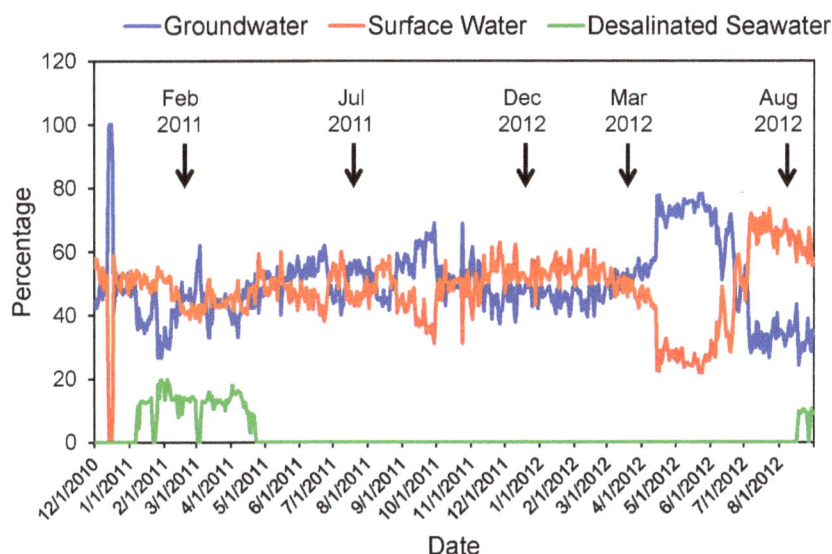

**Figure 1. Relative percentages of source waters within PCU drinking water distribution system.** Biofilm sampling dates are indicated by black arrows.

surface water representing an average of 67% of the source water over that period.

## Water Chemistry

Water temperature varied seasonally in the water main from which the biofilm samples were obtained, being higher in summer months (July and August) than winter months (February, March and December) (Table 1). Some water chemistry parameters varied with sampling date but did not show seasonal trends. For example, there were large fluctuations in concentrations of nitrate (from 0.08 to 0.43 mg $L^{-1}$), free ammonia (from 0.14 to 0.42 mg $L^{-1}$) and iron (from 0.014 to 0.381 mg $L^{-1}$). The August 2012 sampling date had the highest concentration of nitrate, approximately four to five times higher than all other sampling dates, and the lowest concentration of free ammonia (Table 1). Analysis of the source water from TBW confirmed a high concentration of nitrate in the source water immediately prior to the August 2012 sampling date (data not shown). Analysis of the source water from TBW also showed fluctuations in concentrations of total haloacetic acids over the course of the study (Fig 2).

## Plate Count Assay

Sampling date significantly affected the numbers of bacteria within the pipe biofilms as measured by heterotrophic plate count assay, with bacterial counts varying over 3 orders of magnitude between sampling dates (p<0.001; Table 2). Samples from the summer months (July and August) had significantly higher counts than the winter months (February, March and December), and there was a significant positive correlation between plate counts and water temperature ($R^2 = 0.605$; p<0.001). The August 2012 sampling date, which had the highest nitrate concentration, had bacterial counts that were significantly higher than all other sampling dates (p<0.001; Table 2), and there was a significant positive correlation between plate counts and nitrate concentration ($R^2 = 0.799$; p<0.001). There was no significant correlation between plate counts and phosphorous concentration ($R^2 = 0.249$; p = 0.058). There were also no significant correlations between plate counts and total chlorine ($R^2 = 0.005$, p = 0.802) or

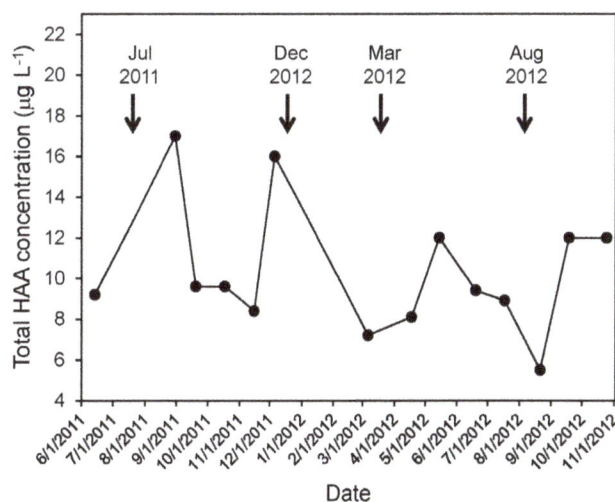

**Figure 2. Concentrations of total haloacteic acids (HAA) in the source water for the PCU drinking water distribution system.** Biofilm sampling dates are indicated by black arrows.

between plate counts and monochloramine ($R^2 = 0.285$, p = 0.548).

## Bacterial Community Analysis

Tag pyrosequencing of 16S rRNA genes was used to profile the bacteria within the biofilms lining the drinking water pipes. Twelve samples representing three replicate biofilm samples from each of four sampling dates (July 2011, December 2011, March 2012 and August 2012) were sequenced successfully. Despite repeated attempts, DNA from the February 2011 pipe samples could not be amplified with the 530F and 1100R primers, so no sequence data were obtained for this sampling date. After processing, the data set included a total of 159,604 high-quality sequence reads. There was significant variation in the number of sequences obtained for each of the samples, from a low of 8,785 to

**Table 2.** Numbers of heterotrophic bacteria in pipe biofilms based on plate count assay.

| Sampling Date | Number of Bacteria (cfu cm$^{-2}$)[1] | |
| --- | --- | --- |
| February 2011 | 215 | a |
| July 2011 | 10,887 | b |
| December 2011 | 2,013 | a |
| March 2012 | 36 | a |
| August 2012 | 23,167 | c |

[1]Data points represent mean values (n = 3) and data points followed by different letters are significantly different based on ANOVA and Tukey's HSD posthoc test (p< 0.05).

a high of 35,854. To avoid biases associated with unequal numbers of sequences, 6,808 sequences were randomly selected from each of the twelve samples using the subsample command in MOTHUR, producing a total of 81,696 high quality sequences that were used for all analyses of community composition. With this subsampled data set, rarefaction curves for all samples had reached plateaus (Fig. 3), suggesting that the sequencing depth obtained in this study was adequate to capture most of the diversity within these communities. Similarly, a comparison of the total number of OTUs observed in each sample and the estimated total number of OTUs present in each sample demonstrated that for all samples more than 50% of the estimated total number of OTUs in each sample were detected (Table 3).

nMDS ordination (Fig. 4) and AMOVA analysis (Table 4) indicated that there were significant differences between the biofilm bacterial communities from the different sampling dates. Biofilm bacterial communities from August 2012 were the most distinct and were significantly different from the communities from all other sampling dates (Fig. 4 and Table 4). The nMDS ordination also revealed relationships between the community composition and water chemistry parameters (Fig 4). The composition of the biofilm bacterial community from August 2012 was positively correlated with nitrate concentration and negatively correlated with calcium concentration and hardness. The August 2012 communities were also positively correlated with

pH and total organic carbon concentration and negatively correlated with alkalinity. In addition, the separation of the biofilm communities from July 2011 and March 2012 on the nMDS ordination was correlated with the concentrations of free ammonia, total chlorine and monochloramine (Fig. 4). Finally, the nMDS ordination demonstrated that there was variation in bacterial community composition between replicates from both the July and March sampling dates, whereas the December and August samples showed a high degree of similarity between replicates (Fig. 4).

### Bacterial Community Diversity

Pyrosequencing analysis identified a total of 677 OTUs (3% distance) within these biofilms, and the number of OTUs per sample ranged from 21 to 199 (Table 3). Despite this large number of OTUs, the diversity of these communities was low, with three of the four sampling dates showing Shannon index scores below 1.2 (Fig 5A). The Shannon diversity index scores for the biofilm bacterial communities varied significantly between sampling dates (p<0.01) with the biofilms from August 2012 being the most diverse (Fig. 5A). There was also a significant positive correlation (p<0.01) between bacterial abundance in the biofilms and the diversity of the bacterial communities (Fig 5B). As indicated by the low diversity index scores, all of the biofilm communities were dominated by a small number of OTUs, with the ten most

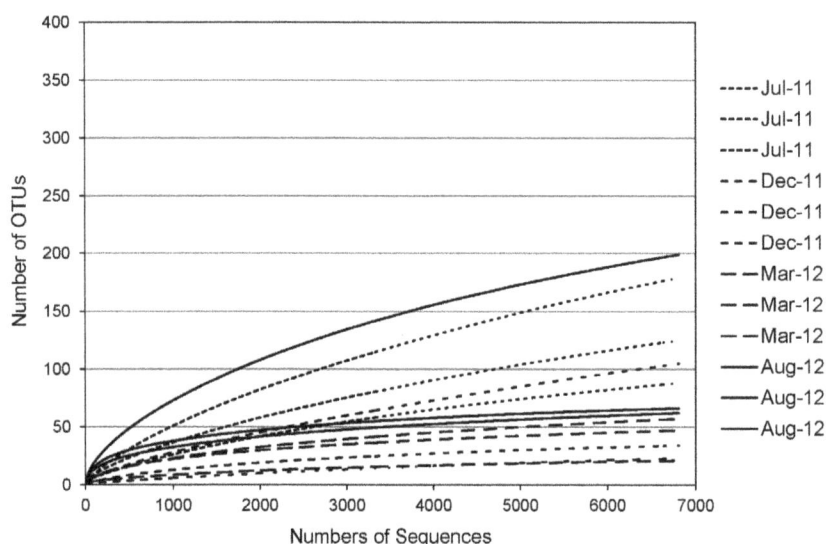

**Figure 3. Rarefaction curves for biofilm bacterial communities based on tag pyrosequencing of 16S rRNA genes.** OTUs were defined based on 3% distance.

**Table 3.** Comparison of the number of observed and estimated bacterial OTUs in pipe biofilm communities based on 16S tag pyrosequencing data.

| Sampling Date | Observed OTUs[1] | Estimated OTUs[2] | Percent Coverage[3] |
|---|---|---|---|
| July 2011 | 88 | 158 | 55.7% |
| July 2011 | 179 | 332 | 53.9% |
| July 2011 | 125 | 239 | 52.3% |
| December 2011 | 34 | 42 | 81.0% |
| December 2011 | 105 | 194 | 54.1% |
| December 2011 | 23 | 38 | 60.5% |
| March 2012 | 47 | 67 | 70.1% |
| March 2012 | 57 | 87 | 65.5% |
| March 2012 | 21 | 26 | 80.8% |
| August 2012 | 199 | 297 | 67.0% |
| August 2012 | 66 | 74 | 89.2% |
| August 2012 | 62 | 75 | 82.7% |

[1]OTUs were defined based on 3% distance.
[2]Total OTUs per sample were estimated based on Chao1 richness estimator.
[3]Percent coverage was calculated by dividing the number of observed OTUs by the number of estimated OTUs.

abundant OTUs accounting for 93% of the total sequences in the data set.

## Bacterial Community Taxonomic Composition

Analysis of pyrosequencing data indicated that sequences corresponding to the genus *Methylomonas* were the most abundant within the biofilm communities, accounting for 41% of the sequences in the total data set (Table 5). Other abundant sequences corresponded to the genera *Acinetobacter*, *Mycobacterium*, *Pseudomonas*, and *Methylobacterium*, as well as an unclassified genus from the family Xanthomonadaceae and an unclassified genus from the class Betaproteobacteria. The relative abundance data

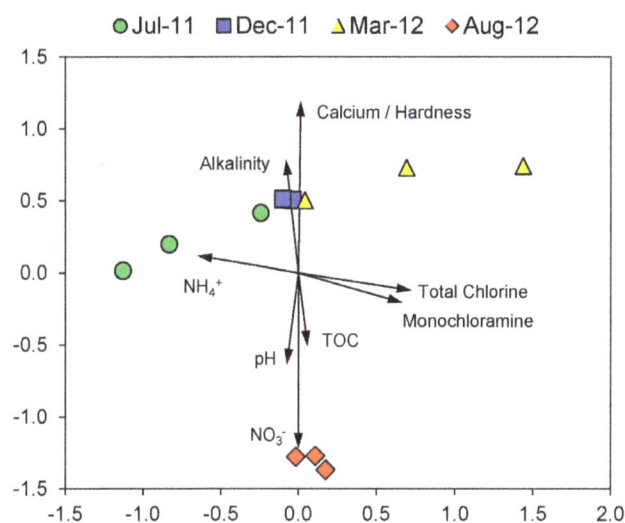

**Figure 4. Non-metric multidimensional scaling ordination of biofilm bacterial communities based on tag pyrosequencing of 16S rRNA genes.** Stress value of ordination is 0.129. Vector lines represent correlations between physical and chemical variables and the ordination axes. Variables with correlation values less than 0.5 for both axes are not shown.

indicated some commonalities between the biofilm bacterial communities from the four sampling dates (Table 5). For example, *Methylomonas* accounted for more than 30% of the sequences for three of the four sampling dates. However, the data also illustrate that there were dramatic differences in the composition of biofilm communities from the different sampling dates, as *Methylomonas* accounted for more than 95% of the sequences in December 2011, but less than 2% in August 2012. A previous study suggested a link between *Methylomonas* bacteria in drinking water and HAA [25], and there was a significant positive correlation between the relative abundance of *Methylomonas* sequences in our biofilms and the concentration of total HAA in the source water ($R^2 = 0.962$; $p = 0.027$). *Acinetobacter* abundance also varied significantly between sampling dates, accounting for 74% of the sequences in August 2012 biofilms, but representing less than 0.1% of the sequences from the other sampling dates.

Sequences corresponding to several genera containing pathogenic species were detected in the biofilms. In a total of 81,696 sequences from all four sampling dates, there were 18 *Escherichia* sequences, 5 *Clostridium* sequences and 3 *Streptococcus* sequences detected, so these genera were extremely rare. *Mycobacterium* represented 59% of the sequences in July 2011, while representing less than 1% of sequences from the other sampling dates. The genus *Mycobacterium* includes two well-known human pathogens, *M. tuberculosis* and *M. leprae*, although these species are generally not found in the environment [26]. The genus *Mycobacterium* also includes a large number of non-pathogenic or occasionally pathogenic species [26]. *Acinetobacter* accounted for 74% of the sequences in August 2012 but less than 1% of sequences from the other sampling dates. Bacteria from the genus *Acinetobacter* are a common cause of nosocomial infections among immunocompromised patients, with the most common example being respiratory infections of ventilated patients [27]. Due to the short length of the sequences obtained in our study, we were unable to discriminate any of the sequences down to the species level, so it is unclear whether the sequences from any of these genera represented pathogenic or non-pathogenic organisms.

The pipes analyzed in this study were ductile iron and did not show significant corrosion or the presence of tubercles. The

**Table 4.** AMOVA analysis of 16S tag pyrosequencing data from pipe samples.

| Comparison | p value |
|---|---|
| August-July | <0.001 |
| August-December | <0.001 |
| August-March | 0.0497 |
| July-December | <0.001 |
| July-March | 0.0500 |
| December-March | 0.0487 |

pyrosequencing data identified only a handful of sequences corresponding to genera known to contain iron-oxidizing species: *Acidovorax* (1 sequence), *Aquabacterium* (1 sequence) and *Thiobacillus* (4 sequences) [28,29]. In addition, no sequences corresponding to any known ammonia oxidizing bacterial genera were detected. However, a few sequences (19 total) from a known nitrite oxidizing genus, *Nitrospira*, were detected, and all of these sequences were found in the July 2011 samples, which also showed the highest concentration of free ammonia (Table 1).

## Discussion

Pipe samples were collected from the same region of a drinking water distribution system in Pinellas County, FL on five dates over an 18-month period between February 2011 and August 2012. Water from these pipes showed seasonal variations in temperature and some large fluctuations in concentrations of nitrate, free ammonia and iron. The August 2012 sample had a much higher concentration of nitrate and a lower concentration of free ammonia relative to other sampling dates. The shut-down of the upstream above ground storage tank prior to the August 2012 sampling date probably contributed to the observed differences in water chemistry between the sampling dates. The above ground storage tank was shut down because of nitrification occurring in the tank, which could have contributed to the high nitrate and low free ammonia concentrations in the August 2012 sample. The August 2012 samples also had a different source-water mixture than other dates: for one month prior to our sampling date the source water was composed of a higher percentage of surface water and a lower percentage of groundwater than typical for this system. Analysis of the TBW source water at its point of entry to the DWDS confirmed the high nitrate and low alkalinity of the source water at the time of our August 2012 sampling, indicating that source water was a key driver of the unique aspects of the August 2012 water chemistry.

The abundance of bacteria in the biofilms varied greatly across the sampling dates, with higher bacterial numbers in the summer months and lower bacterial numbers in the winter months. These differences in abundance may have been driven by the strong seasonal differences in temperature in the system, as the water in the summer months was on average 7.3°C warmer than in the winter months, and a temperature change of this magnitude can significantly increase the growth rates of mesophilic bacteria [30]. The connection between bacterial abundance in the biofilms and water temperature was further supported by a statistically significant correlation between these two parameters. The differences in bacterial abundance may also have been related to the availability of inorganic nutrients in the drinking water, specifically nitrogen. We found a significant correlation between

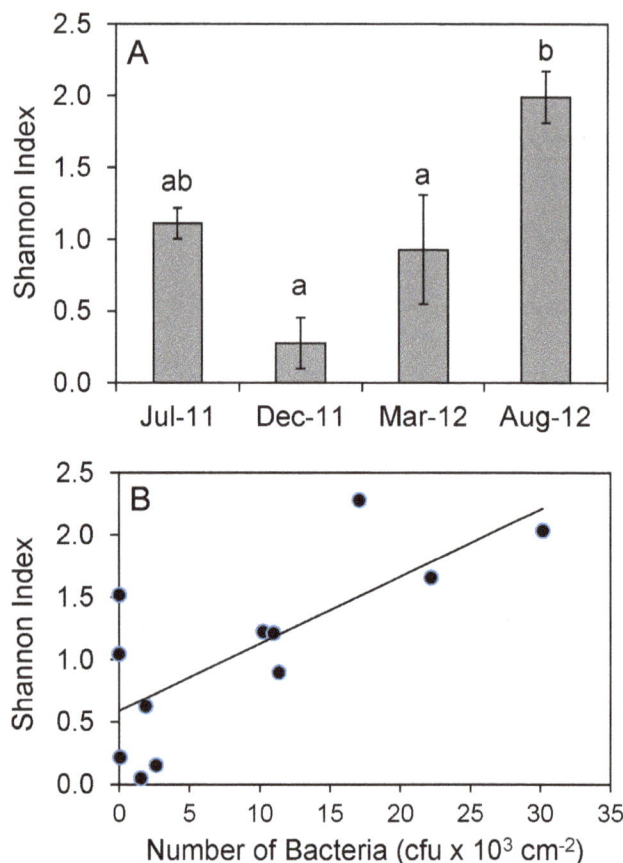

**Figure 5. Diversity of bacterial biofilm communities and relationship between bacterial abundance and diversity.** (A) Shannon index scores calculated using 16S rRNA gene tag pyrosequencing data. Each data point represents mean value (n = 3) with standard error bars. ANOVA indicated a significant effect of sampling date (p<0.01). Data points with different letters are significantly different based on Tukey's posthoc test (p<0.05). (B) Relationship between biofilm bacterial abundance and diversity. Linear regression indicated R2 = 0.541. Pearson correlation analysis indicated a significant correlation between numbers of bacteria and diversity concentration and resistance (p<0.01).

bacterial cell numbers and nitrate concentrations. Other studies have indicated that inorganic nutrients can be a limiting factor for bacterial growth in DWDS [1]. In our samples, nitrate concentrations were higher in the summer months than in the winter months and were highest in the August 2012 sample. Finally, our results indicated no significant relationship between the abundance of bacteria within the biofilms and the concentrations of total chlorine or monochloramine. These results suggest that low concentrations of chlorine disinfectants may not be effective at limiting biofilm growth within DWDS, owing to the protection provided by the biofilm matrix, as has been demonstrated by previous studies [4,5,6]. It should be noted that we assessed bacterial abundance within the pipe biofilms using heterotrophic plate counts. While the limitations of plate counts as estimates of bacterial abundance are well known [31], this method is commonly used to assess bacterial loads in DWDS, so we chose to use this method to make our results comparable to existing data.

Tag pyrosequencing analysis revealed a total of 677 OTUs in the biofilm bacterial communities, with the estimated total numbers of bacterial OTUs per sample ranging from 26 to 332.

**Table 5.** Relative abundance of most numerically dominant bacterial genera[1].

| Bacterial genus | All Samples | July 2011 | December 2011 | March 2012 | August 2012 |
|---|---|---|---|---|---|
| Methylomonas | 40.8 | 33.6±15.6 | 95.5±3.5 | 32.4±21.3 | 1.3±0.6 |
| Acinetobacter | 18.5 | 0.0±0.0 | 0.0±0.0 | 0.0±0.0 | 74.2±2.4 |
| Mycobacterium | 14.7 | 59.1±14.3 | 0.2±0.2 | 0.1±0.1 | 0.0±0.0 |
| Unclass. Xanthomonadaceae | 13.5 | 0.0±0.0 | 2.6±2.4 | 50.4±24.4 | 0.1±0.1 |
| Pseudomonas | 4.7 | 0.6±0.3 | 0.3±0.2 | 15.5±8.6 | 2.1±0.2 |
| Unclass. Betaproteobacteria | 2.3 | 0.3±0.1 | 0.1±0.1 | 0.0±0.0 | 9.0±1.6 |
| Methylobacterium | 1.2 | 3.7±2.3 | 0.2±0.2 | 0.1±0.0 | 0.8±0.1 |
| Unclass. Bacteria | 1.0 | 0.5±0.2 | 0.3±0.1 | 0.0±0.0 | 3.2±2.8 |
| Massilia | 0.9 | 0.0±0.0 | 0.0±0.0 | 0.0±0.0 | 3.4±0.7 |
| Unclass. Gammaproteobacteria | 0.4 | 0.7±0.4 | 0.2±0.2 | 0.7±0.5 | 0.0±0.0 |

[1]Values for each sampling period represent mean values (n = 3) ± standard error.

A previous pyrosequencing survey of drinking water meters revealed a similarly high number of total bacterial OTUs [7]. Despite the large numbers of OTUs observed in our pipe biofilms, these communities were not very diverse, with all samples having Shannon index scores below 2. These low diversity scores reflect the fact that these communities were dominated by a small number of taxa, specifically *Methylomonas*, *Acinetobacter* and *Mycobacterium*. The diversity of the bacterial biofilm communities was significantly correlated with bacterial abundance, suggesting that when conditions within the DWDS were favorable for bacterial growth (e.g. higher temperature and higher nitrate concentrations) a wider range of bacterial taxa were able to proliferate within the biofilms, whereas when conditions were not favorable (e.g. lower temperature and lower nitrate concentrations), a more limited range of bacteria were able to persist. Within our data set the August 2012 samples seemed to represent the most favorable conditions for biofilm growth, as this sampling date had the highest nitrate concentration, one of the highest water temperatures, and supported biofilms with the highest bacterial numbers and the highest bacterial diversity.

*Methylomonas* was the most numerically dominant bacterial genus within the biofilms, with its sequences accounting for more than 40% of all of the sequences recovered. *Methylomonas* is a genus of type I methanotrophic bacteria, which obtain their carbon and energy from the oxidation of methane. *Methylomonas* sequences have been detected previously in drinking water [32], and a recent study of water meter biofilms also detected sequences corresponding to the family *Methylococcaceae*, the bacterial family that includes the genus *Methylomonas*, [7]. *Methylobacterium*, a genus of methylotrophic bacteria that oxidize methyl compounds such as methanol but cannot metabolize methane, was also one of the most commonly detected taxa in our biofilms, although it represented just over 1% of the total sequences. Methylotrophic bacteria, specifically *Methylophilus*, were also detected in biofilms within drinking water meters [7]. The factors favoring high abundance of methanotrophic and methylotrophic taxa within drinking water distribution systems are unclear. We would not expect high concentrations of methane or methanol in drinking water, however these compounds could be produced in anoxic sites within DWDS via anaerobic processes such as methanogenesis or fermentation. Another process that might have supported the growth of methylotrophic bacteria is utilization of haloacetic acid, which is a common by-product of chlorination of drinking water. A recent study isolated a *Methylobacterium* strain from a DWDS

biofilm that was capable of growth with haloacetic acid as the sole carbon source [25]. The results of our study, which showed a significant correlation between the relative abundance of *Methylomonas* sequences and HAA concentration in the source water, lends further support to this hypothesis.

Our results demonstrated that there was significant variation in the taxonomic composition of the biofilm bacterial communities within the pipe sections across our sampling dates. nMDS and AMOVA analyses indicated that the August 2012 bacterial communities were the most distinct in terms of their composition, and further examination of the composition of these biofilms revealed that the August 2012 communities were dominated by *Acinetobacter* sequences (74% of total sequences), while *Acinetobacter* sequences represented less than 0.1% of sequences from the July, December and March samples. *Acinetobacter* is a genus of Gram-negative, heterotrophic bacteria [33] that is commonly found in soils and groundwater [34–36]. *Acinetobacter* is also one of the most common groups of bacteria isolated from drinking water [1], and a number of *Acinetobacter* species have been shown to produce biofilms [37–39]. Therefore the presence of *Acinetobacter* in the pipe biofilms was not surprising. However, the dramatic variation in *Acinetobacter* abundance that we observed between August 2012 and the other sampling dates was remarkable. There were several unique features of the August 2012 sampling date that may have contributed to its distinct biofilm composition. First, the source of the drinking water within the distribution system changed prior to August 2012 to predominantly groundwater for a period of three weeks, with this period of groundwater dominance occurring approximately two months prior to the August 2012 sampling. Since *Acinetobacter* are regularly detected in soil and groundwater, this switch to a groundwater dominated system prior to August 2012 might have provided an additional inoculum of *Acinetobacter* that were able to become established within the pipe biofilms. Another related feature of the August 2012 samples was that the drinking water at that sampling time had a much higher nitrate concentration (at least four times higher) than all of the other sampling dates. The nMDS analysis indicated that nitrate concentration was one of the main drivers of the composition of the bacterial communities within the August 2012 biofilms. The August 2012 biofilms also showed bacterial counts that were more than two times higher than any of the other sampling dates, suggesting more biofilm mass which could have generated more anoxic microsites. Although bacteria from the genus *Acinetobacter* are generally aerobes, there are some species within the genus that can utilize nitrate as an electron acceptor when oxygen is not present [40]. In contrast, *Methylomonas*, which was one of the predominant taxa in July 2011, December 2011 and March 2012 but was less than 2% of total sequences in August 2012, is strictly aerobic [41]. Therefore, the higher drinking water nitrate concentration and possibly more anoxic microsites caused by higher bacterial biofilm growth on the August 2012 sampling date may have provided *Acinetobacter* with a competitive advantage over *Methylomonas*.

*Mycobacterium* abundance also varied considerably over time, as this genus represented 59% of the sequences from the July 2011 sampling date but less than 1% for all other sampling dates. *Mycobacteria* are frequently detected in DWDS and are considered a significant public health issue [42]. The genus *Mycobacterium* consists of approximately 100 species, including a large number of species that are either non-pathogenic or pathogenic under certain situations [26]. For example, nontuberculosis *Mycobacterium* are a major cause of opportunistic infections in immunocompromised hosts [42]. Mycobacteria have been detected previously in this DWDS via culturing, and *M. gordonae* and *M. intracellulare* were the most frequently detected *Mycobacterium* species [43]. *M. gordonae* is among the most frequently reported mycobacteria in drinking water and in DWDS [26] and is generally considered non-pathogenic [44]. *M. intracellulare*, which is part of the *Mycobacterium avium* complex (MAC), has also been detected in DWDS [26]. MAC is the group of non-tuberculosis *Mycobacterium* most commonly associated with human disease, causing primarily pulmonary infections in individuals who are immunocompromised [45]. In this study, we were unable to discriminate the *Mycobacterium* sequences down to the species level, so it is unclear whether the sequences we detected represented potentially pathogenic species.

Several characteristics of *Mycobacteria* enhance their survival in DWDS, including their ability to grow under oligotrophic conditions [46], form biofilms and resist chlorine disinfection [26]. Several recent studies have detected species related to *Mycobacterium* in chlorinated drinking water [47–49]. Previous work at the DWDS considered here indicated that the frequency of detection of *Mycobacteria* increased when the disinfectant was switched from chlorine to chloramine in 2002 [43], suggesting that *Mycobacteria* might be less sensitive to chloramine than chlorine. However, a recent study using a model DWDS observed that the relative sensitivity of *Mycobacterium avium* biofilms to chlorine and monochloramine depended on the pipe material [50]. Specifically, *M. avium* was more sensitive to chlorine than chloramine when biofilms were grown on copper pipe, but the reverse was true for *M. avium* biofilms on iron [50], possibly due to corrosion products interfering with free chlorine [51]. Here we found that *Mycobacterium* sequences were most abundant in biofilms from the July 2011 sampling date, which also had the highest level of free ammonia and the lowest levels of total chlorine and monochloramine. These constituents are related, as monochloramine is reductively dehalogenated to ammonia, and monochloramine is the major component of total chlorine in this system. The results of our nMDS analysis indicate that the monochloramine concentration strongly influenced bacterial community composition for the July 2011 sampling date. Therefore, the fact that *Mycobacterium* was predominant only on the sampling date with the lowest level of monochloramine suggests that within the ductile iron pipes in this DWDS monochloramine significantly reduced the abundance of *Mycobacterium* within the biofilms.

An unclassified genus from the family Xanthomonadaceae also varied considerably in abundance across the sampling dates, representing 50% of the sequences from the March 2012 samples but only a small fraction of the sequences from the other sampling dates. The Xanthomonadaceae are obligate aerobic chemoorganotrophs, and this family includes some well-known plant pathogens [33]. Organisms from the family Xanthomonadaceae family were not detected in a previous pyrosequencing survey of drinking water biofilms [7], but these organisms have been isolated from drinking water and from drinking water pipe biofilms using culture based techniques [52]. In this study, we were unable to discriminate the Xanthomonadaceae sequences down to the genus or species level, and the reason for their high abundance in the March 2012 samples is unclear.

The biofilm communities from July 2011 and March 2012 showed much higher variation in composition among replicates than did biofilm communities from December 2011 and August 2012. The December and August samples were each dominated by a single bacterial genus (*Methylomonas* in December and *Acinetobacter* in August) with very little variation between replicates. These data suggest that the environmental conditions in December and August each favored one specific bacterial genus that dominated all of the biofilms on that sampling date. In contrast, biofilms from July and March had several dominant bacterial genera that showed high variations between the replicates, suggesting that conditions during those months produced greater variability by enabling several genera to compete for dominance within the biofilms.

Nitrification is a significant concern in drinking water distribution systems that use chloramine as the secondary residual, as nitrification can lead to a decrease in the chloramine residual, an increase in the growth of heterotrophic bacteria and an increase in concentrations of nitrate and nitrite, which pose risks to human health [53]. Multiple studies have identified nitrifying bacteria in DWDS [53–55]. No sequences corresponding to any known ammonia oxidizing bacterial genera were detected in the biofilms analyzed in this study. However, a few sequences from a known nitrite oxidizing genus, *Nitrospira*, were detected in the July 2011 samples, but not in samples from any of the other sampling dates. July 2011 also showed the highest concentration of free ammonia (at least two times higher than all other sampling dates) and unpublished data from PCU confirm that there was a peak in nitrification activity in June and July of 2011. The nMDS analysis indicated that ammonia concentration was a significant driver of the composition of the biofilm bacterial communities from July 2011, and previous studies have indicated that the presence of free ammonia is the principal cause of nitrification in DWDS [56]. Therefore, these data suggest that the high free ammonia concentration combined with the high temperature in July 2011 enabled nitrification to occur within the pipes; however, the lack of detection of ammonia oxidizing bacteria suggests that ammonia oxidation within the biofilms may have been driven by ammonia oxidizing archaea (AOA). Previous studies have detected AOA in drinking water distribution systems [54], but archaea would not have been detected by the bacterial primers used in the current study.

In summary, the results of our study demonstrate that the biofilms within the DWDS pipes were dominated by a few bacterial taxa, specifically *Methylomonas*, *Acientobacter* and *Mycobacterium*, and that the dominant taxa within the biofilms varied dramatically between sampling times. It is likely that these differences in dominant taxa were driven by differences in environmental conditions, and our analysis suggests that nitrate, ammonium, total chlorine, and monochloramine concentrations were key drivers of biofilm bacterial community composition.

Another possibility is that these differences in dominant taxa could have been the result of the founder effect, which stipulates that the founding member of a biofilm will have an advantage over subsequent colonizers and will remain dominant. The founder effect has been suggested as a possible driver of biofilm community composition in a variety of habitats [57–60] and could have been a contributing factor to the differences in dominant community members in our biofilms. Further experimental work, which is ongoing in our lab, will be needed to explore the relative contributions of environmental factors and founder effects on the composition of biofilms within drinking water distribution systems.

## Acknowledgments

The authors thank Sharon Waller for coordinating the pipe sampling effort and Kesha Baxi for assistance with data analysis at the start of the project. The authors thank Tim LaPara for his helpful comments on an earlier version of this manuscript.

## Author Contributions

Conceived and designed the experiments: JK MP AP. Performed the experiments: NM AC. Analyzed the data: JK NM AC. Contributed reagents/materials/analysis tools: JK MP AP. Wrote the paper: JK NM AC MP AP.

## References

1. Szewzyk U, Szewzyk R, Manz W, Schleifer K (2000) Microbiological safety of drinking water. Annual Reviews in Microbiology 54: 81–127.
2. Marshall KC (1988) Adhesion and growth of bacteria at surfaces in oligotrophic habitats. Can J Microbiol 34: 503–506.
3. Zacheus OM, Lehtola MJ, Korhonen LK, Martikainen PJ (2001) Soft deposits, the key site for microbial growth in drinking water distribution networks. Water Res 35: 1757–1765.
4. LeChevallier MW, Cawthon CD, Lee RG (1988) Factors promoting survival of bacteria in chlorinated water supplies. Appl Environ Microbiol 54: 649–654.
5. Ridgway H, Olson B (1982) Chlorine resistance patterns of bacteria from two drinking water distribution systems. Appl Environ Microbiol 44: 972–987.
6. Liu Y, Zhang W, Sileika T, Warta R, Cianciotto NP, et al. (2011) Disinfection of bacterial biofilms in pilot-scale cooling tower systems. Biofouling 27: 393–402.
7. Hong P, Hwang C, Ling F, Andersen GL, LeChevallier MW, et al. (2010) Pyrosequencing analysis of bacterial biofilm communities in water meters of a drinking water distribution system. Appl Environ Microbiol 76: 5631–5635.
8. Bois FY, Fahmy T, Block J, Gatel D (1997) Dynamic modeling of bacteria in a pilot drinking-water distribution system. Water Res 31: 3146–3156.
9. Martiny AC, Jørgensen TM, Albrechtsen H, Arvin E, Molin S (2003) Long-term succession of structure and diversity of a biofilm formed in a model drinking water distribution system. Appl Environ Microbiol 69: 6899–6907.
10. Amann RI, Ludwig W, Schleifer K (1995) Phylogenetic identification and in situ detection of individual microbial cells without cultivation. Microbiol Rev 59: 143–169.
11. Kalmbach S, Manz W, Szewzyk U (1997) Isolation of new bacterial species from drinking water biofilms and proof of their in situ dominance with highly specific 16S rRNA probes. Appl Environ Microbiol 63: 4164–4170.
12. Schmeisser C, Stöckigt C, Raasch C, Wingender J, Timmis K, et al. (2003) Metagenome survey of biofilms in drinking-water networks. Appl Environ Microbiol 69: 7298–7309.
13. Parsek MR, Singh PK (2003) Bacterial biofilms: An emerging link to disease pathogenesis. Annual Reviews in Microbiology 57: 677–701.
14. Hoefel D, Monis PT, Grooby WL, Andrews S, Saint CP (2005) Culture-independent techniques for rapid detection of bacteria associated with loss of chloramine residual in a drinking water system. Appl Environ Microbiol 71: 6479–6488.
15. Eichler S, Christen R, Höltje C, Westphal P, Bötel J, et al. (2006) Composition and dynamics of bacterial communities of a drinking water supply system as assessed by RNA-and DNA-based 16S rRNA gene fingerprinting. Appl Environ Microbiol 72: 1858–1872.
16. Marshall K, Blainey BL (1991) Role of bacterial adhesion in biofilm formation and biocorrosion. In: Anonymous Biofouling and Biocorrosion in Industrial Water Systems: Springer. pp. 29–46.
17. Rice EW, Baird RB, Eaton AD, Clesceri LS, editors (2012) Standard methods for the examination of water and wastewater, 22nd ed. Denver, CO: American Water Works Association.
18. Lane DJ (1991) 16S/23S rRNA sequencing. In: Stackebrandt E, Goodfellow M, editors. Nucleic acid techniques in bacterial systematics. Chichester, UK: John Wiley.
19. Youssef N, Sheik CS, Krumholz LR, Najar FZ, Roe BA, et al. (2009) Comparison of species richness estimates obtained using nearly complete fragments and simulated pyrosequencing-generated fragments in 16S rRNA gene-based environmental surveys. Appl Environ Microbiol 75: 5227–5236.
20. Schloss PD, Westcott SL, Ryabin T, Hall JR, Hartmann M, et al. (2009) Introducing mothur: Open-source, platform-independent, community-supported software for describing and comparing microbial communities. Appl Environ Microbiol 75: 7537–7541.
21. Edgar RC, Haas BJ, Clemente JC, Quince C, Knight R (2011) UCHIME improves sensitivity and speed of chimera detection. Bioinformatics 27: 2194–2200.
22. Chao A (1984) Nonparametric estimation of the number of classes in a population. Scandinavian Journal of Statistics 11: 265–270.
23. Shannon CE (2001) A mathematical theory of communication. ACM SIGMOBILE Mobile Computing and Communications Review 5: 3–55.
24. Yue JC, Clayton MK (2005) A similarity measure based on species proportions. Communications in Statistics-Theory and Methods 34: 2123–2131.
25. Zhang P, LaPara TM, Goslan EH, Xie Y, Parsons SA, et al. (2009) Biodegradation of haloacetic acids by bacterial isolates and enrichment cultures from drinking water systems. Environ Sci Technol 43: 3169–3175.
26. Vaerewijck MJ, Huys G, Palomino JC, Swings J, Portaels F (2005) Mycobacteria in drinking water distribution systems: Ecology and significance for human health. FEMS Microbiol Rev 29: 911–934.
27. Forster D, Daschner F (1998) Acinetobacter species as nosocomial pathogens. European Journal of Clinical Microbiology and Infectious Diseases 17: 73–77.
28. Hedrich S, Schlomann M, Johnson DB (2011) The iron-oxidizing proteobacteria. Microbiology 157: 1551–1564.
29. Emerson D, Fleming EJ, McBeth JM (2010) Iron-oxidizing bacteria: An environmental and genomic perspective. Annu Rev Microbiol 64: 561–583.
30. Madigan MT, Martinko JM, Dunlap PV, Clark DP (2009) Brock biology of microorganisms. San Francisco, CA: Pearson/Benjamin Cummings.
31. Staley JT, Konopka A (1985) Measurement of in situ activities of nonphotosynthetic microorganisms in aquatic and terrestrial habitats. Annual Reviews in Microbiology 39: 321–346.
32. Revetta RP, Pemberton A, Lamendella R, Iker B, Santo Domingo JW (2010) Identification of bacterial populations in drinking water using 16S rRNA-based sequence analyses. Water Res 44: 1353–1360.
33. Brenner D, Krieg N, Staley J, Garrity G, Boone D, et al. (2005) The proteobacteria, part B, the gammaproteobacteria. Bergey's manual of systematic bacteriology 2.
34. McKeon DM, Calabrese JP, Bissonnette GK (1995) Antibiotic resistant gram-negative bacteria in rural groundwater supplies. Water Res 29: 1902–1908.
35. Bifulco JM, Shirey J, Bissonnette G (1989) Detection of acinetobacter spp. in rural drinking water supplies. Appl Environ Microbiol 55: 2214–2219.
36. Shirey JJ, Bissonnette GK (1991) Detection and identification of groundwater bacteria capable of escaping entrapment on 0.45-micron-pore-size membrane filters. Appl Environ Microbiol 57: 2251–2254.
37. Hansen SK, Rainey PB, Haagensen JA, Molin S (2007) Evolution of species interactions in a biofilm community. Nature 445: 533–536.
38. Marin M, Pedregosa A, Laborda F (1996) Emulsifier production and microscopical study of emulsions and biofilms formed by the hydrocarbon-utilizing bacteria acinetobacter calcoaceticus MM5. Appl Microbiol Biotechnol 44: 660–667.
39. Tomaras AP, Dorsey CW, Edelmann RE, Actis LA (2003) Attachment to and biofilm formation on abiotic surfaces by acinetobacter baumannii: Involvement of a novel chaperone-usher pili assembly system. Microbiology 149: 3473–3484.
40. Wentzel M, Lötter L, Loewenthal R, Marais G (1986) Metabolic behaviour of acinetobacter spp. in enhanced biological phosphorus removal- a biochemical model. Water S A 12: 209–224.
41. Hanson RS, Hanson TE (1996) Methanotrophic bacteria. Microbiol Rev 60: 439–471.
42. Covert TC, Rodgers MR, Reyes AL, Stelma GN (1999) Occurrence of nontuberculous mycobacteria in environmental samples. Appl Environ Microbiol 65: 2492–2496.
43. Pryor M, Springthorpe S, Riffard S, Brooks T, Huo Y, et al. (2004) Investigation of opportunistic pathogens in municipal drinking water under different supply and treatment regimes. Water Science & Technology 50: 83–90.
44. Weinberger M, Berg SL, Feuerstein IM, Pizzo PA, Witebsky FG (1992) Disseminated infection with mycobacterium gordonae: Report of a case and critical review of the literature. Clinical infectious diseases 14: 1229–1239.
45. Desforges JF, Horsburgh CR Jr (1991) Mycobacterium avium complex infection in the acquired immunodeficiency syndrome. N Engl J Med 324: 1332–1338.
46. Carson LA, Petersen NJ, Favero MS, Aguero SM (1978) Growth characteristics of atypical mycobacteria in water and their comparative resistance to disinfectants. Appl Environ Microbiol 36: 839–846.
47. Beumer A, King D, Donohue M, Mistry J, Covert T, et al. (2010) Detection of mycobacterium avium subsp. paratuberculosis in drinking water and biofilms by quantitative PCR. Appl Environ Microbiol 76: 7367–7370.
48. Falkinham JO, Norton CD, LeChevallier MW (2001) Factors influencing numbers of mycobacterium avium, mycobacterium intracellulare, and other mycobacteria in drinking water distribution systems. Appl Environ Microbiol 67: 1225–1231.

49. Gomez-Alvarez V, Revetta RP, Santo Domingo JW (2012) Metagenomic analyses of drinking water receiving different disinfection treatments. Appl Environ Microbiol 78: 6095–6102.

50. Norton CD, LeChevallier MW, Falkinham JO III (2004) Survival of< i> mycobacterium avium</i> in a model distribution system. Water Res 38: 1457–1466.

51. LeChevallier MW, Lowry CD, Lee RG, Gibbon DL (1993) Examining the relationship between iron corrosion and the disinfection of biofilm bacteria. Journal-American Water Works Association 85: 111–123.

52. Critchley M, Fallowfield H (2001) The effect of distribution system bacterial biofilms on copper concentrations in drinking water. Water Sci Technol Water Supply 1: 247–252.

53. Lipponen MT, Suutari MH, Martikainen PJ (2002) Occurrence of nitrifying bacteria and nitrification in finnish drinking water distribution systems. Water Res 36: 4319–4329.

54. Cunliffe DA (1991) Bacterial nitrification in chloraminated water supplies. Appl Environ Microbiol 57: 3399–3402.

55. Regan JM, Harrington GW, Noguera DR (2002) Ammonia- and nitrite-oxidizing bacterial communities in a pilot-scale chloraminated drinking water distribution system. Appl Environ Microbiol 68: 73–81.

56. Wilczak A, Jacangelo JG, Marcinko JP, Odell LH, Kirmeyer GJ, et al. (1996) Occurrence of nitrification in chloraminated distribution systems. Journal-American Water Works Association 88: 74–85.

57. McKew BA, Taylor JD, McGenity TJ, Underwood GJ (2010) Resistance and resilience of benthic biofilm communities from a temperate saltmarsh to desiccation and rewetting. The ISME journal 5: 30–41.

58. Harrison F (2007) Microbial ecology of the cystic fibrosis lung. Microbiology 153: 917–923.

59. Boomer SM, Noll KL, Geesey GG, Dutton BE (2009) Formation of multilayered photosynthetic biofilms in an alkaline thermal spring in yellowstone national park, wyoming. Appl Environ Microbiol 75: 2464–2475.

60. Ledder R, Gilbert P, Pluen A, Sreenivasan P, De Vizio W, et al. (2006) Individual microflora beget unique oral microcosms. J Appl Microbiol 100: 1123–1131.

# 21

# Sources of Heavy Metals in Surface Sediments and an Ecological Risk Assessment from Two Adjacent Plateau Reservoirs

**Binbin Wu[1], Guoqiang Wang[1]\*, Jin Wu[1], Qing Fu[2], Changming Liu[1]**

1 College of Water Sciences, Beijing Normal University, Key Laboratory of Water and Sediment Sciences, Ministry of Education, Beijing, China, 2 Chinese Research Academy of Environmental Sciences, Beijing, China

## Abstract

The concentrations of heavy metals (mercury (Hg), cadmium (Cd), lead (Pb), chromium (Cr), copper (Cu) and arsenic (As)) in surface water and sediments were investigated in two adjacent drinking water reservoirs (Hongfeng and Baihua Reservoirs) on the Yunnan-Guizhou Plateau in Southwest China. Possible pollution sources were identified by spatial and statistical analyses. For both reservoirs, Cd was most likely from industrial activities, and As was from lithogenic sources. For the Hongfeng Reservoir, Pb, Cr and Cu might have originated from mixed sources (traffic pollution and residual effect of former industrial practices), and the sources of Hg included the inflows, which were different for the North (industrial activities) and South (lithogenic origin) Lakes, and atmospheric deposition resulting from coal combustion. For the Baihua Reservoir, the Hg, Cr and Cu were primarily derived from industrial activities, and the Pb originated from traffic pollution. The Hg in the Baihua Reservoir might also have been associated with coal combustion pollution. An analysis of ecological risk using sediment quality guidelines showed that there were moderate toxicological risks for sediment-dwelling organisms in both reservoirs, mainly from Hg and Cr. Ecological risk analysis using the Hakanson index suggested that there was a potential moderate to very high ecological risk to humans from fish in both reservoirs, mainly because of elevated levels of Hg and Cd. The upstream Hongfeng Reservoir acts as a buffer, but remains an important source of Cd, Cu and Pb and a moderately important source of Cr, for the downstream Baihua Reservoir. This study provides a replicable method for assessing aquatic ecosystem health in adjacent plateau reservoirs.

**Editor:** Jonathan H. Freedman, NIEHS/NIH, United States of America

**Funding:** This research was supported by Beijing Higher Education Young Elite Teacher Project (Grant No. YETP0275), the Program for New Century Excellent Talents in University (Grant No. NCET-12-0058) and the Fundamental Research Funds for the Central Universities (Grant No. 2012LZD10). The funders had no role in study design, data collection and analysis, decision to publish, or preparation of the manuscript.

**Competing Interests:** The authors have declared that no competing interests exist.

\* Email: wanggq@bnu.edu.cn

## Introduction

There is worldwide concern about heavy metal contamination because of the environmental persistence of these elements, biogeochemical recycling and the ecological risks that metals present [1,2]. Large numbers of anthropogenically generated heavy metals from urban areas, agricultural areas and industrial sites are discharged into aquatic environments where they are transported in the water column, accumulated in sediment, and biomagnified through the food chain [3], resulting in significant ecological risk to benthic organisms, fish and humans [4]. Sediments are the main sink for heavy metals in aquatic environments [5], and sediment quality has been recognized as an important indicator of water pollution [6]. However, heavy metals are not permanently bound to sediments [7], and they may be released into the water column when the environmental conditions change (e.g., temperature and pH) or when sediments undergo other physical or biological disturbances [8]. Furthermore, reservoir construction generally leads to an increase in residence time, resulting in high accumulations of heavy metals in sediments. Consequently, it is important to analyze sediments from reservoirs for heavy metals to support environmental management, particularly for sediments from drinking water reservoirs.

Understanding the sources of pollutants in aquatic sediments is important for pollution control. Statistical approaches, such as Pearson correlation analysis, principal components analysis (PCA), and cluster analysis, are considered to be effective tools for uncovering pollution sources and have been used successfully in many studies of heavy metal pollution in sediments [1,2,3,7,9,10,11]. Risk assessments of the environmental pollution are also critical for sediment analysis. The ecological risk of heavy metals in sediments differs for different receptors (e.g., sediment-dwelling organisms, fish or humans). The thresholds in sediment quality guidelines (SQGs) have been used to evaluate the potential adverse effects of heavy metals on sediment-dwelling organisms in freshwater systems [12,13,14,15]. However, few SQGs have been developed to assess the adverse effects of heavy metals in sediment on higher trophic levels (fish or other wildlife) [16,17]. The potential ecological risk index proposed by Hakanson [18] is based on heavy metal concentrations in sediment, and it is the simplest and most popular method for assessing the human health risk from fish consumption.

The rapid growth of urbanization and industrial development has resulted in increasing heavy metal pollution in the aquatic sediment of the Yunnan-Guizhou Plateau [11,19]. Several cascade hydropower stations have been built along the region's large rivers (e.g., Wujiang, Jinshajiang and Nanpanjiang) since the 1950s, and stations are still being built for electricity production today, leading to a continuous series of reservoirs along the rivers [20]. Previous studies have evaluated the carbon (C) cycle [21,22] and the mercury (Hg) balance [20,23] in adjacent plateau reservoirs, and have demonstrated how upstream reservoirs influence downstream reservoirs. However, little research has been conducted on other heavy metals in the adjacent reservoirs on this plateau. In addition, several decades after their construction, the functions of these reservoirs were changed, so they now supply drinking water to the human population, which has grown rapidly due to economic growth. Currently, these reservoirs are the main drinking water sources on the Yunnan-Guizhou Plateau. Although some pollution sources were closed or moved when the reservoir functions were changed, the residue of previous pollutants still remains in reservoir sediments. Furthermore, the Yunnan-Guizhou Plateau is famous for its karst landforms, and the hydrogen carbonate ($HCO_3^-$) concentration and pH are both high in the aquatic environment [24,25,26]. The alkaline environment favors heavy metal accumulation in sediment [27], while the karst landform promotes interactions between groundwater and surface water through fractures (sinkholes, conduits and caves) or carbonate bedrock [26,28,29]. These interactions can complicate heavy metal transport and increase the ecological risk of secondary pollution. Therefore, it is important and necessary to investigate the heavy metal pollution and to assess the associated pollution sources and ecological risks from reservoir sediments on the Yunnan-Guizhou Plateau, particularly for drinking water reservoirs. The objectives of this paper are to (1) identify the pollution sources of heavy metals in the sediment from two adjacent drinking water reservoirs on the Yunnan-Guizhou Plateau, (2) estimate the associated ecological risk by considering different receptors, and (3) discuss the influence of upstream reservoirs (as buffers or sources of heavy metals) on downstream reservoirs.

## Materials and Methods

### Study Areas

The Hongfeng and Baihua Reservoirs are two adjacent reservoirs on the Yunnan-Guizhou plateau, just northwest of Guiyang City, the capital of Guizhou Province, Southwest China (Fig. 1). These two reservoirs were constructed on the main channel of the Maotiao River, a branch of the Wujiang River in the Yangtze River Basin, in 1958 and 1960, respectively. The Maotiao River was one of the first rivers to be used for cascade hydropower in China. The Hongfeng is the first reservoir and the Baihua is the second of seven cascade hydropower stations along the Maotiao River. The Hongfeng Reservoir covers a water surface area of 57.2 km², while the Baihua Reservoir covers an area of 14.5 km². Both reservoirs are very deep, with each having a maximum depth of approximately 45 m. The Hongfeng Reservoir consists of the North and South Lakes (which have different flow directions, Fig. 1), and has five main inflows, two into the North Lake and three into the South Lake. The Maotiao River is the only outlet of the Hongfeng Reservoir, and also serves as the major inlet of the Baihua Reservoir. The Baihua Reservoir has eight additional minor inflows and one outlet. For their first 30 years, the reservoirs were mainly used for electricity generation and flood control. During this period, as industry, agriculture, tourism and fishery production were established and developed in

the basin, the water quality in both reservoirs declined. However, because of an increasing demand for water through the 1990s, the two reservoirs were designated as drinking water sources for Guiyang City. The major function of both reservoirs was changed to drinking water supply in 2000, at which point the government strengthened their environmental protection. The pollution sources have gradually decreased, but the sediments may still hold residue from earlier pollution.

Heavy metal pollution is one of the most prominent environmental problems in the study reservoirs, mainly owing to intense anthropogenic activities. Figure 2 shows the distribution of main point sources in Hongfeng and Baihua Reservoirs basin from the first China pollution source census in 2008 (provided by the Guiyang Research Academy of Environmental Sciences). There are many mining, smelting, mechanical manufacture, chemical and other industries (e.g., building material, food and pharmaceutical factories) in the catchment area of the Hongfeng Reservoir (1596 km²), which are major sources of heavy metals (Fig. 2). There is also a large coal-fired power plant (300 MW) situated on the southeast bank of the Hongfeng Reservoir [30], which is the main source of atmospheric deposition, especially for Hg and other heavy metals associated with coal combustion. The catchment area of the Baihua Reservoir is 1895 km², but pollutants are first transported into the Hongfeng Reservoir, which may serve as a buffer for the Baihua Reservoir. The Baihua Reservoir receives direct inputs from an area of only 299 km². Even so, there are many intense point sources in this small area, including various heavy and light industries (Fig. 2), which have resulted in serious heavy metal pollution in the Baihua Reservoir. In particular, the Baihua Reservoir is noted for its Hg contamination from the Guizhou Organic Chemical Plant (GOCP) [31], which is located in the upper reaches of the Baihua Reservoir and downstream of the Hongfeng Reservoir. The GOCP used Hg-based technology to produce acetaldehyde and discharged Hg-laden wastewater to the Baihua Reservoir via the Dongmenqiao River until 1997. The pollution caused by the GOCP persists to the present day.

### Sampling and Analysis

Two field surveys were conducted in December 2010 and April 2012. Water and sediment samples were collected from 26 sites in the Hongfeng and Baihua Reservoirs (Fig. 1 and Table 1). The field studies were permitted by the Administration of Hongfeng, Baihua and Aha Reservoirs, and did not involve endangered or protected species. There were 13 sites in each reservoir, comprising inlets of main tributaries (sites 1–5 in the Hongfeng Reservoir and sites 14–22 in the Baihua Reservoir) and representative sites within both reservoirs (sites 6–13 in the Hongfeng Reservoir and sites 23–26 in the Baihua Reservoir). Surface water samples were collected in acid-washed polyethylene sample bottles and were acidified with 1:1 nitric acid: deionized water. Water samples were stored at 4°C immediately upon returning from the field. The upper 0–10 cm of sediment was collected, placed into pre-cleaned polyethylene bags, and taken to the laboratory. All sediment samples were freeze-dried and passed through a 2 mm nylon sieve to discard the coarse debris. A pestle and mortar was then used to grind the sieved sediments until all particles were fine enough to pass through a 0.147 mm nylon sieve. Sediment samples were digested in a microwave digestion system with a $HNO_3$-HF-$HClO_4$-HCl acid mixture solution before analysis for total heavy metal content. All water samples and the solutions of the digested sediment samples were analyzed by inductively coupled plasma atomic emission spectroscopy for Cr, Cu and Pb. Cd concentrations were determined by graphite

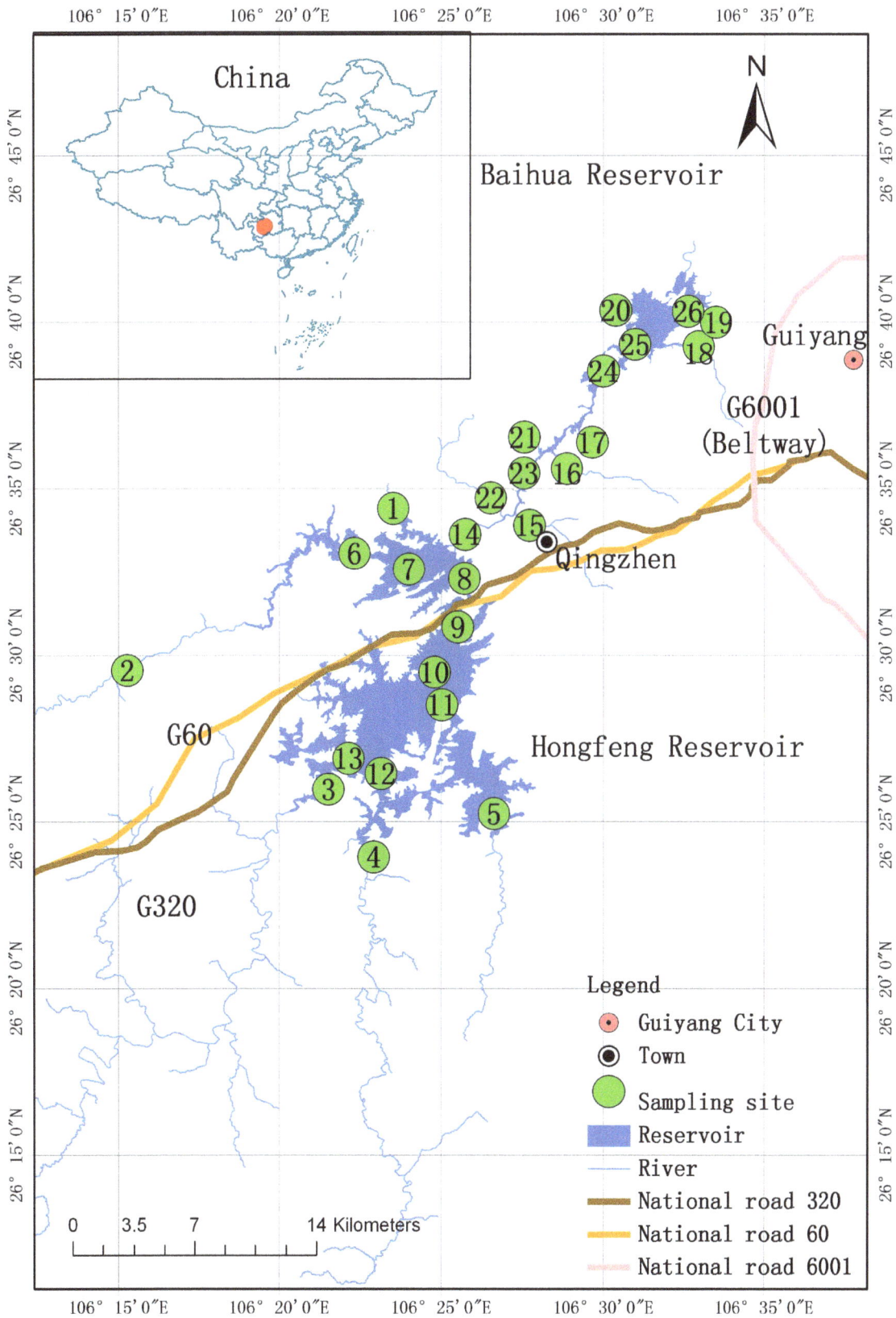

**Figure 1. Map of Hongfeng and Baihua Reservoirs on the Yunnan-Guizhou Plateau, Southwest China.**

furnace atomic absorption spectrophotometry. As and Hg were measured using atomic fluorescence spectrometry. Quality assur- ance and quality control of the analyses processes were assessed by duplicates, method blanks and standard reference materials [11].

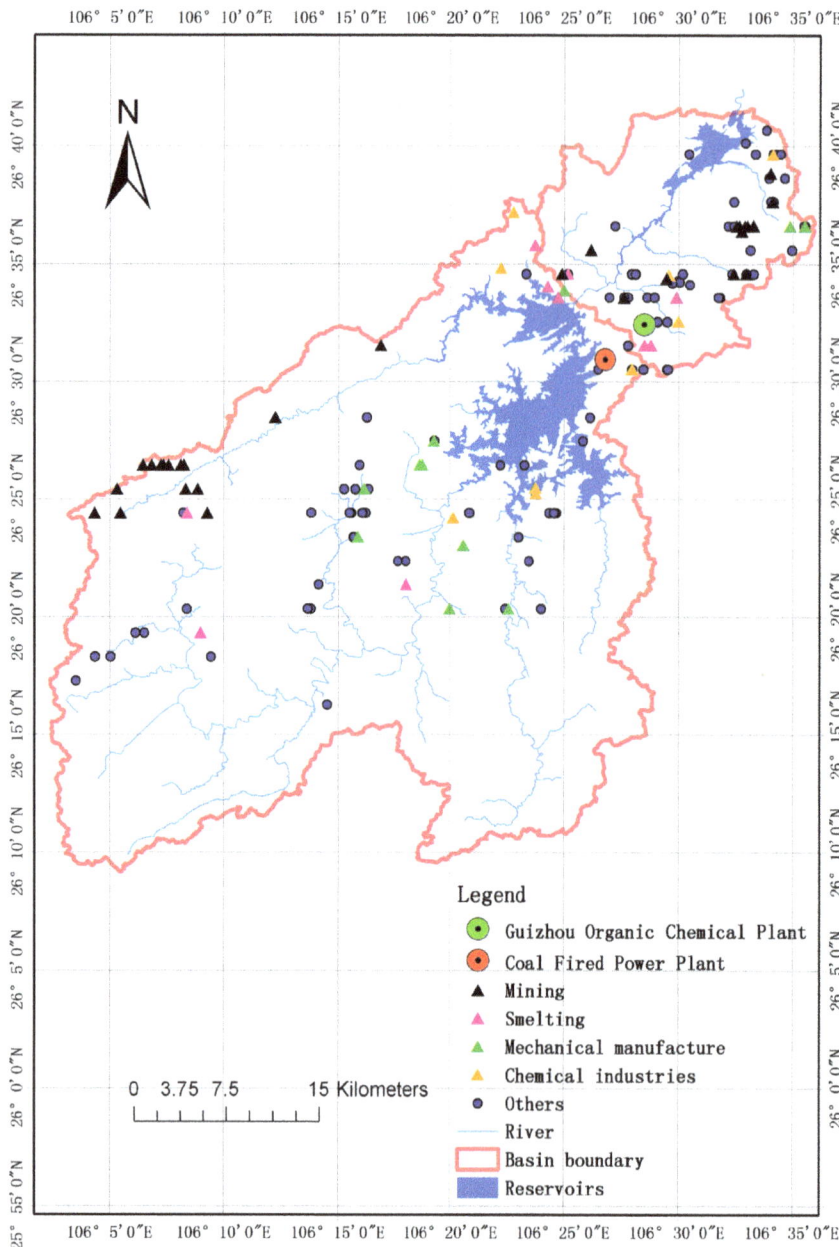

**Figure 2. Distribution of main point sources of pollution in the Hongfeng and Baihua Reservoir basins.**

## Spatial and Statistical Analyses

Spatial and statistical analyses were performed by using Arc GIS 9.3 and SPSS 17.0 for Windows (SPSS Inc., Chicago, IL) to investigate the heavy metal pollution sources separately for the Hongfeng and Baihua Reservoirs. A one-way ANOVA was performed on heavy metal concentrations in sediment to determine whether the differences between the two field surveys were significant. Pearson correlation analysis was used to determine the relationships between the heavy metals in sediment. To obtain more reliable information about the relationships between the heavy metals, a PCA with Varimax normalized rotation was performed separately for the Hongfeng and Baihua Reservoirs. The PCA calculated eigenvectors to determine the common pollution sources, and components with eigenvalues greater than 1 were considered to be relevant [32]. Components

with factor loadings above 0.75, between 0.5 and 0.75, and between 0.3 and 0.5 were considered to be strong, moderate and weak, respectively [33]. Boxplot is a convenient way to depict the full range of data and compare the distributions among different datasets. In this study, the boxplot was used to compare the heavy metal concentrations at site 14, the outlet of the Hongfeng Reservoir and inlet of the Baihua Reservoir, with concentrations at other sites in the tributaries and at sites within the reservoirs, in order to discuss the influence of the upstream Hongfeng Reservoir on the downstream Baihua Reservoir.

## Ecological Risk Assessment

We used two methods to assess the ecological risk of the heavy metals in surface sediments to benthic organisms and humans. First, we used the consensus-based SQGs for freshwater ecosys-

**Table 1.** Locations of sampling sites in the Hongfeng and Baihua Reservoirs on the Yunnan-Guizhou Plateau, Southwest China.

| Reservoir | Site | Longitude | Latitude | Description | |
|---|---|---|---|---|---|
| Hongfeng Reservoir | 1 | 106°23′30.64″ | 26°34′24.52″ | At main tributaries | Maibao River |
| | 2 | 106°15′3.58″ | 26°29′22.80″ | | Maiweng River |
| | 3 | 106°21′30.89″ | 26°25′58.34″ | | Yangchang River |
| | 4 | 106°22′54.88″ | 26°23′56.79″ | | Maxian River |
| | 5 | 106°26′42.01″ | 26°25′4.80″ | | Houliu River |
| | 6 | 106°22′11.10″ | 26°33′44.07″ | Within reservoir | Taipingdi |
| | 7 | 106°24′2.05″ | 26°32′34.80″ | | Center of the North Lake |
| | 8 | 106°23′17.43″ | 26°32′27.13″ | | Junction of the North and South Lakes |
| | 9 | 106°26′07.06″ | 26°30′58.03″ | | Houwu |
| | 10 | 106°24′39.31″ | 26°29′19.07 | | Center of the South Lake |
| | 11 | 106°25′04.17″ | 26°28′33.89″ | | Jiangjundong |
| | 12 | 106°23′01.81″ | 26°26′26.73″ | | Yangjiajun |
| | 13 | 106°22′01.22″ | 26°26′37.57″ | | Sanjiazhai |
| Baihua Reservoir | 14 | 106°25′43.88″ | 26°33′37.65″ | At main tributaries | Outlet of Hongfeng Reservoir |
| | 15 | 106°27′19.60″ | 26°34′7.06″ | | Dongmenqiao River |
| | 16 | 106°28′51.88″ | 26°35′36.16″ | | Maicheng River |
| | 17 | 106°29′39.21″ | 26°36′23.43″ | | Dianzishanggou River |
| | 18 | 106°32′56.81″ | 26°39′15.61″ | | Maixi River |
| | 19 | 106°33′16.38″ | 26°40′6.00″ | | Banpochanggou River |
| | 20 | 106°30′22.33″ | 26°40′21.27″ | | Maolizhaigou River |
| | 21 | 106°27′33.58″ | 26°36′32.72″ | | Xiaohekou River |
| | 22 | 106°26′47.46″ | 26°34′17.11″ | | Changchong River |
| | 23 | 106°27′25.20″ | 26°35′27.60″ | Within reservoir | Huaqiao |
| | 24 | 106°30′4.42″ | 26°38′31.42″ | | Xuantiandong |
| | 25 | 106°30′50.23″ | 26°39′1.37″ | | Tangerpo |
| | 26 | 106°32′48.39″ | 26°40′18.16″ | | Chafan |

tems that were proposed by MacDonald et al. [15], which included a threshold effect concentration (TEC) and a probable effect concentration (PEC). TECs are the concentrations below which adverse effects are not expected on sediment-dwelling organisms, while PECs are concentrations above which adverse effects are expected to occur frequently [34,35]. The mean PEC

**Table 2.** Ecological risk assessment criteria for the sediment quality guidelines (SQGs) and Hakanson index.

| Method | C or $E_r^i$ | Potential ecological risk for single heavy metal | m-PEC-Q or RI | Ecological risk for all Heavy metals |
|---|---|---|---|---|
| SQGs | C<TEC | Low | m-PEC-Q<0.1 | Low (<14%)[a] |
| | TEC <C<PEC | Moderate | 0.1< m-PEC-Q <1.0 | Moderate (15–29%)[a] |
| | C>PEC | High | 1.0<m-PEC-Q<5.0 | Considerable (33–58%)[a] |
| | | | m-PEC-Q >5.0 | Very high (75–81%)[a] |
| Hakanson index | $E_r^i$<40 | Low | RI<95 | Low |
| | 40<$E_r^i$<80 | Moderate | 95<RI<190 | Moderate |
| | 80<$E_r^i$<160 | Considerable | 190<RI<380 | Considerable |
| | 160<$E_r^i$<320 | High | RI>380 | Very high |
| | $E_r^i$>320 | Very high | | |

C: concentration of heavy metal in surface sediment.
TEC: threshold effect level; PEC: probable effect level [15].
m-PEC-Q: mean PEC quotient; [a]incidence of toxicity [36].

**Table 3.** Summary statistics for heavy metal concentrations in surface sediments from the Hongfeng and Baihua Reservoirs.

| | Hg | Cd | Pb | Cr | Cu | As | |
|---|---|---|---|---|---|---|---|
| HongfengReservoir | | | | | | | |
| Mean (Min, Max) | 0.32(0.08, 1.03) | 0.28(0.01, 0.85) | 28.41(0.10, 89.20) | 86.91(34.10, 141.00) | 43.50(15.70, 93.60) | 15.33(0.12, 45.54) | This study (2010.12) |
| S.D. | 0.26 | 0.36 | 27.55 | 29.65 | 25.09 | 14.92 | |
| CV(%) | 82.86 | 130.78 | 96.97 | 34.12 | 57.67 | 97.31 | |
| Mean (Min, Max) | 0.27(0.04, 0.56) | 0.53(0.31, 1.37) | 30. 22(1.21, 89.20) | 82.78(41.00, 141.00) | 45.46(23.20, 73.80) | 23.31(17.31, 29.61) | This study (2012.4) |
| S.D. | 0.17 | 0.27 | 25.63 | 23.45 | 17.72 | 3.98 | |
| CV(%) | 62.33 | 51.27 | 84.78 | 28.32 | 38.99 | 17.07 | |
| 2007.3 | 0.99 | | 34.31 | | 89.11 | 49.90 | Huang et al. [40] |
| 2008.8 | | | | 120.16 | 69.84 | | Zeng et al. [41] |
| 2008.10 | 0.66 | 0.77 | 35.91 | 87.98 | 91.85 | 29.74 | Liu et al. [42] |
| 2009.5 | 0.46 | 0.65 | | 118.00 | 88.00 | 40.80 | He et al. [27] |
| Baihua Reservoir | | | | | | | |
| Mean (Min, Max) | 0.68(-[a], 2.20) | 0.58(0.01, 1.00) | 27.28(0.10, 51.90) | 76.24(30.80, 143.00) | 36.71(0.36, 65.90) | 29.95(7.95, 48.19) | This study (2010.12) |
| S.D. | 0.73 | 0.37 | 16.89 | 31.50 | 19.77 | 14.80 | |
| CV(%) | 108.24 | 64.15 | 61.90 | 41.31 | 53.86 | 49.42 | |
| Mean (Min, Max) | 0.45(0.01, 1.25) | 0.61(0.23, 1.00) | 27.84(0.10, 51.90) | 76.38(30.12, 143.10) | 43.16(9.38, 73.55) | 26.23(7.95, 34.75) | This study (2012.4) |
| S.D. | 0.48 | 0.28 | 18.15 | 31.45 | 21.49 | 7.30 | |
| CV(%) | 107.21 | 46.13 | 65.19 | 41.17 | 49.79 | 27.82 | |
| 2007 | 18.90 | 0.88 | 16.05 | 59.75 | 74.97 | 53.34 | Huang [39] |
| 2010.5 | | 0.95 | 38.90 | 66.00 | 67.50 | | Tian et al. [43] |
| Natural Background Value | 0.08-0.15 | 0.08-0.12 | 18.50-23.90 | 73.90-94.60 | 27.30-36.70 | 27.00-50.00 | NEPA [44] |

All concentrations are in mg/kg dry weight. [a]-: not detected. S.D.: standard deviation; CV: coefficients of variation.

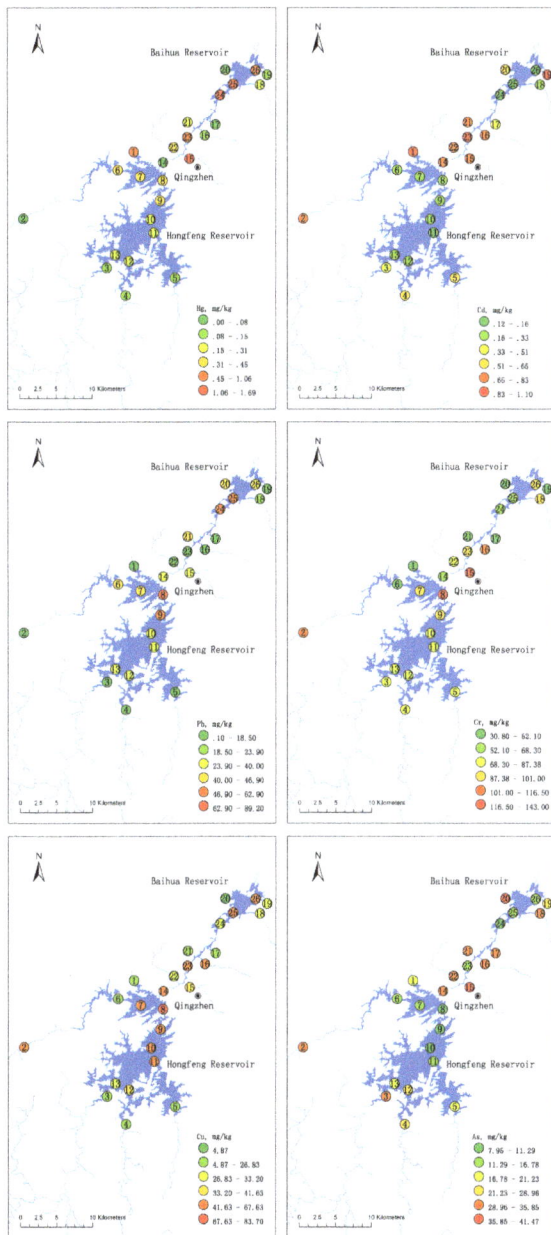

**Figure 3. Heavy metal concentrations in surface sediments from Hongfeng and Baihua Reservoirs.**

quotient (*m-PEC-Q*) [36] was also calculated for each sediment sample to assess the biological significance of the contaminant mixtures as follows:

$$m - PEC - Q = \frac{\sum_{i=1}^{n} (C_i / PEC_i)}{n} \tag{1}$$

where $C_i$ is the sediment concentration of compound $i$, $PEC_i$ is the PEC for compound $i$ and $n$ is the number of compounds $i$. Four ranges of the mean PEC quotient were developed by Long et al. [36] for ranking samples in terms of toxicity incidence (Table 2).

We also used the Hakanson index, which reflects the risk to human health from fish consumption. This index is based on the assumption that the sensitivity of the aquatic system depends on its productivity [3,18]. The potential ecological risk index (*RI*) was introduced to evaluate heavy metal pollution in sediments by considering the toxicity of heavy metals and the environmental response. The *RI* is calculated as follows:

$$RI = \sum E_r^i \tag{2}$$

$$E_r^i = T_r^i C_f^i \tag{3}$$

$$C_f^i = C_0^i / C_n^i \tag{4}$$

where *RI* is the total potential ecological risk index for multiple metals, $E_r^i$ is the potential ecological risk index for a single metal, and $T_r^i$ is the toxic-response factor for a given metal, considering both toxicity and the sensitivity. $C_f^i$ is the contamination factor, $C_0^i$ is the metal concentration in the sediment and $C_n^i$ is a reference value for metals. In this study, because both reservoirs are moderately eutrophic [37], $T_r^i$ was described as Hg (40) > Cd (30) > As (10) > Cu = Pb (5) > Cr (2), based on the assumption that the bioproduction index was 5 [18]. $C_n^i$ was defined as the upper limit of the natural background value for a given metal in the study area (Table 3). Four ranges of the risk factor *RI* were suggested by Hakanson, based on eight metals (polychlorinated biphenyls (PCBs), Hg, Cd, As, Pb, Cu, Cr, and zinc (Zn)). PCBs and Zn were not considered in this study. Based on the different contributions of these elements to the ecological risk index *RI*, the adjusted evaluation criteria for *RI* based on the six metals in this study are listed in Table 2.

**Table 4.** Correlations between heavy metals in surface sediments from the Hongfeng and Baihua Reservoirs.

|      | Hg       | Cd       | Pb       | Cr       | Cu      | As       |
|------|----------|----------|----------|----------|---------|----------|
| Hg   | **1.000**    | **0.090**    | 0.321    | −0.234   | −0.038  | −0.343   |
| Cd   | −0.422*  | **1.000**    | −0.540** | −0.168   | −0.237  | 0.625**  |
| Pb   | 0.511**  | −0.506** | **1.000**    | 0.438*   | 0.595** | −0.524** |
| Cr   | 0.495*   | 0.013    | −0.086   | **1.000**    | 0.785** | 0.046    |
| Cu   | 0.241    | −0.013   | 0.006    | 0.509**  | **1.000**   | −0.183   |
| As   | −0.414*  | 0.476*   | −0.314   | 0.067    | −0.333  | **1.000**    |

Hongfeng Reservoir in the upper right corner (blod); Baihua Reservoir in the lower left corner.
Levels of significance: *p<0.05; **p<0.01.

**Table 5.** Principal Component Analysis (PCA) for heavy metals in surface sediments from the Hongfeng and Baihua Reservoirs.

| Heavy metal | Hongfeng Reservoir | | | Baihua Reservoir | |
|---|---|---|---|---|---|
| | F1 | F2 | F3 | F1 | F2 |
| Hg | | | 0.973 | −0.669 | 0.523 |
| Cd | | 0.924 | | 0.821 | |
| Pb | 0.605 | −0.575 | 0.378 | −0.800 | |
| Cr | 0.929 | | | | 0.902 |
| Cu | 0.930 | | | | 0.799 |
| As | | 0.851 | −0.334 | 0.705 | |
| Variance (%) | 35.29 | 32.24 | 21.35 | 37.84 | 29.32 |
| Cumulative (%) | 35.29 | 67.53 | 88.88 | 37.84 | 67.16 |

Factor loadings smaller than 0.3 have been removed.
Extraction method: PCA, Rotation method: Varimax with Kaiser normalization.

## Results and Discussion

### Heavy Metal Concentrations in Water Samples and Surface Sediments

Of the 6 heavy metals, only Hg and As were detected in the water samples in December 2010, while Hg, Cd, Cr (VI) and As were detected in April 2012. Their concentrations were similar in the Hongfeng and Baihua Reservoirs, with Cd, Cr (VI), and As concentrations lower than Class I as defined in the Chinese Environmental Quality Standards for Surface Water (GB3838-2002, <0.001 mg/L for Cd, <0.01 mg/L for Cr (VI), and < 0.05 mg/L for As) and Hg concentrations ranging from Class I (GB3838-2002, <0.00005 mg/L) to Class IV (GB3838-2002, 0.0001–0.001 mg/L) among different sites. The low heavy metal concentrations in water were primarily due to the accumulation in sediments because the alkaline environment in both reservoirs provides ideal conditions for adsorption and precipitation [27]. Moreover, the sediment accumulation rate in both reservoirs was quite high [38], contributing to the removal of heavy metals from the water column. A prior one-way ANOVA analysis was conducted to examine the variation in heavy metal concentrations in sediment between the two field surveys. None of the heavy metals in Hongfeng and Baihua Reservoirs displayed significant variation in means ($p > 0.05$), although Cd in Hongfeng Reservoir and As in both reservoirs showed significant changes in their variances ($p < 0.05$). The significant differences in variance for Cd and As are mainly because of their higher concentrations in the sites within the reservoirs and the reduced spatial heterogeneity in the second field sampling comparing to the first one. However, the general spatial patterns for Cd and As (with higher concentrations at sites in the tributaries than at sites within the reservoirs) were still similar between the two field surveys. Those results indicate that the pollution sources for the metals were relatively stable between the two surveys. The concentrations of heavy metals in surface sediments of both reservoirs from the two field surveys are summarized in Table 3. Heavy metal concentrations in sediment were much higher than those in water. In general, the mean Hg, Cd and As sediment concentrations in the Baihua Reservoir were higher than those in the Hongfeng Reservoir, while Pb, Cr and Cu were higher in the Hongfeng Reservoir. Comparison with the results of previous studies [27,39,40,41,42,43] shows that most of the heavy metal concentrations in both reservoirs have decreased, though by differing amounts (Table 3). In particular, the Hg concentrations in the Baihua Reservoir have decreased signifi-

cantly, indicating that measures taken in recent years have been effective and resulted in improvements. During the two field surveys, the mean concentrations of Hg, Cd, Pb and Cu, and the maximum concentrations of Cr in both reservoirs exceeded the upper limit of the natural background values for the study area [44], indicating anthropogenic sources. However, the concentrations of As (including the minimum and maximum values) in both reservoirs were well within the range of the natural background values [44], implying no significant anthropogenic impact and primarily lithogenic sources.

### Heavy Metal Pollution Sources

To develop control strategies for environmental pollution, it is very important to identify its source. Spatial and statistical analyses were performed to identify the possible pollution sources for heavy metals in the Hongfeng and Baihua Reservoirs. The average concentrations of heavy metals in the sediments from the two field surveys were used to study the spatial distributions, while all data in the sediments from the two field surveys were used in a Pearson correlation analysis and PCA. The spatial distribution patterns of Hg, Cd, Pb, Cr, Cu and As in surface sediments of both reservoirs are shown in Figure 3. The Pearson correlation coefficients and the results of the PCA for the investigated metals are shown in Table 4 and Table 5, respectively. All of the results were generally consistent with each other.

Specifically, the PCA yielded three significant components for Hongfeng Reservoir and two significant components for Baihua Reservoir, accounting for 88.88% and 67.16% of the cumulative variance, respectively (Table 5). For the Hongfeng Reservoir, the first component (F1), explaining 35.29% of the total variance, had strong positive loadings of Cr and Cu, and moderate positive loading of Pb. Those three heavy metals exhibited similar spatial distributions in the Hongfeng Reservoir, with unexpectedly higher concentrations at reservoir sites than at tributary sites. In particular, site 8 (at the junction of the North and South Lakes) showed the highest concentrations for all of the three heavy metals, and site 9 (near Houwu) also showed relatively high concentrations (Fig. 3). In addition, those three heavy metals were highly correlated (Table 4, $p < 0.01$ for Cr-Cu and Cu-Pb, $p < 0.05$ for Cr-Pb), indicating their similar origins or comparable chemical properties [45]. This phenomenon might be caused by two possible reasons. Firstly, Bai et al. [46] found that traffic pollution was responsible for the high heavy metal concentrations (including comparable Cr, Cu and Pb concentrations with our study) along

**Table 6.** Results of ecological risk assessments for single heavy metal from two methods for the Hongfeng and Baihua Reservoirs.

| | | Hg | Cd | Pb | Cr | Cu | As |
|---|---|---|---|---|---|---|---|
| TEC | | 0.18 | 0.99 | 35.8 | 43.4 | 31.6 | 9.79 |
| PEC | | 1.06 | 4.98 | 128 | 111 | 149 | 33 |
| Hongfeng Reservoir | % samples which exceeded TEC | 69.23 | 7.69 | 38.46 | 92.31 | 61.54 | 84.62 |
| | % samples which exceeded PEC | 0.00 | 0.00 | 0.00 | 15.38 | 0.00 | 15.38 |
| | % samples with $E_r^i<40$ | 30.77 | 7.69 | 100.00 | 100.00 | 100.00 | 100.00 |
| | % samples with $40<E_r^i<80$ | 23.08 | 53.85 | 0.00 | 0.00 | 0.00 | 0.00 |
| | % samples with $80<E_r^i<160$ | 38.46 | 23.08 | 0.00 | 0.00 | 0.00 | 0.00 |
| | % samples with $160<E_r^i<320$ | 7.69 | 15.38 | 0.00 | 0.00 | 0.00 | 0.00 |
| | % samples with $E_r^i>320$ | 0.00 | 0.00 | 0.00 | 0.00 | 0.00 | 0.00 |
| Baihua Reservoir | % samples which exceeded TEC | 53.85 | 7.69 | 38.46 | 84.62 | 46.15 | 92.31 |
| | % samples which exceeded PEC | 30.77 | 0.00 | 0.00 | 7.69 | 0.00 | 38.46 |
| | % samples with $E_r^i<40$ | 38.46 | 23.08 | 100.00 | 100.00 | 100.00 | 100.00 |
| | % samples with $40<E_r^i<80$ | 15.38 | 0.00 | 0.00 | 0.00 | 0.00 | 0.00 |
| | % samples with $80<E_r^i<160$ | 7.69 | 15.38 | 0.00 | 0.00 | 0.00 | 0.00 |
| | % samples with $160<E_r^i<320$ | 15.38 | 61.54 | 0.00 | 0.00 | 0.00 | 0.00 |
| | % samples with $E_r^i>320$ | 23.08 | 0.00 | 0.00 | 0.00 | 0.00 | 0.00 |

All concentrations are in mg/kg dry weight.

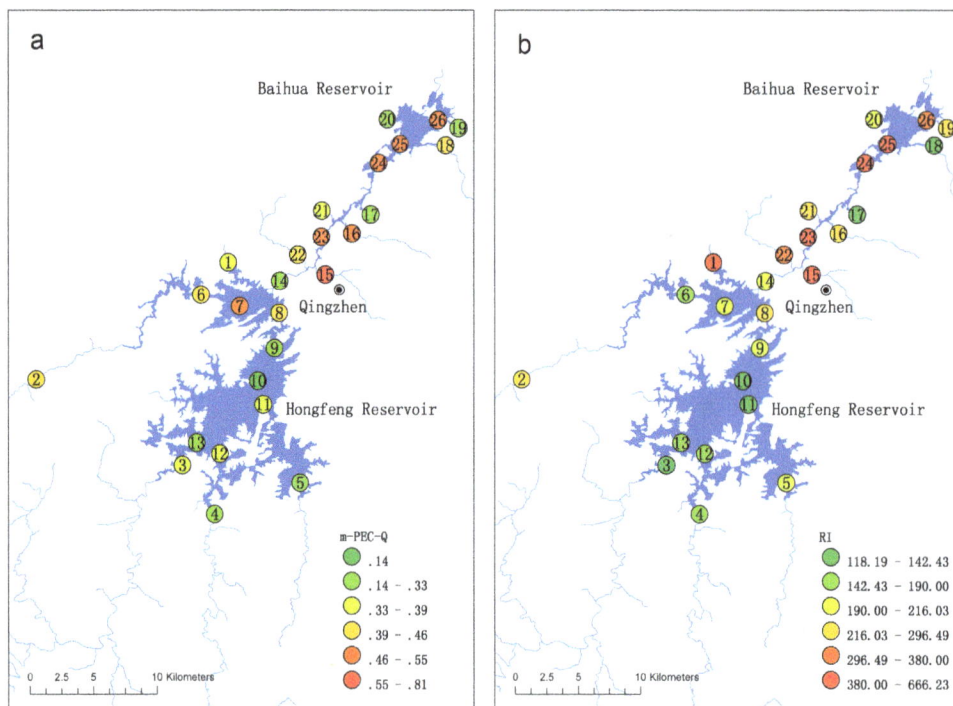

**Figure 4. Mean PEC quotient (a) and potential ecological risk indexes (b) of heavy metals in sediments.**

the roadside of National Road 320 in the Yunnan province (adjacent to Guizhou Province), and Zhu et al. [47] also found that road dust samples were severely polluted by Cr, Cu and Pb in another metal smelting/processing industrial city in Guizhou Province. In this study, the National Road 60 and the National Road 320 pass close to the junction of the South and North Lakes of the Hongfeng Reservoir (Fig. 1), which suggests that the traffic emissions, through atmospheric deposition and road runoff, could result in heavier pollution in sites near the roadway (site 8 and 9). Secondly, the high concentrations of Cr, Cu and Pb at reservoir sites are likely to be related to the residual effect from former industrial activities (e.g., mining, smelting, mechanical manufacture and chemical industry). The metals are more likely retained in the sediment of sites within the reservoirs rather than sites in the tributaries because heavy metal accumulation in lake sediments is generally higher than that in rivers [3]. Additionally, the complex hydrodynamic conditions at the junction of the South and North Lakes may affect the heavy metal distributions in sediment, which requires further research. Therefore, the first component (F1) might reflect mixed sources from traffic pollution and the residual effect of former industrial influence. The second component (F2), explaining 32.24% of the total variance, was dominated by Cd and As. Similar spatial patterns were observed for Cd and As, with higher concentrations at tributary sites than at sites within the reservoirs, indicating that they mainly come from the inflows (Fig. 3). As expected, significantly positive correlations were found between Cd and As (Table 4, $p<0.01$). However, Cd showed apparent anthropogenic origin, with most sites exceeding its natural background values, while As levels suggested natural origins, with all sites well within the natural background values (Table 3). Cd is closely related to industrial activities, such as smelting, electroplating and plastics production in the upstream areas. Hence, F2 may reflect the pollution through inflows from both industrial activities and natural weathering and erosion. The

third component (F3) had strong positive loading on Hg and a weak positive loading on Pb, accounting for 21.35% of the total variance. The highest Hg concentration in Hongfeng Reservoir was found at site 1 in the tributary of Maibao River, which has several smelting and chemical industries in its upstream (Fig. 2). Meanwhile, for both Hg and Pb, the North Lake were more polluted than the South Lake, and three tributaries in the South Lake showed concentrations well within the natural background values. Feng et al. [30] found that runoff due to soil erosion was the main source of Hg in sediment in the South Lake of the Hongfeng Reservoir. Thus, F3 may reflect the pollution from inflows from industrial activities in the North Lake and lithogenic origin in the South Lake. In addition, He [48] found that atmospheric deposition from coal combustion was also an important source of Hg in the Hongfeng Reservoir, which was not clearly distinguished by the PCA.

For Baihua Reservoir, the first component (F1), explaining 37.84% of total variance, showed strong positive loadings on Cd and As. Similar spatial distributions (with higher concentrations at tributary sites than at sites within the reservoirs) (Fig. 3) and positive correlations were also found between Cd and As (Table 4, $p<0.05$). As discussed above, F1 in Baihua Reservoir might be similar to F2 in Hongfeng Reservoir, including the pollution through inflows from both industrial activities and natural origin. The second component (F2) had a moderate positive loading of Hg and strong positive loadings on Cr and Cu. The highest Hg and Cr concentration in the Baihua Reservoir was found at site 15 in the Dongmengqiao tributary, which received wastewater from the GOCP and many other industries (Fig. 2). The highest Cu concentration in the Baihua Reservoir was at site 16 in the tributary of the Maicheng River, which has several industries (especially mining) in its upstream (Fig. 2). Strong associations were found between Cr and Cu ($p<0.01$) and between Cr and Hg ($p<0.05$) (Table 4). F2 obviously represented industrial activity

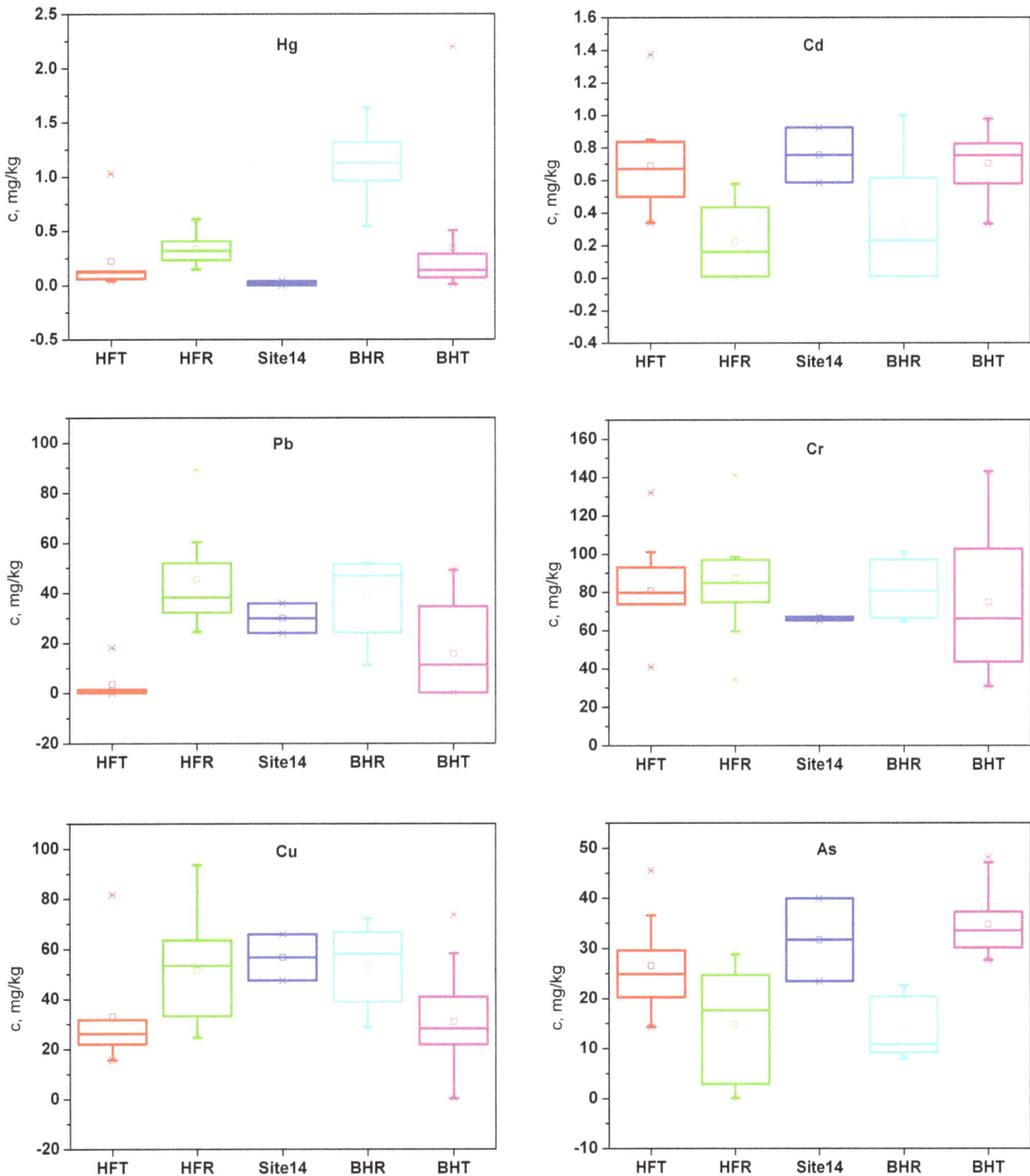

**Figure 5. Comparison of heavy metal concentrations in sediments.** (HFT: sites at inlets of main tributaries in the Hongfeng Reservoir, namely sites 1–5; HFR: representative sites within Hongfeng Reservoir, namely sites 6–13; BHT: sites at inlets of main tributaries in the Baihua Reservoir (except site 14), namely sites 15–22; BHR: representative sites within the Baihua Reservoir, namely sites 23–26).

upstream. On the other hand, the PCA failed to identify Pb sources in Baihua Reservoir, but only showed its strong negative associations with F1, indicating that there may be significant sources other than F1 and F2 for Pb. The spatial pattern of Pb in the Baihua Reservoir also showed much higher concentrations at sites within the reservoir than at sites in the tributaries (Fig. 3).

Although no main road crosses Baihua Reservoir directly, it is close to the urban district of Guiyang City, which has several roads and high traffic density (Fig. 1 only shows the main beltway and there are many other crisscrossed roads inside the beltway). In addition, Pb was added to gasoline in China until June 2000 [49]. Hence, Pb may originate from traffic pollutants deposited atmospherically. Moreover, highly positive correlations were found between Pb and Hg (p<0.01), implying that Hg might also come from atmospheric deposition (coal combustion) in addition to from the factors associated with F2.

## Ecological Risk Assessment of Heavy Metals in Surface Sediments

Because there was no significant change in the mean heavy metal concentrations in sediment between the two field surveys, the average concentrations were used to study the ecological risk of single heavy metals and the combined ecological effects of six heavy metals for the two study reservoirs using both the SQG and Hakanson index methods (Table 6). The SQG method revealed that the metal concentrations were within the TEC and PEC ranges for Hg, Cd, Pb and Cu at 69.23%, 7.69%, 38.46% and 61.54% of the sites in the Hongfeng Reservoir, and for Cd, Pb and Cu at 7.69%, 38.46% and 46.15% of sites in the Baihua Reservoir. The heavy metals at the remaining sites in the corresponding reservoirs fell below the TEC. Cr and As exceeded the PEC at 15.38% of sites in the Hongfeng Reservoir and Hg, Cr and As exceeded the PEC at 30.77%, 7.69% and 38.46%, respectively, of sites in the Baihua Reservoir. Previous studies have shown that ecological risk assessments should consider the regional background values and that exceeding the SQG values does not always lead to adverse ecological effects [50]. Therefore, the As concentrations within background ranges in both reservoirs should be excluded. The Cr in the Hongfeng Reservoir and both Hg and Cr in the Baihua Reservoir may pose significant ecological risks for sediment-dwelling organisms, and they deserve special attention. Heavy metals within the TEC–PEC (viz., Hg in Hongfeng Reservoir and Cd, Pb and Cu in both reservoirs) are also a cause for concern because this uncertain area may be considered to be moderately polluted [10]. The toxicity, derived from mean PEC quotients, that results from the mixture of the six heavy metals at each sampling site in both reservoirs is shown in Figure 4a. Overall, mean PEC quotients for samples in the Baihua Reservoir (range 0.28–0.81) were slightly higher than those in the Hongfeng Reservoir (range 0.14–0.55). However, the mean PEC quotients for all of the samples in both reservoirs were well within the range of 0.1 to 1.0, indicating moderate toxicological risks for sediment-dwelling organisms, with a toxicity incidence of between 15 and 29% in the study areas (Table 2).

The Hakanson method expresses the threat to humans from fish consumption. The results from this index were quite different from those for the SQG method (Table 6). Both Hg and Cd posed high potential ecological risks at 7.69% and 15.38% of sites, considerable risks at 38.46% and 23.08% of sites and moderate risks at 23.08% and 53.85% of sites, respectively, in the Hongfeng Reservoir. The risks were higher in the Baihua Reservoir, in which there was a very high potential ecological risk from Hg at 23.08% of sites, high risks from Hg and Cd at 15.38% and 61.54% of sites, considerable risks from Hg and Cd at 7.69% and 15.38% of sites, and a moderate risk from Hg at 15.38% of sites. However, the other heavy metals (Pb, Cr, Cu and As) posed little potential ecological risks for all sites in both reservoirs, with $E_r^i$ values lower than 40. The high concentrations and toxic-response factors of Hg and Cd in both reservoirs contribute to their posing higher ecological risks than the other metals we examined. $RI$ illustrates the potential ecological risk from heavy metal mixtures, and $RI$ at all sites in both reservoirs were higher than 95 (Fig. 4b). Site 1 showed the highest potential ecological risk ($RI = 484.25$) in the Hongfeng Reservoir, at a level that should cause concern because it poses a very high risk. Sites 2, 5 and 7–9 exhibited considerable ecological risks, while other sites showed moderate ecological risks

in the Hongfeng Reservoir. The combined ecological risk was more severe in the Baihua Reservoir. The $RI$ at site 15 (Dongmenqiao tributary) and 23–25 (within the reservoir) were much higher than 380, which indicates a very high potential ecological risk. All of the other sites exhibited considerable ecological risks (except for the moderate ecological risks at site 17 and 18). Therefore, there are moderate to very high potential ecological risks from heavy metal mixtures in the sediments of both reservoirs. In addition, the contribution of the monomial potential ecological risk to $RI$ for the six heavy metals in both reservoirs decreased in the following order: Hg ≈ Cd > As > Cu > Pb > Cr, with the greatest ecological risk from Hg and Cd.

Overall, the ecological risks from either a single heavy metal or from mixed heavy metals were different for the two receptors (viz., sediment-dwelling organisms and human beings through fish consumption) in both prior contaminants and risk level. However, hot spots with higher ecological risks were similar even though two different methods were used, and they were mainly located in the North Lake and the Houwu area of the Hongfeng Reservoir and in the key tributaries and at all of the sites in the Baihua Reservoir. Therefore, the need for industrial wastewater and mining tailings treatment in upstream watersheds of both reservoirs should be highlighted, especially for the tributaries in the North Lake of the Hongfeng Reservoir and in the key tributaries of the Baihua Reservoir. Additionally, given that the lakes are sources for drinking water, continuous monitoring should be increasingly implemented in areas near their inflows. Finally, there is uncertainty in both the SQG and Hakanson index methods because the SQGs were developed in North America and the toxic-response factor in the Hakanson method is not very sophisticated. Therefore, further on-site or laboratory toxicological experiments should be carried out to ascertain the actual adverse effects on sediment-dwelling organisms and different fish species [10,50], as well as to determine the impacts on human health from consuming fish from the study area.

## Influence of the Hongfeng Reservoir on the Baihua Reservoir

Reservoir construction generally leads to an increase in residence time and a decrease in suspended solids and turbidity. For an alkaline reservoir on the Yunnan-Guizhou Plateau such as the Hongfeng Reservoir, heavy metals tend to be adsorbed to suspended solids, and sediments then settle on the lake bed, resulting in fewer heavy metals in water. Hence, in heavily polluted areas, reservoirs may serve as a sink for pollutants and a buffer for downstream receiving areas. In this study, the catchment area upstream of the Hongfeng Reservoir occupies 84% of the Baihua Reservoir catchment area, and the heavy metal concentrations at site 14, the outlet of the Hongfeng Reservoir and inlet of the Baihua Reservoir, reflect the buffering effect of the Hongfeng Reservoir according to our two field surveys. The metals (Hg, Cd, Cr (VI) and As) in the surface water at site 14 had lower concentrations than they did at the inflow tributary sites of the Hongfeng Reservoir (results not shown). The sediment concentration of Hg was much lower at site 14 than at most sites at the tributaries and within the Hongfeng Reservoir (Fig. 5). He [48] also found that the Hongfeng Reservoir functioned as a net sink for Hg and that it intercepted a large amount of Hg before it was conveyed to the Baihua Reservoir. The sediment concentrations of Cd and As were lower at site 14 than the maximum concentration at the tributaries of Hongfeng Reservoir, although they were generally higher there than at sites within Hongfeng Reservoir (Fig. 5). Due to the different pollution sources (some indirect rather than through inflows), higher concentrations of Pb, Cr and Cu were found at the reservoir sites than at the tributary sites, and the sediment concentrations of Pb, Cr and Cu were generally lower at site 14 than at sites within the Hongfeng Reservoir (Fig. 5). On the other hand, the outflow of the Hongfeng Reservoir has accounted for an average of 70% of the total inflow of the Baihua Reservoir over the last 6 years (2005–2010, provided by the Administration of Hongfeng, Baihua and Aha Reservoirs), implying that the Hongfeng Reservoir may also serve as an important source for total metals in the Baihua Reservoir. For example, the concentrations of the metals (Hg, Cd, Cr (VI) and As) detected in water samples at site 14 fell in the mid-range of the concentrations in the other tributaries of the Baihua Reservoir (results not shown). The concentrations of all of the heavy metals in sediments at site 14 were generally within the concentration ranges of heavy metals in other tributaries of the Baihua Reservoir, with Hg at the low end, Cr in the medium range, and Cd, Pb, Cu and As at the high end (Fig. 5). However, the sediment concentrations of Cd and As at site 14 were generally higher than at sites within the Hongfeng Reservoir. This pattern was not found in the other heavy metals, resulting in uncertainty when considering whether the relatively high concentrations of Cd and As at site 14 came from the Hongfeng Reservoir or not (Fig. 5). Because Cd was mainly from industrial activities and no emission sources exist near site 14 (as determined by the field investigation), the high concentrations at site 14 might result from the pollution of site 1 (Fig. 3) because our sampling sites within the Hongfeng Reservoir did not cover the area near the outlet. In terms of the spatial distribution and the lithogenic source of As, the high concentrations at site 14 might be affected by other factors, such as soil type and land use of the nearby banks. Therefore, for heavy metal concentrations in sediment, the Hongfeng Reservoir might be an important source of Cd, Cu and Pb, and a moderately important source of Cr, but might not be an important source of Hg and As for the Baihua Reservoir. The results also indicate that the Hongfeng Reservoir is not always the most important source for total metals in the Baihua Reservoir, and other tributaries contribute large quantities of pollutants to the Baihua Reservoir, some of which even exceed the levels in the Hongfeng Reservoir. It should also be noted that our field surveys only indirectly reflect the potential long-term impacts from Hongfeng Reservoir on heavy metals in the sediment of the Baihua Reservoir, and continuous monitoring of inflows and outflows of both reservoirs are needed for the specific contribution of the upstream reservoir to the downstream reservoir in future studies.

Other factors contribute to the adverse effects of the Hongfeng Reservoir on the Baihua Reservoir. The Hongfeng Reservoir is a deep reservoir with thermal stratification from May to November [37], and its release water is mainly from the hypolimnion, which has a lower DO concentration, higher $CO_2$ concentration and lower pH than surface water during those months [21,37]. The water chemistry in the hypolimnion favors the release of heavy metals from the sediments and changes the speciation and toxicity of heavy metals. He et al. [37] found that the low DO and pH in hypolimnion accelerated Hg methylation at Houwu (near site 9) and enhanced the release of methylmercury from sediments at Daba (near the outlet) in the Hongfeng Reservoir in summer. He et al. [37] also concluded that the Hongfeng Reservoir was a net source of methylmercury for the Baihua Reservoir. In addition,

the release of hypolimnetic water has a cooling effect in the summer and a warming effect in winter, which may have a significant influence on the temperature downstream, and thus may indirectly influence the heavy metal distribution. Therefore, the outflow of the Hongfeng Reservoir may pose serious risks to ecosystems in the Baihua Reservoir. Further research is needed to help understand the influence of heavy metals and their chemical forms as they are transported downstream from reservoirs.

## Conclusions

This study of heavy metal (Hg, Cd, Pb, Cr, Cu and As) concentrations in surface water and sediments from two adjacent drinking water reservoirs (the Hongfeng and Baihua Reservoirs) on the Yunnan-Guizhou Plateau, Southwest China, showed that surface water was polluted by Hg, and sediments were polluted by Hg, Cd, Pb, Cr and Cu. In both reservoirs, Cd and As mainly came from industrial activities and lithogenic source through inflows, respectively. The Pb, Cr and Cu in Hongfeng Reservoir may have arisen from a mixture of sources (traffic pollution and residual effect of former industrial influence), and they were present at higher concentrations at the junction of the North and South Lakes. Hg sources in the Hongfeng Reservoir might include the sources that contribute Hg through inflows, which were different for the North (industrial activities) and South Lakes (lithogenic origin), and atmospheric deposition resulting from coal combustion. For the Baihua Reservoir, Hg, Cr and Cu were primarily derived from upstream industrial activities, and the Pb originated from traffic pollution. Additionally, the Hg in Baihua Reservoir might have come from atmospheric deposition (coal combustion). Ecological risk was assessed using the SQGs and the Hakanson potential ecological risk index. There were moderate toxicological risks for sediment-dwelling organisms (with the main risks from Hg and Cr) and moderate to very high potential ecological risks for humans from fish consumption (with the main risk coming from Hg and Cd) in both reservoirs. Overall, the risks were higher in the Baihua Reservoir. Improved treatment of industrial wastewater and mining tailings in upstream watersheds would alleviate the pollution and ecological risk in both reservoirs, especially for tributaries of the North Lake of the Hongfeng Reservoir and the key tributaries of the Baihua Reservoir. Ecological restoration could be considered to counteract the residual effects from previous pollution; however, more research is needed in this area. In terms of heavy metal concentrations, the Hongfeng Reservoir acts as a buffer, but it is still an important source of Cd, Cu and Pb and a moderately important source of Cr for the Baihua Reservoir. The Hongfeng Reservoir also had adverse effects on the Baihua Reservoir and merits further research. These findings provide useful information about sediment quality in adjacent reservoirs on the Yunnan-Guizhou Plateau.

## Author Contributions

Conceived and designed the experiments: GQW CML. Performed the experiments: BBW GQW JW QF. Analyzed the data: BBW JW. Contributed reagents/materials/analysis tools: BBW JW QF. Wrote the paper: BBW GQW.

## References

1. Liu WX, Li XD, Shen ZG, Wang DC, Wai OWH, et al. (2003) Multivariate statistical study of heavy metal enrichment in sediments of the Pearl River Estuary. Environ Pollut 121: 377–388.

2. Chabukdhara M, Nema AK (2012) Assessment of heavy metal contamination in Hindon River sediments: A chemometric and geochemical approach. Chemosphere 87: 945–953.

3. Yi YJ, Yang ZF, Zhang SH (2011) Ecological risk assessment of heavy metals in sediment and human health risk assessment of heavy metals in fishes in the middle and lower reaches of the Yangtze River Basin. Environ Pollut 159: 2575–2585.

4. Uluturhan E, Kucuksezgin F (2007) Heavy metal contaminants in Red Pandora (Pagellus erythrinus) tissues from the Eastern Aegean Sea, Turkey. Water Res 41: 1185–1192.

5. Singh KP, Mohan D, Singh VK, Malik A (2005) Studies on distribution and fractionation of heavy metals in Gomti river sediments-a tributary of the Ganges. J Hydrol 312: 14–27.

6. Larsen B, Jensen A (1989) Evaluation of the sensitivity of sediment monitoring stationary in pollution monitoring. Mar Pollut Bull 20: 556–560.

7. Li XD, Wai OWH, Li YS, Coles BJ, Ramsey MH, et al. (2000) Heavy metal distribution in sediment profiles of the Pearl River estuary, South China. Appl Geochem 15: 567–581.

8. Agarwal A, Singh RD, Mishra SK, Bhunya PK (2005) ANN-based sediment yield river basin models for Vamsadhara (India). Water SA 31: 95–100.

9. Loska R, Wiechula D (2003) Application of principal component analysis for the estimation of source of heavy metal contamination in surface sediments from the Rybnik Reservoir. Chemoshere 51: 723–733.

10. Larrose A, Coynel A, Schafer J, Blanc G, Masse L, et al. (2010) Assessing the current state of the Gironde Estuary by mapping priority contaminant distribution and risk potential in surface sediment. Appl Geochem 25: 1912–1923.

11. Bai JH, Cui BS, Chen B, Zhang KJ, Deng W, et al. (2011) Spatial distribution and ecological risk assessment of heavy metals in surface sediments from a typical plateau lake wetland, China. Ecol Model 222: 301–306.

12. Persaud D, Jaagumagi R, Hayton A (1993) Guidelines for the protection and management of aquatic sediment quality in Ontario. Water Resources Branch. Ontario Ministry of the Environment, Toronto, 27.

13. Smith SL, MacDonald DD, Keenleyside KA, Ingersoll CG, Field J (1996) A preliminary evaluation of sediment quality assessment values for freshwater ecosystems. J Great Lakes Res 22: 624–638.

14. Ingersoll CG, Haverland PS, Brunson EL, Canfield TJ, Dwyer FJ, et al. (1996) Calculation and evaluation of sediment effect concentrations for the amphipod Hyalella azteca and the midge Chironomus riparius. J Great Lakes Res 22: 602–623.

15. Macdonald DD, Ingersoll CG, Berger TA (2000). Development and evaluation of consensus-based sediment quality guidelines for freshwater ecosystems. Arch Environ Contam Toxical 39: 20–31.

16. Word JQ, Albrecht BB, Anghera ML, Baudo R, Bay MS, et al. (2002). Predictive ability of sediment quality guidelines. In: Wenning RJ, Batley GE, Ingersoll CG, Moore DW, editors. Use of sediment quality guidelines and related tools for the assessment of contaminated sediments. Pensacola (FL): SETAC. p 121–162.

17. Bhavsar SP, Gewurtz SB, Helm PA, Labencki TL, Marvin CH, et al. (2010) Estimating sediment quality thresholds to prevent restrictions on fish consumption: Application to polychlorinated biphenyls and dioxins–furans in the Canadian Great Lakes. Integr Environ Assess Manag 6: 641–652.

18. Hakanson L (1980) An ecological risk index for aquatic pollution control: A sedimentological approach. Water Res 14: 975–1001.

19. Liu Y, Guo HC, Yu YJ, Huang K, Wang Z (2007) Sediment chemistry and the variation of three altiplano lakes to recent anthropogenic impacts in southwestern China. Water SA 33: 305–310.

20. Feng XB, Jiang HM, Qiu GL, Yan HY, Li GH, et al. (2009a) Mercury mass balance study in Wujiangdu and Dongfeng Reservoirs, Guizhou, China. Environ Pollut 157: 2594–2603.

21. Wang FS, Wang BL, Liu CQ, Wang YC, Guan J, et al. (2011a) Carbon dioxide emission from surface water in cascade reservoirs-river system on the Maotiao River, southwest of China. Atmos Environ 45: 3827–3834.

22. Wang FS, Liu CQ, Wang BL, Liu XL, Li GR, et al. (2011b) Disrupting the riverine DIC cycling by series hydropower exploitation in Karstic area. Appl Geochem 26: S375–S378.

23. Feng XB, Jiang HM, Qiu GL, Yan HY, Li GH, et al. (2009b) Geochemical processes of mercury in Wujiangdu and Dongfeng reservoirs, Guizhou, China. Environ Pollut 157: 2970–2984.

24. Han GL, Liu CQ (2004) Water geochemistry controlled by carbonate dissolution: a study of the river waters draining karst-dominated terrain, Guizhou Province, China. Chem Geo 204: 1–21.

25. Wang B, Liu CQ, Wu Y (2005) Effect of heavy metals on the activity of external carbonic anhydrase of microalga chlamydomonas reinhardtii and microalgae from karst lakes. Bull Environ Contam Toxicol 74: 227–233.

26. Lang YC, Liu CQ, Zhao ZQ, Li SL, Han GL (2006) Geochemistry of surface and ground water in Guiyang, China: Water/rock interaction and pollution in a karst hydrological system. Appl Geochem 21: 887–903.

27. He SL, Li CJ, Pan ZP, Luo MX, Meng W, et al. (2012) Geochemistry and environmental quality assessment of Hongfeng Lake sediments, Guiyang. Geophysical and Geochemical Exploration 36: 273–297 (in Chinese).

28. Wang Y, Luo TMZ (2001) Geostatistical and geochemical analysis of surface water leakage into groundwater on a regional scale: a case study in the Liulin karst system, northwestern China. J Hydrol 246: 223–234.

29. Sophocleous M (2002) Interactions between groundwater and surface water: the state of the science. Hydrogeol J 10: 52–67.

30. Feng XB, Foucher D, Hintelmann H, Yan HY, He TR, et al. (2010) Tracing mercury contamination sources in sediments using mercury isotope compositions. Environ Sci Technol 44: 3363–3368.

31. Yan HY, Feng XB, Shang LH, Qiu GL, Dai QJ, et al. (2008) The variations of mercury in sediment profiles from a historically mercury-contaminated reservoir, Guizhou province, China. Sci Total Environ 407: 497–506.

32. Kaiser HF (1960) The application of electronic computers to factor analysis. Educ Psychol Measure 20: 141–151.

33. Liu CW, Lin KH, Kuo YM (2003) Application of factor analysis in the assessment of groundwater quality in a Blackfoot disease area in Taiwan. Sci Total Environ 313: 77–89.

34. Macdonald DD, Carr RS, Calder FD, Long ER, Ingersoll CG (1996) Development and evaluation of sediment quality guidelines for Florida coastal waters. Ecotoxicology 5: 253–278.

35. Swartz RC (1999) Consensus sediment quality guidelines for PAH mixtures. Environ Toxicol Chem 18: 780–787.

36. Long ER, Ingersoll CG, Macdonald DD (2006) Calculation and uses of mean sediment quality guideline quotients: a critical review. Environ Sci Technol 40: 1726–1736.

37. He TR, Feng XB, Guo YN, Qiu GL, Li ZG, et al. (2008) The impact of eutrophication on the biogeochemical cycling of mercury species in a reservoir: A case study from Hongfeng Reservoir, Guizhou, China. Environ Pollut 154: 56–67.

38. Bai ZG, Wan GJ, Liu TS, Huang RG (2002) A comparative study on accumulation characteristics of $^7$Be and $^{137}$Cs in sediments of Lake Erhai and Lake Hongfeng, China, Geochimica 31: 113–118 (in Chinese).

39. Huang XF (2008) Studies on characteristics of pollution in sediments from Baihua Lake. Guiyang (in Chinese).

40. Huang XF, Qin FX, Hu JW, Li CX (2008) Characteristic and ecological risk of heavy metal polltuion in sediments from Hongfeng Lake. Res Environ Sci 21: 18–23 (in Chinese).

41. Zeng Y, Zhang W, Chen JA, Zhu ZJ (2010) Analysis of heavy metal pollution in the sediment of the inflow-lake rivers of the Hongfeng Lake, Earth Environ 38: 470–475 (in Chinese).

42. Liu F, Hu JW, Wu D, Qin FX, Li CX, et al. (2011) Speciation characteristics and risk assessment of heavy metals in sediments from Hongfeng Lake, Guizhou Province. Environmental Chemistry 30: 440–446 (in Chinese).

43. Tian LF, Hu JW, Luo GL, Ma JJ, Huang XF, et al. (2012) Ecological risk and stability of heavy metals in sediments from Lake Baihua in Guizhou Province. Acta Scientiae Circumstantiae 32: 885–894 (in Chinese).

44. NEPA: National Environmental Protection Agency (Presently known as MEP; Ministry of Environmental Protection) (1994) The Atlas of Soil Environmental Background Value in the People's Republic of China. China Environmental Science Press.

45. Hakanson L, Jasson M (1983) Principles of Lake Sedimentology. Springer Verlag, Berlin.

46. Bai JH, Cui BS, Wang QG, Gao HF, Ding QY (2009) Assessment of heavy metal contamination of roadside soils in Southwest China. Stoch Environ Res Risk Assess 23: 341–347.

47. Zhu ZM, Li ZG, Bi XY, Han ZX, Yu GH (2013) Response of magnetic properties to heavy metal pollution in dust from three industrial cities in China. J Hazard Mater 246–247: 189–198.

48. He TR (2007) Biogeochemical cycling of mercury in Hongfeng Reservior, Guizhou, China. Guiyang (in Chineses).

49. SEPA: State Environmental Protection Administration (Presently known as MEP; Ministry of Environmental Protection) (2000) Report on the State of the Environment in China. Available: http://english.mep.gov.cn/SOE/soechina2000/english/atmospheric/atmospheric_e.htm.

50. Farkas A, Claudio E, Vigano L (2007) Assessment of the environmental significance of heavy metal pollution in surficial sediments of the River Po. Chemosphere 68: 761–768.

# 22

# Development of Composite Indices to Measure the Adoption of Pro-Environmental Behaviours across Canadian Provinces

**Magalie Canuel[1]\*, Belkacem Abdous[2,3], Diane Bélanger[2,4], Pierre Gosselin[1,2,4]**

**1** Institut national de santé publique du Québec (INSPQ), Québec City, Canada, **2** Centre de recherche du Centre hospitalier universitaire de Québec, Québec City, Canada, **3** Département de médecine sociale et préventive de l'Université Laval, Québec City, Canada, **4** Institut national de la recherche scientifique, Centre Eau Terre Environnement, Québec City, Canada

## Abstract

*Objective:* The adoption of pro-environmental behaviours reduces anthropogenic environmental impacts and subsequent human health effects. This study developed composite indices measuring adoption of pro-environmental behaviours at the household level in Canada.

*Methods:* The 2007 Households and the Environment Survey conducted by Statistics Canada collected data on Canadian environmental behaviours at households' level. A subset of 55 retained questions from this survey was analyzed by Multiple Correspondence Analysis (MCA) to develop the index. Weights attributed by MCA were used to compute scores for each Canadian province as well as for socio-demographic strata. Scores were classified into four categories reflecting different levels of adoption of pro-environmental behaviours.

*Results:* Two indices were finally created: one based on 23 questions related to behaviours done inside the dwelling and a second based on 16 questions measuring behaviours done outside of the dwelling. British Columbia, Quebec, Prince-Edward-Island and Nova-Scotia appeared in one of the two top categories of adoption of pro-environmental behaviours for both indices. Alberta, Saskatchewan, Manitoba and Newfoundland-and-Labrador were classified in one of the two last categories of pro-environmental behaviours adoption for both indices. Households with a higher income, educational attainment, or greater number of persons adopted more indoor pro-environmental behaviours, while on the outdoor index, they adopted fewer such behaviours. Households with low-income fared better on the adoption of outdoors pro-environmental behaviours.

*Conclusion:* MCA was successfully applied in creating Indoor and Outdoor composite Indices of pro-environmental behaviours. The Indices cover a good range of environmental themes and the analysis could be applied to similar surveys worldwide (as baseline weights) enabling temporal trend comparison for recurring themes. Much more than voluntary measures, the study shows that existing regulations, dwelling type, households composition and income as well as climate are the major factors determining pro-environmental behaviours.

**Editor:** Judi Hewitt, University of Waikato (National Institute of Water and Atmospheric Research), New Zealand

**Funding:** This study was funded by the Green Fund for Action 21 of the 2006-2012 Climate Change Action Plan of the Quebec government. The funders had no role in study design, data collection and analysis, decision to publish, or preparation of the manuscript.

**Competing Interests:** The authors have declared that no competing interests exist.

\* Email: magalie.canuel@inspq.qc.ca

## Introduction

A significant source of pollution to our natural environment comes from domestic activities and behaviours. For example household-generated waste in Canada accounts for around a third of total waste and household energy use and municipal water consumption for 17% and 57%, respectively [1-3]. Also, 46% of greenhouse gas emissions (GHG), which contribute to climate change, come from direct and indirect household emissions [4]. The impacts of such household pollution can be important.

Municipal waste can impact the environment in various ways including soil and water contamination from leachate in landfills disposal and the production of greenhouse gas emissions (GHG) and air pollution, either from landfills or the incineration process. When solid waste are recycled or composted instead of being landfilled or incinerated, the demand for energy and new-resources can be reduced significantly [3].

The production of energy can impact the environment in various ways, depending on the technology. In Canada, energy production and consumption accounts for around 80% of all GHG emission [5]. A household can reduce its emission of GHG by reducing electric power use. For instance high energy efficiency electronic devices or cleaner energy sources will generate less pollution and GHG.

Water shortages are happening worldwide and one way to limit their occurrence is through water conservation behaviours. In most homes, more than 60% of water use comes from toilet flushing, showers and baths, making water-saving devices like low-flow shower head an efficient way of reducing water consumption. In summer, water use can increase by 50% for yard activities such as watering the lawn. There are behaviours that households can implement to decrease their water consumption in summer time like using sprinklers with a timer or adopting the use of a rain barrel [6].

It thus becomes clear that addressing sustainability concerns has to take into account not only industry or agriculture, but also household behaviours, their impacts on ecosystems and ultimately on human health. Monitoring trends of household behaviours can inform policy and research agendas on the development of incentives or other mechanisms such as information campaigns to reduce domestic pollution and facilitate adaptative measures to minimize related health risks. The adoption of several pro-environmental behaviours, i.e. actions that contribute to the preservation of the environment, should be encouraged to significantly reduce the anthropic impact on the environment.

In Canada, the Households and the Environment Survey (HES) was designed to measure household behaviours with respect to the environment. The HES is a periodic survey conducted by Statistics Canada, the federal government statistical agency, and administered across Canadian provinces. The survey covers 12 broad themes including energy use and heating, water use, transportation decisions, motor vehicle use, recycling and composting (Figure 1) [7]. While this survey provides various estimations of up to 83 Canadian practices (Figure 1) as well as some information on their socio-demographic characteristics, survey reports are limited to analyses of simple cross-tabulation frequencies for some of the 83 separate behaviours [7-13].

It is difficult to follow up on such a wide array of relevant behaviours and their trends over time, unless they are summarized in some way. A composite index is a tool which can be useful to that purpose as it incorporates several aspects of an issue and allow for monitoring across several themes simultaneously, thus facilitating the measurement of trends [14]. While other environmental indices exist, such as the environmental sustainability index [15], to our knowledge no index currently exists to reflect trends of pro-environmental behaviours at the household level in Canada.

This study thus sets out to develop a composite index that summarizes pro-environmental behaviours at the household level across Canadian provinces based on the HES (2007) given the periodicity and geographical coverage of the survey. Pro-environmental behaviours are defined as actions that contribute to the preservation of the environment and can have a positive impact on the health of the population. This study will serve as baseline of the trend of the composite index over time, given the periodicity and geographical coverage of the survey.

## Materials and Methods

### Ethics statement

This research did not require the approval of an ethics review board as we used an existing and anonymized database made available to universities by Statistics Canada. Statistics Canada obtained consent previous to survey administration. No new data was collected for this study.

### Survey

The Households and the Environment Survey (HES) is conducted by Statistics Canada. It was designed to address the needs of the Canadian Environmental Sustainability Indicators project. The project reports on air quality, water quality and greenhouse gas emissions in Canada using indicators to identify areas of importance to Canadians and monitor progress [16].

The survey aimed Canadian households with at least one person aged 18 year or older. The HES covers all 10 of the provinces and excludes the 3 northern territories, Indian reserves and members of the Canadian Armed Forces. The survey was first conducted in 1991 and since 2005 has been carried out biennially. In the present study, the 2007 HES database was used in its Public Use Microdata Files format (PUMF) [16]. As a sub-sample of the dwellings that were part of the Canadian community health survey (CCHS), the sampling allocation for the HES followed that of the CCHS closely. The CCHS used a multistage stratified cluster design in which the dwelling is the final sampling unit. Three sampling frames were used to select the sample of households: 50% of the sample came from an area frame, 49% came from a list frame of telephone numbers and 1% came from a Random Digit Dialing sampling frame [16].

From the 40 584 households selected in the 2007 CCHS, a sub-sample of 29 957 households were selected for the HES. Of those, 21 690 households responded to the survey resulting in an overall response rate of 72%. The survey is representative of 12 932 350 households, corresponding to 97% of all Canadian households [16]. The questionnaire was administered to the 21 690 households by telephone interview spread over a 6-month period, from October 2007 to February 2008.

### Questionnaire

The person with the best knowledge of environmental household practices was asked to respond on behalf of the household. The main questionnaire covered 12 themes and included 121 questions (figure 1) [16]. Among the questions, 83 measured behaviours and 7 measured socio-demographic characteristics. The other 31 questions covered knowledge, reasons for not adopting the behaviour, or served to specify some characteristics (e.g. of a good) or to filter for the next question.

### Database

The PUMF was used for the analysis and unlike the master file, applies privacy measures to protect personal information [16]. In the PUMF, data were mostly coded as categorical variables. Three different labels (don't know, not stated, and refusal) were used to classify households who did not participate despite eligibility or to protect the anonymity of the household. A 'valid skip' label was used when the provision of a response was not appropriate. For example, a household who answered 'no' to the question for car ownership was allocated a 'valid skip' label for subsequent questions on the characteristics of the car.

### Sampling weights

Sampling weights were applied to ensure that any derived composite index is representative of the study population. They were used when proportions and averages were estimated and to weight the relative frequencies of the Burt matrix in the MCA (see Statistical analysis below).

### Variables selection

This study focuses on everyday pro-environmental behaviours, defined as actions that contribute to the preservation of the

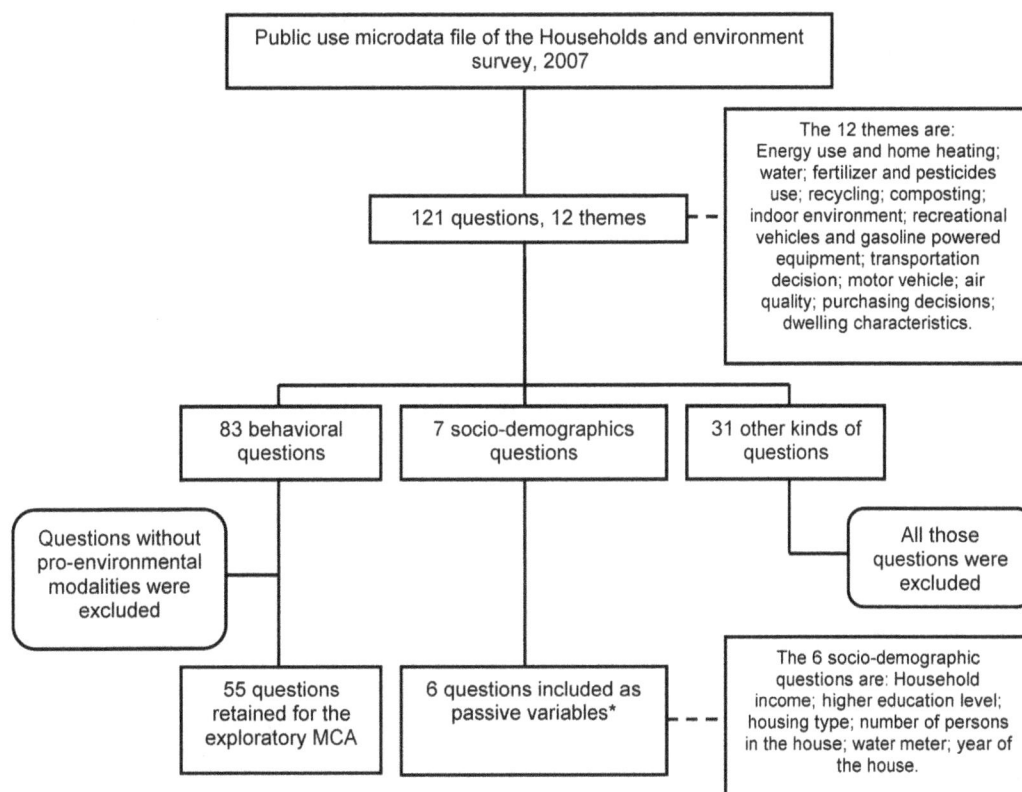

**Figure 1. Number and type of questions selected to develop the composite index. Legend:** *The composite index was also created for the 7th socio-demographic variable, the census metropolitan area (n = 33), but is not presented in this article.

environment and can have a positive impact on the health of the population. For example, air pollutants can be reduced when households adopt behaviours that decrease their energy consumption such as the use of energy-efficient appliances or when they use more sustainable transport options such as public or active transport.

Based on the above definition, a panel of four environmental health experts applied progressive development consensus after iterations, based on a nominal group technique [17] to evaluate HES variables for exclusion. These were either: variables not measuring a behaviour or questions with no clearly pro-environmental response option. Socio-demographic variables were kept as passive variables with zero mass and no influence on the analysis. They support and complement the interpretation of the map representation of the active variables [18].

## Statistical analysis

Given that the data was mostly categorical, the indices in this paper were developed by multiple correspondence analysis (MCA) [18]. Several authors have used Multiple Correspondence Analysis (MCA) as a weighting method for the construction of a composite index [19–23]. MCA is a data reduction procedure for categorical variables (nominal or ordinal) as much as Principal Components Analysis is for quantitative variables [18]. It enables the exploration of associations within a set of variables by transforming the whole data set into dummy variables to form an indicator matrix or upon construction of a matrix from all two-way cross-tabulations among the variables (Burt matrix). This transformed data is treated as a cloud in a space equipped with the classical Chi-square distance. This distance is used in the assessment of

homogeneity and variance (inertia) of rows or columns of the indicator or Burt matrix. The most crucial step of MCA is its use of singular value decomposition and weighted least squares techniques to find low-dimensional best fitting subspaces with minimal inertia and information loss [18].

MCA was conducted using the 'ca' package of the R statistical software [24]. First, the HES database was converted to a Burt matrix taking into consideration the sampling weights. A Burt matrix is a square symmetric categories-by-categories matrix formed from all two-way contingency tables of pairs of variables [18].

Then, an exploratory MCA was performed to project data onto maps where potential outliers were identified and excluded from subsequent analyses. MCA was then applied again to determine the most relevant factorial axes that would serve to build the composite index. There are no universal rules for the determination of the number of dimensions to retain in MCA. However, since the first factorial axis captures the most important part of the total inertia, it plays a central role in the computation of a composite index.

As recommended by Asselin [23], we sought questions having the property of First Axis Ordering Consistency (FAOC). To this end, we projected all the questions on the first axis and tried to identify those having an ordinal structure consistent with respect to this axis, i.e. all questions with pro-environmental responses improving from left to right (or conversely).

The computation of the index score was performed as follows: first, the score of any household was obtained by taking the average of its category-weights generated by the MCA. Then for each province we took the average over all household scores as the value of its composite index. The sampling weight was used in this final step. Coordinates were missing for excluded responses.

The 10 average provincial scores were grouped into categories reflecting different levels of adoption of pro-environmental behaviours. First we applied a cluster analysis and then we used a dendrogram plot using SAS version 9.2 (SAS Institute, Cary, NC) to determine such groups. The categories limits generated for the provincial index were used as reference categories for indices on other socio-demographic variables.

Finally, others indices based on various socio-demographic variables were constructed (Figure 1). Household scores were calculated by taking the average of its category-weights generated by the MCA. Then the index score of the socio-demographic category (e.g. household with annual income less than $40,000) is set as the average of the corresponding household scores.

## Results

### Multiple Correspondence Analysis

Of the 121 questions in the survey, 55 were kept by the Expert Panel for use in the MCA. These represented 285 response possibilities. On the MCA map projection there was a clear opposition between the missing data (don't know, refusal, not stated) located far from the map center and the other responses which gathered close to the center (Figure S1). Excluding the missing data rebalanced the model (179 remaining responses) (Figure 2). However, since pro-environmental behaviours were spread over both sides of the first axis, we failed to find any meaning to this first dimension.

We then screened the projected responses to identify questions following an ordinal structure, (i.e all pro-environmental responses of a question have negative coordinates on the first factorial axis (or conversely)). Twenty-three such questions with pro-environmental responses deteriorating from left to right on the first axis (group A), and 16 questions with opposite ordinal structure (group B) were identified. The remaining 16 (of 55) questions were excluded from the analysis because their responses were not sufficiently discriminating (i.e. the pro- and anti-environmental responses were on the same side of the axis or they were grouped close together on the map). As well the majority of these questions (10/16) had at least two responses with a contribution of zero to the first axis (Table S1).

These exploration steps led us to consider two separate composite indices. Group A included 96 responses but after excluding missing data, 52 responses were used in the MCA. The majority of excluded responses had frequencies lower than 2.0% and two responses had frequencies of 4.6% and 4.7%. After exclusion of missing data, some responses still looked like extreme values on the map (Figure 3). They were kept in the analysis as they are 2 of the 3 responses for all questions concerning recycling. Excluding these responses would have resulted in the exclusion of all recycling questions. Responses used in this analysis had a frequency of 7.5% or higher, except for two responses with frequencies of 2.5% and 3.5% (responses on recycling).

For group A, the first dimension explained 32.6% of the inertia while the second explained 16.1% (Table 1). Given that the first factorial axis plays a central role in the construction of this composite index, only the first dimension was selected to construct the index. This group respects the FAOC as pro-environmental responses are located on the left of the first axis as opposed to others responses deteriorating to the right (Figure 3). Also, we noted that the retained questions were associated with five themes of the survey: energy use and home heating, water, recycling, composting and, purchasing decisions. All 23 questions assessed

behaviours practiced inside the dwelling and thus the first axis measures these behaviours. Twelve of 15 responses contributing the most to the first factorial axis concerned recycling (Table S2).

The second group of 16 questions (group B) consisted of 86 responses, 41 of which were missing values. The 45 remaining responses used for the MCA had frequencies of 7.0% or higher while excluded responses had frequencies lower than 3.5%. For group B, the first dimension explained 62.1% of the inertia while the second explained only 12.4% (Table 2). Again, the first dimension was selected for the construction of the index and pro-environmental responses were located on the right of the first axis with other responses deteriorating from right to left (Figure 4). The 16 questions cover five themes of the survey: water, fertilizer and pesticide use, recreational vehicles and gasoline powered equipment, transport decisions and air quality, all behaviours being practiced outdoors. Of note, 9 of the 15 responses contributing the most to the first factorial axis concern households with no lawn or garden (i.e. the application of fertilizers or pesticides, yard waste and watering of the lawn or the garden) (Table S3).

Because two distinct behavioural categories resulted from the MCA, two composite indices were created instead of one. The first index (group A) is named the 'Indoor Index' and the second one (group B) the 'Outdoor Index'. Questions included for each index are presented in supporting information, Table S4 and Table S5.

### Composite indices by province

The map representations of the final coordinates generated by the MCA are shown in Figure 3 and Figure 4. Coordinates and other results of the MCA are available in supporting information, Table S2 and Table S3. The coordinates of the first dimension were used to construct each of the two composite indices. Coordinates are missing for responses that have been excluded. Only 0.9% and 0.4% of coordinates are missing for the indoor and outdoor indices, respectively.

For the Indoor Index, the households belonging to a province with negative coordinates tend to adopt more pro-environmental behaviours than those of a province with positive coordinates. In contrast, for the Outdoor Index provinces with positive coordinates adopt more outdoor pro-environmental behaviours than those with negative coordinates.

The cluster analysis and dendrogram plot resulted in the classification of each province into one of four categories reflecting different levels of adoption of pro-environmental behaviours: 1) adopting the most; 2) adopting slightly fewer; 3) adopting much fewer and; 4) adopting the fewest. The provincial coordinates and the categories generated from the cluster analysis are shown in Table 3 and Table 4. Maps of the Canadian provinces with their categories of pro-environmental behaviours are shown in Figure 5 and Figure 6.

None of the 10 provinces were classified in both indices as adopting the most pro-environmental behaviours. For the Indoor Index, Ontario (ON), Prince Edward Island (PEI) and Nova Scotia (NS) rated in the top category, British Columbia (BC) and Québec (QC) in the next, the three Prairie provinces and New Brunswick (NB) in the third and Newfoundland and Labrador (NL) in "adopting the fewest" category (Figure 5). For the Outdoor Index, QC scored in the top category with BC, NS, NB and PEI following in second, and Manitoba (MN), ON and NL in third, followed by Alberta (AB) and Saskatchewan (SK) in the bottom category (Figure 6).

Four provinces (BC, QC, NS and PEI) were classified in the top two categories for both indices while four provinces (AB, SK, MN and NL) were classified for both indices, in the two lower categories.

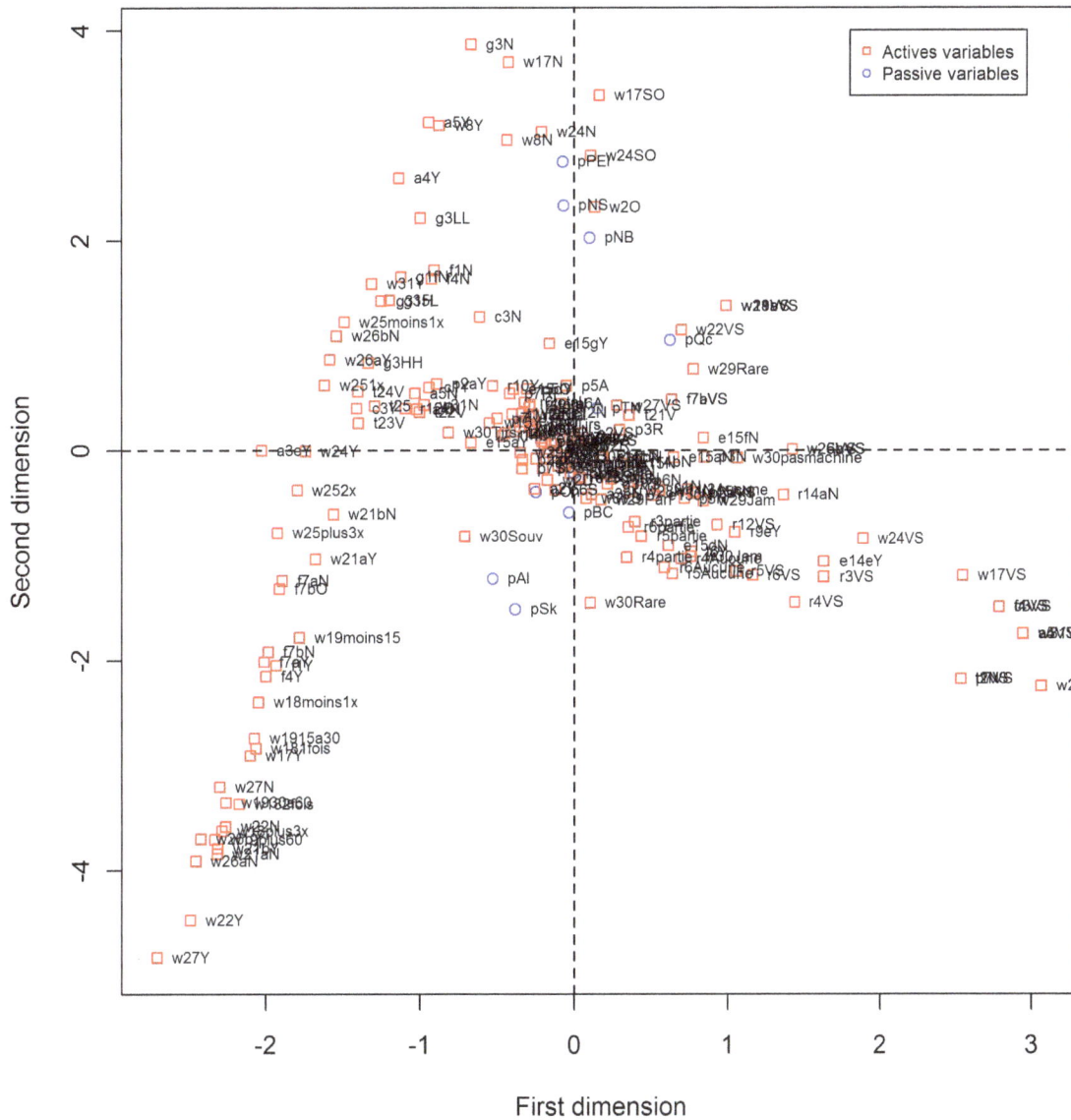

**Figure 2. Map representation of the MCA results on the 55 questions without extreme responses.**

## Composite indices by socio-demographic variables

The coordinates and the classification for the six comparison variables are shown in Table 5. For household income, educational attainment and number of persons in the household, there were oppositions in the classification of the responses. Households with a higher income, or higher educational attainment, or greater number of persons adopted more indoor pro-environmental behaviours, while those with a lower household income, educational attainment, or number of people, adopted more outdoor such behaviours. As well, households with water meters tended to adopt more indoor pro-environmental behaviours than those without, but for outdoors behaviours, the opposite applied – not having a water meter was associated with better adoption of pro-environmental behaviours. And finally, the dwelling's year of construction did not influence the adoption of pro-environmental behaviours as there was no trend on either index (Table 5).

## Discussion

This study sought to develop a composite index which measures the overall adoption of pro-environmental behaviours among Canadian households. MCA, our main analytical technique, was used to aggregate survey data and to provide weights to the responses in the construction of the index. Our approach is similar to other studies in different fields [19–23]. This was followed by a cluster analysis to classify the provinces, as well as an exploration of relationships with socio-demographic factors.

The MCA generated two indices based on 39 of the 55 behavioural questions, an Indoor Index and an Outdoor Index, each reflecting environmental behaviours for 5 of the 12 survey themes. Retaining both indices allowed for better representation of the survey; together they cover 9 themes out of 12 (water use is in both) whereas one single index would have covered only 5, excluding important environmental themes such as fertilizer and pesticide use. As well, because the provincial classifications were different for each index and varied as well in the classification by

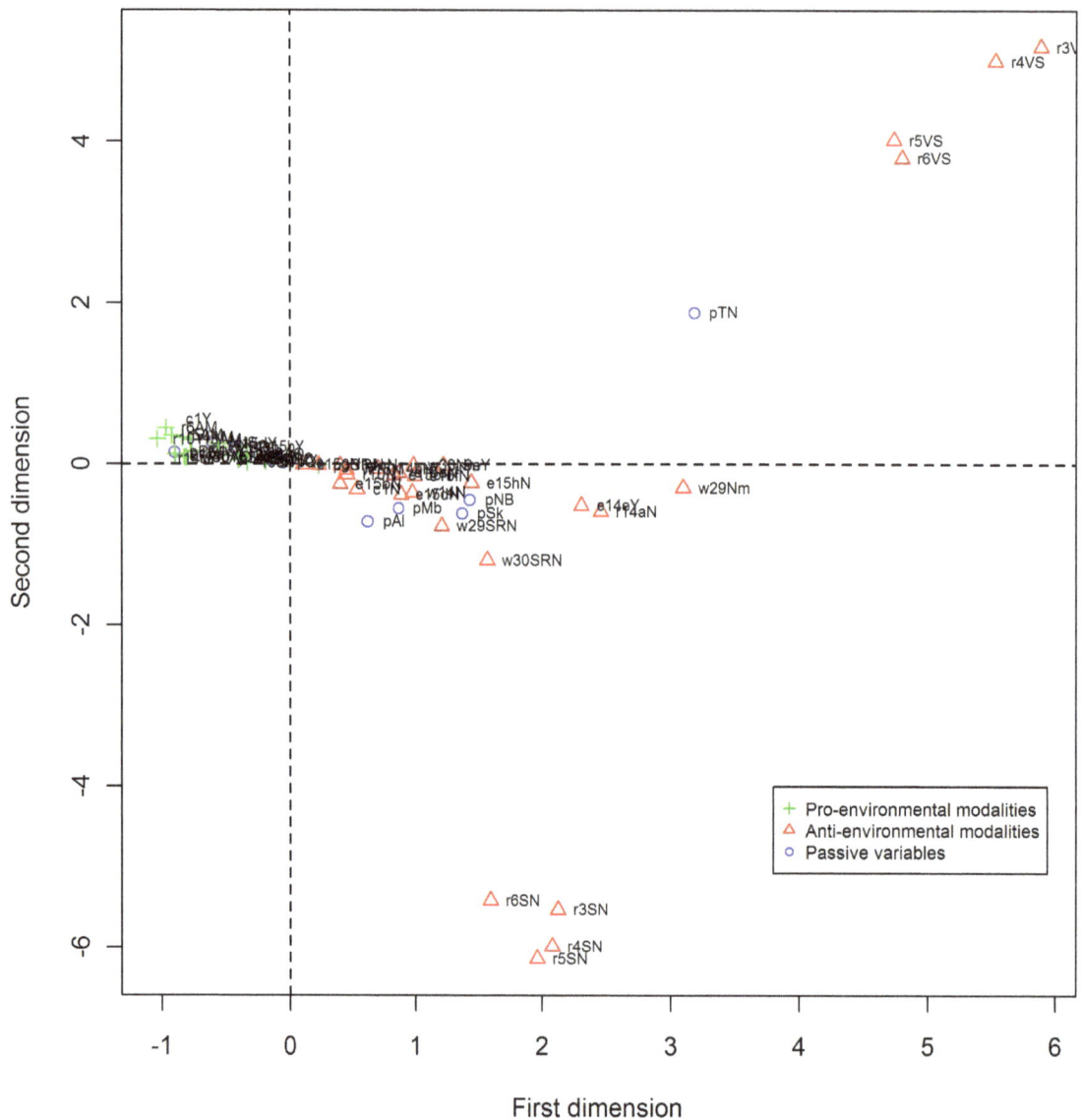

**Figure 3. Map representation of the MCA results on the 23 questions of the group A (Indoor).**

socio-demographics factors for each index (e.g., household income) it was deemed justifiable to keep both indices.

Most (19/23) questions included in the Indoor Index were asked to all households with the exception of questions on recycling where only those households with access to a program were asked to respond. For the Outdoor Index, most questions (11/16) concerned watering of the lawn or the garden, and the use of fertilizers or pesticides. These (11) questions were answered only by households having a yard. However, even if households living in an apartment did not have to answer these questions, they were still recorded in the Index as households adopting pro-environmental behaviours (i.e., most valid skips were classified as pro-environmental responses and some as anti-environmental ones).

### The Indoor Index

One likely explanation for PEI and NS being classified in the top category for the Indoor Index is that nearly 100% of their households recycle and the proportion that compost is substantially above the Canadian average, as reported by Statistics

Canada. In these two provinces, households are obligated by law to recycle and compost [25]. Moreover, questions regarding recycling contributed the most to the Indoor Index.

Recycling and composting are also common in ON but its good ranking is also related to the proportion of households that adopt water conservation behaviours (i.e. use water-efficient shower heads and toilets, run dishwasher and washing machine only when full) [7]. Provinces have been slowly adopting a provincial plumbing code requiring that new buildings use water-saving fixtures, with the exception of NL [26;27]. ON however was the first to adopt such a code in 1996 [26;28], and saw an increase in new residential construction from 1996 to 2002 [29], likely contributing to the higher proportion of households practicing water conservation behaviours [28]. This is an example where building codes may be effective in beneficially influencing the passive uptake of pro-environmental practices.

QC's good classification in the Indoor Index is in part due to its proportion of households adopting recycling behaviours being higher than the Canadian average. There were four questions on

**Table 1.** Explained inertia by each dimension for group A: Indoor Index, 2007.

| Dimension | Inertia | Inertia (%) | cumulative Inertia (%) | scree plot |
|---|---|---|---|---|
| 1 | 0,0270 | 32,6 | 32,6 | ************************* |
| 2 | 0,0133 | 16,1 | 48,8 | ************ |
| 3 | 0,0078 | 9,4 | 58,2 | ******* |
| 4 | 0,0032 | 3,8 | 62,0 | *** |
| 5 | 0,0030 | 3,6 | 65,6 | *** |
| 6 | 0,0026 | 3,2 | 68,8 | ** |
| 7 | 0,0024 | 2,9 | 71,7 | ** |
| 8 | 0,0021 | 2,5 | 74,2 | ** |
| 9 | 0,0019 | 2,3 | 76,5 | ** |
| 10 | 0,0018 | 2,1 | 78,7 | ** |
| 11 | 0,0017 | 2,0 | 80,7 | ** |
| 12 | 0,0016 | 1,9 | 82,6 | * |
| 13 | 0,0015 | 1,8 | 84,4 | * |
| 14 | 0,0014 | 1,7 | 86,1 | * |
| 15 | 0,0014 | 1,7 | 87,7 | * |
| 16 | 0,0013 | 1,6 | 89,3 | * |
| 17 | 0,0012 | 1,5 | 90,8 | * |
| 18 | 0,0012 | 1,4 | 92,2 | * |
| 19 | 0,0011 | 1,3 | 93,5 | * |
| 20 | 0,0011 | 1,3 | 94,8 | * |
| 21 | 0,0010 | 1,2 | 96,1 | * |
| 52 | 0 | 0,0 | 100,0 | |

recycling which contributed significantly to the first dimension, thus contributing to QC's classification. Despite QC having the lowest proportion of households that compost [9] or participate in alternative recycling activities such as donations of furniture and clothing, QC's classification was only slightly affected as these behaviours had only moderate or low contributions to the Index.

In AB, MN, SK and NB, the proportion of households that adopted indoor pro-environmental behaviours is below the Canadian average (data not shown), explaining their lower classification in the Indoor Index. NL had only a few variables above the Canadian average and had most often the lowest proportion of all provinces. For example, the proportion is below the average for all four questions on water conservation and for all questions on recycling. In this province, there is no provincial plumbing code requiring the use of water-saving fixtures in new buildings [26;27]. Also, the proportion of households with access to a recycling program is only 71% [25].

## The Outdoor Index

Results for the Outdoor Index show a pattern with respect to Coastal proximity, with coastal provinces, with the exception of NL, rating in the two higher categories, and the continental provinces in the two lower categories with the two lowest rated provinces situated in the Prairies. The climate of the Prairies grasslands is characterized by hot summers combined with low precipitation and periodic drought. The climatic region of the Maritimes however is the one with the greatest annual precipitation [30-32], a pattern which is likely reflected in the frequency of watering lawn and or garden. Although watering of the lawn or garden is around the Canadian average in NL, its inhabitants own more recreational vehicles, use more gas and burn more yard waste on the property (data not shown) which may explain its lower classification than the other coastal provinces.

Also, there was an important difference in the proportion of households that used fertilizers and pesticides and QC had, by far, the lowest proportion. QC was the first province to adopt a provincial law in 2006 prohibiting the sale of pesticides for cosmetic purposes [7;33]. The Prairies on the other hand had the highest proportions of households that used pesticides or fertilizers in 2007 [7;10]. Subsequently, other jurisdictions have adopted similar laws begging the question of whether their classifications in the Outdoor Index will change over time.

It should also be noted that QC and BC have the highest proportion of households living in an apartment [10]. Given that most households living in an apartment do not have a backyard, they do not water neither lawn nor garden, nor do they use pesticides outdoors. Hence, they passively adopt pro-environmental behaviours and are considered as such by the MCA. In fact, these responses, recorded as 'valid skip', had the highest contribution to the Index, likely contributing to the higher classification for BC and QC on the Outdoor Index. Such passive behaviours or external factors were not excluded from the Index as they significantly contribute to the preservation of environmental resources.

## Indices for socio-demographic variables

For most socio-demographic variables, there were oppositions in the classification of the modalities, which means that it is not the

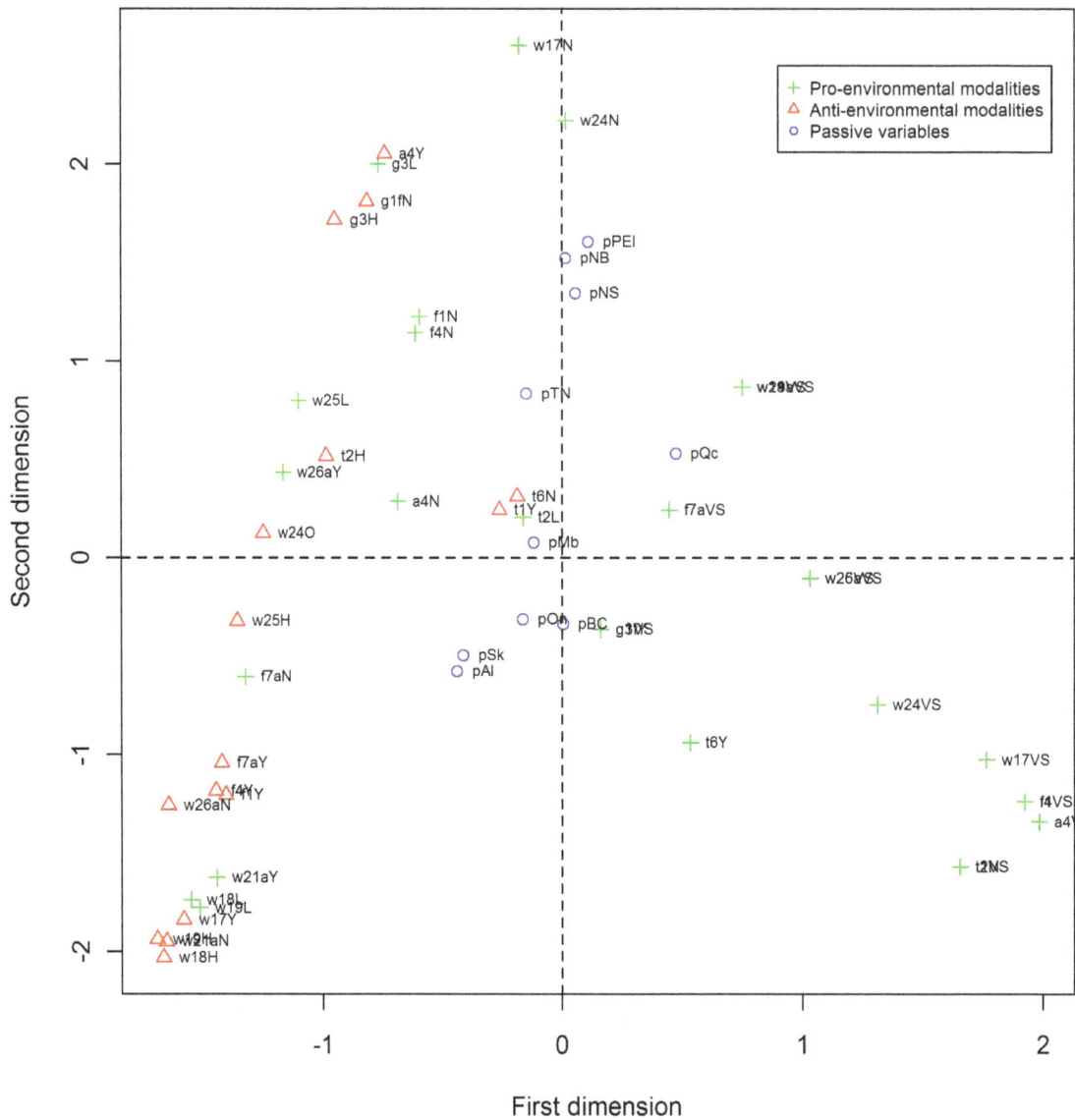

**Figure 4. Map representation of the MCA results on the 16 questions of the group B (Outdoor).**

**Table 2.** Explained inertia by each dimension for group B: Outdoor Index, 2007.

| Dimension | Inertia | Inertia (%) | Cumulative inertia (%) | scree plot |
|---|---|---|---|---|
| 1 | 0,2173 | 62,1 | 62,1 | ************************** |
| 2 | 0,0434 | 12,4 | 74,5 | ***** |
| 3 | 0,0162 | 4,6 | 79,1 | ** |
| 4 | 0,0129 | 3,7 | 82,8 | * |
| 5 | 0,0111 | 3,2 | 86,0 | * |
| 6 | 0,0094 | 2,7 | 88,7 | * |
| 7 | 0,0066 | 1,9 | 90,6 | * |
| 8 | 0,0061 | 1,7 | 92,3 | * |
| 9 | 0,0041 | 1,2 | 93,5 | |
| 45 | 0 | 0,0 | 100,0 | |

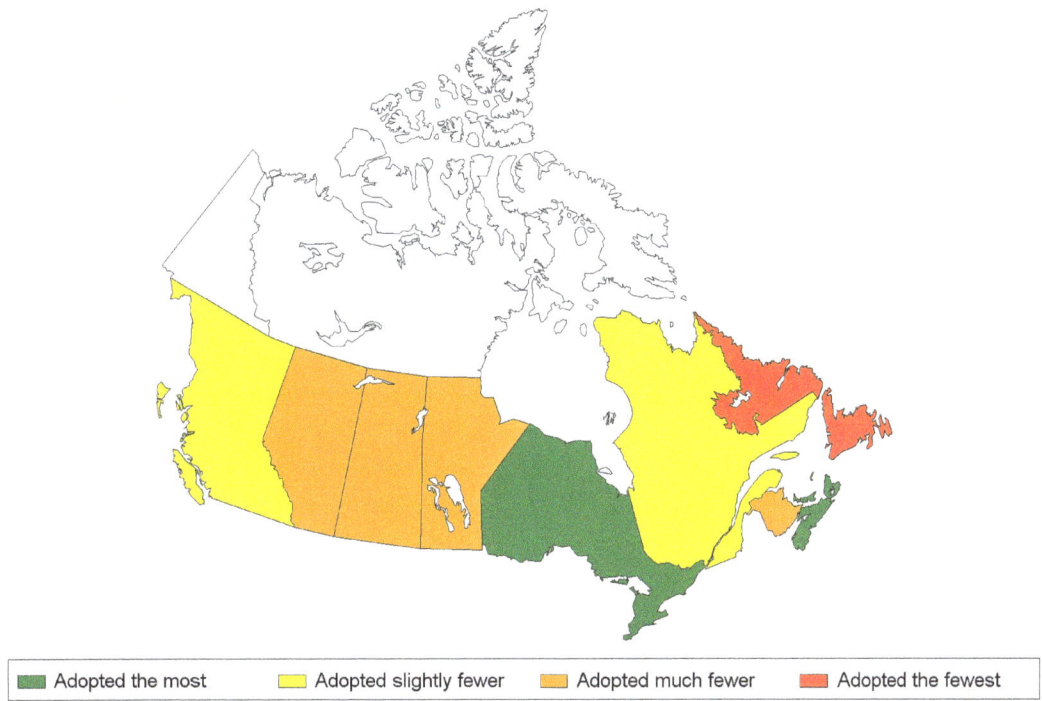

**Figure 5. Provinces' classification according to the four categories of pro-environmental behaviours, Indoor Index, 2007. Legend**: from left to right – British-Columbia, Alberta, Saskatchewan, Manitoba, Ontario, Quebec, New-Brunswick, Nova-Scotia. Prince-Edward-Island is North of the two latter provinces and Newfoundland-and-Labrador is located North-East of Quebec.

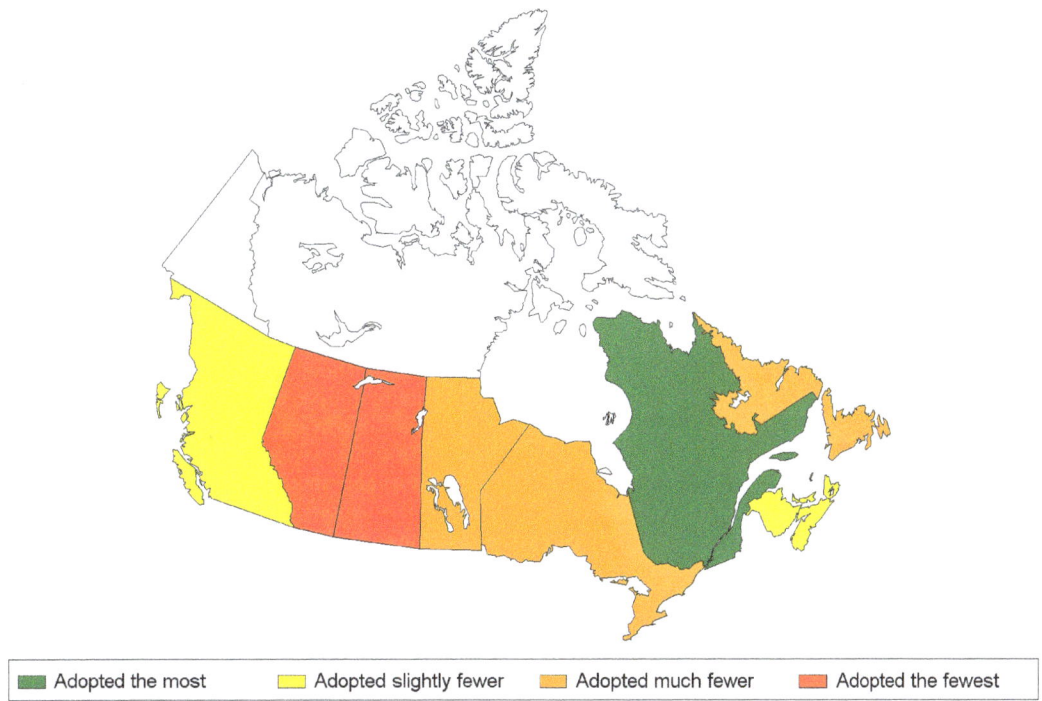

**Figure 6. Provinces' classification according to the four categories of pro-environmental behaviours, Outdoor Index, 2007. Legend**: from left to right – British-Columbia, Alberta, Saskatchewan, Manitoba, Ontario, Quebec, New-Brunswick, Nova-Scotia. Prince-Edward-Island is North of the two latter provinces and Newfoundland-and-Labrador is located North-East of Quebec.

**Table 3.** Provinces' coordinates on the Indoor Index, 2007.

| Provinces | Coordinates | Categories[a] |
|---|---|---|
| Prince-Edward-Island | −0,0262 | ++ |
| Nova-Scotia | −0,0179 | ++ |
| Ontario | −0,0130 | ++ |
| British-Columbia | −0,0055 | + |
| Quebec | −0,0029 | + |
| Alberta | 0,0159 | − |
| Manitoba | 0,0225 | − |
| Saskatchewan | 0,0363 | − |
| New-Brunswick | 0,0381 | − |
| Newfoundland-and-Labrador | 0,0853 | − |

[a]Categories are: adopted the most pro-environmental behaviours (++), adopted slightly fewer (+), adopted much fewer (−) and adopted the fewest (−).

same households that adopt pro-environmental behaviours on both indices. Higher income households may be more able to maintain and repair their housing and also invest in environmentally friendly products such as water and energy efficient appliances or fixtures, which can be more expensive than their regular counterparts [34;35]. Access to such products may contribute to the better classification on the Indoor Index for higher income households. On the other hand, lower income households may be less willing to pay water taxes linked to consumption levels, or to buy chemical products for their lawn or garden. Furthermore, those lower income households live more frequently in apartments where they do not have a yard, and they also own fewer recreational vehicles (data not shown). All these factors likely weigh in on the higher classification attributed to lower versus higher income households on the Outdoor Index.

In Canada, income is usually positively associated to educational attainment [36]. Also, the number of persons in a household will influence the household income. In the HES database, there was a significant correlation between households' income and education level as well as one with the households' income and the number of persons in the households (data not shown). This may explain why the indices by educational level and by number of

**Table 4.** Provinces' coordinates on the Outdoor Index, 2007.

| Provinces | Coordinates | Categories[a] |
|---|---|---|
| Quebec | 0,1038 | ++ |
| Prince-Edward-Island | 0,0261 | + |
| Nova-Scotia | 0,0123 | + |
| New-Brunswick | 0,0052 | + |
| British-Columbia | 0,0012 | + |
| Manitoba | −0,0243 | − |
| Newfoundland-and-Labrador | −0,0273 | − |
| Ontario | −0,0350 | − |
| Saskatchewan | −0,0887 | − |
| Alberta | −0,0962 | − |

[a]Categories are: adopted the most pro-environmental behaviours (++), adopted slightly fewer (+), adopted much fewer (−) and adopted the fewest (−).

persons in a household are similar to the one by household income. Any one of these three socio-demographic variables could potentially be used as a surrogate for the other two for future data collection for following Index trends over time.

Studies have shown that water meters with appropriate pricing are an incentive to reduce water consumption [2]. The US EPA estimated a 20% reduction in water consumption with universal metering [37] and a Canadian study also estimated a similar reduction according to structured water pricing [38]. While our results showed that households with water meters tended to score higher on the Indoor Index, households without a water meter scored higher on the Outdoor Index, which is in contrast to the other studies. We estimated that only 9% of households living in an apartment have water meters as opposed to 58% for all other types of dwellings in Canada (data not shown). As stated earlier, a household living in an apartment passively adopts more outdoor pro-environmental behaviours for lack of a lawn or garden to maintain with only a few having a water meter, possibly explaining the discrepancy between our results and those of other studies.

## Factors that can lead to pro-environmental behaviours

There is a wide variety of measures or instruments than can be introduced by governments to influence households behaviours, from economic instruments to direct regulation, labeling, information campaigns and provision of environment-friendly public goods such as public transportation or bicycle paths [39].

This study has identified factors which seem to influence the uptake of beneficial environmental behaviours at the household level. Investment in infrastructure is one of them. The physical or material possibility to act pro-environmentally must indeed be available [40], such as what might be needed for Newfoundlanders to improve their recycling profile.

Regulation is frequently used to efficiently influence the environmental impacts of household decision-making [39] and in our study it also seems to be an important incentive for the adoption of pro-environmental behaviours. This was seen both in the case of building codes requiring the installation of water efficient shower heads and toilets, and in the case of the ban on pesticides for lawn care. In Ontario, the ban of cosmetic pesticides decreased significantly the concentration of some pesticides, mainly herbicides, in the majority of streams under surveillance near urban areas with limited agriculture activities [41].

To encourage a reduction in water consumption, both price and non-price policies should be used. Volumetric water charges are associated with both water-saving behaviours and adoption of water-efficient devices [39]. However, in a study in several OECD countries, Canada had the highest proportion of households that did not know how they were charged for residential water consumption, thus reducing the price effect on water-saving behaviours [39]. In our study, presence of water meters was an incentive to water-saving behaviours only for indoor behaviours. Climate was also another factor that could be influential. Hence, public information on the environmental impact of water consumption and on measures households can adopt to save water should be combined to economic measures according to the OECD [39] and this study.

Other than governmental measures, household characteristics may play a role in the adoption of environment-related behaviours such as income, household composition and dwelling characteristics [39]. According to the OECD survey, low income households and tenants households make fewer financial investments in water efficiency, as can be expected. Grants targeted at those households to correct the economic imbalance are thus recommended by the agency. Moreover, our study showed

**Table 5.** Coordinates and categories of pro-environmental behaviours for other socio-demographic variables, Indoor and Outdoor Indices, 2007.

| | Indoor Index | | Outdoor Index | |
|---|---|---|---|---|
| | Coordinates | Categories[a] | Coordinates | Categories[a] |
| **Household income** | | | | |
| Less than $40,000 | 0,0253 | − | 0,1533 | ++ |
| $40,000 to less than $80,000 | −0,0080 | + | −0,0133 | − |
| $80,000 and over | −0,0273 | ++ | −0,1567 | − |
| **Highest education level** | | | | |
| Secondary diploma or less | 0,0228 | − | 0,0927 | ++ |
| Postsecondary certificate or diploma | −0,0030 | + | −0,0134 | − |
| University | −0,0159 | ++ | −0,0451 | − |
| **Dwelling type** | | | | |
| Apartment | 0,0370 | − | 0,4335 | ++ |
| Others | −0,0144 | ++ | −0,1501 | − |
| **Number of persons in the dwelling** | | | | |
| One | 0,0260 | − | 0,2058 | ++ |
| Two | −0,0073 | + | −0,0255 | − |
| Three | −0,0104 | ++ | −0,0723 | − |
| Four or more | −0,0163 | ++ | −0,1383 | − |
| **Water meter** | | | | |
| Yes | −0,0201 | ++ | −0,1682 | − |
| No | 0,0123 | − | 0,1497 | ++ |
| **Year the dwelling was built** | | | | |
| Before 1946 | −0,0089 | + | 0,0034 | + |
| Between 1946 and 1960 | −0,0040 | + | −0,0130 | − |
| Between 1961 and 1977 | −0,0009 | + | 0,0025 | + |
| Between 1978 and 1983 | −0,0054 | + | −0,0218 | − |
| Between 1984 and 1995 | −0,0084 | + | −0,0364 | − |
| Between 1996 and 2000 | −0,0088 | + | −0,0434 | − |
| Between 2001 and 2005 | −0,0099 | ++ | −0,0948 | − |
| 2006 or latter | 0,0042 | + | 0,0384 | + |

[a]Categories are: adopted the most pro-environmental behaviours (++), adopted slightly fewer (+), adopted much fewer (−) and adopted the fewest (−).

households from both income groups (high or low) or dwelling type (owned or rented) have to improve their act in different domains and that programs should target them accordingly. In short, Canadians remain very dependent for many such actions on where they live and what the climate brings to their yards, or not.

## Limits of the study

We used data from a survey that has been created to address the needs of Statistics Canada and the federal government. Thus, we were limited to its content. The questionnaire does not cover all behaviours that can impact the environment and public health. Also, the indices developed here measure the behaviours available in the survey and retained after the analysis by an expert group for their potential positive impacts on health, and not all existing pro-environmental behaviours. The classification could have been different if other behaviours had been included.

Three themes of the survey were not covered by the indices, namely dwelling characteristics, motor vehicle and indoor environment. However, we believe they would not have much impact in the indices. First, there were no behaviours measured in the dwelling characteristics theme and some of the characteristics were included as passive variables in the indices (e.g. year the building was built). The same happened for the motor vehicle theme (focused on the characteristic of the car), yet we used another theme to include the number of vehicles owned by the households in the outdoor index. For the indoor environment theme, only 2 of the 5 questions measured behaviours and they both concerned the type of chemical products used to clean windows and the dwelling. Although every small action is important for the environment, those questions were excluded as some other practices, such as agriculture, use similar products in much larger quantities [42].

A good standing in the classification does not mean that there is no place for improvement. Indeed, a high proportion of households that adopt pro-environmental behaviours on one question can compensate for a lower proportion on another question of the same index. Also, the provinces were compared to each other and not classified in relation to a gold standard.

Furthermore, it was the MCA that attributed the weight for each modality. Thus, a modality with a higher weight has more impact in the index. For instance, all four recycling questions had the highest contribution to the indoor index. Further studies should investigate if the inclusion of only one of those recycling question or a composite index of those four questions would be more appropriate. The same reasoning should also be applied to questions related to the watering of the lawn or of the garden. Households without a garden or a lawn are rewarded for every question on that subject which at the end can impact greatly the province classification. For example, they were not only rewarded for not watering their lawn, but they were also rewarded for the question concerning the watering duration and the number of time they water. Because the MCA attributed the weights, those household without a garden or lawn had a higher 'reward' that households with a garden or lawn that did not water them.

One general limit of MCA is that the first dimension usually explains a low proportion of the total inertia in the data set and the other dimensions explain less than the first [18]. In this study, the first dimension explained 33% and 62% of the total inertia for the indoor and outdoor index respectively. By using only the first dimension, these indices might not properly reflect all of the behaviours, especially for the indoor one. However, using more than one dimension to build the index would not considerably increase the total inertia explained but would in return increase its complexity. Composite indices are indeed built to simplify the analysis.

Because of the study design, based on households, it was also not possible to evaluate the impact of personal attributes, like age and gender, on the adoption of environmental behaviours. The association between environmental behaviours and age is not clear. Studies have observed all possible trends, from older people adopting more pro-environmental behaviours to the opposite trends or no trend at all [43–45]. Also, women would be more likely to take pro-environmental actions than men, although some studies have found the opposite depending on behaviour and region [43–46]. In our study we found that socio-demographic characteristics like household income and a higher level of education did not have the same influence on outdoor behaviours compared to indoor behaviours. Hence, some differences could also be expected between indoor and outdoor behaviours for age and gender.

Because the survey was not meant to measure attitudes or values, we cannot associate the classification of the province to any difference in values or perception. However, others studies have showed cultural differences across Canadian provinces [47–51]. In Canada, French speaking people are at majority in the province of Quebec but a minority in the rest of Canada as opposed to English speaking Canadian that are a majority in the rest of Canada [52;53]. Several studies have observed differences of values and attitudes in terms of personality, political perspective, priorities and social issues between English-Canadians and French-Canadians [47–51]. Differences in those values could also explain some differences in the adoption of pro-environmental behaviours but further studies are required to confirm it.

Attitudes and values can also be different in immigrants compared to the native born. The former usually have a smaller ecological footprint [54–57]. For example, several studies, mostly from United States, have observed that immigrants have lifestyles that are less demanding on the environment: they consume less, possess fewer luxury items like SUVs, they carpool or use public transportation more often and live in smaller houses [54–57]. In 2006, around 55% of all Canadian immigrants were in Ontario, followed by 18% in British-Columbia and 14% in Quebec [58]. British-Columbia and Quebec had a good classification on both indices. However, we could not estimate the impact of immigra-

tion on these classifications, as immigration rules and influx have changed significantly over the last decades [58].

Despite those limits, the indices still give a good idea of the global adoption of pro-environmental behaviours with potential positive impacts on health in Canada and remain easy to explain and understand. The main sectors in which households can have an impact are covered by the indices, like air and soil quality as well as water conservation. The weighting methods used (i.e. MCA) are also more appropriate to assign weights as opposed to an equal weights or expert opinion approach that are often criticized for being arbitrary or simplistic [22]. Others similar indices could be created as the survey is performed every two years. The results obtained with the 2007 indices could serve as the baseline for surveillance purposes, as the survey has been more comprehensive since that date.

## Conclusion

MCA was successfully applied in creating Indoor and Outdoor composite Indices of environmental health relevance based on a readily available periodic Statistics Canada dataset. The Indices cover a good range of environmental themes at the household level and the analysis, particularly the indices weights obtained in the MCA, could be applied to similar surveys worldwide (as baseline weights) enabling temporal trend comparisons for recurring themes. Results uncovered provincial patterns of pro-environmental behaviours adoption with certain provinces scoring consistently higher and others consistently lower, as well as the associations between socio-demographic factors and the indices. Much more than voluntary measures, this study shows that existing regulations, dwelling type, household composition and income as well as climate are the major factors determining pro-environmental behaviours.

## Supporting Information

**Figure S1   Map representation of the MCA results on the 55 questions with extreme responses.**

**Table S1**   Results of the MCA on the 55 questions without extreme responses (exploratory analysis). **Legend:** N/A: Results are not available for supplementary variables. Qlt: Quality (i.e. the sum of the squared correlations for the first two dimensions in this case). Inr: Inertias. K: Principal coordinates for the first dimension. Cor: Squared correlation with the first dimension. Ctr: Contributions of the modality to the explained inertia of the first dimension. All cells are multiplied by 1000. Results are the same on rows and on column when a Burt table is used.

**Table S2**   Results of the MCA for the Indoor Index, 2007. **Legend:** N/A: Results are not available for supplementary variables. Qlt: Quality (i.e. the sum of the squared correlations for the first two dimensions in this case). Inr: Inertias. K: Principal coordinates for the first dimension. Cor: Squared correlation with the first dimension. Ctr: Contributions of the modality to the explained inertia of the first dimension. All cells are multiplied by 1000. Results are the same on rows and on column when a Burt table is used.

**Table S3**   Results of the MCA for the Outdoor Index, 2007. **Legend:** N/A: Results are not available for supplementary variables. Qlt: Quality (i.e. the sum of the squared correlations for the first two dimensions in this case). Inr: Inertias. K: Principal coordinates for the first dimension. Cor: Squared correlation with the first dimension. Ctr: Contributions of the modality to the explained inertia of the first dimension. All cells are multiplied by 1000. Results are the same on rows and on column when a Burt table is used.

**Table S4**  Questions and responses selected for the Indoor Index, 2007.

**Table S5**  Questions and responses selected for the Outdoor Index, 2007.

## References

1. Natural Resources Canada (2011) Energy Efficiency trends in Canada, 1990 to 2009Ottawa (On)54 p.
2. Environment Canada (2011) Ottawa (On)Municipal Water Use Report24 p.
3. Mustapha I, Tait M, Trant D (2012) Human Activity and the Environment. Waste management in Canada. Statistics CanadaOttawa (On)Report No: 16-201-x, 46 p.
4. Milito AC, Gagnon G (2008) Greenhouse gas emissions-a focus on Canadian households. EnviroStats 2(4):3–6.
5. Statistics Canada (2008) Human Activity and the Environment: Annual Statistics 2007 and 2008Ottawa (On)159 p.
6. Environment Canada (2013) Wise Water Use. Available: http://www.ec.gc.ca/eau-water/default.asp?lang = En&n = F25C70EC-1. Accessed 6 may 2014.
7. Statistics Canada (2009). Households and the environment, 2007.Ottawa (On)report no: 11-526-x,102 p.
8. Hardie D, Alasia A (2009) Domestic Water Use: The relevance of Rurality in Quantity Used and Perceived Quality. Rural and Small Town Canada Analysis Bulletin 7(5):1–31.
9. Mustapha I (2013) Composting by households in Canada. EnviroStats 7(11):1–6.
10. Lynch MF, Hofmann N (2007) Canadian lawns and gardens: Where are they the greenest? EnviroStats 1(2): 9–14.
11. Birrell C (2008) Energy-efficient holiday lights. EnviroStats 2(4):19–20.
12. Nelligan T (2008) Household's use of water and wastewater services. EnviroStats 2(4):17–8.
13. Babooram A (2008) Canadian participation in an environmentally active lifestyle. EnviroStats 2(4):7–12.
14. Nardo M, Saisana M, Saltelli A, Tarantola S, Hoffman A, et al. (2005) Handbook on Constructing Composite Indicators: Methodology and User GuideParis (Fr)Organisation for Economic Co-operation and Development Publishing162 p.
15. Esty DC, Levy M, Srebotnajk T, de Sherbinin A (2005) 2005 Environmental Sustainability Index: Benchmarking National Environmental Stewardship. New Haven: Yale Center for Environmental Law & Policy, 403 p.
16. Statistics Canada (2010) Microdata User Guide – Households and the environment survey, 2007Ottawa (On)53 p.
17. Stewart DW, Shamdasani PN, Rook DW (2007) Focus Groups: Theory and Practise. 2nd ed.Thousand OaksSAGE Publications200 p.
18. Greenacre M (2007) Correspondence analysis in practice. 2nd EditionNew YorkChapman & Hall/CRC284 p.
19. Dossa LH, Buerkert A, Schlecht E (2011) Cross-Location Analysis of the Impact of Household Socioeconomic Status on Participation in Urban and Peri-Urban Agriculture in West Africa. Hum Ecol Interdiscip J 39(5): 569–581.
20. Charreire H, Casey R, Salze P, Kesse-Guyot E, Simon C, et al. (2010) Leisure-time physical activity and sedentary behaviour clusters and their associations with overweight in middle-aged French adults. Int J Obes (Lond) 34(8):1293–1301.
21. Cortinovis I, Vella V, Ndiku J (1993) Construction of a socio-economic index to facilitate analysis of health data in developing countries. Soc Sci Med 36(8): 1087–1097.
22. Howe LD, Hargreaves JR, Huttly SR (2008) Issues in the construction of wealth for the measurement of socio-economic position in low-income countriesEmerg Themes Epidemiol 5(3): 14 p.
23. Asselin LM (2002) Composite indicator of Multidimensional Poverty - TheoryQuébec (Qc)Institut de Mathématique Gauss33 p.
24. Greenacre M, Nenadic O (2010) Package 'ca' - Simple, Multiple and Joint Correspondence AnalysisR project, 20 p.
25. Munro A (2010) Recycling by Canadian Households, 2007.Statistics Canada, Ottawa (On)34 p.
26. Oaks(2012) Province and Territory Water Efficiency and Conservation Policy Information. Available: http://www.allianceforwaterefficiency.org/2012-Province-Information.aspx. Accessed 13 November 2013.
27. Kinkead J, Boardley A, Kinkead M (2006) An analysis of Canadian and other water conservation practices and initiativesMississauga (On)Canadian Council of Ministers of the Environment274 p.
28. Gibbons WD (2008) Who uses water-saving fixture in the home? EnviroStats 2(3): 8–12.
29. Statistics Canada (2011) CANSIM Table 027-0017: Canada Mortgage and Housing Corporation, mortgage loan approvals, new residential construction and existing residential properties, monthly. Available: http://www5.statcan.gc.ca/cansim/a26?lang = eng&retrLang = eng&id = 0270017&paSer = &pattern = &stByVal = 1&p1 = 1&p2 = 37&tabMode = dataTable&csid = . Accessed 13 November 2013.
30. Bonsal B, Koshida G, O'Brien EG, Wheaton E (2013). Droughts. Available: http://www.ec.gc.ca/inre-nwri/default.asp?lang = En&n = 0CD66675-1&offset = 8&toc = hide . Accessed 13 july 2012.
31. Environment Canada (2010) Water and climate change. Available: http://www.ec.gc.ca/eau-water/default.asp?lang = En&n = 3E75BC40-1. Accessed 13 November 2013.
32. Mekis É, Vincent LA (2011) An overview of the second generation adjusted daily precipitation dataset for trend analysis in Canada. Atmosphere-Ocean 49(2): 163–77.
33. Ministère du Développement durable, de l'Environnement, de la Faune et des Parcs (2011) The pesticides Management Code - Highlights. Available: http://www.mddep.gouv.qc.ca/pesticides/permis-en/code-gestion-en/index.htm. Accessed 13 July 2012.
34. Canada mortgage and Housing Corporation (2013) Reducing energy cost. Available: https://www.cmhc-schl.gc.ca/en/inpr/afhoce/afhoce/afhostcast/afhoid/opma/reenco/index.cfm. Accessed 22 November 2013.
35. BChydro (2013) Buy, build, or rent an efficient home. Available: http://www.bchydro.com/powersmart/residential/guides_tips/green-your-home/whole_home_efficiency/energy_efficient_home.html. Accessed 22 November 2013.
36. Human Resources and Skills Development Canada (2007) What difference does learning make to financial security. Indicators of Well-Being – Special ReportGovernment of Canada14 p.
37. U.S. Environmental Protection Agency (1998) Washington (DC)Water Conservation Plan Guidelines208 p.
38. Reynaud A, Renzetti S, Villeneuve M (2005) Residential water demand with endogenous pricing: The Canadian CaseWater Resour Res 41(w11409)11 p.
39. OECD (2013) Greening Household Behaviour: Overview from the 2011 Survey, OECD Studies on Environmental Policy and Household Behaviours.OECD Publishing306 p.
40. Kollmuss A, Agyeman J (2002) Mind the Gap: Why Do People Act Environmentally and What Are the Barriers to Pro-Environmental Behaviour? Environmental Education Research Aug;8(3):239.
41. Todd A, Struger J (2014) Changes in acid herbicide concentrations in urban streams after a cosmetic pesticides ban. Challenges, 5:138–151.
42. Environment Canada (2013) Ammonia Emissions. Available: https://www.ec.gc.ca/indicateurs-indicators/default.asp?lang = en&n = FE578F55-1. Accessed 30 January 2014.
43. Mainieri T, Barnett EG, Valdero TR, Unipan JB, Oskamp S (1997) Green buying: The influence of environmental concern on consumer behavior. The Journal of Social Psychology Apr;137(2):189–204.
44. Melgar N, Mussio I, Rossi M (2013) Environmental Concern and Behavior: Do Personal Attributes Matter?Facultad de Ciencias Sociales, Universidad de la Republica; 21 p.
45. Xiao C, Hong D (2010) Gender differences in environmental behaviours in China. Population and Environment Sep;32(1):88–104.
46. Lopez A, Torres CC, Boyd B, Silvy NJ, Lopez RR (2007) Texas Latino College Student Attitudes Toward Natural Resources and the Environment. Journal of Wildlife Management Jun;71(4):1275–80.
47. Baer DE, Curtis JE (1984) French Canadian-English Canadian Differences in Values: National Survey Findings. Canadian Journal of Sociology/Cahiers canadiens de sociologie 9(4):405–27.
48. Baillargeon JP (1994) The Cultural Practices of Anglophones in Quebec. Recherches Sociographiques May;35(2):255–71.
49. Gibson KL, McKelvie SJ, Man AF (2008) Personality and Culture: A Comparison of Francophones and Anglophones in Québec. The Journal of Social Psychology Apr;148(2):133–65.

## Acknowledgments

The authors thank Mr. Yves Lafortune of Statistics Canada for relevant comments on a preliminary version of this study and Ms Sandra Owens for her contribution to the redaction of this article. Also, thanks to Mr. Gaston Quirion of Laval University Library for facilitating access to the Statistics Canada survey database.

## Author Contributions

Conceived and designed the experiments: MC BA DB PG. Analyzed the data: MC BA. Wrote the paper: MC BA DB PG.

50. Wu Z, Baer DE (1996) Attitudes toward family and gender roles: A comparison of English and French Canadian women. Journal of Comparative Family Studies 27(3):437–52.
51. Young N, Dugas E (2012) Comparing climate change coverage in Canadian English and French-language print media: environmental values, media cultures, and the narration of global warming. Canadian journal of sociology 37(1):25–54.
52. Corbeil JP (2012) Ottawa (On)French and the francophonie in Canada. Census in brief no. 1, Statistics Canada12 p.
53. Corbeil JP (2012) Linguistic Characteristics of Canadians. Language, 2011 Census of Population, Statistics CanadaOttawa (On)22 p.
54. Atiles JH, Bohon SA (2003) Camas Calientes: Housing Adjustments and Barriers to Social and Economic Adaptation among Georgia's Rural Latinos. Southern Rural Sociology 19(1):97–122.
55. Blumenberg E, Smart M (2010) Getting by with a little help from my friends and family: immigrants and carpooling. Transportation May;37(3):429–46.
56. Bohon SA, Stamps K, Atiles JH (2008) Transportation and Migrant Adjustment in Georgia. Population Research and Policy Review Jun;27(3):273–91.
57. Price CE, Feldmeyer B (2012) The Environmental Impact of Immigration: An Analysis of the Effects of Immigrant Concentration on Air Pollution Levels. Population Research and Policy Review Feb;31(1):119–40.
58. Statistics Canada (2011) Immigration in Canada: A portrait of the Foreign-born Population, 2006 Census: Data tables, figures and maps. Available: http://www12.statcan.ca/census-recensement/2006/as-sa/97-557/tables-tableaux-notes-eng.cfm. Accessed 30 January 2014.

# 23

# Metabolomic Response of *Calotropis procera* Growing in the Desert to Changes in Water Availability

**Ahmed Ramadan**[1,2], **Jamal S. M. Sabir**[1], **Saleha Y. M. Alakilli**[1], **Ahmed M. Shokry**[1,2], **Nour O. Gadalla**[1,3], **Sherif Edris**[1,4], **Magdy A. Al-Kordy**[1,3], **Hassan S. Al-Zahrani**[1], **Fotouh M. El-Domyati**[1,4], **Ahmed Bahieldin**[1,4], **Neil R Baker**[6], **Lothar Willmitzer**[5]*, **Susann Irgang**[5]

**1** Department of Biological Sciences, Faculty of Science, King Abdulaziz University (KAU), Jeddah, Saudi Arabia, **2** Agricultural Genetic Engineering Research Institute (AGERI), Agriculture Research Center (ARC), Giza, Egypt, **3** Genetics and Cytology Department, Genetic Engineering and Biotechnology Division, National Research Center, Dokki, Egypt, **4** Department of Genetics, Faculty of Agriculture, Ain Shams University, Cairo, Egypt, **5** Max-Planck-Institut für Molekulare Pflanzenphysiologie, Potsdam-Golm, Germany, **6** Department of Biological Sciences, University of Essex, Colchester, United Kingdom

## Abstract

Water availability is a major limitation for agricultural productivity. Plants growing in severe arid climates such as deserts provide tools for studying plant growth and performance under extreme drought conditions. The perennial species *Calotropis procera* used in this study is a shrub growing in many arid areas which has an exceptional ability to adapt and be productive in severe arid conditions. We describe the results of studying the metabolic response of wild *C procera* plants growing in the desert to a one time water supply. Leaves of *C. procera* plants were taken at three time points before and 1 hour, 6 hours and 12 hours after watering and subjected to a metabolomics and lipidomics analysis. Analysis of the data reveals that within one hour after watering *C. procera* has already responded on the metabolic level to the sudden water availability as evidenced by major changes such as increased levels of most amino acids, a decrease in sucrose, raffinose and maltitol, a decrease in storage lipids (triacylgycerols) and an increase in membrane lipids including photosynthetic membranes. These changes still prevail at the 6 hour time point after watering however 12 hours after watering the metabolomics data are essentially indistinguishable from the prewatering state thus demonstrating not only a rapid response to water availability but also a rapid response to loss of water. Taken together these data suggest that the ability of *C. procera* to survive under the very harsh drought conditions prevailing in the desert might be associated with its rapid adjustments to water availability and losses.

**Editor:** Wagner L. Araujo, Universidade Federal de Vicosa, Brazil

**Funding:** The project was funded by the Deanship of Scientific research (DSR), King Abdulaziz University, Jeddah, under Grant no 2-3-1432/HiCi. The authors therefore acknowledge with thanks DSR technical and financial support. The funders had no role in study design, data collection and analysis, decision to publish, or preparation of the manuscript.

**Competing Interests:** The authors have declared that no competing interest exists.

* E-mail: Willmitzer@mpimp-golm.mpg.de

## Introduction

Drought is one of the most serious limitations for agriculture limiting plant growth, photosynthesis and, thus, productivity in many areas of this planet. Given the anticipated climate change, it is expected to worsen in the future thus becoming an even more important threat for global food supply [1]. Due to the extraordinary importance of drought stress, many studies have been performed aiming at an improved understanding of the mechanisms underlying drought tolerance with the hope to ultimately translate this knowledge into improved crop varieties.

Essentially, a variety of different approaches have been followed in different plant systems. Starting from the observation that response to drought/drought tolerance is a multigenic trait, quantitative genetic studies taking advantage of natural diversity with respect to drought tolerance have been performed. Many of these studies have been performed in crop species such as corn or rice [2,3] and improved varieties, e.g. for corn have been developed [4,5]. These studies have furthermore led to the identification of QTL's associated with improved performance

under water limiting conditions and recently also to the molecular cloning of underlying genes such as DRO1 from rice [6].

As to the molecular responses, numerous studies have been performed in different plant systems such as *A. thaliana* or corn [7,8], resurrection plants, e.g. *Craterostigma plantagineum*, displaying extreme drought/desiccation tolerance [9,10] or following a sister group contrast approach by comparing two closely related species differing significantly in drought tolerance [11,12]. These studies showed that responses to limiting water availability at the organismal and cellular levels include inhibition of growth, stomatal closure and reduced photosynthesis. Further responses include the accumulation of osmoprotectants such as sugars, sugar alcohols and amino acids, such as proline, and a change in the glutathione/ascorbate cycle to probably combat oxidative stress. These changes are accompanied by a plethora of further biochemical and gene expression changes aiming at keeping the negative consequences of the limited water availability at a minimum [2,9,13–19]

Membranes are very sensitive to dehydration and in consequence lipid composition changes to cope with this stress [20].

Major changes observed consistently as a result of dehydration are a decline in galactolipids, an increase in digalactosyldiacyglycerol (DGDG) as compared to monogalactosyldiacylglycerol (MGDG), a decline in the degree of fatty acid desaturation [19–21] and an increase in triacylglycerols (TAG) [19].

In this study we describe the metabolomic and lipidomic response of a wild plant species, *Calotropis procera*, growing in the desert near Jeddah in Saudi Arabia to sudden water availability.

*Calotropis procera* belongs to the spurge family (*Euphorbiaceae*). This family includes about 280 genera and 8000 species which occur in tropical and in temperate regions all over the world. The genus *Calotropis* is distributed around sub-tropical and tropical regions of Asia and Africa and is used in traditional medicine.

In addition *Calotropis procera* displays an impressive drought tolerance given the fact that it grows in extremely dry areas such as the Saudi-Arabian desert thus making it an interesting model for studying the response to extreme drought respectively scarce water supply.

We here describe the metabolomic and lipidomic response of *C. procera* with respect to rehydration. To this end wild plants growing in the desert near to Jedda, Saudiarabia were sampled at different time intervals before and after giving a onetime water supply thus providing a first insight into the metabolic response under true field conditions. Even given that the metabolomic and lipidomic data presented are only semiquantitative the results obtained demonstrate that *C. procera* has the ability to respond very fast (within one hour) to changes in water availability by reorienting its metabolism. With equally impressive speed, the plant responds to the loss of water by transforming its metabolism back to pre-watered state. This fast response may be a crucial parameter to survive under the harsh conditions prevailing in the Saudiarabian desert.

## Materials and Methods

### Plant material, location of sampling and watering

The experiment was conducted using plants grown wild in the desert near to Jedda. No specific permission was required to perform these experiments as this area is used routinewise by the Colleagues from the King Abdullaziz University. The plant used in these experiments Calotropis procera is not an endangered or protected species.

This experiment was conducted during September 2012 in the desert about 30 km away from Jeddah (latitude 21°26′6.00, longitude 39°28′3.00). The average temperature during the time the experiments were conducted varied from 28–37°C, humidity was 70–75%. In this area, *Calotropis procera* shrubs grow as single plants in the wild (cf. figure 1). For the conduction of this experiment, three *C. procera* plants of equal size were selected for watering (25 liters dH$_2$O) and leaf sampling. Average rainfall in the Jeddah province in winter is ~25 cm distributed over 10 individual days. Thus the the expected amount of water received by individual plant in one day in 1 m$^2$ equals 25,000 cm$^3$, i.e., 25 liters dH$_2$O. Single similar-sized plants were given this amount of water in the evening. In the next day, we determined by visual inspection how far the water has penetrated. Water was no longer detectable 36 hours after watering as determined by a frequency domain probe CS615-FDR (Campbell Scientific Inc., Utah, USA). Plot edges were raised to avoid water flowing away and watering was done gradually over a period of 5 minutes to avoid spillage.

### Execution of the experimental treatment

**a. Determination of relative water content (RWC).** Leaves of the three *C. procera* plants selected for this experiment were sampled one day before and three consecutive days after watering. Leaves at 50 cm from the ground were the targets for all samples. Leaf samples for RWC were immediately weighed and fresh weight (FW) determined. Leaves were transferred to sealed amber flasks, rehydrated in one L of water for five hours until fully turgid at 48°C, surface dried, and reweighed (turgid weight, TW). The leaf samples were then oven-dried at 72°C for 48 hours and reweighed (dry weight, DW) (Silva *et al.*, 1996). RWC was calculated by the following formula:

RWC (%) equals (FW – DW) divided by ( TW – DW)×100.

Multiple comparisons were performed following the procedure outlined by Duncan's New Multiple Range test.

**b. Metabolomics analysis.** Samples for metabolomic studies were taken one hour (at dawn), six (at midday) and 12 hours (one hour pre-dusk) after water treatment. In order to be able to identify possible changes in metabolism due to diurnal fluctuations, samples were in addition taken one day before watering at the same three time points. Samples taken were frozen in liquid nitrogen and kept at −80 C until extraction. Three independent but comparable plants were used for this experiment, thus representing three biological replicates.

Leaf samples were extracted and processed for metabolomics analysis as detailed below [22,23].

Approximately 100 mg of the frozen plant tissue was homogenized in 2-ml Eppendorf tubes twice for 1 min at maximum speed within a Retschmill. The metabolites were extracted from each aliquot in 1 ml of a homogenous mixture of −20°C methanol: methyl-tert-butyl-ether: water (1:3:1), with shaking for 30 min at 4°C, followed by another 10 min of incubation in an ice cooled ultrasonication bath. After adding 650 μl of UPLC-grade methanol: water 1:3, the homogenate was vortexed and spun for 5 min at 4°C in a table-top centrifuge. The addition of methanol: water leads to a phase separation, providing the upper organic phase, containing the lipids, a lower aqueous phase, containing the polar and semipolar metabolites, and a pellet of starch and proteins at the bottom of Eppendorf tube. The separate phases are isolated and dried down in a speed vac and stored at −80°C until use in the different metabolomic or lipidomic analyses.

### UPLC-FT-MS measurement of lipids and semipolar metabolites and GC-TOF analysis of primary metabolites

UPLC separation of the semipolar fraction of the fractionated metabolite extract is performed using a Waters Acquity UPLC system, using an HSS T3 C18 reversed-phase column (100 mm ×2.1 mm ×1.8 μm particles; Waters). The mobile phases are 0.1% formic acid in H$_2$O (Buffer A, ULC MS grade; BioSolve, http://www.biosolve-chemicals.com) and 0.1% formic acid in acetonitrile (Buffer B, ULC MS grade; BioSolve). A 2-μl sample (the dried-down aqueous fraction was re-suspended in 100 μl of UPLC grade water) is loaded per injection, and the gradient, which is taken out with a flow rate of 400 μl min−1, is: 1 min 99% A, 13-min linear gradient from 99% A to 65% A, 14.5-min linear gradient from 65% A to 30% A, 15.5-min linear gradient from 30% A to 1% A, hold 1% A until 17 min, 17.5-min linear gradient from 1% A to 99% A, and re-equilibrate the column for 2.5 min (20-min total run time).

The lipid fraction of the fractionated metabolite extract is performed on the same UPLC system using a C8 reversed-phase column (100 mm ×2.1 mm ×1.7 μm particles; Waters). The mobile phases are water (UPLC MS grade; BioSolve) with 1% 1 M NH4Ac, 0.1% acetic acid (Buffer A,) and acetonitrile: isopropanol (7:3, UPLC grade; BioSolve) containing 1% 1 M NH4Ac, 0.1% acetic acid (Buffer B). A 2-μl sample (the dried-down organic fraction was re-suspended in 500 μl of UPLC-grade acetonitrile: isopropanol 7:3) is loaded per injection, and the

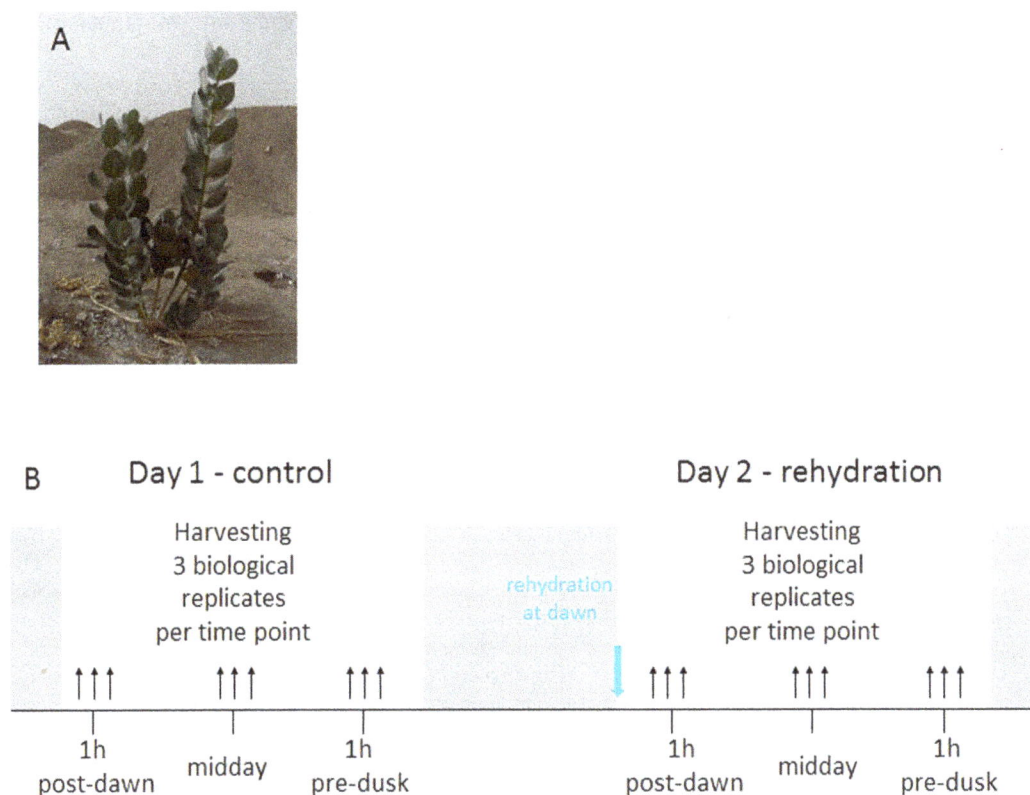

**Figure 1. Experimental Setup and plants chosen.** (a) Representative photo of the plants chosen for this experiment growing in its natural habitat in Saudi Arabia near to Jeddah. For this study representative species of similar size and performance were chosen. (b) Experimental set-up. At day 1 (control) leaves of three independent plants were harvested 1 h post-dawn, at midday and 1 h pre-dusk. One day later (Day 2), plants were watered at dawn and leaves were harvested 1 h post dawn, at midday and 1 h pre-dusk. Harvested leaves were frozen immediately in liquid –N and processes as described in Experimental procedures.

gradient, which was taken out with a flow rate of 400 µl min−1, is: 1 min 45% A, 3-min linear gradient from 45% A to 35% A, 8-min linear gradient from 25% A to 11% A, 3-min linear gradient from 11% A to 1% A. After washing the column for 3 min with 1% A, the buffer is set back to 45% A and the column is re-equilibrated for 4 min (22-min total run time).

The mass spectra are acquired using an Exactive mass spectrometer. The spectra are recorded alternating between full-scan and all-ion fragmentation-scan modes, covering a mass range from 100 to 1500 m/z. The resolution is set to 10 000, with 10 scans per second, restricting the loading time to 100 ms. The capillary voltage is set to 3 kV with a sheath gas flow value of 60 and an auxiliary gas flow of 35 (values are in arbitrary units). The capillary temperature is set to 150°C, whereas the drying gas in the heated electrospray source is set to 350°C. The skimmer voltage is set to 25 V, whereas the tube lens is set to a value of 130 V. The spectra are recorded from 1 min to 17 min of the UPLC gradients.

The polar phase is analyzed for primary metabolites using an established GC-TOF ms protocol [22,23].GC-TOF chromatograms were extracted and annotated as described by [24]. Lipid annotation was based on retention times, exact molecular mass and comparison to an inhouse database. For statistical analysis and visualization (ANOVA, Bonferroni correction, PCA, boxplots) the R-software was used (http://cran.r-project.org/). ANOVAs were conducted using the harvesting time and condition (control and watering) as factors and resulting p-values were corrected for multiple testing using the stringent Bonferroni method. For

principal component analysis (PCA), the "bpca"- algorithm of the "pcaMethod"- package was used [25]. Heatmaps were visualized using the Multi experiment Viewer software (MeV) version 4.8.1.

## Results and Discussion

### Experimental set-up

In a preliminary experiment leaf samples were taken from three independent plants of similar stature and developmental stage (cf. Figure 1 for a representative plant) at the end of the day (one hour pre-dusk), at seven days and one day before watering; and two and seven days after watering and subjected to our metabolomics platforms. An ANOVA analysis of the metabolomic data showed that the watering had no influence on the metabolite composition (data not shown).

The absence of a significant effect of watering on the metabolism of the treated plants could have two explanations: either metabolism of *Calotropis procera* is highly buffered and does not respond to water treatment or the effect on metabolism is much more transient and already lost two days after water treatment. To distinguish between these possibilities, we devised a second experiment where we collected samples within a more narrow time window, i.e., one (at dawn), six (at midday) and 12 hours (one hour pre-dusk) after water treatment. In order to be able to identify possible changes in metabolism due to diurnal fluctuations, samples were taken one day before watering at the same three time points. Three independent but comparable plants were used for this experiment, thus representing three biological replicates. Samples were, again, immediately frozen and processed for analysis on our three metabolomics platforms (cf. Table S1 and Table S2 for all metabolomics data).

## Changes in primary, secondary and lipid metabolite contents of *Calotropis* plants due to watering

In order to see whether changes in metabolism are detectable at the earlier time-points, metabolite data from samples harvested within 12 hours after watering and obtained at the corresponding time points before watering were subjected to a principal component analysis.

The results are shown in Figure 2, a–c revealing several interesting features:

- There is a very clear and significant effect of watering on metabolism detectable on the level of primary, secondary and lipid metabolite content.

- This effect is of a highly transitory nature. Thus, samples taken one and six hours after watering clearly differ from the non-treated samples, on one hand, and from each other, on the other hand. Besides, the effect of watering on metabolism vanishes after 12 hours (sample: at pre-dusk), thus confirming the results from the pilot experiment.

- Some metabolites measured by the GC-MS platform (primary metabolism) vary with sampling time with the samples taken at dawn being different from those taken at the pre-dusk and midday. This was also evidenced by ANOVA, whereas no such influence is seen for the lipids (Figure 2d).

- Figure 2a–c, furthermore, indicates a high reproducibility of the experiment as evidenced by the clustering of the control samples and the "return" of the samples 12 hours after water treatment to the control level. This is remarkable given the fact that the entire experiment was performed under field conditions in the desert with individual plants grown and rainfed in the wild.

## Changes in amino acids, TCA cycle intermediates and sugar alcohols

In total, 357 primary metabolites could be detected via GC-TofMS analysis of which 118 could be annotated (cf. Table S1). Significant changes were revealed by ANOVA and PCA (figure 2). It should be mentioned that due to the fact that the relative water content increased in response to watering (cf. below) the relative metabolite content does change as well. However this effect is marginal (less than 10%) and the changes observed were as a rule much higher. To identify metabolites that force the separation of samples due to watering within the PCA, we used PCA-loading scores for each metabolite (cf. Table S1). Analysis of the main metabolites driving the separation in the PCA, respectively, the results of ANOVA of primary metabolites (Figure 2) reveal significant changes in amino acids, TCA cycle intermediates, sugars and sugar alcohols. Figure 3a–c presents box-plots of a number of typical examples and the pathways view is shown in Figure 4. The clearest trend is observed in behavior of the majority of amino acids. Thus, most amino acids respond to watering by a fast and significant increase in steady-state concentration. This effect is very pronounced for the branched-chain amino acids such as leucine, isoleucine or valine, in addition to phenylalanine, lysine, methionine, proline and asparagine. Notably, glutamine is not vastly exceeding the levels found in non-watered plants, whereas glutamic acid rather does not change or at midday is even lower as compared to the non-watered control. The reason for the observed increase in amino acids must remain unclear. One possible explanation would be an increased demand due to increased protein synthesis, another (opposite) explanation would be the degradation of (storage) proteins.

The increase in concentration for most amino acids seen here in case of *C. procera* as a result of watering is in contrast to most other metabolomics studies where amino acids were observed to increase in parallel to applying drought stress [8,11,12] Also in a more recent study using the resurrection lycophyte *S. lepidophylla*, more than half of the amino acids were more abundant in the dry as compared to the hydrated state [18].

We do not know the reason underlying these differences however except the use of a different plant system the experiments

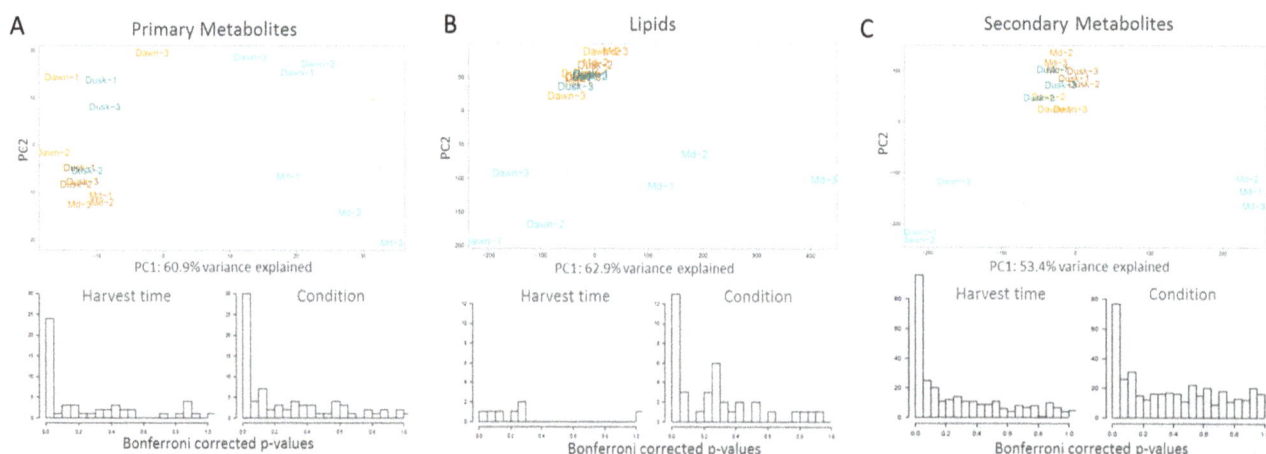

**Figure 2. Principal Component Analysis (PCA) and ANOVA for metabolomic analysis of leaf samples from control and watered plants.** (a) PCA (upper part) and ANOVA (lower part) for primary metabolites. (b) PCA (upper part) and ANOVA (lower part) for complex lipids. (c) PCA (upper part) and ANOVA (lower part) for secondary metabolites. Shown are always three independent samples per time point (dawn (1 hour post dawn/after watering), midday (6 hours after dawn/after watering) and pre-dusk (12 hours after dawn/after watering). Watered samples are shown in blue, non-watered in red. The lower part shows the results of a Bonferroni corrected ANOVA displaying the influence of treatment (watering) for all samples and of harvesting time for primary and secondary metabolites.

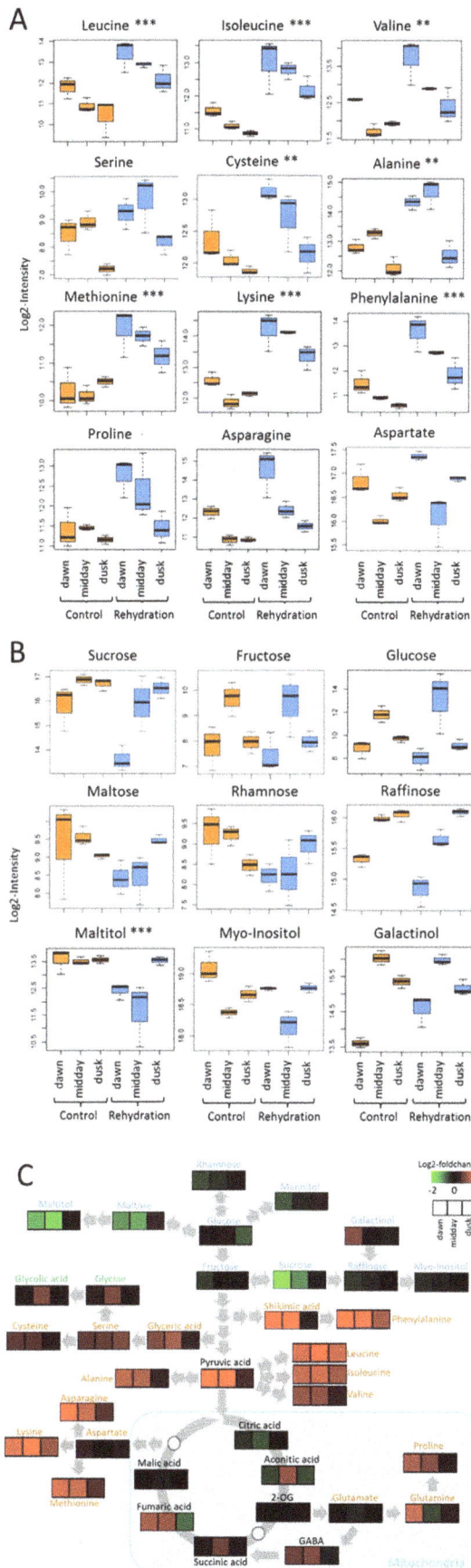

**Figure 3. Boxplots and pathway visualization of representative primary metabolites. (a)** and **(b)**: Boxplot-visualizations for a subset of amino acids (A) and sugars and sugar alcohols(B) as determined for the three independent samples for the different time points and treatments as indicated on the x-axis. **(b)** Pathway mapping of a number of primary metabolites visualized as their averaged log2-foldchange ratio of rehydration versus control (green = decrease; red = increase).

described here were performed under natural conditions in the field and furthermore the time intervals at which samples were taken after drought respectively rehydration treatment were as a rule longer in these studies as compared to our study. As described above all changes are of a highly transient nature.

Another aspect of our study worth mentioning is that the pattern of the amino acid concentration with respect to the three time points is very similar in the control and the watered conditions, i.e., the highest level is reached 1 hour after dawn in both control and watered samples with the concentrations subsequently decreasing towards midday and pre-dusk. Specifically the fact that the amino acid abundance in the samples taken at dawn before watering is also higher as compared to the midday and dusk samples is an independent confirmation of the response of the amino acids to watering.

Concerning TCA cycle intermediates the situation is similar though less pronounced. Whereas pyruvate and fumarate and, to a lesser extent, succinate display a significant increase as a result of watering, oxo-glutaric and malic acids essentially remain unchanged, whereas citric acid actually shows a decrease (cf. Table S1 and Figure 4). An increase in pyruvate and succinate has also been observed during rehydration of *S. lepidophylla* [18].

With respect to sugars and sugar alcohols, a more complex picture emerges. Glucose and fructose largely remain unchanged with only the one hour value being lower in the watered as compared to the non-watered control. Maltose is initially reduced in the water control which could be due to either an increase in maltose consumption or a decreased starch degradation. Sucrose, raffinose and maltitol are believed to serve as osmoprotectant. All three compounds display a significant reduction in watered as compared to non-watered control at the first two time points. This observation could be taken as indication that *C. procera*, senses during these early time points after watering as a relief from drought and, thus, osmotic stress and in consequence reduces the amount of compatible solutes. The members of the raffinose pathway, myoinositol and galactinol decrease transiently. As described above, proline like most other amino acids increases after rehydration. Proline is an accepted osmolyte and thus would be expected to decrease in parallel with the sugars and sugar alcohols. The increase observed could be either due to an increased need of proline for processes such as increased protein biosynthesis or due to a blockage of proline-consuming processes.

With respect to osmoprotecant sugars and sugar alcohols most studies analyzed their behavior in response to drought stress and not surprisingly an increase has been described in most studies [8,11,12,18]. It should be noted however that in case of *S. lepidophylla*, some sugar alcohols were observed to increase after rehydration [18].

The significant transitory decrease in malonic acid described in our study is interesting when connecting this observation with the lipidomics data where we observed a transitory decrease in storage lipids (triacylglycerides). Taken together, this might suggest a reduced flux into storage lipids as an early response to watering.

Finally, it is noteworthy to comment on the behavior of glycolate and glycine (Figure 4 and Table S1). Glycine, in contrast

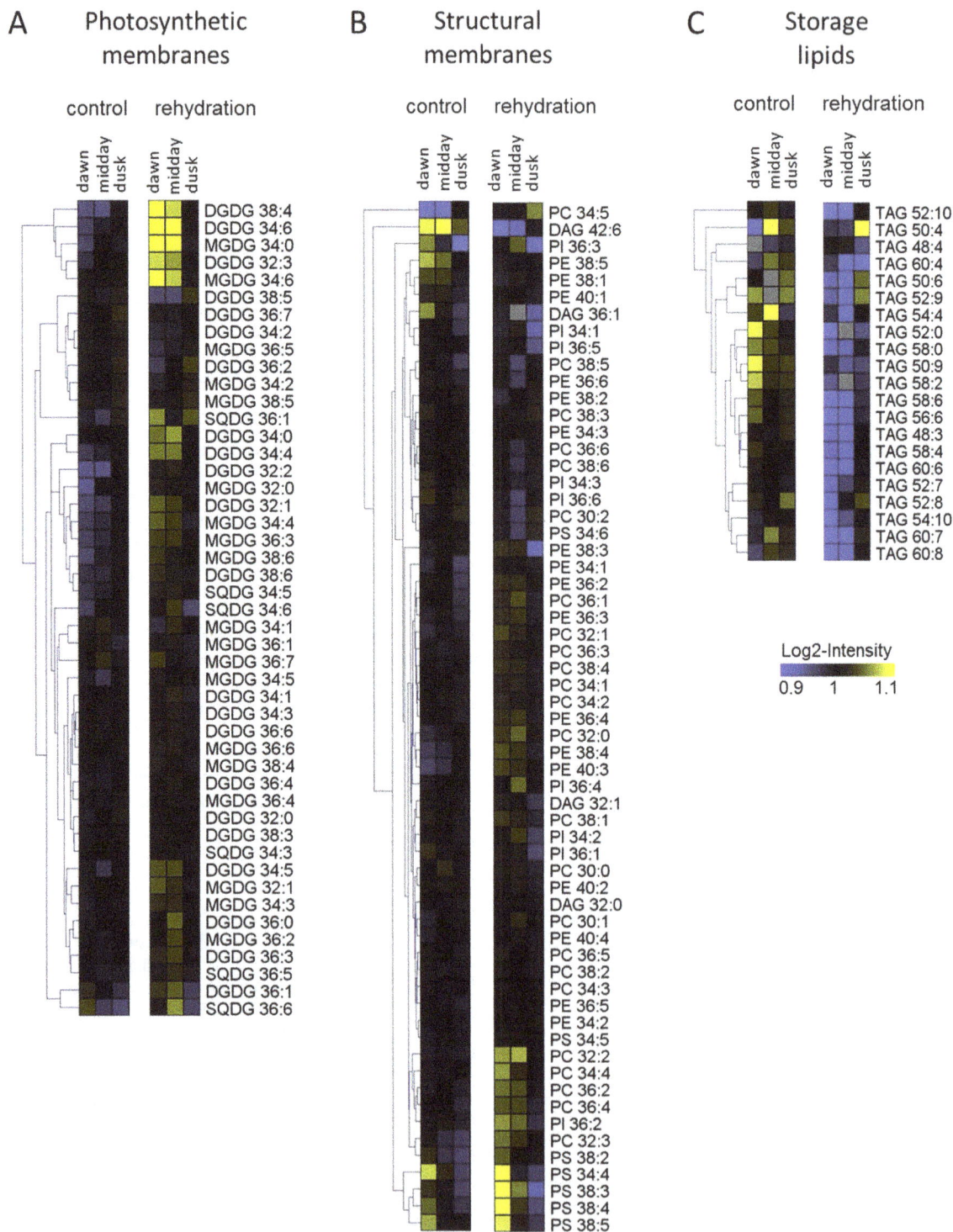

**Figure 4. Clustered heatmap visualization of different lipid classes.** Shown is the average abundance of several complex lipids visualized in a false-color heatmap at the three time points before and after watering ordered according to their presence in photosynthetic membranes, in cellular membranes or representing storage lipids.

to most other amino acids, reached its highest level at midday both in the watered and the non-watered control though the level is much higher in the watered sample. As we see the same pattern for glycolate, one possible though speculative explanation is that this increase is due to increased photorespiration which would also be in agreement with kinetics of serine accumulation (Figure 4a).

## Storage and membrane lipids

The non-polar phase of the extracts was subjected to UPLC-MS measurements and we identified 133 lipids belonging eight different classes: Diacylglycerol (DAG), Mono-galactosyl-diacylglycerol (MGDG), Di-galactosyl-diacylglycerol (DGDG), Sulfoquinovosyl-diacylglycerol (SQDG), Phosphatidylcholine

(PC), Phosphatidylserine (PS), Phosphatidylinositol (PI), Phsopha-tidylethanolamine (PE), Triacylglycerols TAG). Lipid species within each class are characterized by the number of C-atoms and by the number of double bonds in the acyl-chains.

As visible from Figure 2b, watering has a strong influence on complex lipid composition of Calotropis plants. A more detailed analysis shows that membrane lipids, in general and more specifically lipids of the photosystem, increase after watering. Most prominent examples comprise all MGDG's and the vast majority of the DGDG's (cf. Figure 5 a–d). A similar picture is observed for the majority of the SQDG's.

A contrasting picture emerges for the storage lipids, namely TAG's. Here for all classes, we observe a fast and significant decrease for the first two time points after watering and a reversion to the non-watered condition at the third time point.

These results largely agree with data described for other plant systems. Thus galactolipids have been described consistently to be

reduced as a result of dehydration (thus mirroring the decline after the first transient increase after watering) although a change in the ratio between MGDG's and DGDG's is not obvious in our case. Also the transient decrease observed for TAG's as a result of watering is in agreement with the described data (an increase in TAG's as a result of drought stress;[19]). With respect to membrane lipids specifically phospholipids however the data are only in partial agreement with data reported for other systems [18,19] which again might be contributed to the different plant system and/or the different experimental set-up.

## Secondary metabolites

Extracts of Calotropis are well known to display numerous pharmaceutical activities. Thus next to GC-MS measurements for primary metabolites, the polar phase was also subjected to secondary metabolite measurements by UPLC-MS. However, the secondary metabolism of Calotropis is only scarcely defined,

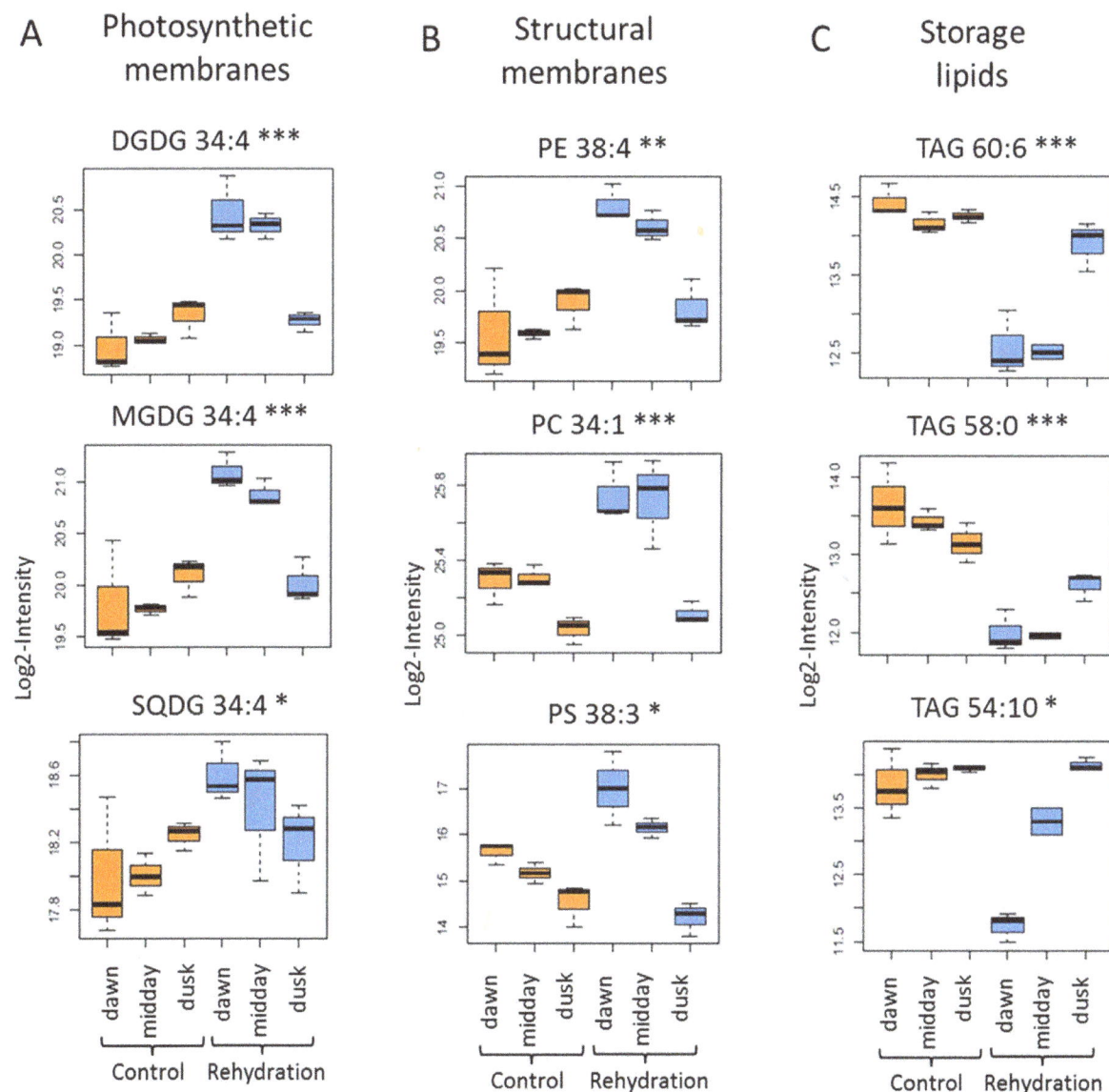

**Figure 5. Boxplots of representative species of photosynthetic, structural and storage lipids.** Boxplot-visualizations for a subset of complex lipids as determined for the three independent samples for the different time points and treatments as indicated on the x-axis of three replicates that were harvested at 1 h post-dawn, midday and 1 h pre-dusk for control and rehydrated plants.

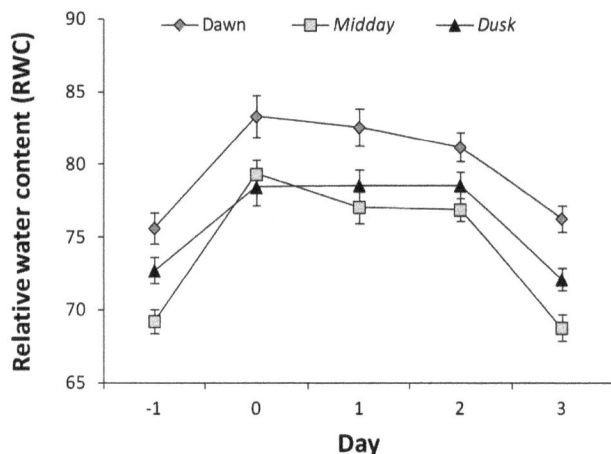

**Figure 6. Relative water content of leaves of *C. procera* plants taken at dawn, midday and one hour pre-dusk one day before watering and up to three days after watering.**

of metabolism back to prewatering conditions is seemingly disconnected from the RWC which goes back to prewatering conditions much later, i.e. 3 days after watering. We are not in a position to explain this puzzling observation however one speculation is that metabolism at least in case of Calotropis does not directly respond to RWC as measured in the leaves but to other obviously still nonidentified parameters. One possibility could be water uptake as measured in the root system however as said this is first a pure speculation and second difficult to prove/disprove. Thus we are at present left with the description of this phenomenon.

## Conclusion

A time-resolved metabolomics and lipidomics response of *Calotropis procera*, a shrub growing in arid regions, towards the sudden supply of a limited amount of water is described. To increase the relevance of this study, the entire experiment was performed in the field respectively desert using wild grown *Calotropis* plants as experimental object.

Key observations are the transitory decrease in maltitol and raffinose (indicating a reduced drought stress), an increase in essentially all amino acids which might suggest that increased protein biosynthesis takes place, an increase in all structural lipids of the photosynthetic membranes (DGDG's, MGDG's, SQDG's) which may suggest that the plant prepares itself for increasing its photosynthetic capacity as well as an increase in most other membrane lipids.

Thus most changes observed and specifically their kinetics suggest that water availability in the natural habitat of *C. procera*, i.e. the desert, is such a scarce event that it has developed the capacity to respond fast and massively by remodeling it metabolic machinery towards growth. Understanding the molecular mechanisms behind this response may open new approaches for adapting crop plants to arid conditions.

thus, making annotation of compounds difficult. In order to have a first look into changes occurring on the secondary metabolite level, we identified m/z features, which is based on their exact mass and retention time behavior could putatively be assigned to some of the described secondary metabolites of *Calotropis*. No clear trend can be observed for secondary metabolites analyzed some of them increasing, others decreasing. The most significant increase amongst the putatively annotated compounds is observed for Uscharin, a compound with strong molluiscidal effects found in the latex of *Calotropis procera* (data not shown).

## Changes in relative water content (RWC) before and after watering

As shown in Figure 6, relative water content at dawn was shown to increase from 75% one day before watering to 83% one hour after watering, withheld for two more days before it started to fall back (76%) to control (before watering) levels at day three after watering. The same trend of results was reached at midday and pre-dusk, as RWC increased from 69% and 72%, respectively, one day before watering to 79% and 78%, respectively, in the day of watering, then fell back to pre-watering control level three days after watering. It is obvious that RWC was lowest at midday and highest at dawn across the five days, except at the day of watering, where RWC was higher at midday than at pre-dusk. The overall results indicate that *Calotropis* plant was able to hold water efficiently for two days, then returned to its original level of RWC prior to watering.

As evident from Figure 6 metabolism responds parallel to changes in RWC with respect to induction however the reversion

## Supporting Information

**Table S1** GC measured metabolite list and ANOVA values for individual metabolites.

**Table S2** Lipid list and ANOVA values for individual lipids.

## Author Contributions

Conceived and designed the experiments: NRB AB LW AR JSMS. Performed the experiments: SYA NOG SE MAA HSA SI AMS FME. Analyzed the data: SI AB NRB LW AR JSMS. Wrote the paper: AR JSMS AB NRB LW SI.

## References

1. Schmidhuber J, Tubiello FN (2007) Global food security under climate change. Proceedings of the National Academy of Sciences of the United States of America 104: 19703–19708.
2. Mir RR, Zaman-Allah M, Sreenivasulu N, Trethowan R, Varshney RK (2012) Integrated genomics, physiology and breeding approaches for improving drought tolerance in crops. Theoretical and Applied Genetics 125: 625–645.
3. Swamy BVPDSAHKA (2011) Meta-analysis of grain yield QTL identified during agricultural drought in grasses showed consensus. BMC Genomics, 12: 319.
4. Lopes MS, Araus JL, van Heerden PDR, Foyer CH (2011) Enhancing drought tolerance in C-4 crops. Journal of Experimental Botany 62: 3135–3153.

5. Messina CD, Podlich D, Dong Z, Samples M, Cooper M (2011) Yield-trait performance landscapes: from theory to application in breeding maize for drought tolerance. Journal of Experimental Botany 62: 855–868.
6. Uga Y, Sugimoto K, Ogawa S, Rane J, Ishitani M, et al. (2013) Control of root system architecture by DEEPER ROOTING 1 increases rice yield under drought conditions. Nature Genetics 45: 1097–1102.
7. Hirayama T, Shinozaki K (2010) Research on plant abiotic stress responses in the post-genome era: past, present and future. Plant Journal 61: 1041–1052.
8. Witt S, Galicia L, Lisec J, Cairns J, Tiessen A, et al. (2012) Metabolic and Phenotypic Responses of Greenhouse-Grown Maize Hybrids to Experimentally Controlled Drought Stress. Molecular Plant 5: 401–417.

9. Gechev TS, Dinakar C, Benina M, Toneva V, Bartels D (2012) Molecular mechanisms of desiccation tolerance in resurrection plants. Cellular and Molecular Life Sciences 69: 3175–3186.

10. Alcazar R, Bitrian M, Bartels D, Koncz C, Altabella T, et al. (2011) Polyamine metabolic canalization in response to drought stress in Arabidopsis and the resurrection plant Craterostigma plantagineum. Plant signaling & behavior 6: 243–250.

11. Yobi A, Wone BWM, Xu W, Alexander DC, Guo L, et al. (2012) Comparative metabolic profiling between desiccation-sensitive and desiccation-tolerant species of Selaginella reveals insights into the resurrection trait. Plant Journal 72: 983–999.

12. Oliver MJ, Guo L, Alexander DC, Ryals JA, Wone BWM, et al. (2011) A Sister Group Contrast Using Untargeted Global Metabolomic Analysis Delineates the Biochemical Regulation Underlying Desiccation Tolerance in Sporobolus stapfianus. Plant Cell 23: 1231–1248.

13. Marshall A, Aalen RB, Audenaert D, Beeckman T, Broadley MR, et al. (2012) Tackling Drought Stress: RECEPTOR-LIKE KINASES Present New Approaches. Plant Cell 24: 2262–2278.

14. Skirycz A, Inze D (2010) More from less: plant growth under limited water. Current Opinion in Biotechnology 21: 197–203.

15. Obata T, Fernie AR (2012) The use of metabolomics to dissect plant responses to abiotic stresses. Cellular and Molecular Life Sciences 69: 3225–3243.

16. Shao H-B, Chu L-Y, Jaleel CA, Manivannan P, Panneerselvam R, et al. (2009) Understanding water deficit stress-induced changes in the basic metabolism of higher plants - biotechnologically and sustainably improving agriculture and the ecoenvironment in arid regions of the globe. Critical Reviews in Biotechnology 29: 131–151.

17. Bhargava S, Sawant K (2013) Drought stress adaptation: metabolic adjustment and regulation of gene expression. Plant Breeding 132: 21–32.

18. Yobi A, Wone BWM, Xu W, Alexander DC, Guo L, et al. (2013) Metabolomic Profiling in Selaginella lepidophylla at Various Hydration States Provides New Insights into the Mechanistic Basis of Desiccation Tolerance. Molecular Plant 6: 369–385.

19. Gasulla F, vom Dorp K, Dombrink I, Zähringer U, Gisch N, et al. (2013) The role of lipid metabolism in the acquisition of desiccation tolerance in Craterostigma plantagineum: a comparative approach. The Plant Journal 75: 726–741.

20. Torres-Franklin M-L, Gigon A, de Melo DF, Zuily-Fodil Y, Pham-Thi A-T (2007) Drought stress and rehydration affect the balance between MGDG and DGDG synthesis in cowpea leaves. Physiologia Plantarum 131: 201–210.

21. Gigon A, Matos AR, Laffray D, Zuily-Fodil Y, Pham-Thi AT (2004) Effect of drought stress on lipid metabolism in the leaves of Arabidopsis thaliana (ecotype Columbia). Annals of Botany 94: 345–351.

22. Giavalisco P, Li Y, Matthes A, Eckhardt A, Hubberten H-M, et al. (2011) Elemental formula annotation of polar and lipophilic metabolites using C-13, N-15 and S-34 isotope labelling, in combination with high- resolution mass spectrometry. Plant Journal 68: 364–376.

23. Lisec J, Schauer N, Kopka J, Willmitzer L, Fernie AR (2006) Gas chromatography mass spectrometry-based metabolite profiling in plants. Nature Protocols 1: 387–396.

24. Cuadros-Inostroza A, Caldana C, Redestig H, Kusano M, Lisec J, et al. (2009) TargetSearch - a Bioconductor package for the efficient preprocessing of GC-MS metabolite profiling data. Bmc Bioinformatics 10.

25. Stacklies W, Redestig H, Scholz M, Walther D, Selbig J (2007) pcaMethods - a bioconductor package providing PCA methods for incomplete data. Bioinformatics 23: 1164–1167.

# Permissions

All chapters in this book were first published in PLOS ONE, by The Public Library of Science; hereby published with permission under the Creative Commons Attribution License or equivalent. Every chapter published in this book has been scrutinized by our experts. Their significance has been extensively debated. The topics covered herein carry significant findings which will fuel the growth of the discipline. They may even be implemented as practical applications or may be referred to as a beginning point for another development.

The contributors of this book come from diverse backgrounds, making this book a truly international effort. This book will bring forth new frontiers with its revolutionizing research information and detailed analysis of the nascent developments around the world.

We would like to thank all the contributing authors for lending their expertise to make the book truly unique. They have played a crucial role in the development of this book. Without their invaluable contributions this book wouldn't have been possible. They have made vital efforts to compile up to date information on the varied aspects of this subject to make this book a valuable addition to the collection of many professionals and students.

This book was conceptualized with the vision of imparting up-to-date information and advanced data in this field. To ensure the same, a matchless editorial board was set up. Every individual on the board went through rigorous rounds of assessment to prove their worth. After which they invested a large part of their time researching and compiling the most relevant data for our readers.

The editorial board has been involved in producing this book since its inception. They have spent rigorous hours researching and exploring the diverse topics which have resulted in the successful publishing of this book. They have passed on their knowledge of decades through this book. To expedite this challenging task, the publisher supported the team at every step. A small team of assistant editors was also appointed to further simplify the editing procedure and attain best results for the readers.

Apart from the editorial board, the designing team has also invested a significant amount of their time in understanding the subject and creating the most relevant covers. They scrutinized every image to scout for the most suitable representation of the subject and create an appropriate cover for the book.

The publishing team has been an ardent support to the editorial, designing and production team. Their endless efforts to recruit the best for this project, has resulted in the accomplishment of this book. They are a veteran in the field of academics and their pool of knowledge is as vast as their experience in printing. Their expertise and guidance has proved useful at every step. Their uncompromising quality standards have made this book an exceptional effort. Their encouragement from time to time has been an inspiration for everyone.

The publisher and the editorial board hope that this book will prove to be a valuable piece of knowledge for researchers, students, practitioners and scholars across the globe.

# List of Contributors

**Georgina O'Farrill**
Ecology and Evolutionary Biology Department, University of Toronto, Toronto, Ontario, Canada

**Raja Sengupta**
Geography Department, McGill University, Montreal, Quebec, Canada

**Kim Gauthier Schampaert**
Département de géomatique (KGS), Département de biologie (SC), Université de Sherbrooke, Sherbrooke, Québec, Canada

**Bronwyn Rayfield and Andrew Gonzalez**
Biology Department, McGill University, Montreal, Quebec, Canada

**Örjan Bodin**
Stockholm Resilience Centre, Stockholm University, Stockholm, Sweden

**Sophie Calmé**
Département de géomatique (KGS), Département de biologie (SC), Université de Sherbrooke, Sherbrooke, Québec, Canada
Departamento de conservación de la biodiversidad, El Colegio de la Frontera Sur, Chetumal, Quintana Roo, Mexico

**Fanchao Meng**
Institute of Eco-Environment and Agro-Meteorology, Chinese Academy of Meteorological Sciences, Beijing, China
College of Atmospheric Science, Nanjing University of Information Science & Technology, Nanjing, China

**Jiahua Zhang**
Institute of Eco-Environment and Agro-Meteorology, Chinese Academy of Meteorological Sciences, Beijing, China
Key Laboratory of Digital Earth Science, Institute of Remote Sensing and Digital Earth, Chinese Academy of Sciences, Beijing, China

**Fengmei Yao**
Key Laboratory of Computational Geodynamics, Chinese Academy of Sciences, Beijing, China

**Cui Hao**
Institute of Eco-Environment and Agro-Meteorology, Chinese Academy of Meteorological Sciences, Beijing, China

**Claudio Biscaro**
Department of Economics, Ca' Foscari University of Venice, Venice, Italy
Department of Management, Ca' Foscari University of Venice, Venice, Italy
Institut für Organization und Globale Managementstudien, Johannes Kepler Universität, Linz, Austria

**Carlo Giupponi**
Department of Economics, Ca' Foscari University of Venice, Venice, Italy

**Rosana Ferrero**
Departamento Protección de Cultivos, Instituto de Agricultura Sostenible, Consejo Superior de Investigaciones Científicas (CSIC), Córdoba, Spain
Center of Applied Ecology and Sustainability (CAPES), Santiago, Chile

**Mauricio Lima**
Departamento de Ecología, Pontificia Universidad Católica de Chile, Santiago, Chile
Laboratorio Internacional de Cambio Global, LINCG (CSIC-PUC), Santiago, Chile
Center of Applied Ecology and Sustainability (CAPES), Santiago, Chile

**Jose Luis Gonzalez-Andujar**
Departamento Protección de Cultivos, Instituto de Agricultura Sostenible, Consejo Superior de Investigaciones Científicas (CSIC), Córdoba, Spain
Laboratorio Internacional de Cambio Global, LINCG (CSIC-PUC), Santiago, Chile

**Afifuddin Latif Adiredjo**
Université de Toulouse, INP-ENSAT, UMR 1248 AGIR (INPT-INRA), Castanet-Tolosan, France
Brawijaya University, Faculty of Agriculture, Department of Agronomy, Plant Breeding Laboratory, Malang, Indonesia

**Olivier Navaud and Thierry Lamaze**
Université de Toulouse, UPS-Toulouse III, UMR 5126 CESBIO, Toulouse, France

**Stephane Muños and Nicolas B. Langlade**
INRA, Laboratoire des Interactions Plantes-Microorganismes (LIPM), UMR 441, Castanet-Tolosan, France
CNRS, Laboratoire des Interactions Plantes-Microorganismes(LIPM), UMR 2594, Castanet-Tolosan, France

**Philippe Grieu**
Université de Toulouse, INP-ENSAT, UMR 1248 AGIR
(INPT-INRA), Castanet-Tolosan, France

**Tomonori Tsunoda, Naoki Kachi and Jun-Ichirou
Suzuki**
Department of Biological Sciences, Tokyo Metropolitan
University, Hachioji, Tokyo, Japan

**Joep F. Schyns and Arjen Y. Hoekstra**
Twente Water Centre, University of Twente, Enschede,
The Netherlands

**Xiao-Yan Chen**
College of Resources and Environment/Key Laboratory
of Eco-environment in Three Gorges Region (Ministry
of Education), Southwest University, Chongqing,
China
State Key Laboratory of Soil Erosion and Dryland
Farming on the Loess Plateau, Institute of Soil and
Water Conservation, CAS and MWR, Yangling, China

**Yu Zhao, Bin Mo and Hong-Xing Mi**
College of Resources and Environment/Key Laboratory
of Eco-environment in Three Gorges Region (Ministry
of Education), Southwest University, Chongqing,
China

**Andrew J. Cole, Rocky de Nys and Nicholas A. Paul**
MACRO — the Centre for Macroalgal Resources
and Biotechnology, and School of Marine and
Tropical Biology, James Cook University, Townsville,
Queensland, Australia

**Susanne Petzold and Tobias Polte**
UFZ – Helmholtz Centre for Environmental Research
Leipzig-Halle, Department of Environmental
Immunology, Leipzig, Germany
Department of Dermatology, Venerology and
Allergology, Leipzig University Medical Center,
Leipzig, Germany

**Marco Averbeck and Jan C. Simon**
Department of Dermatology, Venerology and
Allergology, Leipzig University Medical Center,
Leipzig, Germany

**Irina Lehmann**
UFZ – Helmholtz Centre for Environmental Research
Leipzig-Halle, Department of Environmental
Immunology, Leipzig, Germany

**Feng Du and Xingchang Zhang**
Institute of soil and Water Conservation, Northwest
Sci-Tech University of Agriculture and Forestry,
Yangling, Shaanxi, China

Institute of soil and Water Conservation, Chinese
Academy of Science, Yangling, Shaanxi, China
State key laboratory of soil erosion and dryland
farming on Loess Plateau, Yangling, Shaanxi, China

**Huijun Shi**
Institute of soil and Water Conservation, Northwest
Sci-Tech University of Agriculture and Forestry,
Yangling, Shaanxi, China

**Xuexuan Xu**
Institute of soil and Water Conservation, Northwest
Sci-Tech University of Agriculture and Forestry,
Yangling, Shaanxi, China
Institute of soil and Water Conservation, Chinese
Academy of Science, Yangling, Shaanxi, China

**Meha Jain, Yili Lim and María Uriarte**
Department of Ecology, Evolution and Environmental
Biology, Columbia University, New York, New York,
United States of America

**Javier A. Arce-Nazario**
Department of Biology, University of Puerto Rico in
Cayey, Cayey, Puerto Rico, United States of America

**Julianne L. Baron**
Department of Infectious Diseases and Microbiology,
University of Pittsburgh, Graduate School of Public
Health, Pittsburgh, Pennsylvania, United States of
America
Special Pathogens Laboratory, Pittsburgh,
Pennsylvania, United States of America

**Amit Vikram**
Department of Civil and Environmental Engineering,
University of Pittsburgh, Swanson School of
Engineering, Pittsburgh, Pennsylvania, United States
of America

**Scott Duda**
Special Pathogens Laboratory, Pittsburgh, Pennsylvania,
United States of America

**Janet E. Stout**
Special Pathogens Laboratory, Pittsburgh,
Pennsylvania, United States of America
Department of Civil and Environmental Engineering,
University of Pittsburgh, Swanson School of Engineering,
Pittsburgh, Pennsylvania, United States of America

**Kyle Bibby**
Department of Civil and Environmental Engineering,
University of Pittsburgh, Swanson School of
Engineering, Pittsburgh, Pennsylvania, United States
of America

Department of Computational and Systems Biology, University of Pittsburgh Medical School, Pittsburgh, Pennsylvania, United States of America

**Takeki Hamasaki, Noboru Nakamichi, Kiichiro Teruya and Sanetaka Shirahata**
Department of Bioscience and Biotechnology, Faculty of Agriculture, Kyushu University, Higashi-ku, Fukuoka, Japan

**Marcelo Zeri, Celso von Randow and Gilvan Sampaio**
Centro de Ciência do Sistema Terrestre, Instituto Nacional de Pesquisas Espaciais, Cachoeira Paulista, SP, Brazil

**Leonardo D. A. Sá**
Centro Regional da Amazônia, Instituto Nacional de Pesquisas Espaciais, Belém, PA, Brazil

**Antônio O. Manzi**
Instituto Nacional de Pesquisas da Amazénia (INPA), Manaus, Amazonas, Brazil

**Alessandro C. Araújo**
Embrapa Amazônia Oriental, Belém, Pará, Brazil

**Renata G. Aguiar**
Universidade Federal de Rondônia, Porto Velho, Rondônia, Brazil

**Fernando L. Cardoso**
Universidade Federal de Rondônia, Ji-Paraná, Rondônia, Brazil

**Carlos A. Nobre**
Secretaria de Políticas e Programas de Pesquisa e Desenvolvimento, Ministério da Ciência, Tecnologia e Inovação, Brasília, DF, Brazil

**Xia Yao, Wenqing Jia, Haiyang Si, Ziqing Guo, Yongchao Tian, Xiaojun Liu, Weixing Cao and Yan Zhu**
National Engineering and Technology Center for Information Agriculture, Jiangsu Key Laboratory for Information Agriculture, Nanjing Agricultural University, Nanjing, Jiangsu, P. R. China

**Yangquanwei Zhong and Zhouping Shangguan**
State Key Laboratory of Soil Erosion and Dryland Farming on the Loess Plateau, Northwest A & F University, Yangling, Shaanxi, P.R. China

**Sruthi Narayanan and P. V. Vara Prasad**
Department of Agronomy, Kansas State University, Manhattan, Kansas, United States of America

**Amita Mohan and Kulvinder S. Gill**
Department of Crop and Soil Sciences, Washington State University, Pullman, Washington, United States of America

**John J. Kelly and Nicole Minalt**
Department of Biology, Loyola University Chicago, Chicago, Illinois, United States of America

**Alessandro Culotti and Aaron Packman**
Department of Civil and Environmental Engineering, Northwestern University, Evanston, Illinois, United States of America

**Marsha Pryor**
Pinellas County Utilities Laboratory, Largo, Florida, United States of America

**Binbin Wu, Guoqiang Wang, Jin Wu and Changming Liu**
College of Water Sciences, Beijing Normal University, Key Laboratory of Water and Sediment Sciences, Ministry of Education, Beijing, China

**Qing Fu**
Chinese Research Academy of Environmental Sciences, Beijing, China

**Magalie Canuel**
Institut national de santé publique du Québec (INSPQ), Québec City, Canada

**Belkacem Abdous**
Centre de recherche du Centre hospitalier universitaire de Québec, Québec City, Canada
Département de médecine sociale et préventive de l9Universite´ Laval, Québec City, Canada

**Diane Bélanger**
Centre de recherche du Centre hospitalier universitaire de Québec, Québec City, Canada
Institut national de la recherche scientifique, Centre Eau Terre Environnement, Québec City, Canada

**Pierre Gosselin**
Institut national de santé publique du Québec (INSPQ), Québec City, Canada
Centre de recherche du Centre hospitalier universitaire de Québec, Québec City, Canada
Institut national de la recherche scientifique, Centre Eau Terre Environnement, Québec City, Canada

**Ahmed Ramadan and Ahmed M. Shokry**
Department of Biological Sciences, Faculty of Science, King Abdulaziz University (KAU), Jeddah, Saudi Arabia
Agricultural Genetic Engineering Research Institute (AGERI), Agriculture Research Center (ARC), Giza, Egypt

**Jamal S. M. Sabir, Saleha Y. M. Alakilli and Hassan S. Al-Zahrani**
Department of Biological Sciences, Faculty of Science, King Abdulaziz University (KAU), Jeddah, Saudi Arabia

**Nour O. Gadalla and Magdy A. Al-Kordy**
Department of Biological Sciences, Faculty of Science, King Abdulaziz University (KAU), Jeddah, Saudi Arabia
Genetics and Cytology Department, Genetic Engineering and Biotechnology Division, National Research Center, Dokki, Egypt

**Fotouh M. El-Domyati, Ahmed Bahieldin and Sherif Edris**
Department of Biological Sciences, Faculty of Science, King Abdulaziz University (KAU), Jeddah, Saudi Arabia
Department of Genetics, Faculty of Agriculture, Ain Shams University, Cairo, Egypt

**Neil R Baker**
Department of Biological Sciences, University of Essex, Colchester, United Kingdom

**Lothar Willmitzer and Susann Irgang**
Max-Planck-Institut für Molekulare Pflanzenphysiologie, Potsdam-Golm, Germany

# Index

www.ingramcontent.com/pod-product-compliance
Lightning Source LLC
Chambersburg PA
CBHW080248230326
41458CB00097B/4086